PROGRESS IN CLINICAL AND BIOLOGICAL RESEARCH

Series Editors
Nathan Back
George J. Brewer

Vincent P. Eijsvoogel
Robert Grover
Kurt Hirschhorn

Seymour S. Kety
Sidney Udenfriend
Jonathan W. Uhr

RECENT TITLES

Vol 50: **Rights and Responsibilities in Modern Medicine: The Second Volume in a Series on Ethics, Humanism, and Medicine,** Marc D. Basson, *Editor*

Vol 51: **The Function of Red Blood Cells: Erythrocyte Pathobiology,** Donald F. H. Wallach, *Editor*

Vol 52: **Conduction Velocity Distributions: A Population Approach to Electrophysiology of Nerve,** Leslie J. Dorfman, Kenneth L. Cummins, and Larry J. Leifer, *Editors*

Vol 53: **Cancer Among Black Populations,** Curtis Mettlin and Gerald P. Murphy, *Editors*

Vol 54: **Connective Tissue Research: Chemistry, Biology, and Physiology,** Zdenek Deyl and Milan Adam, *Editors*

Vol 55: **The Red Cell: Fifth Ann Arbor Conference,** George J. Brewer, *Editor*

Vol 56: **Erythrocyte Membranes 2: Recent Clinical and Experimental Advances,** Walter C. Kruckeberg, John W. Eaton, and George J. Brewer, *Editors*

Vol 57: **Progress in Cancer Control,** Curtis Mettlin and Gerald P. Murphy, *Editors*

Vol 58: **The Lymphocyte,** Kenneth W. Sell and William V. Miller, *Editors*

Vol 59: **Eleventh International Congress of Anatomy,** Enrique Acosta Vidrio, *Editor-in-Chief.* Published in 3 volumes:
Part A: **Glial and Neuronal Cell Biology,** Sergey Fedoroff, *Editor*
Part B: **Advances in the Morphology of Cells and Tissues,** Miguel A. Galina, *Editor*
Part C: **Biological Rhythms in Structure and Function,** Heinz von Mayersbach, Lawrence E. Scheving, and John E. Pauly, *Editors*

Vol 60: **Advances in Hemoglobin Analysis,** Samir M. Hanash and George J. Brewer, *Editors*

Vol 61: **Nutrition and Child Health: Perspectives for the 1980s,** Reginald C. Tsang and Buford Lee Nichols, Jr., *Editors*

Vol 62: **Pathophysiological Effects of Endotoxins at the Cellular Level,** Jeannine A. Majde and Robert J. Person, *Editors*

Vol 63: **Membrane Transport and Neuroreceptors,** Dale Oxender, Arthur Blume, Ivan Diamond, and C. Fred Fox, *Editors*

Vol 64: **Bacteriophage Assembly,** Michael S. DuBow, *Editor*

Vol 65: **Apheresis: Development, Applications, and Collection Procedures,** C. Harold Mielke, Jr., *Editor*

Vol 66: **Control of Cellular Division and Development,** Dennis Cunningham, Eugene Goldwasser, James Watson, and C. Fred Fox, *Editors*. Published in 2 Volumes.

Vol 67: **Nutrition in the 1980s: Constraints on Our Knowledge,** Nancy Selvey and Philip L. White, *Editors*

Vol 68: **The Role of Peptides and Amino Acids as Neurotransmitters,** J. Barry Lombardini and Alexander D. Kenny, *Editors*

Vol 69: **Twin Research 3, Proceedings of the Third International Congress on Twin Studies,** Luigi Gedda, Paolo Parisi, and Walter E. Nance, *Editors*. Published in 3 volumes:
  Part A: **Twin Biology and Multiple Pregnancy**
  Part B: **Intelligence, Personality, and Development**
  Part C: **Epidemiological and Clinical Studies**

Vol 70: **Reproductive Immunology,** Norbert Gleicher, *Editor*

Vol 71: **Psychopharmacology of Clonidine,** Harbans Lal and Stuart Fielding, *Editors*

Vol 72: **Hemophilia and Hemostasis,** Doris Ménaché, D. MacN. Surgenor, and Harlan D. Anderson, *Editors*

Vol 73: **Membrane Biophysics: Structure and Function in Epithelia,** Mumtaz A. Dinno and Arthur B. Callahan, *Editors*

Vol 74: **Physiopathology of Endocrine Diseases and Mechanisms of Hormone Action,** Roberto J. Soto, Alejandro De Nicola, and Jorge Blaquier, *Editors*

Vol 75: **The Prostatic Cell: Structure and Function,** Gerald P. Murphy, Avery A. Sandberg, and James P. Karr, *Editors*. Published in 2 volumes:
  Part A: **Morphologic, Secretory, and Biochemical Aspects**
  Part B: **Prolactin, Carcinogenesis, and Clinical Aspects**

Vol 76: **Troubling Problems in Medical Ethics: The Third Volume in a Series on Ethics, Humanism, and Medicine,** Marc D. Basson, Rachel E. Lipson, and Doreen L. Ganos, *Editors*

Vol 77: **Nutrition in Health and Disease and International Development: Symposia From the XII International Congress of Nutrition,** Alfred E. Harper and George K. Davis, *Editors*

Vol 78: **Female Incontinence,** Norman R. Zinner and Arthur M. Sterling, *Editors*

Vol 79: **Proteins in the Nervous System: Structure and Function,** Bernard Haber, Joe Dan Coulter, and Jose Regino Perez-Polo, *Editors*

Vol 80: **Mechanism and Control of Ciliary Movement,** Charles J. Brokaw and Pedro Verdugo, *Editors*

Vol 81: **Physiology and Biology of Horseshoe Crabs: Studies on Normal and Environmentally Stressed Animals,** Joseph Bonaventura, Celia Bonaventura, and Shirley Tesh, *Editors*

Vol 82: **Clinical, Structural, and Biochemical Advances in Hereditary Eye Disorders,** Donna Daentl, *Editor*

Vol 83: **Issues in Cancer Screening and Communications,** Curtis Mettlin and Gerald P. Murphy, *Editors*

Vol 84: **Progress in Dermatoglyphic Research,** Christos S. Bartsocas, *Editor*

Vol 85: **Embryonic Development,** Max M. Burger and Rudolf Weber, *Editors*. Published in 2 volumes:
  Part A: **Genetic Aspects**
  Part B: **Cellular Aspects**

See pages following the index for other titles in this series.

EMBRYONIC DEVELOPMENT

PART B
CELLULAR ASPECTS

EMBRYONIC DEVELOPMENT

PART B
CELLULAR ASPECTS

IX Congress of the
International Society
of Developmental Biologists
August 28–September 1, 1981
Basel, Switzerland

Editors

Max M. Burger
Biozentrum der Universität Basel
Basel, Switzerland

Rudolf Weber
Zoologisches Institut
Universität Bern
Bern, Switzerland

ALAN R. LISS, INC., NEW YORK

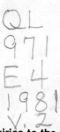

Address all Inquiries to the Publisher
Alan R. Liss, Inc., 150 Fifth Avenue, New York, NY 10011

Copyright © 1982 Alan R. Liss, Inc.

Printed in the United States of America.

Under the conditions stated below the owner of copyright for this book hereby grants permission to users to make photocopy reproductions of any part or all of its contents for personal or internal organizational use, or for personal or internal use of specific clients. This consent is given on the condition that the copier pay the stated per-copy fee through the Copyright Clearance Center, Incorporated, 21 Congress Street, Salem, MA 01970, as listed in the most current issue of "Permissions to Photocopy" (Publisher's Fee List, distributed by CCC, Inc.), for copying beyond that permitted by sections 107 or 108 of the US Copyright Law. This consent does not extend to other kinds of copying, such as copying for general distribution, for advertising or promotional purposes, for creating new collective works, or for resale.

Library of Congress Cataloging in Publication Data
Main entry under title:

Embryonic development.

   (Progress in clinical and biological research; 85)
   Includes bibliographies and indexes.
   Contents: pt. A. Genetic aspects — Pt. B. Cellular aspects.
   1. Embryology — Congresses. 2. Developmental cytology — Congresses. 3. Developmental genetics — Congresses.
I. Burger, Max M. II. Weber, Rudolf, 1922 —
III. International Society of Developmental Biologists.
IV. Series. [DNLM: 1.Embryology — Congresses. 2. Genetics — Congresses. 3. Cytology — Congresses. W1 PR66E v. 85/QS 604 161 1981e]
QL971.E4   574.3′3   82-15351

ISBN 0-8451-0085-8 (set)   AACR2
ISBN 0-8451-0164-1 (pt. B)

Contents

Contributors	xiii
Contents of Part A: Genetic Aspects	xxi
Foreword	
A.A. Moscona	xxiii
Preface	
Max M. Burger and Rudolf Weber	xxv
IN MEMORY OF MICHAEL ABERCROMBIE	
J.A. Weston	1
IN MEMORY OF J.C. DAN	
T.S. Okada	5
IN MEMORY OF VALERIO MONESI	
A. Monroy	7

FERTILIZATION AND EARLY DEVELOPMENT OF VERTEBRATE AND INVERTEBRATE EGGS

Studies on the Role of Exogenous Calcium in Fertilization, Activation and Development of the Sea Urchin Egg
  Tobias Schmidt 11
Cell-Cell Interactions and the Role of Micromeres in the Control of the Mitotic Pattern in Sea Urchin Embryos
  P. Andreuccetti, S. Filosa, A. Monroy and E. Parisi 21
Some Effects of Heat Shock on the Regulation of Macromolecular Syntheses in Sea Urchin Embryos
  M.C. Roccheri, G. Spinelli, G. Sconzo, C. Casano and G. Giudice .. 31
Immunolocalization of α-Actinin in an Amphibian Egg (Discoglossus pictus)
  C. Campanella, E. Rungger-Brändle, and G. Gabbiani 45

FERTILIZATION AND DEVELOPMENT OF THE MAMMALIAN EGG

Physiological Basis for Human In Vitro Fertilization
  Luigi Mastroianni, Jr. 55
The Fine Structure of Human Blastocysts Developed in Culture
  A. Lopata, D.J. Kohlman and G.N. Kellow 69
Experimental Genetics of the Mouse Embryo
  Karl Illmensee 87

Transplantation of the Human Insulin Gene Into Fertilized Mouse Eggs
Kurt Bürki and Axel Ullrich 103
Germ Line Transmission in Transgenic Mice
Jon W. Gordon and Frank H. Ruddle 111

MEMBRANES, CELL INTERACTION AND PATTERN FORMATION

A: Biogenesis and Function of Cell Membranes and the Cell Periphery

Proteins Involved in the Vectorial Translocation of Nascent Peptides Across Membranes
David I. Meyer 125
Susceptibility to Merocyanine 540-Mediated Photosensitization as a Differentiation Marker in Murine Hematopoietic Stem Cells
Fritz Sieber, Richard C. Meagher and Jerry L. Spivak 135
Human Lymphocyte Membrane Proteins in Activation by Phytohemagglutinin
P. Lerch, H. Gmünder and W. Lesslauer 145
Basement Membrane Glycoproteins in the Extracellular Matrix of a Teratocarcinoma-Derived Differentiated Cell Line
Ilmo Leivo and Jorma Wartiovaara 155

B: Cell Interaction and Cell Recognition

Cell-Cell Contact, Cyclic AMP, and Gene Expression During Differentiation of the Cellular Slime Mold Dictyostelium discoideum
Daphne D. Blumberg, Stephen Chung, Scott M. Landfear, Harvey F. Lodish 167
Adhesion of Dictyostelium discoideum Cells to Sugar Derivatized Polyacrylamide Gels
Salvatore Bozzaro and Saul Roseman 183
The Molecular Basis of Species Specific Cell-Cell Recognition in Marine Sponges, and a Study on Organogenesis During Metamorphosis
G.N. Misevic and M.M. Burger 193
Aggregation Factors of Sea Urchin Embryonic Cells
M.L. Vittorelli, V. Matranga, M. Cervello, and H. Noll 211
The Cephalopod Egg, A Suitable Material for Cell and Tissue Interaction Studies
Hans Jürg Marthy 223
Adhesion and Polarity of Amphibian Embryo Blastomeres
Peter B. Armstrong, Marie M. Roberson, Richard Nuccitelli and Douglas Kline 235
Epithelial-Derived Basal Lamina Regulation of Mesenchymal Cell Differentiation
Harold C. Slavkin, Elaine Cummings, Pablo Bringas and Lawrence S. Honig 249

Mesenchyme-Dependent Differentiation of Intestinal Brush-Border Enzymes in the Gizzard Endoderm of the Chick Embryo
   K. Haffen, M. Kedinger, P.M. Simon-Assmann and B. Lacroix 261

Acid Phosphatase of Schizosaccharomyces pombe Is Involved in Morphogenetic and Behavioral Differentiation
   M.E. Schweingruber, A.M. Schweingruber, M.E. Schüpbach and F. Schönholzer 271

C: Pattern Formation

Pattern Formation in the Development of Dictyostelium discoideum
   Ikuo Takeuchi, Masao Tasaka, Masakazu Oyama, Akitsugu Yamamoto and Aiko Amagai 283

Intraspecific Tissue Incompatibilities in the Metagenetical Podocoryne carnea M. SARS (Cnidaria, Hydrozoa)
   Pierre Tardent and Max Bührer 295

Effects of Retinoic Acid on Developmental Properties of the Foot Integument in Avian Embryo
   Danielle Dhouailly 309

DEVELOPMENT OF MUSCLE, CELL AND MORPHOGENETIC MOVEMENTS

Regulation of the Assembly of the Z-Disc in Muscle Cells
   Elias Lazarides, David L. Gard, Bruce L. Granger, Clare M. O'Connor, Jennifer Breckler, and Spencer I. Danto 317

Organisations of Actin and Fibroblast Locomotion
   J.V. Small 341

Neural Crest Cell Development
   James A. Weston 359

Developmental History of the Two M-Line Proteins MM-Creatine Kinase and Myomesin During Myogenesis
   H.M. Eppenberger, E.E. Strehler, Th.C. Doetschman, U.B. Rosenberg, J.C. Perriard, and Th. Wallimann 381

Analysis of the Expression of Myosin and Actin Genes During Mouse Myoblast Differentiation Using Specific Recombinant Plasmids
   B. Robert, M. Caravatti, A. Minty, A. Weydert, S. Alonso, A. Cohen, P. Daubas, F. Gros and M. Buckingham 391

Early Alteration Induced by Tumor Promoters on Chick Embryo Muscle Cells in Culture
   B.M. Zani and M. Molinaro 403

A New Procedure to Analyse the Role of Adhesion to Solid Substrata and of Motility in Chemokinesis of Neutrophil Granulocytes
   H.U. Keller and H. Cottier 415

Restricted Developmental Options of the Metanephric Mesenchyme
   Hannu Sariola, Peter Ekblom and Lauri Saxén 425

In Situ Recording of the Mechanical Behaviour of Cells in the Chick Embryo
    P. Kucera and Y. de Ribaupierre 433

DEVELOPMENT OF THE NEURAL SYSTEM

Cell Contact-Dependent Regulation of Hormonal Induction of Glutamine Synthetase in Embryonic Neural Retina
    P. Linser, A.D. Saad, B.M. Soh and A.A. Moscona 445

Carbonic Anhydrase-C in Neural Retina of the Chick Embryo: Developmental Changes in Cellular Localization
    Paul J. Linser and A.A. Moscona 459

Progress in the Purification of a Factor Involved in the Neuro-Transmitter Choice Made by Cultured Sympathetic Neurons
    Michel J. Weber and Agathe Le Van Thai 473

Clonal Analysis of the Quail Neural Crest: Pluripotency, Autonomous Differentiation, and Modulation of Differentiation by Exogenous Factors
    Maya Sieber-Blum 485

Mechanism of Avian Crest Cell Migration and Homing
    J.L. Duband, A. Delouvee, R.A. Rovasio and J.P. Thiery 497

Biochemistry of Granule Cell Migration in Developing Mouse Cerebellum
    M.E. Hatten, D.B. Rifkin, M.B. Furie, C.A. Mason and R.K.H. Liem .. 509

PLANT DEVELOPMENT

Studies on Plant Cell Lines Showing Temperature-Sensitive Embryogenesis
    M. Terzi, G. Giuliano, F. Lo Schiavo, and V. Nuti Ronchi 521

Comparison of Endogenous Lectins and Glycosidases in Wheat Grains
    Rosemary C. Miller and Dianna J. Bowles 535

Globulins of Developing Maize Seeds: Preliminary Characterization
    Christa Dierks-Ventling 545

Studies on the Chalcone Synthase Gene of Two Higher Plants: Petroselinum hortense and Matthiola incana
    V. Hemleben, M. Frey, S. Rall, M. Koch, M. Kittel, F. Kreuzaler, H. Ragg, E. Fautz and K. Hahlbrock 555

DEVELOPMENT AND CANCER

Tumor Reversal in Plants
    Frederick Meins, Jr. 567

DNA Integrity in Plant Embryos and the Importance of DNA Repair
    Daphne J. Osborne 577

The Blastocyst in Control of Embryonal Carcinoma
    G. Barry Pierce and Robert S. Wells 593

Recent Observations on the Pathogenesis of Cancer Metastasis
    Isaiah J. Fidler and Ian R. Hart................................ 601
Gonial Cell Neoplasm of Genetic Origin Affecting Both Sexes of Drosophila melanogaster
    Elisabeth Gateff 621
Proliferating Rat Hepatocytes in Culture Contain mRNA Coding for Alpha-Feto-Protein and Albumin
    Hans H. Arnold, H.J. Ruthe and Dieter Paul 633
Differences in the Adhesive Properties of Osteosarcoma-Derived Clonal Variants Enhanced by Protease Treatment
    D. De Martelaere, J.J. Cassiman, F. Van Leuven and H. Van den Berghe 647
Index 659

Contributors

S. Alonso, Pasteur Institute, Department of Molecular Biology, Paris, France [391]

Aiko Amagai, Department of Botany, Faculty of Science, Kyoto University, Kyoto, Japan [283]

Piero Andreuccetti, Istituto di Istologia ed Embriologia, Universita di Napoli, Napoli, Italy [21]

Peter B. Armstrong, Department of Zoology, University of California, Davis, California [235]

Hans H. Arnold, , Department of Toxicology, University of Hamburg Medical School, Hamburg, Federal Republic of Germany [633]

Daphne D. Blumberg, Laboratory of Molecular Oncology, NCI-FCRF, Frederick, Maryland [167]

Dianna J. Bowles, Department of Biochemistry, University of Leeds, West Yorkshire, United Kingdom [535]

Salvatore Bozzaro, Department of Biology and McCollum-Pratt Institute, The Johns Hopkins University, Baltimore, Maryland [183]

Jennifer Breckler, Division of Biology, California Institute of Technology, Pasadena, California [317]

Pablo Bringas, Laboratory for Developmental Biology, Graduate Program in Craniofacial Biology, University of Southern California, Los Angeles, California [249]

M. Buckingham, Pasteur Institute, Department of Molecular Biology, Paris, France [391]

Max Bührer, Zoological Institute, University of Zürich, Zürich, Switzerland [295]

Max M. Burger, Department of Biochemistry, Biocenter, University of Basel, Basel, Switzerland [xxv, 193]

Kurt Bürki, Department of Animal Biology, University of Geneva, Geneva, Switzerland [103]

Chiara Campanella, Istituto di Istologia ed Embriologia, Universita di Napoli, Napoli, Italy, [45]

M. Caravatti, Pasteur Institute, Department of Molecular Biology, Paris, France [391]

The bold face number in brackets following each contributor's affiliation indicates the opening page of that author's paper.

Contributors

C. Casano, Institute of Comparative Anatomy, University of Palermo, Palermo, Italy [31]

J.J. Cassiman, Division of Human Genetics, University of Leuven, Leuven, Belgium [647]

Melchiorre Cervello, Istituto di Biologia dello Sviluppo, Consiglio Nazionale delle Ricerche, Palermo, Italy [211]

Stephen Chung, Department of Biological Chemistry, University of Illinois Medical Center, Chicago, Illinois [167]

A. Cohen, Pasteur Institute, Department of Molecular Biology, Paris, France [391]

H. Cottier, Institute of Pathology, University of Berne, Berne, Switzerland [415]

Elaine Cummings, Laboratory for Developmental Biology, Graduate Program in Craniofacial Biology, University of Southern California, Los Angeles, California [249]

Spencer I. Danto, Division of Biology, California Institute of Technology, Pasadena, California [317]

P. Daubas, Pasteur Institute, Department of Molecular Biology, Paris, France [391]

Annie Delouvee, Institut d'Embryologie du CNRS et du College de France, Nogent sur Marne, France [497]

D. De Martelaere, Division of Human Genetics, University of Leuven, Leuven Belgium [647]

Y. de Ribaupierre, Institute of Physiology, University of Lausanne, Lausanne, Switzerland, [433]

Danielle Dhouailly, Laboratoire de Zoologie et Biologie Animale, Université Scientifique et Médicale de Grenoble, Grenoble, France [309]

Christa Dierks-Ventling, Friedrich Miescher Institut, Basel, Switzerland [545]

Thomas C. Doetschman, Department of Cell Biology, Swiss Federal Institute of Technology, Zurich, Switzerland [381]

Jean Loup Duband, Institut d'Embryologie du CNRS et du College de France, Nogent sur Marne, France [497]

Peter Ekblom, Department of Pathology, University of Helsinki, Helsinki, Finland [425]

Hans M. Eppenberger, Department of Cell Biology, Swiss Federal Institute of Technology, Zurich, Switzerland [381]

E. Fautz, Institute of Biology II, University of Freiburg, Freiburg, Federal Republic of Germany [555]

Isaiah J. Fidler, Cancer Metastasis and Treatment Laboratory, NCI Frederick Cancer Research Center, Frederick, Maryland [601]

Silvana Filosa, Istituto di Istologia ed Embriologia, Universita di Napoli, Napoli, Italy [21]

M. Frey, Institute of Biology II, University of Tübingen, Tübingen, Federal Republic of Germany [555]

Martha B. Furie, Department of Cellular Physiology and Immunology, The Rockefeller University, New York, New York [509]

Contributors / xv

G. Gabbiani, Department of Pathology, University of Geneva, Geneva, Switzerland [45]
David L. Gard, Division of Biology, California Institute of Technology, Pasadena, California [317]
Elisabeth Gateff, Biologisches Institut I, Albert-Ludwigs-Universität, Freiberg, Federal Republic of Germany [621]
G. Giudice, Institute of Developmental Biology, C.N.R., Palermo, Italy [31]
Giovanni Giuliano, Istituto di Mutagenesi e Differenziamento del CNR, Pisa, Italy [521]
H. Gmünder, Department of Biochemistry, University of Bern, Bern, Switzerland [145]
Jon W. Gordon, Department of Biology, Yale University, New Haven, Connecticut [111]
Bruce L. Granger, Division of Biology, California Institute of Technology, Pasadena, California [317]
F. Gros, Pasteur Institute, Department of Molecular Biology, Paris, France [391]
K. Haffen, Unité de Recherche INSERM 61, Strasbourg-Hautepierre, France [261]
K. Hahlbrock, Institute of Biology II, University of Freiburg, Freiburg, Federal Republic of Germany [555]
Ian R. Hart, Cancer Metastasis and Treatment Laboratory, NCI Frederick Cancer Research Center, Frederick, Maryland [601]
Mary E. Hatten, Department of Pharmacology, New York University School of Medicine, New York, New York [509]
V. Hemleben, Institute of Biology II, University of Tübingen, Tübingen, Federal Republic of Germany [555]
Lawrence S. Honig, Laboratory for Developmental Biology, Graduate Program in Craniofacial Biology, University of Southern California, Los Angeles, California [249]
Karl Illmensee, Department of Animal Biology, University of Geneva, Geneva Switzerland [87]
M. Kedinger, Unité de Recherches, INSERM 61, Strasbourg-Hautepierre, France [261]
H.U. Keller, Institute of Pathology, University of Berne, Berne, Switzerland [415]
G.N. Kellow, Department of Obstetrics and Gynaecology and Reproductive Biology Unit, University of Melbourne, Royal Women's Hospital, Carlton, Victoria, Australia [69]
M. Kittel, Institute of Biology II, University of Tübingen, Tübingen, Federal Republic of Germany [555]
Douglas Kline, Department of Zoology, University of California, Davis, California [235]
M. Koch, Institute of Biology II, University of Tübingen, Tübingen, Federal Republic of Germany [555]
D.J. Kohlman, Department of Obstetrics and Gynaecology and Reproductive

Biology Unit, University of Melbourne, Royal Women's Hospital, Carlton, Victoria, Australia [69]
F. Kreuzaler, Institute of Biology II, University of Freiburg, Freiburg, Federal Republic of Germany [555]
P. Kucera, Institute of Physiology, University of Lausanne, Lausanne, Switzerland [433]
B. Lacroix, Unitè de Recherches, INSERM 61, Strasbourg-Hautepierre, France [261]
Scott M. Landfear, Department of Biology, Massachusetts Institute of Technology, Cambridge, Massachusetts [167]
Elias Lazarides, Division of Biology, California Institute of Technology, Pasadena, California [317]
Ilmo Leivo, Department of Pathology, University of Helsinki, Helsinki, Finland [155]
P. Lerch, Sidney Farber Cancer Institute, Boston, Massachusetts [145]
W. Lesslauer, Department of Biochemistry, University of Bern, Bern, Switzerland [145]
Agathe Le Van Thai, Laboratoire de Pharmacologie et Toxocologie Fondamentales, CNRS, Toulouse, France [473]
Ronald K.H. Liem, Department of Pharmacology, New York University School of Medicine, New York, New York [509]
Paul J. Linser, Department of Biology, University of Chicago, Chicago, Illinois [445,459]
Harvey F. Lodish, Department of Biology, Massachusetts Institute of Technology, Cambridge, Massachusetts [167]
A. Lopata, Department of Obstetrics and Gynaecology and Reproductive Biology Unit, University of Melbourne, Royal Women's Hospital, Carlton, Victoria, Australia [69]
Fiorella Lo Schiavo, Istituto di Mutagenesi e Differenziamento del CNR, Pisa, Italy [521]
Hans Jürg Marthy, Universite Pierre et Marie Curie, C.N.R.S., Laboratoire Arago, Banyuls-sur-Mer, France [223]
Carol A. Mason, Department of Pharmacology, New York University School of Medicine, New York, New York [509]
Luigi Mastroianni, Jr., Department of Obstetrics and Gynecology, University of Pennsylvania School of Medicine [55]
Valeria Matranga, Istituto di Biologia dello Sviluppo, Consiglio Nazionale delle Ricerche, Palermo, Italy [211]
Richard C. Meagher, Department of Medicine, The Johns Hopkins University School of Medicine, Baltimore, Maryland [135]
Frederick Meins, Jr., Friedrich Miescher-Institut, Basel, Switzerland [567]
David I. Meyer, Department of Cell Biology, European Molecular Biology Laboratory, Heidelberg, Germany [125]
Rosemary C. Miller, Department of Biochemistry, University of Leeds, West Yorkshire, United Kingdom [535]

A. Minty, Pasteur Institute, Department of Molecular Biology, Paris, France [391]

G.N. Misevic, Department of Biochemistry, Biocenter, University of Basel, Basel, Switzerland [193]

Mario Molinaro, Institute of Histology and General Embryology, University of Rome, Rome, Italy [403]

A. Monroy, Zoological Station, Naples, Italy [7,21]

A.A. Moscona, Department of Biology, University of Chicago, Chicago, Illinois [xxiii,445,459]

Hans Noll, Department of Biochemistry and Molecular Biology, Northwestern University, Evanston, Illinois [211]

Richard Nuccitelli, Department of Zoology, University of California, Davis, California [235]

Clare M. O'Connor, Division of Biology, California Institute of Technology, Pasadena, California [317]

T.S. Okada, Institute for Biophysics, Faculty of Science, University of Kyoto, Kyoto, Japan [5]

Daphne J. Osborne, Department of Developmental Botany, Weed Research Organization, Oxford, United Kingdom [577]

Masakazu Oyama, Department of Botany, Faculty of Science, Kyoto University, Kyoto, Japan [283]

Elio Parisi, Istituto di Embriologia Molecolare, CNR, Arco Felice, Italy [21]

Dieter Paul, Department of Toxicology, University of Hamburg Medical School, Hamburg, Federal Republic of Germany [633]

Jean-Claude Perriard, Department of Cell Biology, Swiss Federal Institute of Technology, Zurich, Switzerland [381]

G. Barry Pierce, Department of Pathology, University of Colorado School of Medicine, Denver, Colorado [593]

H. Ragg, Institute of Biology II, University of Freiburg, Freiburg, Federal Republic of Germany [555]

S. Rall, Institute of Biology II, University of Tübingen, Tübingen, Federal Republic of Germany [555]

Daniel B. Rifkin, Department of Cell Biology, New York University School of Medicine, New York, New York [509]

Marie M. Roberson, Department of Zoology, University of California, Davis, California [235]

B. Robert, Pasteur Institute, Department of Molecular Biology, Paris, France [391]

M.C. Roccheri, Institute of Comparative Anatomy, University of Palermo, Palermo, Italy [31]

V. Nuti Ronchi, Istituto di Mutagenesi e Differenziamento del CNR, Pisa, Italy [521]

Saul Roseman, Department of Biology and McCollum-Pratt Institute, The Johns Hopkins University, Baltimore, Maryland [183]

Urs B. Rosenberg, Department of Cell Biology, Swiss Federal Institute of

Technology, Zurich, Switzerland [381]
Roberto Americo Rovasio, Institut d'Embryologie du CNRS et du College de France, Nogent sur Marne, France [497]
Frank H. Ruddle, Department of Biology and Human Genetics, Yale University, New Haven, Connecticut [111]
E. Rungger-Brändle, Department of Pathology, University of Geneva, Geneva, Switzerland [45]
H.J. Ruthe, Department of Toxicology, University of Hamburg Medical School, Hamburg, Federal Republic of Germany [633]
A.D. Saad, Department of Biology, University of Chicago, Chicago, Illinois [445]
Hannu Sariola, Department of Pathology, University of Helsinki, Helsinki, Finland [425]
Lauri Saxén, Department of Pathology, University of Helsinki, Helsinki, Finland [425]
Tobias Schmidt, Stanford University, Hopkins Marine Station, Pacific Grove, California [11]
F. Schönholzer, Institute for General Microbiology, Bern, Switzerland [271]
M.E. Schüpbach, Biological and Pharmaceutical Research Department, F. Hoffmann-La Roche and Co., Ltd., Basel, Switzerland [271]
A.M. Schweingruber, Institute for General Microbiology, Bern, Switzerland [271]
M.E. Schweingruber, Institute for General Microbiology, Bern, Switzerland [271]
G. Sconzo, Institute of Comparative Anatomy, University of Palermo, Palermo, Italy [31]
Fritz Sieber, Department of Medicine, The Johns Hopkins University School of Medicine, Baltimore, Maryland [135]
Maya Sieber-Blum, Department of Cell Biology and Anatomy, The Johns Hopkins University School of Medicine, Baltimore, Maryland [485]
P.M. Simon-Assmann, Unité de Recherches, INSERM 61, Strasbourg-Hautepierre, France [261]
Harold C. Slavkin, Laboratory for Developmental Biology, Graduate Program in Craniofacial Biology, University of Southern California, Los Angeles, California [249]
J.V. Small, Institute of Molecular Biology of the Austrian Academy of Sciences, Salzburg, Austria [341]
B.M. Soh, Department of Biology, University of Chicago, Chicago, Illinois [445]
G. Spinelli, Institute of Comparative Anatomy, Univesity of Palermo, Palermo, Italy [31]
Jerry L. Spivak, Department of Medicine, The Johns Hopkins University School of Medicine, Baltimore, Maryland [135]
Emanuel E. Strehler, Department of Cell Biology, Swiss Federal Institute of Technology, Zurich, Switzerland [381]

Contributors / xix

Ikuo Takeuchi, Department of Botany, Faculty of Science, Kyoto University, Kyoto, Japan [283]
Pierre Tardent, Zoological Institute, University of Zürich, Zürich, Switzerland, [295]
Masao Tasaka, Department of Botany, Faculty of Science, Kyoto University, Kyoto, Japan [283]
Mario Terzi, Istituto di Mutagenesi e Differenziamento del CNR, Pisa, Italy [521]
Jean Paul Thiery, Institut d'Embryologie du CNRS et du College de France, Nogent sur Marne, France [497]
Axel Ullrich, Genentech, Inc., South San Francisco, California [103]
H. Van Den Berghe, Division of Human Genetics, University of Leuven, Leuven, Belgium [647]
F. Van Leuven, Division of Human Genetics, University of Leuven, Leuven, Belgium [647]
Maria Letizia Vittorelli, Istituto di Anatomia Comparta, Universita di Palermo, Palermo, Italy [211]
Theo Wallimann, Department of Cell Biology, Swiss Federal Institute of Technology, Zurich, Switzerland [381]
Jorma Wartiovaara, Department of Electron Microscopy, University of Helsinki, Helsinki, Finland [155]
Michel V. Weber, Laboratoire de Pharmacologie et Toxicologie Fondamentales — CNRS, Toulouse, France [473]
Rudolf Weber, Zoologisches Institut, Universität Bern, Bern Switzerland [xxv]
Robert S. Wells, Department of Pathology, University of Colorado School of Medicine, Denver, Colorado [593]
James A. Weston, Department of Biology, University of Oregon, Eugene, Oregon [1,359]
A. Weydert, Pasteur Institute, Department of Molecular Biology, Paris, France [391]
Akitsugu Yamamoto, Department of Botany, Faculty of Science, Kyoto University, Kyoto, Japan [283]
Bianca M. Zani, Institute of Histology and General Embryology, University of Rome, Rome, Italy [403]

Contents of Part A: Genetic Aspects

GENE STRUCTURE AND FUNCTION
Developmental Control of Histone Gene Expression/*Max L. Birnstiel*
The Conalbumin "TATA" Box Sequence and the SV40 72 Base-Pair Repeat Region Influence Expression of a Chimeric Gene *in vivo*/*B. Wasylyk, M.P. Gaub, A. Dierich, C. Wasylyk, and P. Chambon*
Phenotypic Analysis of Globin Gene Expression: The Thalassaemias/*R.A. Flavell, N.K. Moschonas, E. deBoer, G.C. Grosveld, M. Busslinger, H.H.M. Dahl and F.G. Grosveld*
Moving Genes/*Philip Leder*
Examination of the Regulation of the Actin Multigene Family in *Dictyostelium discoideum*/*Michael McKeown, Klaus-Peter Hirth, Cynthia Edwards, and Richard A. Firtel*
Transcription and Processing of a tRNA Gene Containing an Intervening Sequence, G.P. Tocchini-Valentini/*G. Carrara, A. De Paolis, G. Di Segni, P. Fabrizio, D. Gandini Attardi and A. Otsuka*
Left-Handed Z-DNA: A DNA Structure With Biological Function?/*Alfred Nordheim, Eileen M. Lafer, Achim Möller, B. David Stollar, Mary Lou Pardue, Alexander Rich*
Expression and Organization of the Globin Genes in *Xenopus laevis*/*Hans A. Hosbach, Heinrich J. Widmer, Anne-Catherine Andres, and Rudolf Weber*
Molecular Cloning and Sequence Analysis of Highly Repetitive DNA Sequences Contained in the Eliminated Genome of *Ascaris lumbricoides*/*F. Müller, P. Walker, P. Aeby, H. Neuhaus, E. Back and H. Tobler*
Characterization and Sequence Analysis of Interspersed Repetitive DNA Sequences Transcribed in *X. laevis* Embryos/*Walter Reith, Irmi Sures and Georges Spohr*
Developmentally Regulated Poly(A) Containing RNA Sequences in *Xenopus laevis*/ *Maliyakal E. John, Wolfgang Meyerhof and Walter Knöchel*
DNA Mediated Transfer and Fate of a Proviral Gene of Mouse Mammary Tumor Virus in Cultured Cells/*N.E. Hynes, U. Rahmsdorf, N. Kennedy, P. Herrlich, B. Hohn and B. Groner*

CHROMOSOMES
Structure of the Active Nucleolar Chromatin of *Xenopus laevis* Oocytes/*Paul Labhart and Theo Koller*
Improvements in the Ultrastructural Approach of Lampbrush Chromosomes of Amphibians/*Nicole Anglier and Annie Lavaud*
Visualization of *In Vivo* Transcription Patterns In *Xenopus* rDNA Spacer Chromatin/ *Michael F. Trendelenburg*

DETERMINATION AND DIFFERENTIATION
Amphibian Oocytes and Gene Control in Development/*J.B. Gurdon*
Analogues of 5-Methyl Cytosine and Early Embryonic Development/*V. Maharajan, L. Tosi, M. Pratibha and E. Scarano*
Identification of a Nuclear Polypeptide ("Cyclin") Whose Relative Proportion is Sensitive to Changes in the Rate of Cell Proliferation and to Transformation/*Rodrigo Bravo, Stephen J. Fey, Jaime Bellatin, Peter Mose Larsen and Julio E. Celis*

Transdetermination and Transdifferentiation of Neural Retinal Cells into Lens in Cell Culture/*T.S. Okada, Kunio Yasuda, Hisato Kondoh, Kazuya Nomura, Shin Takagi and Keiji Okuyama*
Regeneration Occurring In Vitro by Cellular Transdifferentiation/*Volker Schmid, Monikay Wydler, Hansjürg Alder and Andreas Bally*

INSECT DEVELOPMENT
Control of Body Segment Differentiation in *Drosophila* by the Bithorax Gene Complex/ *E.B. Lewis*
The Developmentally Regulated Chorion Gene Families of Insects/*Fotis C. Kafatos*
Studies on Fibroin Gene Transcription by *In Vitro* Genetics/*Yoshiaki Suzuki*
Tissue-Specific and Gene-Specific Sites of Hemoglobin Synthesis in *Chironomus*/ *H. Laufer, X. Vafopoulou, R. Kuliawat, and G. Gundling*
Theory of Regulatory Functions of the Genes in the Bithorax Complex/*Hans Meinhardt*
Cell Determination in the *Drosophila melanogaster* Embryo/*A.A. Simcox and J.H. Sang*
Loss of the Ability to Form Pole Cells in *Drosophila* Embryos with Artificially Delayed Nuclear Arrival at the Posterior Pole/*Masukichi Okada*
Hormonal Regulation of a Larval Haemolymph Protein in *drosophila*/*T. Jowett, J.H. Postlethwait and D.B. Roberts*

MATURATION OF GAMETES
RNA and Protein Synthesis During Oogenesis of *Ciona intestinalis*/*F. Cotelli, F. Andronico, R. De Santis, A. Monroy and F. Rosati*
Uracil-DNA Glycosylase in Meiotic and Post Meiotic Male Germ Cells of the Mouse/*P. Grippo, P. Orlando, G. Locorondo and R. Geremia*
Multinucleate Oogenesis/Embryonic Development and Other Adaptations for Reproduction on Land in Egg-Brooding Hylid Frogs/*Eugenia M. del Pino*
Nucleo-Cytoplasmic Interactions in Oogenesis and Early Embryogenesis in the Mouse/ *Andrzej K. Tarkowski*
Effects of the Organophosphate Pesticide Dichlorvos on Quail Embryo Germ Population. Numerical Study and Ultrastructural Cytochemistry/*M.T. Bruel and D. David*

DEVELOPMENTAL ASPECTS OF IMMUNOLOGY
Differentiation of T Cells: Thymic Selection of Specificity for Self/*Rolf M. Zinkernagel*
Surface Markers of Avian T Lymphocytes as Defined by a Monoclonal Antibody and Antisera/*B. Péault, R. Pink, M. Coltey, B. Bruner and N.M. Le Douarin*
Ontogeny of Murine Thymus Cells: Correlation of Surface Phenotype with the Appearance of Cytolytic T Lymphocyte Precursors/*Rhodri Ceredig and H. Robson MacDonald*
Cellular Aspects of the Graft-Versus-Host Reaction in the Chick Embryo/ *J. Desveaux-Chabrol and F. Dieterien-Lievre*
Allotypic Trophoblast-Lymphocyte Cross Reactive (TLX) Antigens/*John A. McIntyre and W. Page Faulk*
The Role of Plasminogen Activator in Cell Migration and Morphogenesis in the Bursa of Fabricius/*Jay E. Valinsky, E. Reich, and Nicole M. Le Douarin*

Foreword

These volumes contain the symposia papers and a selection of contributed papers presented at the IXth International Congress of the International Society of Developmental Biologists. The Congress was held in Basel during August 28 thru September 1, 1981. Traditionally, the quadrennial Congresses of ISDB have served as landmarks of progress in developmental biology and as springboards for future advances in this field. The IXth Congress very successfully met these aims, as clearly evidenced by the content of these volumes. The wealth of new information, ideas and directions reflects the rapid progress of modern developmental biology and conveys the sense of excitement and challenge shared by those working in this field.

Organization of a major international conference is a formidable task. The success of the IXth Congress of ISDB was largely due to the superb work of the Swiss Organizing Committee which spared no effort in its dedication to this enterprise. The Committee was headed by Dr. Max Burger and consisted of Drs. M. Birnstiel, W. Gehring, G. Gerisch, K. Illmensee, H. Ursprung and R. Weber. The initial planning of the scientific program was done jointly with the officers and board members of ISDB whose functions were coordinated by Dr. M. Spiegel, the past Secretary-Treasurer of our Society. Drs. Burger and Weber accepted the additional responsibility of preparing and editing these volumes. To all these colleagues goes the gratitude of the International Society of Developmental Biologists. Thanks are also due to the speakers for contributing papers of unusual interest and significance, and for submitting their manuscripts on time. The Editors received the full cooperation of Alan R. Liss, Inc., assuring prompt publication of these timely volumes.

A.A. Moscona, President
International Society of
Developmental Biologists

Preface

For the first time the traditional proceedings of the ISDB congress are published in two volumes. This is due to the fact that the programme of the IXth Congress in Basel 1981, originally meant to focus on developmental genetics, was expanded to include some related topics. In doing so the organizers hoped to promote interaction between the rapidly expanding field of molecular embryology and the more classical areas of developmental biology. For this reason the usual number of four symposia was increased to six and presented under the heading: "Embryonic Development: Genes and Cells".

We felt that a selection of manuscripts from congress participants presenting posters and oral communications might enhance the value of the articles from the symposia speakers. Some 70 contributors were selected in such a manner as to widen the scope, to provide some of the latest data in the particular field and focus and adjust thereby the information in the symposia articles. The response was highly successful despite an early deadline after the congress.

We have decided to publish the topically related and invited communications together with the review lectures of the symposia in volumes separated according to topics. Volume I deals with the genetic aspects of development with particular emphasis on the structure and function of eurokaryotic genes. Volume II is concerned with cellular aspects of development and includes the more classical areas, such as fertilization, pattern formation and emergence of cell phenotypes.

It is our hope that these proceedings provide adequate insight into the current state of developmental biology in general and reflect at the same time in which areas the exchange of ideas and experience between developmental biologists was most intense at the congress.

The editors are grateful to all the contributors for their cooperation in the preparatory phase of these volumes. All but one symposium speaker delivered their manuscripts and all congress participants selected and invited to send a manuscript did so promptly. We also wish to thank Alan R. Liss, Inc. for logistic support and speedy publication.

M.M. Burger

R. Weber
Basel/Bern
January 1982

IN MEMORY OF MICHAEL ABERCROMBIE

J.A. Weston

University of Oregon

Eugene, Oregon, 97403

Michael Abercrombie died in May, 1979, as he was about to retire from the Directorship of the Strangeways Research Laboratory in Cambridge. An obituary written by his colleague Graham Dunn has appeared in Nature (1). A Biographical Memoir was also published by Michael's close friend Sir Peter Medawar (2). We celebrate his memory here for the lasting positive influence he had on his many admiring colleagues and friends, and on the fields of Cell and Developmental Biology.

Michael was the son of a distinguished poet and literary critic, and his childhood environment must surely have shaped his widely recognized ability to formulate ideas precisely and to write them tersely. Michael received his B.Sc. degree from Queen's College, Oxford, in 1934, after reading Zoology under Gavin deBeer, and doing research with C.H. Waddington at the Strangeways Laboratory. It was 43 years later, however, before he received a Doctorate-- an honorary degree from Uppsala University.

Characteristically, Michael always eschewed honorific titles. If it couldn't be "Michael", "Mr. Abercrombie" would have to do. Even after he had been appointed Professor of Embryology in J.Z. Young's Anatomy Department at University College London, in 1959, or the Jodrell Professor of Zoology at U.C., in 1962, it was still "Michael" or "Mr. Abercrombie".

His early scientific work centered on aspects of cell behavior. These included studies on gastrulation in avian embryos (begun with Waddington); his work with his wife, colleague, and constant friend, Jane (M.L. Johnson) on the migratory and adhesive behavior of Schwann sheath cells during neuron degeneration in vivo and in vitro; and his work with D.W. James and others on liver regeneration and wound healing. In these studies, Michael insightfully recognized the need to characterize the social behavior of cells in an objective and quantitative way. With his colleagues at U.C.--especially Joan Heaysman--he introduced the concept of "contact inhibition of movement" that made an immediate and lasting impact on cell biology. He recognized that this behavior of normal cells in vitro--the inhibition of a cell's locomotor activity and movement in the direction of contact with another cell--could be the basis for understanding morphogenetic processes such as directed cell migration in vivo. Michael also argued that defective contact inhibition of movement could account for some of the invasive behavior of cancer cells, and he thoughtfully summarized these arguments in a Nature article he was completing at the time of his death (3).

In fact, his work focussed attention for the first time on the immensely difficult, but now yielding problems of the locomotory and adhesive mechanisms that regulate normal and abnormal motile and social behavior of cells. In doing so, he and his coworkers helped to identify some of the cytoskeletal and membrane specializations that underlie these mechanisms. These and other contributions were the bases for his election as a Fellow of the Institut International d'Embryology in 1956; and in 1958 into the Fellowship of the Royal Society.

Michael was also an important force in scientific publishing. He edited the New Biology Series for Penguin Paperback Books, and with Jane, his wife, he edited the

Penguin Dictionary of Biology, the 7th edition of which has recently been published. For many years, he was an editor of Experimental Cell Research and Journal of Cell Science. He was a cofounder and editor of the Journal of Embryology and Experimental Morphology, whose Editorial Board meetings were often occasions for international congresses of developmental biologists. With Brachet, he conceived and edited Advances in Morphogenesis, which, between 1961 and 1970, focussed attention on various problems in developmental biology, and served as a prototype for a number of subsequent serial publications of reviews in cell and developmental biology.

Michael was an accomplished and universally admired scientist. Those who worked with him, however, would acknowledge that he was also a very complex man. Such complexity defies brief characterization, and it would be presumptuous of me even to try. Noting a few apparent contradictions, however, may help me convey Michael's special qualities.

--Although he was not gregarious (in fact, he claimed to be neurotically shy (1)), he had great personal charm and was able to make people who met him--especially younger people--feel accepted and valued. Considering his manifestly high standards, this was a great source of encouragement for many of us!

--Although I think he generally disliked large congregations, he was a cofounder of the London Embryology Club, which later became the British Society for Developmental Biology.

--Although he was scrupulously polite and fair, he held very strong opinions that he would occasionally share, when it seemed necessary and appropriate.

--Although he often claimed to know very little about a subject under discussion (and, indeed, by his standards, this might have been so), it was clear to most of us that he had extraordinary breadth of knowledge and analytical insights.

--Although at meetings, he spoke little and rarely argued, in most discussions, he actually took the offensive--he listened!

--Although he was internationally recognized for his scientific contributions, he remained self-effacing. This diffidence also hid an extraordinary knowledge of the arts and literature. His libraries of books and musical recordings were enormous. In London, he would often manifest his dedication to the performing arts, for example, by queueing long hours for tickets to ballet and opera performances at Covent Garden.

In London's magnificent St. Paul's Cathedral, there is a memorial to the building's distinguished architect, Christopher Wren. Considering Michael's contributions to Cell and Developmental Biology, in general, and to this Society in particular, the words, as paraphrased from that plaque, suit him too:

"If you seek his monument, look around you."

(1) Nature 281, 163 (1979).

(2) Biographical Memoirs of Fellows of the Royal Society 26, 1 (1980).

(3) Nature 281, 259 (1979).

IN MEMORY OF J.C. DAN (1910-1978)

T.S. OKADA

Institute for Biophysics, Faculty of Science,
University of Kyoto
Kyoto, Japan, 606

Jean Clark Dan, former Professor in Biology at Ochanomizu Women's University, Tokyo, passed away on November 13, 1978, at Tateyama near Tokyo, Japan. Her numerous important contributions to the study of fertilization in marine animals are very well known. Particularly, her name will remain immortal in the field of Developmental Biology as a discoverer of the acrosomal reaction, which is indeed the critical moment that raises the curtain of the most exciting drama of animal development.

She was born in the United States, married with a distinguished Japanese biologist, Prof. Katsuma Dan, and spent her later life as a Japanese lady. Many people think of her as a bridge between Japanese biology and international biology, particularly when Japanese scientists still had great difficulty in their international activities. Among her many achievements demonstrating this, I should like to mention, for example, the fact that the first phase-contrast microscope which the Japanese ever had, was introduced to Japanese scientists as a result of her visit to her mother country immediately after the war. You can easily imagine that many young Japanese biologists were then enormously stimulated by a new world which was now open before them through this "ultra-modern" microscope of that time. A number of observations on fertilization were, of course, made by herself and they became an indispensable prelude to her great discovery of the acrosomal reaction.

She never stopped being kind and optimistic throughout her life, even during the most difficult times of serious conflicts between Japan and other countries. We really see her as a great lady. However, she hated most to be referred in this manner. She was always modest and reserved. As Prof. Mazia (1) wrote in his article dedicated to the memory of J.C. Dan, she remains ever as a "Sister" to all of us.

(1) Mazia D (1980). Dr. J.C. Dan. Dev Growth Differ (Memorial Issue of J.C. Dan): xiii.

IN MEMORY OF VALERIO MONESI

A. Monroy

Zoological Station

Naples, Italy

Valerio Monesi died, while he was on holiday in Sicily with his family, of a heart attack, on 29 December 1979, at the age of 51.

He was Professor of Histology and Embryology at the Medical School of the University of Rome. After obtaining his M.D. in 1953 at the University of Pavia, he spent some years at the Oak Ridge National Laboratory working on the cell cycle and cellular radiosensitivity, which he approached by using mammalian spermatogenesis as a model.

In 1968 he became Professor of Histology and Embryology at the University of Siena and the following year he moved to the Medical School of the University of Rome. There he set up a new Institute of Histology and Embryology which under his stimulating and friendly leadership soon became a major centre for studies on reproductive and developmental biology.

He was fascinated by the study of cell differentiation and of cell cycle control; indeed, these problems were his main interest during twentyfive years of active research. The main contributions of Monesi's research deal with the regulation of mammalian spermatogenesis and in particular with the characterization of the spermatogonial cell cycle, the study of meiotic and post-meiotic macromolecular synthesis, and the role of cyclic nucleotides in the differentiation of the male gamete. In the early 60's, using autoradiographic techniques, he was able to evaluate, in the mouse, the number of spermatogonial generations occurring between A_1 and B spermatogonia. He showed that spermatogonial proliferation is characterized by a constant generation time, a gradual increase in the duration of the S phase and a parallel shortening of G_2. He showed that the radiosensitivity of spermatogonia increases with differentiation, and correlated these data with the gradual increase of the S phase during spermatogonial proliferation. To explain spermatogonial stem cell renewal, he proposed a model widely accepted until quite recently, which postulated that in the mouse A_1 spermatogonia there are stem cells capable of producing stem cells and differentiated A_2 spermatogonia by bivalent mitosis.

Of fundamental interest are his early autoradiographic studies, which showed that in the mouse the stages of midlate pachytene spermatocyte and of round spermatid are the most active in the synthesis of ribosomal and messenger RNA. He showed that at least part of the RNA synthesized during meiosis and early spermiogenesis is still present in the cytoplasm at more advanced stages of differentiation, thus suggesting that haploid germ cell differentiation may be controlled by the diploid as well as by the haploid genome. During these studies he found that the XY bivalent is invariably genetically inactive throughout meiotic prophase.

His most recent work was concerned with the study of cyclic nucleotide-dependent pathways in differentiating germ cells. The early results of his involvement in this field led to the demonstration of changes in iso-enzymes of cAMP-dependent protein kinase during spermatid maturation.

He was a founding member and past President of the Italian Society of Reproductive and Developmental Biology and a member of the Directory Board of ECBO and ISDB and of the Executive Committees of EDBO.

His untimely death has deprived Developmental Biology of a brilliant scientist and of a man whose charming personality made him highly respected by a wide circle of friends all over the world.

STUDIES ON THE ROLE OF EXOGENOUS CALCIUM IN FERTILIZATION,
ACTIVATION AND DEVELOPMENT OF THE SEA URCHIN EGG.

Tobias Schmidt

Stanford University, Hopkins Marine Station

Pacific Grove, California 93950

INTRODUCTION

Large changes in ion permeability accompany fertilization of the sea urchin egg. The earliest ionic change detected after successful interaction of both gametes is an influx of Ca^{2+} from the seawater (Paul & Johnston, 1978; Chambers & deArmendi, 1979), but the role of this influx in fertilization and activation is unknown. Since activation of the sea urchin egg results from a transient rise in intracellular Ca^{2+}, most likely from intracellular stores, one might expect a role of this Ca^{2+} influx in the activation which might be a case of Ca^{2+}-induced Ca^{2+} release (Steinhardt & Epel, 1974; Chambers et al., 1974; Gilkey et al., 1977). It could also be part of the fast block to polyspermy (Jaffe, 1976).

An assessment on the role of this Ca^{2+} influx, by attempting fertilization in Ca^{2+} free seawater, has so far given ambiguous or conflicting results (Takahashi & Sugiyama, 1973; Sano & Kanatani, 1980; Chambers, 1980). An attempt to reconcile these results is reported in this paper, which shows that exogenous Ca^{2+} (and hence the Ca^{2+} influx) is not essential for fertilization and the subsequent events in S. purpuratus and L. pictus. It is, however, required for the stability of the unfertilized egg and for the structural integrity of the embryo. The data support the hypothesis that sperm induce release of Ca^{2+} from intracellular stores. They suggest that under certain conditions, which will be discussed, Ca^{2+} influx might be important in augmenting this sperm induced intracellular Ca^{2+} release.

RESULTS & DISCUSSION

Fertilization and Activation

The fertilization protocol. It is not possible to assess fertilization in Ca^{2+} free medium since the sperm must first undergo the acrosome reaction and this requires exogenous Ca^{2+}. The problem can be circumvented, however, by first inducing the acrosome reaction in Ca^{2+}-containing seawater and then adding acrosome reacted sperm to eggs in Ca^{2+} free/ EGTA buffered seawater (OCaSW).
Immediately before fertilization 20 μl of eggs were washed 3x in OCaSW and then resuspended in 2 ml of OCaSW containing 1 mM EGTA & 10 mM Glycylglycine (Gg), pH 8. Then 5 μl of dry sperm were suspended in 50 μl of seawater (ASW) and acrosome reacted 10 sec thereafter by adding 50 μl of egg jelly. The reaction was stopped after 5 sec by adding 500 μl OCaSW containing 10 mM EGTA & 50 mM Gg, pH 8.2. The spermsuspension was then immediately added to the egg suspension. An elevated Gg buffer concentration as well as a higher external pH were used for the chelation of free Ca^{2+} after the acrosome reaction since this process would otherwise strongly acidify the seawater. The final free Ca^{2+} concentration was calculated to be $<10^{-11}$M (Steinhardt et al., 1977). Fertilization in *S. purpuratus* was determined by counting the % of Carnoy's fixed eggs with a cortical reaction, using phase contrast microscopy on the fixed eggs previously suspended in 45% acetic acid. This method is more reliable than counting fertilization membranes since these are fragile in OCaSW and can hardly be seen under phase optics. In *L. pictus* this method can not be used and sperm entry as well as development were used as criteria.
The parameters of fertilization: temperature, sperm titer and timing of acrosome reaction. In preliminary experiments fertilization was obtained in the complete absence of Ca^{2+} using the described protocol. The initial results, however, varied widely from day to day, with a fertilization efficiency between 0 and 60%. An examination of the experimental conditions revealed that there was an apparent correlation with the temperature at which fertilization was attempted.
Therefore the effects of temperature on fertilization were studied. As shown in Tab. 1, fertilization rates of 90-95% were obtained only at temperatures below 10-12°C. At higher temperatures the rate dropped rapidly. Egg lysis

increased, indicating that the elevated temperature was affecting membrane stability. The temperature effect was on the eggs since the temperature at which the sperm were acrosome reacted did not alter the fertilization rate.

% cortical reaction at	5	10	15	20	°C
in ASW	100	99	100	100	
in OCaSW	92	90	58	27	

Tab.1: Temperature effect on the fertilizability of S. purpuratus eggs. Fertilization was performed at indicated temperatures. Sperm were acrosome reacted for 5 sec at 10°C.

To determine whether the rate of fertilization was comparable to that in ASW, eggs were inseminated at 10°C with different sperm concentrations. The sperm volume was kept constant to assure identical conditions. Five minutes after adding the acrosome-reacted sperm, the eggs were fixed in Carnoy's fixative and the percentage of fertilization determined by assessing cortical granule breakdown. This experiment revealed that in OCaSW fertilization requires 20 times more sperm than in ASW a 1:5,000 sperm dilution is required for 50% fertilization in OCaSW as opposed to a 1:50,000 dilution in normal seawater. Transmission electron microscopy on formaldehyde fixed and uranyl acetate stained sperm revealed that 85% of the sperm used in the experiments were acrosome reacted.

A problem in using acrosome reacted sperm is that their fertilizability decreases rapidly (see e.g. Vacquier, 1979). In order to define optimal conditions for fertilization in OCaSW, the time of acrosome reaction was varied. Maximal fertilization in OCaSW was obtained with 5 sec acrosome reacted sperm. Prolonged jelly treatment decreased the percentage of fertilization and a 20 sec treatment allowed only 50% of the eggs to be fertilized (which therefore corresponds to the fertilization "half-life" of these sperm; a similar half-life was also seen by Vacquier (1979), using sperm reacted with jelly in ASW).

Interpretation of the data. The present study indicates that eggs of two different species, S. purpuratus and L. pictus can be fertilized in the complete absence of exogenous calcium ($<10^{-11}$M). The previous contradictory findings most probably resulted from the reduced fertilizability in OCaSW, which can be circumvented by using higher sperm concentrations, and from the unexpected sensitivity to temperature in OCaSW, which can be circumvented by fertilizing at temperatures lower than 12°C.

It is surprising that sperm-egg fusion takes place in the absence of Ca^{2+} since many studies on cell fusion indicate that fusion normally depends on this cation (see e.g., Lucy, 1977). Fish eggs can also be fertilized in Ca^{2+} free media (Nuccitelli, 1980) so perhaps sperm-egg fusion is an exception to this rule. Another possibility is that the sperm carries a high concentration of calcium at the site specialized for fusion or that the egg has high affinity calcium binding proteins at the putative fusion sites which are not removed under the Ca^{2+} free conditions. Finally, a Ca^{2+} involvement in fusion could account for the observed twenty fold reduction in fertilization. In this view, Ca^{2+} is involved in fusion but is not essential for fusion.

The experiments indicate that the Ca^{2+} influx normally seen during fertilization is not essential. This supports the concept that the activation of the egg results from release of Ca^{2+} from intracellular stores (Steinhardt & Epel, 1974; Chambers et al., 1974) which had been suggested since divalent ionophores activate eggs in Ca^{2+}- free media. However, the argument could be made that the ionophores released Ca^{2+} from a store normally not used during fertilization. It had also been postulated that this Ca^{2+} release is Ca^{2+}- mediated and results either from the influx of Ca^{2+} from the seawater or from Ca^{2+} introduced by the sperm after fusion of both gametes. The data indicate that if there is a Ca^{2+}- mediated Ca^{2+} release it can only be introduced into the egg via the sperm (Gilkey et al, 1979).

Successful fertilization in OCaSW can only be obtained at low temperature. This temperature effect appears to result from changes in the egg since sperm do not lose their fertilizing capacity at the higher temperature and since sperm binding is the same in OCaSW as compared to ASW. This was shown by fixing eggs 10 sec after insemination with glutaraldehyde and counting the number of sperm around the circumference of these eggs (Kato & Sugiyama, 1978). The detrimental effect of the higher temperature on the eggs may relate to membrane stability; unfertilized eggs lysed in OCaSW and this effect was promoted at the higher temperatures. Interestingly, the OCaSW effect on the stability of the cells is not seen in fertilized eggs.

A second explanation for the lower temperature needed to attain fertilization in OCaSW is that the sperm somehow changes a critical equilibrium between rates of Ca^{2+} release and Ca^{2+} sequestration to induce the massive Ca^{2+} increase seen at fertilization, and that this equilibrium is highly temperature sensitive. If so, one would hypothesize that the sperm-induced change in equilibrium at lower temperatures (<15° C) is by itself sufficient to increase the Ca^{2+} concentration high enough to initiate a propagated Ca^{2+} release. At higher temperatures (>15°C), however, the change in equilibrium is not sufficient to induce Ca^{2+} release and an additional influx of Ca^{2+} from the seawater that normally accompanies fertilization is now necessary. Once stimulated, however, the propagated Ca^{2+} release can now continue in the absence of any exogenous source. This view, if correct, would indicate that there is a role for the Ca^{2+} influx in normal fertilization, but only at the upper limits of the egg's temperature range.

An alternative view, also based on this equilibrium concept, is that the temperature effect is simply an artefact of the unnatural conditions of fertilization in OCaSW. It is probable that the Ca^{2+} efflux of the cell is increased in OCaSW since the concentration gradient is now reversed. If so, the sperm would have to induce a greater Ca^{2+} release than normal in order for the Ca^{2+} to achieve levels sufficient to activate the egg. Higher temperatures could increase this efflux even more and thus prevent activation; lower temperatures could decrease this efflux to more natural levels and allow activation to occur.

This equilibrium view is also consistent with the phenomenon of partial activation of eggs, such as seen when ionophore A 23187-coated rods or beads contact egg surfaces. These eggs exhibit a localized cortical reaction (Chambers & Hinkley, 1979; Epel, 1980) and metabolism is only partially activated (Whittaker & Steinhardt, 1981). Several workers (Chambers & Hinkley, 1979; Jaffe, 1980) have suggested that these results can be explained if the rate of Ca^{2+} release is not great enough to trigger further Ca^{2+} release. A related view would be that a Ca^{2+} increase, sufficient to completely activate the egg, also requires a shift in equilibrium between sequestration, efflux and release.

Development

The formation of the fertilization envelope is the first visible event after the union of both gametes. This envelope elevates upon fertilization in OCaSW and Carnoy's fixation indicates that propagation and complete exocytosis of the cortical granules has taken place in S. purpuratus. The fertilization coat was thinner than normal and less resistant to mechanical stress.

Development was followed through the first 12 hours. Therefore the egg suspension was diluted to a final concentration of 1 mM EGTA at 5 min after insemination. Although the timing of development, as judged by cleavage, was similar to the ASW control, blastomere adhesion was affected under Ca^{2+}- free conditions. The cells were unorganized and only in loose contact with each other. Fertilization membranes were seldomn detectable. Fig.1 depicts a 7- hour old

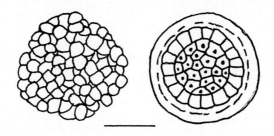

Fig.1: Seven hour old L. pictus embryo raised in OCaSW and ASW. Bar equals 50 um.

L. pictus embryo fertilized and raised in OCaSW as well as one raised in ASW. This indicates that Ca^{2+} is not required for cleavage but for the structural integrity of the developing embryo.

Fertilization Related Events

Fast block to polyspermy. Fertilization, activation and cleavage do not seem to be dependent on exogenous Ca^{2+} and the Ca^{2+} influx. Could there be a role for the influx at fertilization other than in activation. Polyspermy prevention might be one role for the Ca^{2+} influx. The initial depolarization is related to the Ca^{2+} influx (Chambers & deArmendi, 1979) and Jaffe (1976) has presented data showing (1) a relationship between the amplitude of depolarization and susceptibility to polyspermy and (2) a direct relationship between membrane potential and polyspermy since depolarizing eggs (to greater than 0 mV) prevents their fertilization. One would therefore predict that polyspermy should be increased in OCaSW, since the initial depolarization should not occur (Chambers & deArmendi, 1979). This increase in polyspermy was not observed, but might have been impossible to see since fertilizability was reduced by 20-fold (i.e., 20x more sperm were required).

Cortical reaction. As mentioned earlier the cortical reaction is complete after fertilization in OCaSW. Thus Ca^{2+} influx does not seem to be involved in the induction or propagation of the cortical reaction. It still, however, could influence the kinetics of this reaction. This was tested by comparing the rate of cortical reaction in OCaSW versus ASW. S. purpuratus eggs were fixed at different times after fertilization and the extent of cortical reaction was scored as a function of time. Tab.2 shows that, when using identical sperm concentrations, the kinetics of this process is faster in ASW. However, when the amount of sperm used for

% cortical reaction at		20	40	60	90	120	150 sec
ASW;	1:500 sperm dil.	4	9	97	99	99	99
ASW;	1:10000 sperm dil.	0	2	7	56	88	96
OCaSW;	1:500 sperm dil.	0	4	6	59	89	93

Tab.2: Kinetics of the cortical reaction of S. purpuratus eggs following fertilization in OCaSW at 5°C. The % of eggs was determined having a 'partial' to 'total' cortical reaction.

fertilization in ASW was reduced by a factor of 20 to achieve the same fertilizing capacity as sperm in OCaSW, there was no difference in the kinetics of the cortical reaction. This indicates that the presence of extracellular Ca^{2+}, and presumably the related Ca^{2+} influx, does not induce or influence the propagation or kinetics of the cortical reaction.

SUMMARY

Both isotopic and microelectrode studies reveal a significant Ca^{2+} influx at fertilization. The role, if any, of this influx is disputed. An attempt to reevaluate contradictory findings by others on this role was made and the results with S. purpuratus and L. pictus eggs, using acrosome reacted sperm and EGTA- buffered media (free $[Ca^{2+}]$

$<10^{-11}$M), indicate that exogenous Ca^{2+}, and hence the Ca^{2+} influx, is not required for fertilization, activation and subsequent cleavage. Blastomere adhesion, however, is affected. The contradictory findings by others may have resulted from reduced fertilizability in Ca^{2+}- free seawater, which can be circumvented by higher sperm concentration and by a sensitivity to temperature in Ca^{2+}- free medium, which can be bypassed by carrying out fertilization at lower temperatures. Under the assay conditions, there is also no effect of Ca^{2+}- free media on the kinetics of the cortical reaction or polyspermy. The data support the hypothesis that sperm induce release of Ca^{2+} from intracellular stores, perhaps by affecting an equilibrium between Ca^{2+} sequestration and Ca^{2+} release.

Acknowlegement: This work was done in collaboration with Prof. D. Epel and C. Patton whose contributions are greatfully acknowledged. Supported by DFG and NSF.

REFERENCES

Azarnia R, Chambers EL (1976). The role of divalent cations in activation of the sea urchin egg. I. Effect of fertilization on divalent cation content. J Exp Zool 198:65.
Chambers EL, de Armendi (1979). Membrane potential, action potential and activation potential of eggs of the sea urchin Lytechinus variegatus. Exp Cell Res 122:203.
Chambers EL, Hinkley RE (1979). Non-propagative cortical reactions induced by the divalent ionophore A 23187 in eggs of the sea urchin Lytechinus variegatus. Exp Cell Res 124:441.
Chambers EL (1980). Abstracts of the 2nd Int. Congress of Cell Biology, Berlin. European J Cell Biol 22:476.
Chambers EL, Pressman BC, Rose B (1974). The activation of sea urchin eggs by the divalent ionophores A 23187 and X 537 A. Biochem Biophys Res Comm 60:126.
Epel D (1980). Ionic triggers in the fertilization of sea urchin eggs. Annals of the New York Acad Sci 339:74.
Gilkey JC, Jaffe LF, Ridgeway EB, Reynolds GT (1978). A free calcium wave traverses the activating egg of the medaka, Oryzias latipes. J Cell Biol 76:448.
Jaffe LA (1976). Fast block to polyspermy is electrically mediated. Nature 261:68.
Jaffe LA, Robinson KR (1978). Membrane potential of the unfertilized sea urchin egg. Dev Biol 62:215.

Jaffe LF (1980). Calcium explosions as triggers of development. Annals of the New York Acad Sci 339:86.
Kato KH, Sugiyama M (1978). Species specific adhesion of spermatozoa to the surface of fixed eggs in sea urchins. Devel Growth and Differ 20:337.
Lucy JA (1978). Mechanisms of chemically induced cell fusion. In Poste G, Nicholson GL (eds):"Membrane Fusion", Cell Surface Reviews, Amsterdam: North Holland Publishing Co 5:268.
Nuccitelli R (1980). The electrical changes accompanying fertilization and cortical vesicle secretion in the medaka egg. Dev Biol 76:483.
Paul M, Johnston RN (1978). Uptake of Ca^{2+} is one of the earliest responses to fertilization of sea urchin eggs. J Exp Zool 203:143.
Sano K, Kanatani H (1980). External Ca^{2+} ions are requisite for fertilization of sea urchin eggs by spermatozoa with reacted acrosomes. Devel Biol 78:242.
Steinhardt RA, Epel D (1974). Activation of sea urchin eggs by a Ca^{2+}- ionophore. Proc Nat Acad USA 71:1915.
Steinhardt RA, Zucker R, Schatten G (1977). Intracellular calcium release at fertilization in the sea urchin egg. Devel Biol 58:185.
Takahashi Y, Sugiyama M (1973). Relation between the acrosome reaction and fertilization in the sea urchin egg. I. Fertilization in Ca^{2+}- free seawater with egg-water treated spermatozoa. Dev Growth and Differ 15:261.
Vacquier VD (1979). The fertilizing capacity of sea urchin sperm rapidly decreases after induction of the acrosome reaction. Dev Growth and Differ 21,1:61.
Whittaker MJ, Steinhardt RA (1981). The relationship between the increase in reduced nicotinamide nucleotide and the initiation of DNA synthesis in sea urchin eggs. Cell 25,1:95.

CELL-CELL INTERACTIONS AND THE ROLE OF MICROMERES IN THE CONTROL OF THE MITOTIC PATTERN IN SEA URCHIN EMBRYOS

P.Andreuccetti,[1] S.Filosa,[1] A.Monroy[2] & E.Parisi[3]

[1]Istituto di Istologia ed Embriologia, Università di Napoli; [2]Stazione Zoologica, Napoli; [3]Istituto di Embriologia Molecolare, CNR, Arco Felice, Italy.

INTRODUCTION

Cleavage of the egg is the process whereby the zygote acquires the multicellular state. It is conceivable that spatial and temporal coordination of cell divisions is a prerequisite for embryo growth.

In the sea urchin the first four cleavage cycles are characterized by the synchronous divisions of the blastomeres; however, at the fourth cleavage, with the segregation of the micromeres, four small cells arising from the unequal division of the vegetal blastomeres, a slight asynchrony of cell divisions begins to appear in the embryo. Indeed, from this stage and up to the early blastula stage, cell divisions occur in waves moving from the vegetal to the animal pole; the blastomeres closer to the micromeres enter mitotis before those further from them (Parisi et al. 1978). This and the observation that the "mitotic gradient" sets in at the time of segregation of the micromeres, led us to the hypothesis that these cells may act as "pacemakers" of mitotic activity (De Petrocellis et al. 1977). A direct corollary to this hypothesis is that interfering with the function of the micromeres should upset the mitotic pattern. Indeed, the mitotic gradient is disrupted in sea urchin embryos exposed to Actinomycin D prior to the segregation of the micromeres; while this does not occur when the Actinomycin D treatment is started <u>after</u> their segregation (Parisi et al. 1979). This suggests that although cell division is independent of RNA synthesis, the coordination of mitosis is under direct genetic control. In particular we have proposed that Actinomycin D inhibits a transcriptional event which specifically controls the pacemaker function of the micromeres (Parisi et al. 1979). This hypothesis is consistent with the observation that the micromeres are the site of synthesis of an RNA not completely homologous with the RNA transcribed in the other blastomeres (Mizuno et al. 1974).

The propagating waves of the mitotic activity may be explained in terms of a chemical signal which has the characteristics of a relaxation oscillator which is generated within each cell. If the frequency of the periodic signal in the micromeres is increased, the signal fires slightly before that of the other cells, thus giving rise to a pacemaker wave which propagates upward towards the animal pole (Parisi et al. in press). Hence, the spatial-temporal distribution of the mitotic signal is mediated by chemical diffusion and it can only propagate along the embryonic axis as a travelling wave at critical values of the diffusion coefficients of the chemical species involved in the process (Catalano et al. 1981). For this to occur, cells arising from the cleavage of the egg must be functionally coupled via intercellular communications.

Although the existence of cellular junctions in animal cells is a well-established fact (Staehelin, 1974), little information is available concerning the early sea urchin embryo (Gilula, 1973; Millonig and Giudice, 1967).

We present evidence of intercellular communications in the sea urchin embryo during cleavage. In addition, we present the results of experiments in which, by preventing the segregation of the micromeres, no mitotic gradient arises at the 16-cell stage; i.e. in these embryos cleavage proceeds synchronously in all blastomeres just as in the embryo prior to the formation of the micromeres.

RESULTS

Intercellular Communications

The ultrastructural analysis performed on embryos of Paracentrotus lividus at stages preceeding the segregation of the micromeres shows that the membranes of adjacent blastomeres run parallel for long stretches separated by an intervening space 50-60Å wide. After the fourth cleavage the stretches between adjacent blastomeres become shorter and some cell-cell connections are formed: mainly intercellular bridges and desmosome-like junctions (Figs. 1 and 2).

Intercellular bridges appear between micromeres and macromeres (Fig. 1) and between mesomeres (Fig. 2a); they seem to result from the fusion of two adjacent cell membranes and contain electron-dense material (mid-body) and residues of microtubules, the latter departing from the mid-body and diffusing into the cytoplasm of the interconnected blastomeres (Fig. 2 a and b). These are clearly in interphase as the nuclear envelopes are completely reconstituted and the nuclei contain the typical nucleoli (Fig. 2a).

Desmosome-like junctions are also present in the proximity of the intercellular bridges; electron dense material is attached to the cytoplasmic surface of the adjacent plasma membranes and there is an intercellular space measuring about 150Å (Fig. 2 a and c). In addition to these two classes of cellular connections we have identified regions where the membranes appear so closely apposed that any intervening space cannot be resolved (Fig. 2 a and d).

In the freeze-fracture preparations we have also detected on the fracture face of the membrane, arrays of packed particles (Fig. 3), which we interpret as gap junctions given their similarity with those described in other systems (Gilula et al. 1978).

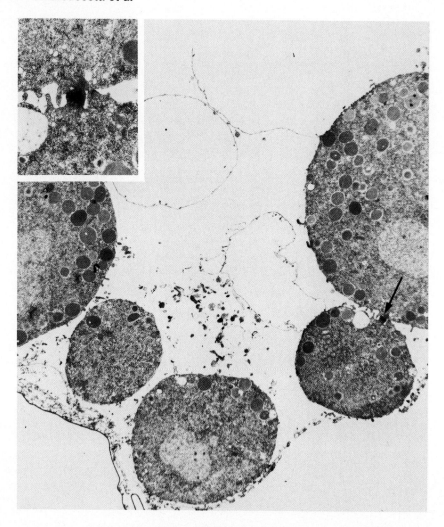

Fig. 1. Intercellular bridges in an embryo of <u>Paracentrotus lividus</u> at the 32-cell stage. A micromere and a macromere are connected by a cytoplasmic bridge (arrow) (X 4000). Insert: higher magnification of the bridge (X 120,000).

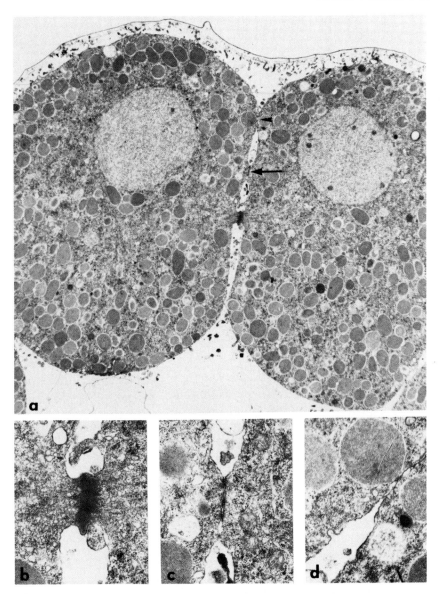

Fig. 2. Intercellular junctions in an embryo at the 32-cell stage. a) Two mesomeres showing a cytoplasmic bridge (see also panel b), a desmosome-like junction (arrow and panel c) and stretches of very close apposition of the membranes (arrow head and panel d). a) X4000, b) X18,000; c) X18,000; d) X30,000.

Fig. 3. Gap junctions in an embryo at the 32-cell stage.
a) Freeze cleaved appearance of the plasma membrane showing many intramembraneous particles, both packed (arrow) and randomly distributed (X 22,000). Insert: higher magnification of the aggregated (X 45,000).

Role of the Micromeres

Tanaka (1976) has shown that when sea urchin embryos at the mid 4-cell stage are briefly treated with certain detergents at the fourth cleavage, the micromeres fail to segregate and an embryo made up of 16 equal-sized cells is formed. The embryos are viable and develop into a blastula which lacks the primary mesenchyme.

About 70% of these embryos exhibited a complete synchrony of cell divisions (Fig. 4a). In the remaining 30% some asynchrony among the blastomeres was observed which, however, never resulted in a mitotic gradient. This effect cannot be attributed to the detergent per se since embryos treated at the 16-cell stage, i.e. when the micromeres have already segregated, display the typical mitotic gradient (Fig. 4b).

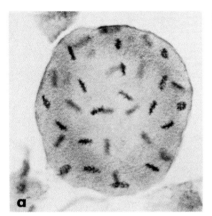 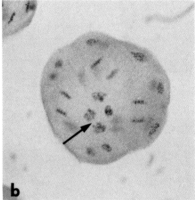

Fig. 4. Whole mounts of flattened embryos treated with sodium dodecyl sulphate (SDS). a) Embryo at the 64-cell stage treated with SDS at the mid 4-cell stage. The micromeres are lacking and all the nuclei are in metaphase. b) Embryo at the mid 32-cell stage treated with SDS at the onset of the 16-cell stage. The micromeres are present and the mitotic gradient (macromeres in anaphase, mesomeres in pro-metaphase) is clearly seen. Feulgen X 510.

DISCUSSION

Our observations show that the blastomeres of the early sea urchin embryo are joined by cellular connection which may provide a suitable pathway for cell-to-cell communication.

The presence of cytoplasmic bridges between interphase cells suggests that the blastomeres may form a kind of syncitial structure for a significant period of the cleavage cycle. This is supported also by some electrophysiological and ultrastructural studies by Dale and his colleagues (personal communication) on the first cleavage of the sea urchin embryo. It is likely that these cytoplasmic bridges may result from incomplete cytokinesis; their presence in early embryos has been inferred (Lo, 1980) but rarely observed (Schmekel, 1970).

The finding that cleaving sea urchin eggs contain gap junctions whose presence has also been detected in embryos of other species (see Lo 1980 for a review), suggests that these cell contacts may play an important role during embryogenesis. Indeed, gap junctions have been considered responsible for the formation of morphogenetic gradients (Lo, 1980) and for the control of growth (Loewenstein, 1979).

It is possible that both cytoplasmic bridges and gap junctions may provide a tool for the transmission of mitotic signals from cell to cell, although perhaps at different rates. Indeed, the bridges may provide a low-resistance pathway through which signals can rapidly diffuse; while gap junctions may be responsible for the transmission of time-delayed signals.

The results obtained with embryos where micromeres segregation was prevented further support the hypothesis that these cells act as coordinators of mitotic activity. This implies that the micromeres produce some kind of regulator of mitotic pace.

REFERENCES

Catalano G, Eilbeck JC, Monroy A, Parisi E (1981). A mathematical model for pattern formation in biological systems. Physica D: Non linear phenomena 3: 439.

De Petrocellis B, Filosa-Parisi S, Monroy A, Parisi E (1977). Cell Interactions and DNA replication in the sea urchin embryo. In Lash JW, Burger MM (eds): "Cell and Tissue Interactions", New York: Raven Press, p 269.

Gilula NB (1973) Development of cell junctions. Amer Zool 13: 1109.

Gilula NB, Epstein ML, Beers WH (1978). Cell-to-cell communication and ovulation. A study of the cumulus-oocyte complex. J Cell Biol 78: 58.

Lo WC (1980). Gap junctions and development. In Johnson MH (ed): "Development in Mammals, vol9", New York: Elsevier North-Holland Inc, p 39.

Loewenstein WR (1979). Junctional intercellular communication and the control of growth. Biochim Biophys Acta 560: 1.

Millonig G, Giudice G (1967). Electron microscopic study of the reaggregation of cells dissociated from sea urchin embryos. Dev Biol 15: 91.

Mizuno S, Lee YR, Whiteley AH, Whiteley HR (1974). Cellular distribution of RNA populations in the 16-cell stage embryos of the sand dollar, Dendraster excentricus. Dev Biol 37: 18.

Parisi E, Filosa S; De Petrocellis B, Monroy A (1978). The pattern of cell division in the early development of the sea urchin, Paracentrotus lividus. Dev Biol 65: 38.

Parisi E, Filosa S, Monroy A (1979). Actinomycin D-disruption of the mitotic gradient in the cleavage stages of the sea urchin embryo. Dev Biol 72; 167.

Parisi E, Filosa S, Monroy A (in press). Regulation of cell divisions in the sea urchin embryo. In Ricciardi LM, Scott A (eds) "Proceedings of biomathematics: current status and future perspectives", Amsterdam: North-Holland Publishing Co.

Schmekel L (1970). Elektronmikroskopie der Makromeren-Mikromerengreuze des Seeigelkeimes. Zool Amz Suppl 334: 141.

Staehelin LA (1974). Structure and function of intercellular junctions. Int Rev Cytol 39: 191.

Tanaka Y (1976) Effect of the surfactants on the cleavage and further development of the sea urchin embryo. Develop Growth and Differ 18: 113.

Some effects of heat shock on the regulation of macromolecular syntheses in sea urchin embryos.

M.C.Roccheri[o], G.Spinelli[o], G.Sconzo[o], C.Casano[o]
and G.Giudice[o+]
Institute of Comparative Anatomy, University[o]
Institute of Developmental Biology C.N.R.[+]
Via Archirafi 20-22 Palermo.

Heat treatment has been found to induce in the cells from a variety of organisms a profound inhibition of the bulk protein synthesis, together with the stimulation of the synthesis of a few proteins, the so called "heat shock proteins" (h.s.p.) (Ritossa,1962; Ashburner and Bonner,1979). This phenomenon presents a twofold interest, first because it offers a practical means for studying the regulation of the synthesis of a small class of proteins by strongly reducing the interference brought about by the enormous background of the synthesis of the bulk proteins, and second because, as it is also suggested by the experiments to be reported here, it may represent a mechanism, widespread in nature for the defence of the eucariotic cells by the stress brought about a change of temperature. For these reason we felt important to investigate the effect of a heat shock on entire embryos, like those of sea urchins, whose developmental mechanisms have been and are being investigated under so many respects (Giudice,1973).

If one submits for 45 minutes a culture of developing sea urchins at the stage of gastrula to increasing temperatures and then pulse labels them for 15 minutes with radioactive leucine, and looks at the pattern of protein synthesis by phluorography of an SDS-polyacrylamide gel electrophoresis, as shown in in fig.1, observes that at the temperature of 31ºC the synthesis of the bulk proteins, which form as expected a smear, in the control embryos (i.e.cultured at 20°C), is strongly inhibited, while two closely migrating very intense bands corresponding to the molecular weights of 72,500 and 70,000 daltons, the h.s.p.,appear. Also other 5 minor h.s.p.are observed; we will however refer to the two major bands throughout the paper for the sake of simplicity, since they are especially evident under all the experimental conditions to be reported.

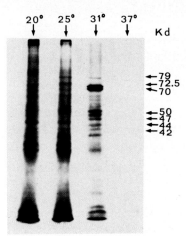

Fig.1 Effect of heat treatment at different temperatures on the pattern of protein synthesis of Paracentrotus lividus gastrulas. (From Roccheri et al.,1981, Dev.Biol.83: 173).

Having found that gastrulas respond to heat shock with the production of h.s.p.,we have asked a series of questions about the mechanism of h.s.p.production as well as about the mechanism of inhibition of the bulk protein synthesis. The first question is whether all the embryonic tissues or only some of them synthesize the h.s.p.. The only method available to separate embryonic sea urchin tissues in bulk amount is that described by Mc Clay (1978), which allows to separate the ectoderm of plutei from the other non ectodermic tissues.

If one then heats the plutei, pulse labels them and then looks at the pattern of protein synthesis of the ectoderm and of the non ectodermic tissues, one finds, as shown in figure 2, that the ectoderm is certainly able to synthesize the h.s.p., whereas the other tissues do not seem to be so, although the possibility of digestion of the h.s.p. by the intestinal enzymes cannot be ruled out with certainty.

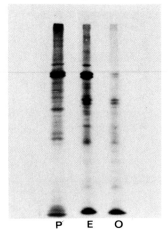

Fig.2 Electrofluorograms of the proteins of plutei labeled after heat shock; P= entire plutei; E= ectoderm; o= other tissues.

A second immediate question is where in the cell are the h.s.p. located. A cell fractionation performed shortly after heat shock shows (fig.3) that the two major h.s.p. are concentrated in the cytoplasm and that at least one of them is also present in the nucleus. If on the other hand a chase is performed, by incubating the embryos at normal temperature in the absence of labeled aminoacids, while the h.s.p. remain highly concentrated in the cytoplasm for at least further 22 hours, those of the nuclei disappear after 5 hours.

The meaning of this transient appearance of the h.s.p. in the nucleus is at present not known, but it is intriguing in view of a possible role of the h.s.p. in regulating transcription. Where in the cytoplasm are they located? A centrifugation of the cytoplasm of the heat treated embryos such as to enlarge the RNP particle region, by pushing the ribosomes to the bottom of the tube shows (fig.4) that the highest concentration of the h.s.p. is in the soluble cytoplasm. Very little if any of them is found the RNP particle region, therefore making the hypothesis improbable that the h.s.p. protein themselves inhibit the bulk protein synthesis by complexing their mRNAs.

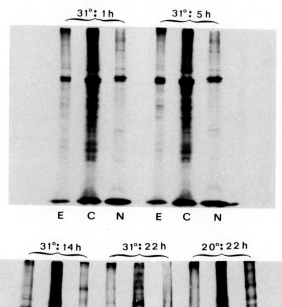
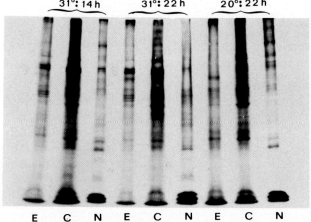

Fig.3. Electrofluorograms of proteins from subcellular fractions of early gastrulas at various times after heating:
E= entire embryos; C= cytoplasms; N= nuclei.

Heat Shock Proteins in Sea Urchin Embryos / 35

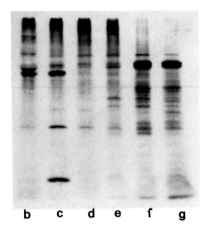

Fig.4. Electrofluorograms of the proteins of different zones of a sucrose gradient of the post-ribosomal supernatant; compartments from b trough e = RNP particles; f and g = soluble cytoplasm.

We have seen that the h.s.p. made at the stage of gastrula (and even blastula) persist in the cytoplasm at least till the pluteus stage, but for how long after the beginning of the heat treatment are they made? We have found that 15 minutes of heating is enough to elicit h.s.p. synthesis, inhibition of the bulk protein synthesis, and that this effect is enhanced by protracting heat treatment up to 2 hours, but if heating is further elongated to 4 hours or more, a reversal of the effect is observed and a complex pattern of protein synthesis is restored. Heating therefore appears to cause a wave of h.s.p. synthesis followed by a return to what looks like a normal pattern of protein synthesis. The return to a normal pattern of protein synthesis is even better observed if the embryos are returned to normal temperature after 1 hour of heating (fig.5).

We have up to now observed the effect of heating on embryos at the gastrula stage, but what happens if we heat the embryos at other stages?

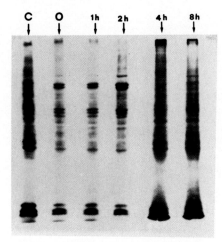

Fig.5. Reversal of the pattern of protein synthesis at various time lengths after 1 hr of heating of Paracentrotus gastrulas. C, control, nonheated embryos; O:1 hr of heating; 1hr,2 hr,4 hr,and 8 hr; 1 hr of heating followed by 1,2,4 and 8 hr, respectively, at normal temperature. (From Roccheri et al,1981, Dev.Biol.83:173).

Fig.6 shows that heating at any stage earlier than hatching blastula, although resulting in the inhibition of the bulk protein synthesis, does not elicit production of h.s.p.. Heating at any stage later that hatching, on the contrary causes both effects: inhibition of protein synthesis and production of the h.s.p.

It is very interesting in this context to observe the effect on development of heating at different developmental stages:

"Heating at any stage between fertilization and hatching blastula does not elicit h.s.p.production and causes immediate arrest of development followed by death.
Heating at any stage after hatching elicits h.s.p. production and development proceeds as normally".

This fact suggests a role for the h.s.p. in protecting the embryos from the heat stress.

There are indications also from other systems, e.g. Drosophila embryos (Graziosi et al.,1980) and yeast (Mc Alister and Finkelstein, 1980) that h.s.p. may have a role in protecting eucaryotic cells from temperature increases.

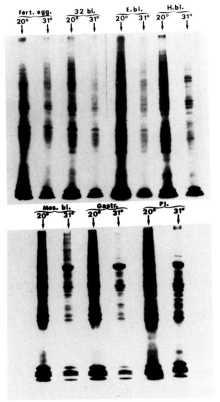

Fig.6. Effect on the pattern of protein synthesis of heating at 31°C at different developmental stages. Fert. egg.: fertilized eggs; 32 bl.: 32 blastomeres; E.bl.: early blastulas; H.bl.: hatched blastulas; Mes.bl.: mesenchyme blastulas; Gastr.: gastrulas; Pl.: plutei (From Roccheri et al.,1981, Dev. Biol. 83: 173).

The fact is also of interest that the ability to respond to heat shock with h.s.p. synthesis arises at a certain developmental stage. What is the mechanism which elicits in the embryos the ability to respond with such a synthesis? This is at present unknown and is being investigated in our laboratory.

As to the mechanism of h.s.p. production at the competent (postblastular) stages, we can conclude from our results that their synthesis is regulated through the production of a specific mRNA for the following reasons: first, if embryos are heated in the presence of actinomycin, no h.s.p. are produced, although the bulk protein synthesis is inhibited;

second, if the poly A+ or poly A- RNAs are extracted from polysomes and assayed for their biological activity in a reticulocyte cell-free system, they clarly (especially the poly A+) stimulate the synthesis of the two major h.s.p., whereas no such a stimulation is obtained with the RNAs from non heated embryos. The results shown in figure 7, which strongly suggest that the synthesis of h.s.p. is regulated at a transcriptional level, deserve some further comment. The RNA for the two major h.s.p. in the cell-free is not translated as preferentially with respect to the minor h.s.p. and to other proteins as it is in vivo. This can be explained with the findings of Scott and Pardue (1981) that, mRNAs for h.s.p. are preferentially used only by cell-free systems obtained by cells which are making h.s.p. in vivo.

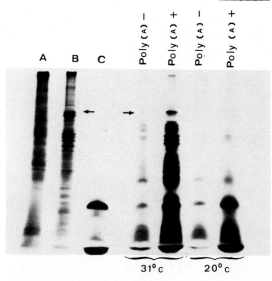

Fig.7. Electrofluorograms of proteins synthesized in vivo by gastrulas at normal temperature (A) and after 1 hour heating (B), and synthesized by a reticulocyte cell free system with no addition (C) or in the presence of 10 ug of poly (A)- or of 3 ug of poly (A)+ RNA deriving from polysomes of heat treated embryos or of embryos reared at normal temperature. The arrows indicate the position of the major h.s.p.

Whereas the mechanism of production of the h.s.p. has to be looked for at a transcriptional level, the inhibition of the bulk protein synthesis and its reversal seem to be post transcriptionally regulated. In fact, if we look at the polysome profile of normal and heated embryos, we see (fig.8) that a remarkable decrease of the polysomes, accompanied by a correspondent monosome increase, occurs after heating which is enough to explain all or most of the inhibition of the bulk protein synthesis. This disengagement of mRNA from polysomes is due, at least in Drosophila, to a decreased affinity of the ribosomes for the bulk mRNA.

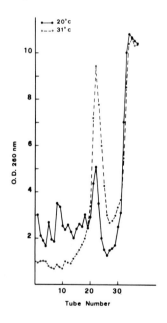

Fig.8. Comparison of the polysomal profiles from embryos reared at normal temperature ●----● or after 1 hour heating o----o.

That also the reversal is post transcriptionally regulated is suggested by the observation that it occurs as normally in the presence of actinomycin. The attractive hypothesis that the h.s.p. themselves bind the bulk mRNA thus preventing it from being degraded during the bulk protein synthesis seems to have to be discarded because no h.s.p. are found bound to the RNA particles.

The fact that the inhibition of protein synthesis is

exerted at a post transcriptional level does not necessarily imply that heating has not other effect on transcription than that of stimulating the synthesis of the RNA coding for the h.s.p.. The experiments to be described here suggest indeed an inhibition of the maturation of the mRNA precursors and of the RNA transport to the cytoplasm. This phenomenon appears especially clear with respect to the precursor of a specific mRNA class that we have investigated in detail, that of histone mRNA.

If one looks indeed at the electrophoretic profile, under fully denaturing conditions of the nuclear RNA of mesenchyme blastulae, synthesized shorthy after heating, one finds an accumulation of the radioactivity from H^3 labeled uridine in the large RNAs and a decrease of radioactivity in the small sized RNA classes, which are on the contrary largely predominant in the non heated control embryos (fig.9 A,B). These results are consistent with the idea that heating has slowed down the maturation processes of the nuclear RNAs, a hypothesis this, which is also supported by the results of a chase experiment carried out after returning the embryos to normal temperature, i.e., under conditions when a normal pattern of protein synthesis is restored and normal development is resumed. Figure 9C shows that in this case: the electrophoretic profile of the nuclear RNA becomes normal, with a predominance of the small sized RNA classes, into which an amount of radioactivity equal to that formerly present into the heavy RNAs has been transferred.

As a consequence of the inhibition of the maturation of the large nuclear RNA classes, one would expect a decrease of the radioactivity of the cytoplasmic RNA during heating. Experiments not reported here, clearly fulfil this expectation. In order to obtain more precise informations about our hypothesis, we have followed the effect on the synthesis and maturation of the histone mRNA precursor. Previous experiments by one of us (Spinelli et al.,1980) have suggested that at least at developmental stages when there is little or no cell division, histone mRNA is synthesizes in the form of a polycistronic precursor containing the mRNAs for all five histone classes, which is thereafter matured into the

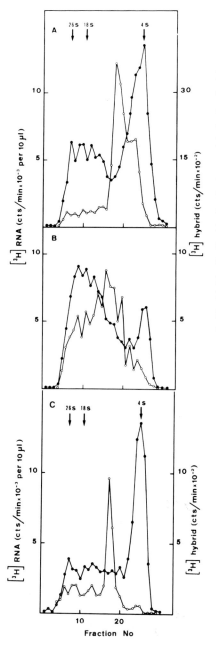

Fig.9. Radioactivity profile of total nuclear RNA ●--●--● and of histone RNA-DNA hybrid o--o--o. Labelled nuclear RNAs extracted from control (A), heat shocked (B) and shocked and chased (C) mesenchyme blastula embryos were denatured with glyoxal and fractionated by electrophoresis on horizontal 1.2% agarose gel in 10 mM Phosphate buffer pH 7.0. Part of each RNA fraction was hybridized with pPH70, containing histone genes.

single mRNAs. If this is the case, one might expect heating to slow down the maturation of such a precursor and to cause its accumulation within the nuclei.

Fig.9A shows that the newly synthesized histone nuclear RNA, revealed by hybridization with a cloned histone DNA probe, is found mostly in the region of about 9S in the control mesenchyme blastulas; if on the other hand the embryos had been heated before labeling, then most of the radioactivity identifiable as nuclear histone RNA is found in a polydisperse area heavier than 9S (fig.9B). At the same time, as shown in fig.10, very little if any newly synthesized histone RNA is transported into the cytoplasm.

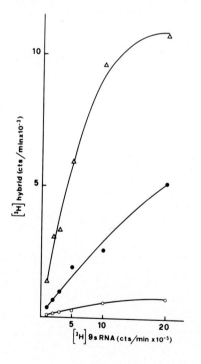

Fig.10. Transfer of histone RNA to cytoplasm in:
△——△ normal embryos
o——o embryos subjected to heat shock
●——● embryos subjected to heat shock followed by chase. Cytoplasmic labelled RNAs of the same embryos used for the experiment described in fig.9,were fractionated on 15-30% sucrose density gradients and the 7-12S RNA hybridized to saturation with pPH70,containing histone genes.

If at this point the heated embryos are returned to normal **temperature** and a chase is operated in the presence of an excess of cold uridine and cytidine, the nuclear histone RNA becomes again predominantly of the 9S size (fig.9C), and histone RNA is transported again into the cytoplasm (fig.10). The latter upon electrophoresis shows a size of about 9S as for the mature histone messenger.

REFERENCES

Ashburner M, Bonner II (1979). The induction of gene activity in Drosophila by heat shock. Cell 17: 241.

Giudice G (1973). "Developmental Biology of the Sea Urchin Embryo". New York and London: Academic Press, p.1.

Graziosi G, Micali F, Mazzari R, De Cristini F, Savoini A (1980). Variability of response of early Drosophila embryos to heat shock. J.Exp.Zool. 214: 141.

McAlister L, Finkelstein V (1980). Heat shock proteins and thermal resistance in yeast. Biochem.Biophys.Res.Commun. 93: 819.

McClay DR, Champers AF (1978). Identification of four classes of cell surface antigens appearing at gastrulation in sea urchin embryos. Develop.Biol. 63: 179.

Ritossa F (1962). A new puffing pattern induced by temperature shock and DNP in Drosophila. Experientia 18: 571

Scott MP, Pardue ML (1981). Tralsational control in lysates of Drosophila melanogaster cells. Proc.Natl.Acad.Sci.USA. 78: 3353.

Spinelli G, Melli M, Arnold E, Casano C, Gianguzza F, Ciaccio M (1980). High Molecular weight RNA containing histone messanger in the sea urchin Paracentrotus lividus. J.Molec.Biol. 139: 111.

IMMUNOLOCALIZATION OF α-ACTININ IN AN AMPHIBIAN EGG (DISCOGLOSSUS PICTUS).

Campanella,C.[°],E.Rungger Brändle,G.Gabbiani[°°]

[°]Istituto di Istologia ed Embriologia,
Via Mezzocannone 8, 80134 Napoli, Italy.
[°°]Department de Pathologie, Bol.de la Cluse 40, 1211, Genève, Switzerland.

INTRODUCTION

In the uterine eggs of Discoglossus the cortex may be divided into two distinct areas. The animal dimple which is a depression of the animal hemisphere and the site of fertilization, containing microfilament bundles in finger-shaped microvilli and extending into the cytoplasm and the remaining cortex which lacks microvilli and where microfilaments are scarce (Campanella,1975; Campanella and Gabbiani,1980).

It has been shown previously that a contractile system, involving actin and myosin, is present throughout the egg cortex. Moreover, in the animal dimple this system is particularly well developed, where an intense rod-like staining pattern following incubation with anti-actin antibodies indicates the actinF nature of the microfilament bundles in the microvilli (Campanella and Gabbiani, 1980). This orderly arrangement of the microfilament bundles with respect to the plasma membrane is similar to that in the microvilli of the intestinal brush border. α-actinin occurs in close association with actin filaments, in the Z-lines of striated muscle, in dense plaques of smooth muscle cells and in the adhesion plaques of fibroblasts, thus it is located where the microfilaments join the plasma membrane. From this, it was assumed that in the intestinal brush order, α-actinin would be found at the plasma membrane of microvilli (Mooseker and Tilney, 1975), however, recent studies have shown that α-actinin is not present at the microvillar plasma membrane, but instead is found in the terminal web , particularly in the zonula adherens of the mucosal epithelium(Bretscher and Weber, 1978; Geiger et al., 1979). Its specific role is not yet understood.

In the present report we asked the questions 1) is α-actinin present in the Discoglossus egg, and 2) if so, what is its distribution. This amphibian egg offers the opportunity to compare in the same cell the distribution of this protein in a brush border-like structure - the animal dimple - and in cortical regions where microvilli are absent.

We give immunochemical evidence that α-actinin is a constituent of the cortex, however throughout the entire egg surface and we have compared its location to that of actin and myosin.

MATERIALS AND METHODS

Adult Discoglossus pictus females, captured in April near Palermo (Italy), were induced to ovulate by the injection of 250 U Pregnyl (Organon, Holland) in Amphibian Ringer's. The uterine eggs were collected and utilized within 18 hrs following hormone treatment.

Characterization of the Antiserum

α-actinin was isolated from swollen myofibrils of pig skeletal muscle and purified by DE 52 and hydroxyapatite chromatography (Singh et al., 1977). Approx. 200 μg of purified protein was mixed with Freund's adjuvant and injected intradermally into rabbits at multiple sites. Booster injections were given 4 and 6 weeks later. Serum was taken 7 weeks after the initial injection. By immunofluorescence the antiserum strongly stains the Z-lines of striated muscle and brush borders of intestinal cells, but stains smooth muscle only weakly. In homogenates from lymphocytes, fibroblasts and platelets, separated by one-dimensional SDS-polyacrylamide gel electrophoresis, the antibodies react with a polypeptide of approx. 100,000 daltons, slightly larger than purified skeletal α-actinin.

Immunotransfer Technique

For the Discoglossus "egg extract", eggs were dejellied in 5mM DTT in 0.1 NaCl-5mM Tris-HCl (pH 8.0) homogenized and centrifuged at 10,000 x g for 10 min. 150-200 µg of the supernatant was subjected to gradient gel (5-20% acrylamide) electrophoresis. Proteins were transferred electrophoretically (Towbin et al., 1979) on nitrocellulose paper sheets (Schleicher and Schüll, pore size 0.45 um) and incubated with either the immune serum or the preimmune serum from the same animal (dilution 1:80). As a second antibody, goat anti-rabbit IgG-fraction coupled with peroxidase (Nordic, Tilburg, Netherlands; 3 µl/5 ml) was used. Blots were stained with dianisidine (Fluka, Buchs, Switzerland).

Immunofluorescent Staining

Before freezing for cryostat sectioning, unfertilized eggs were fixed for 15 min with 3.7% formaldehyde prepared from paraformaldehyde and diluted in phosphate buffered saline (PBS). This procedure does not significantly alter the immunological reaction and improves the quality of the sections. Frozen 4 µm sections were fixed in acetone at 20°C for five minutes, dried, and incubated with anti-α-actinin for 15 min. After incubation the sections were washed in PBS and stained for 15 min with fluorescein-conjugated IgG fraction of sheep antiserum to rabbit TgG (Code Nr 65-173-1, Miles). After rewashing with PBS, the level of fluorescence was compared with

that in control sections treated in parallel with
normal rabbit serum. Ten percent Discoglossus serum
was routinely added to the sera used for the second
step of indirect staining in order to eliminate
nonspecific background.
    Photographs were taken on a Zeiss UV photo-
microscope (Ilford, Basel, Switzerland).

Electron Microscopy
    Uterine eggs were removed manually from their
jelly coats and fixed in 0.2M phosphate buffer at
pH 7.3-7.4, postfixed in 1% phosphate buffered
osmium tetroxide, dehydrated and flat embedded in
Araldite Epon. Ultrathin sections were stained with
uranyl acetate followed by lead citrate and observed with a Siemens Elmiskop 1A.

RESULTS
    As shown in Fig.1, the serum containing α-ac-
tinin antibodies reacts with a component of Discoglossus egg extract with the same mobility as the
α-actinin from striated muscle. α-actinin is often
split in two bands as is the component from the
amphibian egg extract. As α-actinin occurs in minor
quantities in the egg extract, it was necessary to
overload the gel in order to transfer a sufficient
amount of the particular protein to the nitrocellulose sheet. This might explain partly the high
background observed on the blot. Alternatively, a-
specific crossreaction between amphibian and mammalian sera might also contribute to this background,
since rabbit antibodies were neither affinity-purified nor absorbed with Discoglossus serum (cf.immu-
nofluorescence microscopy).
    Fig.2a shows the ultrastructural arrangement
of the animal dimple which is unique to this part
of the egg cortex. Microfilament bundles are present in the peripheral cytoplasm and are arranged
in parallel in the finger-shaped microvilli (for

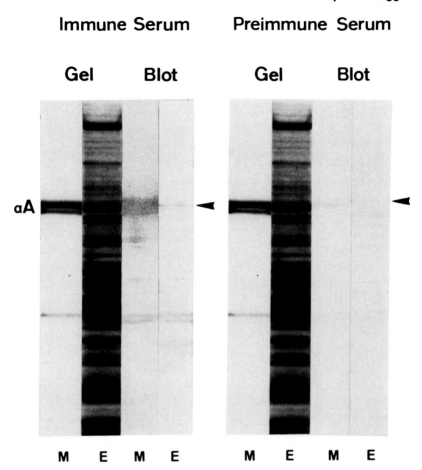

Fig.1. Identification of α-actinin in egg extract of <u>Discoglossus pictus</u>. Proteins were electrophoretically separated, blotted on nitrocellulose paper sheets and incubated with either a rabbit antiserum containing antibodies against α-actinin or with the preimmune serum from the same animal as described in Materials and Methods. <u>Gel</u>: Coomassie blue stained SDS polyacrylamide gel slots after transfer. <u>Blot</u>: corresponding nitrocellulose blots stained for peroxidase. The immune serum reacts with a component (arrow)

of Discoglossus egg extract (E) having the same mobility as α-actinin (αA) in a crude marker preparation (M) of striated muscle.

more details see Campanella and Gabbiani, 1980). The microvilli extend for about 1-1.5 μm at the surface of the animal dimple.

Following exposure to anti α-actinin, the peripheral cytoplasm of the animal dimple is positively stained (Fig.2a), and appears as very thin rims of fluorescence in the cortex, at the bases of the microvilli.

The cortex of the rest of the egg surface also stains for α-actinin and the location and intensity of the immunostaining is practically indistinguishable from that found in the animal dimple (Fig.2c).

Control sections incubated with normal rabbit serum were negative (Fig.2d).

DISCUSSION

Using an immunotransfer technique from SDS-polyacrylamide gel electrophoresis, we have found in uterine eggs of Discoglossus pictus a polypeptide comigrating with α-actinin from pig skeletal muscle which reacts with rabbit antiserum containing α-actinin antibodies. Thus, together with a number of other cytoskeletal and contractile proteins, already identified in eggs of different animals (actin, myosin, dynein, tubulin, tropomyosin), α-actinin appears to be a component - although minor - of uterine eggs, and thus it is present before the onset of embryonic development.

Moreover, we are able to localize α-actinin as a thin band over the whole egg cortex, both in the animal dimple and in the remaining cortical area. It appears that α-actinin is present at the site where the acto-myosin system was localized (Campanella and Gabbiani, 1980). However in contrast to

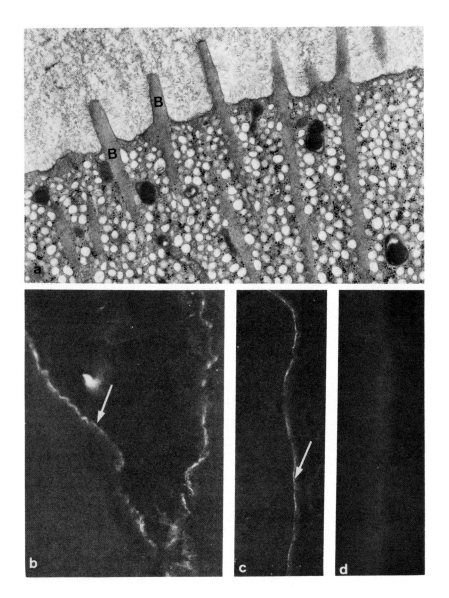

Fig.2 - a. The ultrastructure of the animal dimple: bundles (B) of microfilaments are present in the peripheral cytoplasm; they are located in the finger-shaped microvilli. X 16,000.

b. Immunofluorescent staining of α-actinin in the periphery of a Discoglossus egg: animal dimple incubated with anti-α-actinin antiserum. A thin rim of fluorescence (arrow) is present in the peripheral cytoplasm of the dimple. c. Similarly, in the rest of the egg cortex the immunostaining is restricted to a thin band beneath the egg plasma membrane (arrow). d. Control section of an egg cortex incubated with normal rabbit serum. X 320.

actin, α-actinin does not extend into the microvilli of the animal dimple. This is in agreement with the distribution of α-actinin in the intestinal brush borders, where it was found exclusively in the terminal web (Bretscher and Weber, 1978; Geiger et al., 1979).

REFERENCES

Bretscher A, Weber K (1978). Localization of actin and microfilament-associated proteins in the microvilli and terminal web of the intestinal brush border by immunofluorescence microscopy. J Cell Biol 79: 839-845.

Campanella C (1975). The site of spermatozoon entrance in the unfertilized egg of Discoglossus pictus (Anura): An electron microscope study. Biol Reprod 12: 439-447.

Campanella C, Gabbiani G (1980). Cytoskeletal and contractile proteins in coelomic oocytes, unfertilized and fertilized eggs of Discoglossus pictus (Anura). Gamete Res 3: 99-114.

Fujiwara K, Porter ME, Pollard TD (1978). Alpha-actinin localization in the cleavage furrow during cytokinesis. J Cell Biol 79: 268-275.

Geiger B, Tokuyasu KT, Singer SG (1979). Immunocytochemical localization of α-actinin in intestinal epithelial cells. Proc Nat Acad Sci USA 76: 2833-2837.

Mooseker MS, Tilney LG (1975). Organization of an actin filament-membrane complex. Filament polarity and membrane attachment in the microvilli of intestinal epithelial cells. J Cell Biol 67: 725-743.

Towbin H, Staehelin T, Gordon J (1979). Electrophoretic transfer of proteins from polyacrylamide gels to nitrocellulose sheets: procedure and some applications. PNAS (USA) 78: 27-32.

Singh I, Goll DE, Robson RM, Stromer MH (1977). N- and C-terminal amino acids of purified α-actinin. BBA: 29-45.

PHYSIOLOGICAL BASIS FOR HUMAN IN VITRO FERTILIZATION

Luigi Mastroianni, Jr., M.D.

Department of Obstetrics and Gynecology

University of Pennsylvania School of Medicine

Mammalian fertilization normally occurs in the lumen of the fallopian tube. At that rather inaccessible site, direct observation of the events associated with fertilization cannot be carried out effectively. Through in vitro fertilization, however, the process can be moved from the oviductal lumen to the laboratory bench, where sequential events associated with conditioning of spermatozoa and penetration of the oocyte can be more readily appreciated. In vitro fertilization of mammalian eggs now serves as a standard model for fertilization research. Of at least equal moment has been the recent application of in vitro fertilization and subsequent transfer of embryos to the uterus to problems of animal husbandry and human infertility.

In vitro systems in mammals were developed largely through trial and error, and a review of mammalian fertilization is an exercise in comparative biology. In vitro fertilization has been carried out successfully in a number of species including the rabbit, mouse, rat, hamster, cow and human. Although there are many characteristics in common, distinct species differences are appreciated. Observations on marine forms have provided a wealth of information which has served as a basis for study of mammalian fertilization.

As efforts have been made to develop systems for in vitro fertilization and embryo culture, attention was also accorded to the intralumenal tubal environment. Conditions under which fertilization normally occurs have been assessed, and this information has proved useful in the development of in vitro techniques

METHODS FOR GAMETE COLLECTION

Gametes used for in vitro fertilization studies have been recovered from the reproductive tract in a number of different ways. In the rabbit, ova had been recovered from the ovarian follicle (Suzuki and Mastroianni 1966), the surface of the ovary following ovulation (Mills et al, 1973; Brackett and Olifant, 1975) and from the oviduct shortly after ovulation (Chang, 1959). When follicular oocytes are used, they are aspirated from the mature preovulatory follicle. In the rabbit, when the fimbrae are surgically removed, ova remain attached to the surface of the follicle in a sticky cumulus cell mass and can easily be recovered from the ovarian surface. Since this mechanism does not apply in all species, by and large oocytes are either aspirated from follicles or recovered from the oviduct. Because of the size of the rabbit fallopian tube, oocytes can readily be flushed from the lumen. In smaller species such as the mouse or rat, the oviduct is excised surgically and the oocytes then released from the small coiled oviduct. (Yanagimachi and Chang, 1963; Yanagimachi, 1969, 1972; Toyoda and Chang, 1974). Only a few human oviductal ova have been recovered.

Perhaps one of the most significant breakthroughs in human in vitro fertilization, first applied successfully by Dr. Patrick Steptoe, was recovery of ova from ripe follicles at the time of laparoscopy. Laparoscopy is not a simple office procedure, but requires a general anesthetic and a certain degree of surgical dexterity. The oocytes are collected under direct visualization by aspiration with recovery rates 50% or better (Lopata, 1980). In the spontaneously ovulating patient, but one oocyte is recovered. If recovery of additional oocytes at the same sitting is desired, one may elect to stimulate ovarian production of follicles by treatment with ovulation inducing agents such as clomiphene citrate or human menopausal gonadotropins. There is still a difference of opinion as to the usefulness of preovulatory ovarian stimulation, but substantial success has been obtained by two Australian groups among patients who were pretreated with gonadotropins or other ovulation inducing agents (Lopata, 1978; Trounson, 1981).

TIMING OF OOCYTE RECOVERY

In order to recover fertilizable oocytes, aspiration of the follicle must be timed to coincide within a short interval of anticipated ovulation. This appears to be true in spite of the fact that oocytes recovered from immature follicles often proceed through metaphase II and polar body release in vitro.

The observation that oocytes will mature in vitro, at least at the nuclear level, was made as early as 1935 by Pincus and Enzman in the rabbit and human. More recently, Edwards carried out a comparative study on the rate of oocyte maturation in vitro in a number of species (Edwards, 1965). In vitro maturation of monkey follicular oocytes and their fertilizability have also been assessed (Suzuki and Mastroianni, 1966). Early experiments in the rabbit suggest, however, that oocytes recovered too early from the developing follicle are not completely fertilizable (Suzuki and Mastroianni, 1968). Thibault, (1970), appears to have pinpointed the difficulty at the level of pronucleus formation. Apparently, any such immature oocytes are penetrated, but pronuclear formation is arrested or inhibited, perhaps by some cytoplasmic factors or alternatively, the absence of a pronuclear formation stimulating factor (Thibault, 1975). In any case, the human experience suggests that timing of oocyte recovery in the menstrual cycle is critical.

A number of approaches have been used to monitor human follicular development. The system presently most popular involves monitoring the patient with sequential determinations of urinary gonadotropins at 4 hour intervals. The use of blood determinations for total estrogens and luteinizing hormone (LH) has also been suggested.

Follicular development has been assessed with ultrasound (Mastroianni, 1980; Lachlan, et al 1981; O'Herlehy, et al 1980). Development of the dominant follicle can be appreciated, and serial ultrasonographic studies have been used to predict the time of ovulation. Ultrasonography has also been most useful in patients who are being treated with gonadotropins to determine the number of follicles which are maturing in a given menstrual cycle (Queenan, 1980). In a recent report, ultrasonography has been used to identify a follicle which was then aspirated percutaneously without benefit of laparoscopy (Lenz et al, 1981). Further refinement of the various approaches used for determination of "ripeness" of the follicle will predictably result in improvement in overall success rates. Timing of ovum recovery is one of the principal issues surrounding the techniques of human in vitro fertilization.

Follicular ova have been aspirated from maturing follicles for purposes of ovum transfer in the rhesus monkey. Their fertilizability in vivo is materially affected by the timing of recovery in relation to the anticipated time of ovulation. In the rhesus monkey, follicular development is monitored by determination of peripheral blood levels of gonadotropins and estrogen. The presence of a mature ovulatory follicle is implied when serum estradiol levels exceed 150 pk/ml, preceded by an LH surge 12 to 20 hours earlier (Kreitmann, O and Hodgen, G.D. 1980). Methods of timing ovum recovery are still imprecise in the rhesus monkey. Only 39% of recovered oocytes in one series exhibited evidence of full maturation in spite of meticulous hormonal monitoring (Kreitmann and Hodgen, 1980).

Of particular importance are the reported observations on the function of the monkey corpus luteum following ovum extraction (Kreitmann, 1981). Although the length of the postovulatory luteal phase of the cycle is not altered, initial progesterone levels are significantly lower than normal. It has been suggested that aspiration of the follicle may be associated with removal of substantial numbers of progesterone producing luteinized granulosa cells from the follicular wall. Human corpus luteum function is also altered following oocyte aspiration with significant modifications in both estradiol and progesterone levels (Garcia et al, 1981).

RECOVERY AND TREATMENT OF SPERMATOZOA

For human in vitro fertilization, ejaculated spermatozoa have been used uniformly. In other species, spermatozoa have been harvested from the epididymis. There are substantial differences between epididymal and ejaculated spermatozoa, the latter having been exposed to seminal plasma. Regardless of the source of the spermatozoa, they must undergo further maturation in a process referred to as capacitation, defined as acquisition of fertilizing ability through exposure to the female reproductive tract. The phenomenon of capacitation was appreciated for the first time in the rabbit (Austin, 1951-1952; Chang, 1951). In that species, several hours of exposure to the female reproductive tract are required. The time required for capacitation displays marked species variation. Human spermatozoa are relatively easily capacitated in vitro, and in this regard display characteristics similar to those of the hamster and mouse.

Capacitation is associated with changes in both metabolism and sperm surface components. For investigational purposes, these are often considered separately, although both are part of the sperm maturation process which occurs prior to fertilization. Alterations in the sperm membrane have been extensively investigated, and components which appear or disappear from the sperm surface have been evaluated. The likely sequence of events is initiated by dissociation of acrosomal stabilizing factors (ASF) from the sperm membrane (Eng and Oliphant, 1978; Koehler et al, 1980). ASF is a large molecular weight glycoprotein acquired from the seminal plasma. When epididymal sperm are used for in vitro fertilization, this factor is absent, and therefore, in vitro systems which would remove it are not required. In human systems, however, in which ejaculated spermatozoa are used, a significant feature of sperm conditioning involves dissociation of ASF (Kanwar et al, 1979). Under the continued influence of female reproductive tract fluids, the membrane altered spermatozoan exhibits increased metabolism. The latter is associated with an influx of calcium (Oliphant and Eng, 1981).

An additional feature of sperm conditioning involves alterations in the sperm acrosome, an enzyme containing envelope located on the sperm head. Such alterations are referred to as the "acrosome reaction" (Austin and Bishop, 1958). This occurs as the outer acrosomal membrane is dissociated and is an absolute requisite to final penetration of the oocyte by the spermatozoan.

Although the acrosome reaction may justifiedly be considered apart from capacitation, it does represent a continuum in the process which culminates in fertilizing ability. Culture media which would support in vitro fertilization must necessarily function to bring about capacitation and to condition the spermatozoan such that the acrosome reaction can be completed prior to penetration.

ENVIRONMENTAL CONDITIONS WITHIN THE FALLOPIAN TUBE

Tubal fluid provides appropriate in vivo conditions for capacitation, fertilization, and early embryonal cleavage. A chemically balanced intralumenal environment is provided by the tubal secretions, which represent a combination of a transudate and actively secreted products. Constituents of tubal fluid have been evaluated in a number of species, and methods for continuous collection have been utilized in the rabbit (Hamner and Williams, 1965; Holmdahl and Mastroianni, 1965) and monkey (Mastroianni et al, 1961-69-70). The tubal environment has also been assessed with the use of intralumenal probes (Mastroianni and Jones, 1965). In the monkey, probes have been used to evaluate changes in pH, carbon dioxide and oxygen tension within the oviductal lumen (Maas et al, 1976, 1977). In the mouse, microprobes have been devised to evaluate electrolyte content (Borland et al, 1977). Variations in content as effected by hormonal status have been observed. The rate of fluid production increases dramatically just prior to ovulation in both the rhesus monkey and human (Mastroianni et al, 1970; Lippes et al, 1972; Sham et al, 1977). Hence, fluid is present in greatest quantity at the time of sperm migration and fertilization.

Electrolytes present in tubal fluid provide the in vivo environment for capacitation (Brackett and Mastroianni 1974). An osmolarity higher than that of plasma facilitates capacitation in vitro (Brackett and Oliphant, 1975). The calcium concentration in tubal fluid increases after ovulation in the monkey (Mastroianni and Stambaugh, 1974) and rabbit (Holmdahl and Mastroianni, 1965). Ca^{2+} has been shown to be necessary for capacitation (Singh et al, 1978) for the acrosome reaction (Rogers and Yanagimachi, 1976; Yanagimachi and Usui, 1974) and for binding of sperm to egg (Saling et al, 1978). Experiments examining the competition of Mg^{2+} with Ca^{2+} for binding by sperm suggest that the Mg^{2+}/Ca^{2+} ratio in the reproductive tract may be an important modulator for the timing of the acrosome reaction (Rogers and Yanagimachi, 1976). The Mg^{2+}/Ca^{2+} ratio is very low in the oviduct fluid of species in which the acrosome reaction occurs in the oviduct.

The role that the oviductal milieu plays in the conservation of sperm motility has been investigated. The possibility that macromolecules, such as albumin, may be important in this process has been suggested. Serum albumin exerts a preservative and protective action on sperm motility in vitro (Harrison et al, 1978). Albumin is a major protein of tubal fluid (Mastroianni et al, 1970; Moghissi, 1970; Urzua et al, 1970; Feigelson and Kay, 1972).

Clearly, oviduct fluid contains the necessary components to confer fertilizing ability on sperm. No single component of tubal fluid alone can effect capacitation. Rather, the oviduct contributes to this critical phase by providing the proper conditions of osmolarity, pH, metabolic substrates, and electrolytes in appropriate ratios.

The metabolic substrates produced by the tube include glucose, lactate and pyruvate. Lactate and pyruvate concentration increase following ovulation (Mastroianni et al, 1973) and presumably serve as essential substrates during early cleavage.

Inhibitors of enzymes have been described in oviduct fluid. Acrosin, a proteolytic enzyme bound to the inner membrane of the acrosome, is implicated in the fertilization process. It is suggested that acrosin facilitates sperm penetration through the zona pellucida (Stambaugh and Buckley, 1969; Stambaugh and Smith, 1976). Several inhibitors of acrosin and trypsin have been isolated in the oviduct fluid of rhesus monkey (Stambaugh et al, 1974). The concentrations of these inhibitors vary during the menstrual cycle. They are present at high levels before ovulation, decrease to significantly lower levels around the time of ovulation, and rise again within one to three days after ovulation. A total of six proteinase inhibitors is found, five to which inhibit acrosin from rhesus monkey sperm. These include $alpha_1$-antitrypsin, $alpha_2$-macroglobulin, $alpha_1$-antichymotrypsin. Two others with isoelectric points of 5.8 and 3.3 remained unidentified. Recently, acrosin inhibitors with acidic isoelectric points have been identified as secretory IgA in the tubal fluid of both rhesus monkey and rabbit (Go et al, 1978, 1979).

Rabbit oviduct fluid contains at least four inhibitors of trypsin (McLaughlin and Hamner, 1975). Total inhibitor concentration varies with hormonal state of the female, with low levels at estrus and high levels several days after ovulation.

These enzyme inhibitors may play a role in the regulation of fertilization by preventing the fertilization of an ovum at other than the optimum time in the life of the gamete. The inhibitors may also afford the oviductal lumen protection from the released proteolytic enzymes from degenerating spermatozoa.

The oviduct, by a combination of active secretion and transudation, elaborates a complex fluid. The tubal fluid contains a spectrum of biologically active constituents, enabling the oviduct to prepare the gametes for fertilization and sustain the developing zygote during its transport to the uterus. Study of the environment provided by the oviduct has brought us to a better understanding of the conditions which must be reproduced to support in vitro fertilization and culture.

CULTURE CONDITIONS FOR IN VITRO FERTILIZATION

The culture media which have been used for in vitro fertilization have certain ingredients in common. By and large, they represent modifications of a basic Kreb's-Ringer bicarbonate solution or Tyrode's solution. In general, their ionic concentrations mimic that of mammalian blood and differ only slightly from that which has been observed in oviductal fluid.

The osmotic pressures of the numerous media which have been used vary substantially (Bavister, 1981). The commonly employed energy sources provide appropriately for both in vitro fertilization and embryo development. Protein content appear to be important, and it has been shown that protein not only sustains the viability of spermatozoa in culture, but is also important for the acrosome reaction (Meizel, 1978). Albumin satisfies this requisite and is used uniformly. The source of the albumin may be serum derived from the species in question or bovine serum albumin. In many of the culture systems used for human in vitro fertilization, the latter is added, although human serum, derived from the ovum donor has also been employed successfully. For regulation of pH, a bicarbonate-CO_2 buffer system yields the best overall results. In vitro fertilization is accomplished best under a gaseus mixture of 5% CO_2, supporting the bicarbonate-CO_2 buffer system. Although satisfactory culture media which would sppport human in vitro fertilization are provided by a balanced salt solution supplemented with appropriate energy sources and protein, it is likely that the "ideal" environmental conditions have not as yet been defined.

CRITERIA FOR IN VITRO FERTILIZATION

Over the years, claims for successful in vitro fertilization have been made in the literature which had not stood the test of time. By and large, the criteria applied to support these claims have been either inadequate or inappropriate. Inasmuch as parthenogenetic cleavage occurs without benefit of spermatozoa, we must look to evidence other than cleavage to be sure that fertilization has, in fact, occurred. Disintegration of nonfertilized ova into structures which superficially resemble cells occurs with some frequency. Disintegrated forms have, in fact, been published as supporting evidence for human in vitro fertilization, and indeed, a disintegrated follicular ovum has actually been represented a parthenogenetically cleaved human egg. Some authors have suggested that the presence of a slit in the zona suggests that that egg has been penetrated by spermatozoan. Such a finding does not, in fact, indicate sperm penetration as it clearly can be artifactual. Presence of two polar bodies per se does not indicate that fertilization has occurred. The first polar body can cleave in the absence of sperm penetration, and artifacts resembling polar bodies are observed frequently in the perivitelline space. The morphological criteria which support the conclusion that fertilization has occurred include the presence of components of the fertilizing sperm in the vitellus, two well developed pronuclei with characteristic nuclei and two polar bodies all observed in the same specimen. Such findings, however, do not offer assurance that the process will be completed and simply represent the status of the specimen at a given time during the process of fertilization. The ultimate criterion for in vitro fertilization lies, of course, in the successful transfer of the embryo to a recipient with subsequent establishmen of a pregnancy, followed by delivery of young exhibiting normal karyotypes.

Successful application of in vitro fertilization and embryo transfer techniques to the human has significant implications in terms of our ability to treat the infertile patient. There is little doubt that the information which has accumulated thus far from in vitro fertilization studies and embryo culture in both laboratory and domestic animals and humans has contributed greatly to our understanding of early reproductive processes. In spite of herculian efforts there are still significant gaps in our knowledge. These clearly justify ongoing emphasis on laboratory experimentation if we are to continue to define systems which will lead to successful clinical applications in the management of human reproductive problems.

REFERENCES

Austin, CR (1951). Observations on the penetration of the sperm into the mammalian egg. Aust J Sci Res Sci B4:581-596.

Austin, CR (1952). The "capacitation" of the mammalian sperm. Nature (London) 170:326.

Austin, CR and Bishop, MWH (1958). Role of the rodent acrosome and perforatorium in fertilization. Proc R Soc London Ser B149:241-248.

Bavister, BD (1981) Analysis of culture media for in vitro fertilization and criteria for success. In Mastroianni, LM, Biggers J and Sadler W (eds): "Fertilization and Embryonic Development In Vitro," New York: Plenum Press.

Borland RM, Hazra S, Biggers JD and Lechene CP (1977). The elemental composition of the gametes and preimplantation embryo during the initiation of pregnancy. Biol Reprod 16:147-157.

Brackett, BG and Mastroianni L (1974). Composition of oviductal fluid. In Johnson AD, Foley CW (eds): "The Oviduct and Its Functions," New York: Academic Press pp 133-159.

Brackett BG, Oliphant G (1975). Capacitation of rabbit spermatozoa in vitro. Biol Reprod 12:260-274.

Chang MC (1951). Fertilizing capacity of spermatozoa deposited into fallopian tubes. Nature (London) 168:697-698.

Chang MC (1959). Fertilization of rabbit ova in vitro. Nature (London) 184:466-467.

Edwards RG (1965). Maturation in vitro of mouse, sheep, cow, pig, rhesus monkey and human ovarian oocytes. Nature 208:349.

Eng LA, Oliphant G (1978). Rabbit sperm reversable decapacitation by membrane stabilization with a high purified glycoprotein from seminal plasma. Biol Reprod 19:1083-1094.

Feigelson M, Kay E (1972). Protein patterns of rabbit oviductal fluid. Biol Reprod 6:244-252.

Garcia JE, Georgeanna SJ, Acosta A (1981). The effect of follicular aspiration on corpus luteum function (Abstract) Fertil Steril 35:235.

Go KJ, Mastroianni L, Stambaugh R (1978). The presence of secretory IgA specific for acrosin in rhesus monkey (macaca mulatta) oviduct fluid. Fed Proc (Abstract) 37: 1474.

Go KJ, Mastroianni L (1979). Tubal secretions. In Beller, FK and GFB Schumacher (eds): "The Biology of the Fluids of the Female Genital Tract," New York: Elsevier/North Holland.

Hamner, CE, Williams, WL (1965). Composition of rabbit oviduct secretions. Fertil Steril 16:170-176.

Harrison RAP, Cott HM, Foster CG (1978). Effect of ionic strength, serum albumin and other macromolecules on the maintenance of motility and the surface of mammalian sperm in a simple medium. J Reprod Fert 52:65-73.

Holmdahl TH, Mastroianni L (1965). Continuous collection of rabbit oviduct secretions at low temperature. Fertil Steril 16:587-595.

Kanwar KC, Yanagimachi R, Lopata A (1979). Effects of human seminal plasma on fertilizing capacity of human spermatozoa. Fertil Steril 31:321-327.

Koehler JK, Nudelman ED, Hakomoris (1980). A collagenbinding protein on the surface of ejaculated rabbit spermatozoa. J Cell Biol 86:529-536.

Kreitmann O, Nixon WE, Hodgen GD (1981). Induced corpus luteum dysfunction after aspiration of the preovulatory follicle in monkeys. Fertil Steril 35:671-675.

Kreitmann O, Hodgen GD (1980). Low tubal ovum transfer: an alternative to in vitro fertilization. Fertil Steril 34:375.

Lachlan CH, O'Herlehy C, Holt IJ, Robinson HP (1981). Ultrasound in an in vitro fertilization program. Fertil Steril 35:25.

Lenz S, Lauritsen GJ, Kjellow M (May 1981). Collection of human oocytes for in vitro fertilization by ultrasonically guided follicular puncture. Lancet 1163-1164.

Lippes J, Enders RG, Pragay DA, Bartholomew WR (1972). The collection and analysis of human fallopian tube fluid. Contraception 5:85-103.

Lopata A, Brown JB, Leeton JF, Talbot J, Wood C (1978). In vitro fertilization of preovulatory oocytes and embryo transfer in infertile patients treated with clomiphene and human chorionic gonadotropin. Fertil Steril 30:27-35.

Lopata A, (1980). Successes and failures in human in vitro fertilization. Nature 288:18-25.

Maas DHA, Storey BT, Mastroianni L (1977). Hydrogen ion concentration and carbon dioxide content of the oviduct fluid of the rhesus monkey. Fertil Steril 28:981-985.

Mastroianni L, Wallach RC (1961). Effect of ovulation and early gestation on oviduct secretions in the rabbit. Amer J Physiol 200:815:818.

Mastroianni L, Jones R (1965). Oxygen tension within the rabbit fallopian tube. J Reprod Fert 9:99-102.

Mastroianni L, Urzua M, Avalos M, Stambaugh R (1969). Some observations on fallopian tube fluid in the monkey. Amer J Obstet Gynecol 103:703-709.

Mastroianni L, Urzua M, Stambaugh R (1970). Protein patterns in monkey oviductal fluid before and after ovulation. Fertil Steril 21:817-820.

Mastroianni L, Urzua M, Stambaugh (1973). The internal environmental fluids of the oviduct. In Segal SJ, Crozier R, Corfman PA, Condliffe RG (eds): "The Regulation of Mammalian Reproduction," Springfield: Charles C. Thomas, pp 376-381.

Mastroianni L, Stambaugh R (1974). The secretory function of the primate oviduct. In Coutinho EM, Fuchs F (eds): "Physiology and Genetics of Reproduction," New York: Plenum Press, pp 25-34.

Mastroianni L (1980). The role and value, present and future, of ultrasound in the detection of ovulation. Fertil Steril 34:177-178.

McLaughlin C, Hamner CE (1975). Preliminary characterization of rabbit oviduct fluid trypsin inhibitors. Biol Reprod 12:556-565.

Meizel S (1978). The mammalian sperm acrosome reaction; a biochemical approach. In Johnson MH (ed): "Development in Mammals," Amsterdam: Elsevier/North-Holland, 3:1-64.

Mills JA, Jeitles GG, Brackett BG (1973). Embryo transfer following in vitro and in vivo fertilization in rabbit ova. Fertil Steril 24:602-608.

Moghissi KS (1970). Human fallopian tube fluid. I. Protein composition. Fertil Steril 21:821-829.

O'Herley C, Lachlan CH, Lopata A, Johnston I, Hoult I, Robinson H (1980). Preovulatory follicular size: a comparison of ultrasound and laparoscopic measurement.

Fertil Steril 34:24.

Oliphant G, Eng LA (1981). Collection of gametes in laboratory animals and preparation of sperm for in vitro fertilization. In Mastroianni L, Biggers J, Sadler W (eds): "Fertilization and Embryonic Development In Vitro," New York: Plenum Press.

Queenan JT, O'Brien GD, Bains LM, Simpson J, Collins WP, Campbell S (1980). Ultrasound scanning of ovaries to detect ovulation in women. Fertil Steril 34:99.

Rogers BJ, Yanagimachi R (1976). Competitive effect of magnesium on the calcium-dependent acrosome reaction in guinea pig spermatozoa. Biol Reprod 15:614-619.

Saling PM, Storey BT, Wolf DP (1978). Calcium-dependent binding of mouse epididymal spermatozoa to the zona pellucida. Develop Biol 65:515-525.

Shams A, Rizk MA, Toppozada HK, Khowessah MM, Abul-Enin M, Said S, Habib YA, Kira LH (1977). Human tubal fluid collection via vagina and its quantitative variations during the menstrual cycle. J Reprod Med 18:61-65.

Singh JP, Babcock DF, Lardy DH (1978). Increased calcium-ion influx is a component of capacitation of spermatozoa. Biochem J 172:549-556.

Smith DH, Picker RH, Sinosich M, Saunders DM (1980). Assessment of ovulation and estradiol levels during spontaneous and induced cycles. Fertil Steril 33:387.

Stambaugh R, Buckley J (1969). Indentification and subcellular localization of the enzymes effecting penetration of the zona pellucida by rabbit spermatozoa. J Reprod Fertil 19:423-432.

Stambaugh R, Seitz HM, Mastroianni L (1974). Acrosomal proteinase inhibitors in rhesus monkey oviduct fluid. Fertil Steril 25:352-357.

Stambaugh R, Smith M (1976). Sperm proteinase release during fertilization of rabbit ova. J Exp Zool 197:121-125.

Shettles LB (1955). Further observations on living human oocytes and ova. Am J Obstet Gynecol pp 365-371.

Suzuki S, Mastroianni L (1965). In vitro fertilization of rabbit ova in tubal fluid. Amer J Obstet Gynecol 93:465.

Suzuki S, Mastroianni L (1966). Maturation of monkey follicular oocytes in vitro. Am J Obstet Gynecol 96:723.

Suzuki S, Mastroianni L (1968). In vitro fertilization of rabbit follicular oocytes in tubal fluid. Fertil Steril 19:500.

Thibault C, Gerard M (1970). Factear cytoplasmique necessaire a la formation du pronucleus male dans l'ovocyte

de lapine, CR. Acad Sci 270:2025-2076.
Thibault C, Gerard M, Menezo Y (1975). Acquisition par l'ovocyte de lapine at de veaudufacteur de decondensation du noyau du spermatozoide fecondant (MPGF). Ann Biol Anim Biochim Biophys 15:705-714.
Toyoda Y, Chang MC (1974). Fertilization of rat eggs in vitro by epididymal spermatozoa and the development of eggs following transfer. J Reprod Fertil 36:9-22.
Trounson AO, Leeton JF, Wood C, Webb J (1981). Pregnancies in humans by fertilization in vitro and embryo transfer in the controlled ovulatory cycle. Science 212:681-683.
Urzua MA, Stambaugh R, Flickinger G, Mastroianni L (1970). Uterine and oviduct fluid patterns in the rabbit before and after ovulation. Fertil Steril 21:860-865.
Yanagimachi R, Chang MC (1963). Fertilization of hamster eggs in vitro. Nature (London) 200:281-283.
Yanagimachi R (1969). In vitro capacitation of hamster spermatozoa by follicular fluid. J Reprod Fertil 18:275-286.
Yanagimachi R (1972). Fertilization of guinea pig eggs in vitro. Anat Rec 174:9-24.
Yanagimachi R, Usui N (1974). Calcium dependence of the acrosome reaction and activation of guinea pig spermatozoa. Exp Cell Res 89:161-174.

THE FINE STRUCTURE OF HUMAN BLASTOCYSTS DEVELOPED IN CULTURE

A. Lopata, D.J. Kohlman and G.N. Kellow

Dept. of Obstetrics & Gynaecology and Reproductive Biology Unit, University of Melbourne, Royal Women's Hospital, Carlton, Victoria, Australia

INTRODUCTION

The recent development of successful methods for human in vitro fertilization and embryo culture (Lopata et al, 1980; Edwards, 1981) has provided a unique opportunity to investigate the ultrastructure of early cleavage and blastocyst stages in homo sapiens. Although human embryos resulting from natural conceptions are rarely available for electron microscopy (Pereda and Croxato, 1978), it is now known that embryos produced in vitro do give rise to viable blastocysts capable of producing normal offspring (Edwards, 1981; Lopata et al, 1981; Trounson et al, 1981; Wood et al, 1981). Only a sparse amount of information is available on the fine structure of non-human primate blastocysts developed in the genital tract (Panigel et al, 1975; Hurst et al, 1978). There is also a lack of information on the development of blastocysts in culture media in these species. In other mammals a limited number of studies have been carried out to compare the fine morphology of blastocysts grown in vivo and in vitro (McReynolds and Hadek, 1972; Van Blerkom et al, 1973; Anderson et al, 1975). In the present study electron microscopy has been used to evaluate human blastocysts developed from eggs fertilized in vitro.

METHODS

Preovulatory oocytes were obtained for in vitro fertilization and embryo culture from infertile patients treated with clomiphene citrate and human chorionic gonadotrophin (Lopata et al, 1981). When 3 or more preovulatory oocytes were aspirated from the stimulated ovaries, all of the eggs were inseminated in vitro, and in some patients 3 apparently normal embryos were obtained. In these cases, 2 were inserted into the infertile patient's uterus at the 4-cell to 8-cell stage of development, and the third was cultured to the blastocyst stage (table 1).

Table 1. Culture of Eggs Fertilized in Vitro to Blastocysts

Blastocyst	Potassium content in medium	Culture duration (hr)
Zonal degenerated	10 mM	122
Zonal degenerated	10 mM	136
Zonal degenerated	5 mM	139
Hatching	10 mM	143
Zonal unexpanded	5 mM	145
Zonal + endoderm	10 mM	165
Zonal expanded	5 mM	170

Two types of culture media, both based on Ham's F10 solution, were used for fertilizing the preovulatory eggs and for growing blastocysts. The media were prepared as described previously, except that the pH was adjusted to 7.35 in both the insemination medium (IM) and the embryo growth medium (GM), and the IM was supplemented with 10% inactivated human serum instead of crystalline albumin. In addition, the potassium concentration in some of the media was raised to 10 mM/l. All cultures were done in loosely capped Falcon tubes in a humidified atmosphere of 5% CO_2 + 5% O_2 + 90% N_2 at 37° C. If an embryo became available, it would be cultured to the blastocyst stage in the normal (approximately 5 mM potassium) or in the 10 mM potassium medium, at random, depending entirely on the medium being used during that week.

At the completion of culture, the blastocysts were fixed in 3.5% glutaraldehyde in 0.1 M cacodylate buffer, pH 7.3, post fixed in 1% osmium tetroxide in the same buffer, dehydrated in alcohol and embedded in Spurr Resin. Survey sections of embryos were cut serially, mounted on glass slides in sets of 8 to 10, and stained with Toluidine Blue. After each set, 5 to 10 ultrathin sections were cut and mounted on copper grids. These were stained with uranyl acetate and lead citrate and examined with a Phillips EM400T electron microscope.

RESULTS

Table 1 summarises some of the blastocysts obtained, the media used and the duration of culture. Electron microscopy revealed degenerative changes in the first 3 blastocysts. These will not be described in this article. It was of interest that a morphologically normal blastocyst developed in the 10 mM potassium medium and that it was hatching from its zona pellucida after 143 hours of culture, while another blastocyst developed a distinct layer of endoderm in the high potassium medium. The present report will deal with the ultrastructure of 3 zonal blastocysts cultured for 145 to 170 hours.

Light Microscopy

Serial 0.5 µm sections were cut close to the thickest part of the inner cell mass (ICM) in each of the zona-intact blastocysts. The section through the zonal blastocyst grown for 145 hours showed 4 main components (Fig. 1): 1. The zona pellucida formed the outermost layer. 2. Vacuolated cells between the zona pellucida and the trophectoderm. 3. A continuous layer of trophectoderm cells and an ICM protruding from one pole. 4. The blastocoele cavity. In addition to these 4 components, the blastocyst grown for 165 hours contained a layer of endoderm cells (Fig. 2). This sheet of cells partly covered the ICM and extended across the blastocoele to reach the trophectoderm. In the expanded blastocyst (Fig. 3) the zona pellucida was considerably attenuated due to extensive expansion of the blastocoele cavity. Discarded cells and debris were present in the cavity of each blastocyst (Figs. 2 and 3).

Electron Microscopy

Trophectoderm and ICM. The nucleus and cytoplasm of trophectoderm and inner mass cells of the human blastocyst

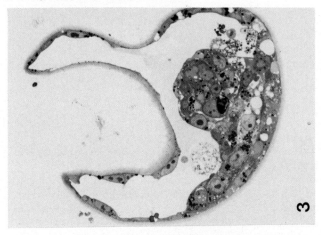
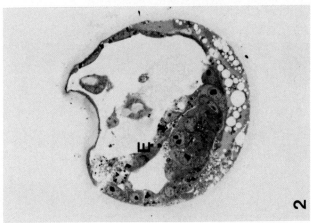
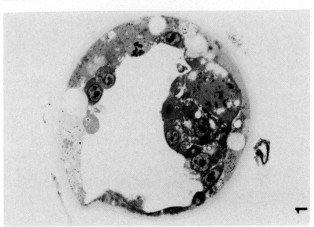

Figures 1 to 3: Sections through the inner cell mass of zonal blastocysts.

Fig. 1. Unexpanded blastocyst grown 145 hours.
Fig. 2. Endoderm (E) in blastocyst grown for 165 hours.
Fig. 3. Expanded blastocyst grown for 170 hours. Note that indentation of the trophectoderm has occurred during processing.

had similar ultrastructural features. In both cell types the following specific features are noteworthy about the nuclei: 1. Clumps of heterochromatin were often attached to the inner nuclear membrane and scattered in the nucleoplasm (Fig. 4). 2. The differentiated nucleoli which were not excessively reticulated (Fig. 4). 3. The uniform perinuclear space continuous with cisternae of the rough endoplasmic reticulum (Fig. 6). 4. Ribosomes aligned along the cytoplasmic surface of the outer nuclear membrane (Fig. 6). 5. Annulate lamellae were not observed in the nucleoplasm. They were, however, present in the cytoplasm of both ICM and trophectoderm cells.

Polyribosomes and free ribosomes were abundant in the cytoplasm of all differentiating blastocyst cells (Fig. 4). Elements of the rough endoplasmic reticulum (RER) comprised a prominent feature in the cytoplasm of trophectoderm and ICM cells (Figs. 4 and 5). As in other mammalian blastocysts, lengths of RER were closely associated with the surfaces of mitochondria (Fig. 4), generally running along their long axis, or bridging the cytoplasm between adjacent mitochondria. The RER were particularly long in the inner mass cells underlying endoderm cells (Fig. 5). In some cells a cisternum of RER was closely apposed to the parallel membranes of Golgi complexes. Continuity between the RER and annulate lamellae was observed in some regions.

The mitochondria were often elongated or curved, contained transversely running cristae (Fig. 4) which in some cases were closely stacked together. Frequently the cristae were slightly dilated, giving the mitochondria a vacuolated appearance. The matrix of these enlarged mitochondria was electron lucent compared to that of mitochondria in the earlier cleavage stages.

Golgi complexes were generally located in a juxtanuclear region (Fig. 7). Smooth surfaced vesicles (Fig. 7) and coated vesicles were usually found in the vicinity of the Golgi zones of both cell types. Vesicles of various size containing electron dense material (Fig. 9), probably representing lysosomes, were observed in some cells. Larger multivesicular structures were also seen.

Bundles of microfilaments (Fig. 4) were observed in the cytoplasm of ICM, trophectoderm and endoderm cells. These bundles of fibrils were particularly prominent in the troph-

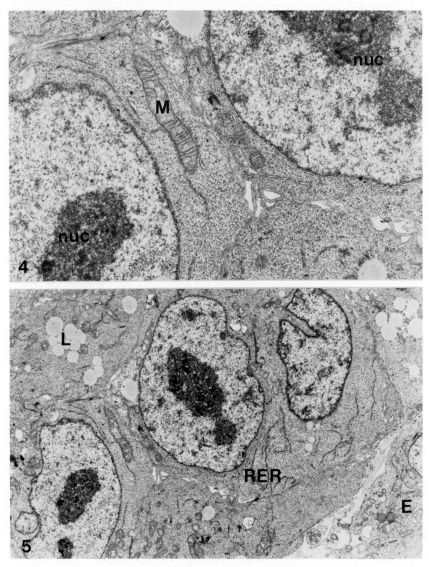

Fig. 4. Inner mass cells. Note reticulated nucleoli (Nuc) and well-developed mitochondria (M). x 10,800.
Fig. 5. Region of inner cell mass facing endoderm (E). Note lipid (L) droplets and rough endoplasmic reticulum (RER). x 4,800.

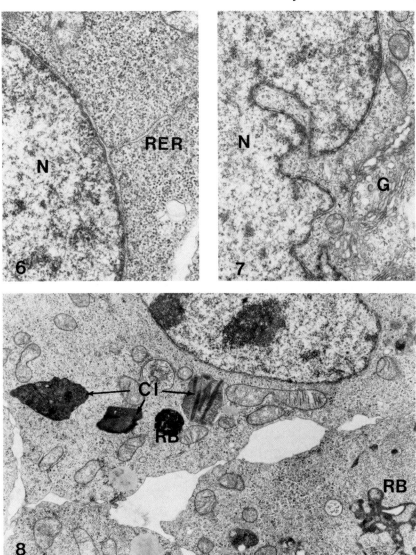

Fig. 6. Inner mass cell showing portion of nucleus (N). Note continuity of rough endoplasmic reticulum (RER) with the outer nuclear membrane. x 27,200.
Fig. 7. Inner mass cell nucleus (N) and Golgi complex (G). x 17,000.
Fig. 8. Inner cell mass region showing residual bodies (RB) and crystalline inclusions (CI). x 9,000.

ectoderm of the hatching blastocyst.

<u>Microvilli and cell junctions</u>. Plasma membrane specialisations in the trophectoderm distinguished it from the ICM. Highly developed microvilli and complex intercellular junctions gave the trophectoderm an appearance of a specialised epithelium.

Microvilli were present at two main sites: 1. They were most numerous on the outer surface of the trophectoderm where they projected toward the zona pellucida (Fig. 9). However, if the trophectoderm cells were covered with discarded cells the microvilli either abutted against the inner surface of the sequestered cells, or alternatively pushed their way between the cells toward the zona pellucida (Fig. 11). The latter phenomenon was observed only in the polar trophectoderm region. In some instances extensive cytoplasmic processes of polar trophectoderm extended toward the zona pellucida between the sequestered cells (Fig. 12). 2. Microvilli were also highly developed in regions of intercellular junctional contact between trophectoderm cells. In these areas, generally on the surface facing the blastocoele cavity, but also on the outer surface, microvilli of adjacent cells curved toward one another, forming arched interdigitations over the junctions (Fig. 10).

Intercellular junctional complexes, closely resembling those found in blastocysts of non-human primates and other mammals, linked trophectoderm cells in human blastocysts grown in culture. Typically, an apical complex consisting of a tight junction, a desmosome and gap junction, sealed the zonal borders of adjacent trophectoderm cells (Figs. 13 and 14). Less well developed desmosomes, gap junctions and regions of multiple plasma membrane contact were observed between trophectoderm and inner mass cells and between neighbouring inner mass cells.

<u>Endoderm</u>. Endoderm cells lined the ICM, extended to the trophectoderm and were characterised by: 1. Dilated branches of RER cisternae containing a flocculent material (Fig. 16). 2. Material released into the extracellular space which appeared to form a basement membrane-like structure lining the inner mass cells (Fig. 17). 3. Lipid globules which were more electron-dense than the lipid in

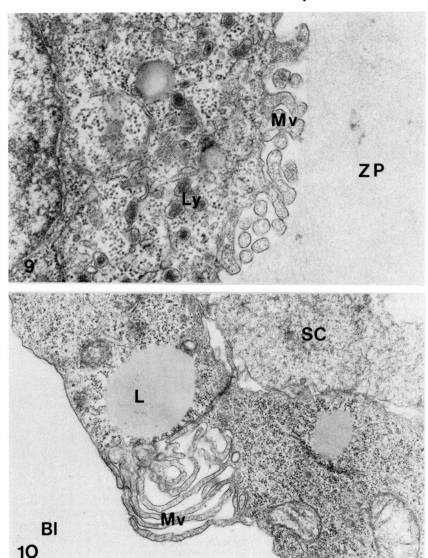

Fig. 9. Trophectoderm cell. Note microvilli (Mv), lysosome-like vesicles (Ly) and zona pellucida (ZP). x 34,200.
Fig. 10. Junctional region of two trophectoderm cells showing interdigitating microvilli (Mv), lipid (L), sequestered cell (SC) and blastocoele (Bl). x 18,000.

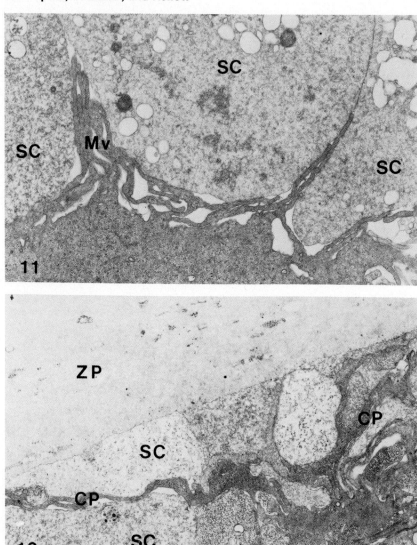

Fig. 11. Polar trophectoderm region showing microvilli (Mv) extending between sequestered cells (SC). x 10,800.
Fig. 12. Polar trophectoderm region showing cell processes (CP) extending between sequestered cells (SC) towards the zona pellucida (ZP). x 12,000.

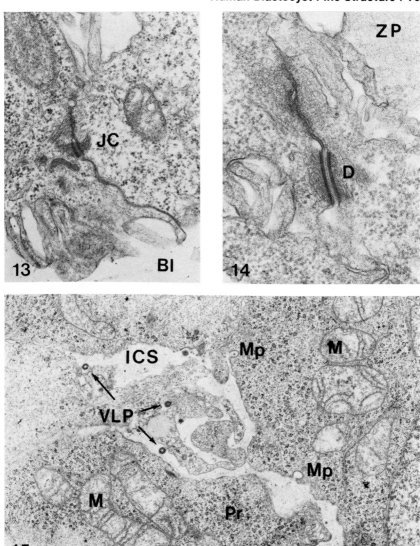

Fig. 13. Junctional complex (JC) between trophectoderm cells. Blastocoele (Bl). x 27,000.
Fig. 14. Junctional complex between trophectoderm cells. Note desmosome (D), zona pellucida (ZP). x 50,400.
Fig. 15. Viral-like particles (VLP) in the intercellular space (ICS) between inner mass cells. Also note mitochondria (M), polyribosomes (Pr) and micropinocytotic vesicles (Mp). x 18,000.

inner mass cells. 4. Aggregates of dark granules, considerably larger than ribosomes, which probably represent glycogen granules.

Cytoplasmic inclusions: 1. Lipid globules were common (Fig. 5). Some were closely associated with mitochondria. 2. Residual bodies (Fig. 8) and lysosome-like structures were present in both inner mass and trophectoderm cells. In some cases the contents of residual bodies appeared to have been released into the blastocoele. 3. Polymorphic electron dense material of unknown significance and autophagic vacuoles. 4. Crystalloid material (Fig. 8) showing linear periodicity was found in some inner mass and trophectoderm cells. 5. Virus-like particles (type C) were observed in the intercellular space of 2 blastocysts (Fig. 15).

Discarded cells. Two types of cells were apparently discarded by the developing embryo: 1. Sequestered cells. These were a group of cells located between the zona pellucida and trophectoderm (Fig. 18). Nuclei were not detected in serial sections of these cells. However, extensive arrays of annulate lamellae were present in the cytoplasm (Fig. 19). In addition, aggregates of dense granular material, resembling chromatin, were present in the cytoplasm. Microtubules were also observed. Polyribosomes and RER were not detected in the sequestered cells. However, free ribosomes were present.

Cisternae and vesicles of SER were abundant in the cytoplasm of the sequestered cells (Fig. 18). These elements were filled with a granular material and were often moderately to grossly dilated, giving the cytoplasm a vacuolated appearance (Fig. 18). In some cells whorls of parallel membranes and structures resembling dilated Golgi complexes were observed.

Undifferentiated and poorly differentiated mitochondria containing an electron dense matrix (Fig. 19) and relatively few cristae, were a prominent feature of the sequestered cells. Cisternae of SER were closely associated with the surface of these mitochondria (Fig. 19).

Some surface activity appeared to be occurring in the sequestered cells. In some cases a granular material was

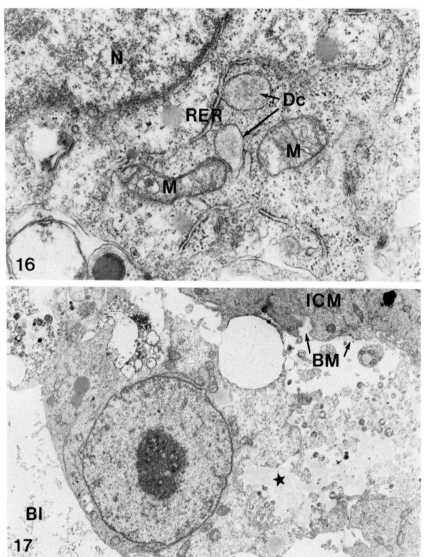

Fig. 16. Endoderm cell showing dilated cisternae (DC) of rough endoplasmic reticulum (RER). Also note nucleus (N) and mitochondria (M). x 30,600.
Fig. 17. Endoderm cell with extruded material (*) in the extracellular space. Note inner cell mass (ICM), basement membrane (BM) and blastocoele (Bl). x 5,000.

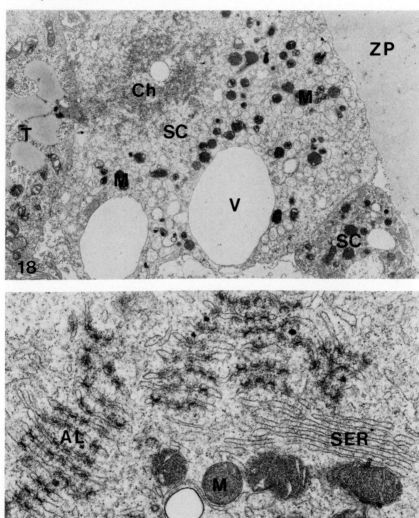

Fig. 18. Sequestered cells (SC) in polar trophectoderm region. Note chromatin-like material (Ch), vacuoles (V), mitochondria (M), zona pellucida (ZP) and trophectoderm cell (T). x 5,400.
Fig. 19. Sequestered cell. Note mitochondria (M), smooth endoplasmic reticulum (SER) and annulate lamellae (AL). x 34,200.

apparently being released into the sub-zonal space. In others micropinocytotic activity appeared to be occurring. Microvilli were absent. Intercellular junctional specialisations were not present between the sequestered cells, nor between these cells and the underlying trophectoderm.

Cytoplasmic inclusions, such as lipid, autophagic vacuoles and residual bodies, were rare or absent in sequestered cells.

2. Isolated cells. The second group of cells discarded from the blastocyst were found mainly in the blastocoele cavity. They occurred as isolated cells, rounded in shape and considerably more electron dense than cells of the trophectoderm or inner mass. In some cases the isolated cells were undergoing advanced degradation.

Unlike the sequestered cells, the cells shed into the blastocoele contained a nucleus and an extensively reticulated nucleolus. In addition, elements of RER and numerous ribosomes were present in the cytoplasm. The cytoplasm also contained small mitochondria which were reasonably well differentiated.

COMMENTARY

Blastocysts containing a cytologically normal trophectoderm, ICM and endoderm were developed from preovulatory eggs fertilized and cultured in Ham's F10 medium. Although hatching of the blastocyst was observed in this culture medium at about 140 hours after insemination, the present report deals with the ultrastructure of blastocysts which failed to hatch between 145 and 170 hours. These embryos contained a layer of discarded cells interposed between the trophectoderm and zona pellucida. The origin and significance of these sequestered cells is uncertain. However, in view of their absence from the hatching blastocyst, and their prominence in the zonal blastocysts, it would be interesting to know whether their presence is related to the failure to hatch in vitro.

The outgrowth of highly developed microvilli and long cell processes, from the surface of polar trophectoderm, is also worth noting. These cellular extensions penetrated between the sequestered cells to reach the zona pellucida, in contrast to microvilli confined by sequestered cells in regions of cavity trophectoderm. The questions raised by these observations are whether the microvilli of polar trophectoderm normally interact with the zona pellucida to initiate hatching and whether the interposition of discarded cells delays this process.

The absence of nuclei in the cells sequestered between the trophectoderm and zona pellucida, and the presence of annulate lamellae and chromatin-like material in their cytoplasm, suggests that these cells arrested during division. Moreover, the presence of undifferentiated mitochondria, absence of RER and polyribosomes and lack of intercellular junctions, suggests that the sequestered cells originated during early cleavage stages. Their vacuolated and degenerating appearance resembled the devitalised cells observed in mouse blastocysts from $A^y/a \times A^y/a$ matings (Cizadlo and Granholm, 1978). Our observations indicate, however, that despite the loss of a proportion of its cells the early human embryo continued to develop to the blastocyst stage.

Unlike the sequestered cells, the cells shed into the blastocoele derived from the ICM and/or trophectoderm. This was indicated by the presence of advanced differentiation in isolated cells which were not grossly disintegrated. El-Shershaby and Hinchliffe (1974) found that in mouse blastocysts recovered from the genital tract, devitalised or necrotic cells, similar to those observed in human blastocysts, were either isolated in the blastocoele cavity or phagocytosed by the inner mass and trophectoderm cells. This type of cell loss was attributed to abnormal differentiation in some ICM cells.

It may be postulated, therefore, that cell death in human embryos growing in a culture medium results from abnormalities in a proportion of the dividing cells, unfavourable conditions in vitro, or deranged cell differentiation at the more advanced stages of development.

ACKNOWLEDGEMENTS

We are grateful to Mr. Ian Johnston, Dr. Andrew Speirs and Dr. John McBain for providing us with preovulatory eggs and to Professor Graeme Ryan, University of Melbourne, for the use of electron microscopes in his department.

REFERENCES

Anderson E, Hoppe PC, Whitten WK, Lee GS (1975). In vitro fertilization and early embryogenesis: A cytological analysis. J Ultrastruct Res 50:231.

Cizadlo GR, Granholm NH (1978). Ultrastructural analysis of preimplantation lethal yellow (A^y/A^y) mouse embryos. J Embryol exp Morphol 45:13.

Edwards RG (1981). Test-tube babies, 1981. Nature 293:253.

El-Shershaby AM, Hinchliffe JR (1974). Cell redundancy in the zona-intact preimplantation mouse blastocyst: A light and electron microscopic study of dead cells and their fate. J Embryol exp Morphol 31:643.

Hurst PR, Jefferies K, Eckstein P, Wheeler AG (1978). An ultrastructural study of preimplantation uterine embryos of the rhesus monkey. J Anat 126:209.

Lopata A, Johnston WIH, Hoult IJ, Speirs AL (1980). Pregnancy following intrauterine implantation of an embryo obtained by in vitro fertilization of a preovulatory egg. Fertil Steril 33:117.

Lopata A, Kellow GN, Johnston WIH, Speirs AL, Hoult IJ, Pepperell RJ, du Plessis YP (1981). Human embryo transfer in the treatment of infertility. Aust NZ J Obstet Gynaecol 21:156.

McReynolds HD, Hadek R (1972). A comparison of the fine structure of late mouse blastocysts developed in vivo and in vitro. J Exp Zool 182:95.

Panigel M, Kraemer DC, Kalter SS, Smith GC, Heberling RL (1975). Ultrastructure of cleavage stages and preimplantation embryos of the baboon. Anat Embryol 147:45.

Pereda J, Croxato HB (1978). Ultrastructure of a seven-cell human embryo. Biol Reprod 18:481.

Trounson AO, Leeton JF, Wood C, Webb J, Wood J (1981). Pregnancies in humans by fertilization in vitro and embryo transfer in the controlled ovulatory cycle. Science 212:681.

Van Blerkom J, Manes C, Daniel JCJr (1973). Development of preimplantation rabbit embryos in vivo and in vitro. I. An ultrastructural comparison. Dev Biol 35:262.

Wood C, Trounson A, Leeton J, Talbot JMc, Buttery B, Webb J, Wood J, Jessup D (1981). A clinical assessment of nine pregnancies obtained by in vitro fertilization and embryo transfer. Fertil Steril 35:502.

EXPERIMENTAL GENETICS OF THE MOUSE EMBRYO

Karl Illmensee

Department of Animal Biology,
University of Geneva
CH-1224 Geneva, Switzerland

In recent years, several approaches suitable for the analysis of gene activity during mammalian embryogenesis have been realized in the mouse and I should like to briefly summarize these new results and discuss their relevance and applicability to experimental embryology. Firstly, microsurgical removal of one of the two pronuclei from the fertilized mouse egg enables to independently study the maternal or paternal genome for its protein synthesis capacity during development. Secondly, enucleation by removing either both pronuclei from the fertilized egg or the female pronucleus from the unfertilized egg provides an experimental means to analyze the protein pattern derived from maternal mRNA stored during oogenesis. Thirdly, in nuclear transplantations the egg genome is replaced by a somatic cell nucleus in order to reveal the developmental potential of nuclei originating from different cell types and lineages. Fourthly, in gene transfers cloned recombinant DNA molecules are introduced into the developing organism via injection into the pronucleus of the fertilized egg to examine their presence, expression and regulation during cellular differentiation.

1. MATERNAL AND PATERNAL GENE ACTIVITY

Early preimplantation development of the mouse embryo seems to be supported by both the translation of mRNA made during oogenesis and the activation of the embryonic genome (reviewed by Sherman, 1979). During this transient period when oogenetic and embryonic gene products seem to function in a temporally coordinated way, it is rather difficult to separate these two processes under normal developmental conditions. Although there is accumulating evidence for early activation and expression of the embryonic genome, the paternal and maternal gene products can usually not be distinguished from each other. Only in a few instances, paternal gene contributions have been documented in preimplantation mouse embryos (reviewed by Chapman et al., 1977). It would therefore be desirable to study the paternal and maternal genome independently as far as their genetic activity is concerned.

Recently, it has been demonstrated that both parental genomes can microsurgically be separated soon after fertilization, thereby producing androgenetic and gynogenetic mouse embryos (Markert and Petters, 1977) which after transfer into foster mothers have developed into normal isodiploid mice (Hoppe and Illmensee, 1977). Following this experimental scheme (Fig. 1), androgenetic and gynogenetic embryos were subsequently analyzed for their protein synthesis using high resolution 2-dimensional gel electrophoresis. We have found that both types of embryos irrespective of the parental genetic origin synthesized very similar patterns of polypeptides throughout different stages of preimplantation development (Petzoldt et al., 1981). In this respect, the paternal as well as the maternal nuclear genome seems to be equally competent in controlling translation during cleavage and blastocyst formation. While certainly some proteins are derived from maternal mRNA (as discussed later), a significant proportion of newly synthesized proteins has to come from embryonic gene activity and it is therefore important to emphasize that the paternal genome by itself is fully capable of promoting normal early embryogenesis. In this context, further studies on X-chromosomally linked gene expression during subsequent postimplantation stages should provide new insights into the process of maternal versus paternal sex-chromosomal determination (reviewed by Takagi, 1978).

Contrary to the similarities in protein synthesis between androgenetic and gynogenetic embryos, a remarkable discrepancy in stage-specific protein patterns occurred in these uniparental embryos when compared with control embryos. The diploidized eggs expressed a polypeptide pattern which very much resembled that of normal 2-cell embryos, and when the experimental eggs cleaved to the 2-cell stage they synthesized proteins typical of the normal 4-cell embryos. These differences between stage-specific protein synthesis and the actual cell stages are most likely caused by the

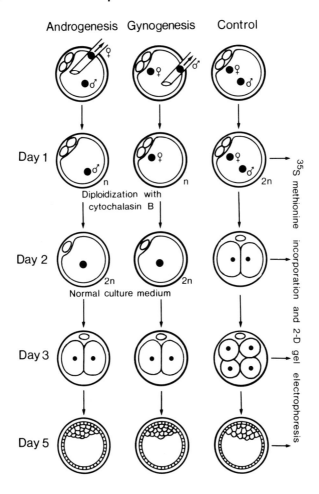

Figure 1. Experimental scheme for the production of andro- and gynogenetic mouse embryos. Fertilized C57BL/6 eggs were placed in culture medium containing cytochalasin B (CB). Subsequently either the maternal or the paternal pronucleus was removed by sucking it into a small glass pipette. The resulting haploid eggs were diploidized with CB and then cultured *in vitro* in normal medium. As in nonoperated controls they developed to blastocysts at day 5. Androgenetic, gynogenetic and control embryos of day 1, 2, 3 and 5 were used for incorporation with ^{35}S-methionine and further processed for protein analysis using 2-D polyacrylamide gel electrophoresis (after Petzoldt et al., 1981).

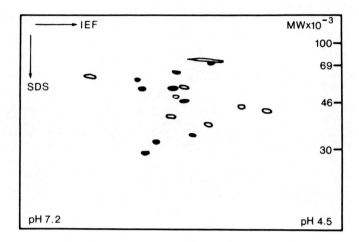

Figure 2. Schematic presentation of proteins whose synthesis is correlated with nuclear replication at the first and second division but independent of cytokinesis. After having eliminated all proteins presumably translated from maternal mRNA (see Petzoldt et al., 1980) or being present ubiquitously during preimplantation, only those proteins showing developmental stage specificity have been retained in the diagram. Following the first nuclear replication, 9 proteins appear in andro- and gynogenetically diploidized eggs that are also specific for normal 2-cell embryos (filled spots). After the second nuclear replication, 8 proteins including groups of polypeptides are found in androgenetic and gynogenetic 2-cell embryos which are typical for normal 4-cell embryos (open spots). Electrophoresis directions and approximate molecular weights are indicated in the scheme (Petzoldt et al., 1981).

temporal discordance between nuclear and cellular division, since CB treatment of the haploid eggs prevented the first cleavage but did allow chromosomal replication to proceed normally to an isodiploid genome (Hoppe and Illmensee, 1977). After having removed these eggs from CB, their asynchrony to normal embryos with respect to cytokinesis was maintained during the second and subsequent cleavage divisions. Are these changes in protein synthesis causally related to nuclear replication rather than cellular division ? Do they possibly result, at least to some extent, from a pool of maternal mRNA likewise involved in early embryonic protein synthesis and/or originate from newly transcribed mRNA after fertilization ?

To clarify these questions, we first omitted from the protein pattern of normal 2-cell embryos all polypeptides presumably derived from maternal mRNA (see Petzoldt et al., 1980) or being ubiquitously present during preimplantation. In doing this, we were indeed left with a few characteristic proteins which appeared stage-specifically and were also correlated with nuclear replication (Fig. 2). Following the next cleavage division, another set of 8 particular proteins was regarded as characteristic for normal 4-cell embryos as well as uniparental 2-cell embryos.

From our data we conclude that the appearance or increased synthesis of some stage-specific polypeptides is correlated with the replication of the genome independent of whether cytokinesis is occuring. To more intensively investigate the effect of chromosomal and DNA replication on stage-specific protein synthesis, it will be important to further characterize these proteins and eventually reveal their function during early mouse embryogenesis.

2. PROTEIN SYNTHESIS FROM MATERNAL MESSAGE

Although it has been known for some time that mouse eggs and early embryos synthesize proteins (reviewed by van Blerkom, 1981), only recently there have been attempts to determine whether the mRNA coding for these proteins is derived from transcription during oogenesis or from embryonic gene activity (Braude et al., 1979; reviewed by Schultz et al., 1981). However, it is not yet entirely clear, to which extent these proteins coded by maternal mRNA contribute to early embryogenesis and interact with proteins derived from newly synthesized mRNA of the embryonic genome (reviewed by Sherman, 1979).

In this way, we tried to further investigate the presence of a maternally derived mRNA pool by analyzing the translational capacity of the cytoplasm of unfertilized and fertilized mouse eggs in the absence of the nucleus. For this purpose the eggs were microsurgically enucleated and the surviving cytoplasts analyzed quantitatively and qualitatively for protein synthesis at different times during culture *in vitro* (Petzoldt et al., 1980). Some remarkable differences were observed between the polypeptide pattern synthesized by fertilized and unfertilized eggs following enucleation.

Furthermore, cytoplasts of fertilized eggs showed a more rapid decrease of incorporation than those of unfertilized eggs. Assuming that the pool of total mRNA is originally rather similar between the two egg types, our data may imply that a higher depletion rate of active RNA takes place after fertilization and/or less active RNA is available for translation before fertilization. Both fertilized and unfertilized enucleated eggs reached approximately the same low level of protein synthesis after three days in culture. Newly synthesized proteins could only be detected in enucleated fertilized eggs when cultured for 2 days and appeared similarly to those polypeptides expressed in normal 2-cell embryos (Fig. 3). The differences in protein synthesis between fertilized and unfertilized eggs might result either from newly synthesized mRNA (Piko and Clegg, 1981) or from "masked" maternal mRNA being activated by specific developmental events, e.g. fertilization. Maternally derived mRNA is shown to be present in fertilized eggs, gradually decreases during subsequent cleavage divisions and eventually disappears during the blastocyst stage (Bachvarova and De Leon, 1980).

For some of the polypeptides newly appearing in fertilized eggs, activation of maternally derived mRNA seems a plausible mechanism. Braude et al. (1979) extracted mRNA from unfertilized mouse eggs and translated it *in vitro*. Besides the pattern normally expressed by unfertilized eggs, they also obtained some polypeptides typical for late fertilized eggs and early 2-cell embryos. The mRNA coding for these proteins might remain inactive in our enucleated unfertilized eggs due to the lack of appropriate "activation signals". Recently, Schultz et al. (1978) demonstrated that artificial maturation of cytoplasmic oocyte fragments could be initiated with special media. Under these conditions the protein synthesis pattern changed qualitatively, and some meiosis-specific polypeptides appeared. It therefore seems reasonable to conclude that activation of maternal mRNA is feasible in oocytes and eggs if the proper stimulus is used.

In future experiments, we shall attempt to artificially activate enucleated unfertilized eggs in order to investigate whether equivalent changes of the polypeptide pattern occur in these cytoplasts. It will also be important to isolate and characterize the different populations of maternal mRNA stored in the egg and, eventually, the genes coding for these messages in order to approach at the molecular level the problem of gene activity and its control during mammalian oogenesis.

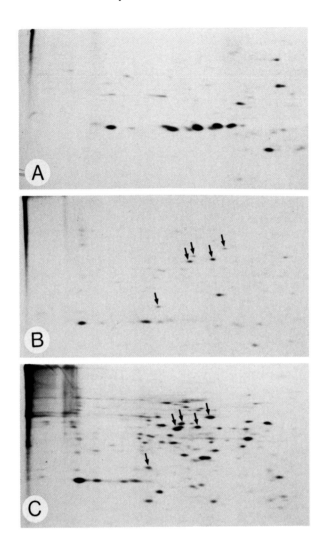

Figure 3. Protein synthesis in enucleated fertilized eggs (A), enucleated unfertilized eggs (B), and nonoperated 2-cell embryos (C), using 2-D polyacrylamide gel electrophoresis. Both kinds of enucleated eggs continue to synthesize proteins, though qualitative and quantitative differences are observed after two days of culture *in vitro*. Additionally, in enucleated fertilized eggs only, proteins appear that correspond to those of normal 2-cell embryos (after Petzoldt et al., 1980).

3. DEVELOPMENTAL POTENTIAL OF TRANSPLANTED NUCLEI

Nuclear transplantations carried out in mammals have met so far with rather limited success. Initial attempts to introduce somatic nuclei into mouse eggs via inactivated Sendai virus-mediated cell fusion resulted at best, in a few abnormal cleavage divisions of the treated eggs (Graham, 1969; Baranska and Koprowski, 1970; Lin et al., 1973). Failure to develop beyond the early cleavage stage could probably be attributed to several technical and biological factors such as damaging effects during virus induced fusion, inadequate culture conditions, or introduction of large amounts of donor cell cytoplasm, particularly during fusion of eggs with embryonic blastomeres. In recent studies, the possible damage due to the viral fusion procedure has been prevented by microsurgically injecting early embryonic nuclei into unfertilized rabbit eggs (Bromhall, 1975). The transfer of radioactively labeled nuclei from cells of morulae into unfertilized and parthenogenetically activated mature oocytes resulted in the development of preimplantation embryos in which weakly labeled nuclei were identified in about one-half of their cells. Because of the lack of chromosomal analysis and donor specific markers, it was not possible to determine whether these embryos were derived exclusively from the implanted nuclei or from the residing egg nucleus together with the injected somatic nucleus which had participated in early preimplantation development. In order to identify clearly the transplanted nucleus from the egg nucleus both biologically and genetically, Modlinski (1978, 1981) transferred chromosomally marked nuclei from cells of the CBA/H-T6 mouse morula and blastocyst into fertilized mouse eggs of a different strain. Some of the developing morulae and blastocysts revealed a tetraploid karyotype including the T6 translocation chromosomes of the donor nuclei, thus indicating functional participation of the transferred nuclei in preimplantation development. However, the injection of a somatic nucleus into a fertilized egg does not allow one to trace the donor nucleus through normal embryogenesis, which is obscured by the developmental contributions of the residing zygote nucleus of the recipient egg, thereby giving rise to a tetraploid embryo.

In collaboration with Dr. P.C. Hoppe (Jackson Laboratory, Bar Harbor, USA), we attempted to circumvent these complications by injecting a genetically marked nucleus into a fertilized but enucleated mouse egg and have subsequently probed this bioassay on various cell nuclei of the late preimplantation embryo (Fig. 4). About four days after fertilization, the mouse embryo develops into a blastocyst composed of the trophectoderm (TE) and the inner cell mass (ICM). The molecular basis for this temporal and spacial morphogenetic program leading to two different cell lineages has not yet been unravelled (reviewed by Johnson et al., 1981). During normal development, ICM cells will give rise to the embryo proper (reviewed by Gardner, 1978). TE cells, on the other hand, do not contribute to the embryo but to an extraembryonic cell lineage (reviewed by Gardner and Rossant, 1976), cease to divide and become polyploid during implantation (Barlow and Sherman, 1972), synthesize particular proteins (van Blerkom et al., 1976) and form intermediate filaments characteristic of epithelial cells (Jackson et al., 1980). Do these various cellular changes also restrict the nuclear potential to cell lineage specific pathways ?

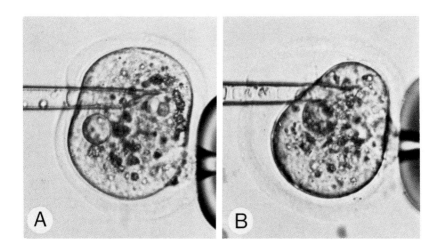

Figure 4. Nuclear transplantation into a fertilized mouse egg. A. The egg is attached to a blunt holding pipette (right) in the proper position. The nucleus of an embryonic cell is transferred into the egg using a thin injection pipette. B. Subsequently, the male and female pronucleus are removed from the egg by sucking them into the same pipette (after Illmensee and Hoppe, 1981a).

The question of whether irreversible alterations in gene function occur during differential segregation into TE and ICM cells has been answered in nuclear transplantations. After injection of TE nuclei into fertilized but enucleated eggs, the resulting early embryos showed abnormalities and their development was usually arrested during the early preimplantation period. On the contrary, nuclei from ICM cells initiated development of the recipient eggs and gave rise to normally appearing blastocysts. Biochemical and karyotypic analyses were carried out on the nuclear transplant embryos in order to determine whether the transplanted nuclei had actually supported development of the injected eggs. Allelic variants of the enzyme glucosephosphate isomerase (GPI) and chromosomal markers of the T6 translocation enabled us to distinguish between the genome of the transplanted nucleus and the recipient egg. Preimplantation embryos derived from TE nuclei exhibited predominantly the egg specific GPI pattern and only occasionally, in addition, enzyme activity originating from the nuclear donor. Apparently, nuclei from TE cells failed to promote normal development and appeared already limited in their differentiation capacity. On the other hand, enzyme tests carried out on ICM nuclear transplant embryos showed that a considerable proportion of them expressed exclusively enzyme activity characteristic of the donor genome. In addition, chromosomal analysis of the ICM-derived embryos revealed a diploid karyotype with the two short T6 translocation chromosomes of the donor nucleus (Illmensee and Hoppe, 1981a).

Are these nuclear transplant blastocysts able to continue in development and, if so, reach the adult stage ? For this reason, we transferred some of them together with nonoperated control embryos into the uteri of pseudopregnant females to allow development to term. Live-born mice were derived genetically from the transplanted ICM nuclei as judged by their coat color, karyotype and GPI enzyme pattern (Fig. 5). In additional breeding tests, the nuclear transplant mice proved to be fertile and transmitted the ICM genome to their progeny, all of which exhibited the nuclear donor phenotype (Illmensee and Hoppe, 1981a). From these results we conclude that the nuclear genome of ICM cells can still express all the genes necessary to code for the entire animal. Nuclear transplantation in the mouse may therefore provide a vigorous bioassay to functionally determine the genomic capacity of the entire nucleus, to uncover the progressive loss of nuclear potential in differentiating cells,

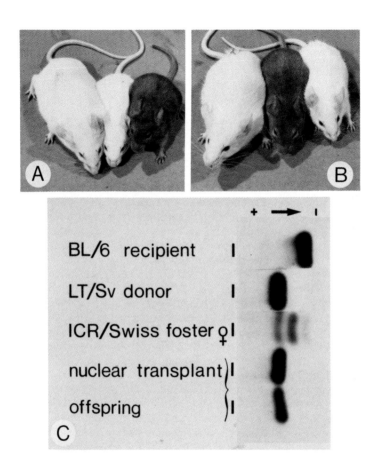

Figure 5. Nuclear-transplant mice derived from ICM nuclei of the LT/Sv blastocyst after their injection into fertilized but enucleated BL/6 eggs. A and B. Two ICR/Swiss foster females each of which is shown with one of their regular (white coat) and nuclear-transplant (grey coat) offspring. C. Biochemical analysis using cellulose acetate electrophoresis of glucosephosphate isomerase (GPI) from blood cell lysates of the two nuclear-transplant mice only shows the LT/Sv specific enzyme variant characteristic of the nuclear donor strain (after Illmensee and Hoppe, 1981a).

and to search for the mechanisms of gene inactivation during cell diversification. In this respect, it will be important to extend nuclear transplantations to more advanced embryonic stages (for preliminary results see Illmensee, 1981; Illmensee and Hoppe, 1981b) in order to gain insight into the biological and genetic consequences of nuclear changes during mammalian cell differentiation.

GENE TRANSPLANTATION

Alternatively to DNA-mediated transformation of cultured cells (reviewed in Celis et al., 1980), the expression and regulation of cloned genes may be analyzed in the developing organism, which would require the injection of those genes into eggs or early embryos. In mammals, preimplantation embryos of the mouse seem to be suitable recipients for gene transplantation because of their well investigated genetic background and the relative ease with which they can be collected from pregnant females, manipulated and cultured *in vitro*, and then retransferred into foster mothers for further development. In this way, the integration and possible expression of the foreign genes can be most effectively studied at various stages of embryonic, fetal and adult life and, eventually, transmitted to the next generation for genetic mapping of their location in the host genome. This *in vivo* approach was first used by injecting purified simian virus 40 DNA into the cavity of mouse blastocysts and by demonstrating the presence of injected DNA copies in some organs of surviving mice (Jaenisch and Mintz, 1974). More recently, recombinant DNA plasmids composed of segments of herpes simplex virus, simian virus 40 and the bacterial vector pBR322 were injected directly into fertilized mouse eggs in order to increase the probability of integration of donor DNA into all cells of the developing embryos. Two of the newborn mice derived from plasmid-injected eggs were found to contain donor DNA specific sequences, though in abnormally rearranged form (Gordon et al., 1980). Similarly, cloned plasmids containing the human β-globin gene and thymidine kinase gene of herpes simplex virus were transferred into mouse eggs and retained in the developing fetuses of which one was found to express the viral gene (Wagner et al., 1981).

In collaboration with Dr. A. Ullrich and his associates (Genentech, South San Francisco, USA), we attempted to introduce recombinant DNA molecules containing the human insulin gene into mouse eggs (Bürki and Ullrich, 1981; Illmensee et al., 1981). Following DNA injection, the developing embryos were then transferred into the uteri of pseudopregnant foster females. Fetuses and their corresponding placentas were isolated from the uteri at late pregnancy and their DNA was extracted and screened for the presence of the injected DNA sequences by using a radioactively labeled 310 bp probe that contains most of the human preproinsulin (Ullrich et al., 1980). With the aid of restriction endonuclease digestion in conjunction with Southern blot hybridization we found that in two normally developed fetuses at day 18 of pregnancy, both fetal and placental tissues contained the human insulin gene including the flanking regions and bacterial plasmid sequences (for details see Bürki and Ullrich, this volume). Although we have demonstrated the presence of the human insulin gene within the fetal mouse genome, it still remains to be shown whether this particular gene is functionally expressed during cellular differentiation, organ-specifically controlled at the adult stage, and eventually transmitted through the germline to the progeny.

ACKNOWLEDGEMENTS

I should like to thank Drs. P.C. Hoppe and A. Ullrich for continuous collaboration and my associates Drs. K. Bürki and U. Petzoldt, M.F. Blanc and G.R. Illmensee for their invaluable help and excellent assistance. Inbred mice were kindly donated by the Füllinsdorf Institute of Biomedical Research, Switzerland. I should like to acknowledge support from the Swiss Science Foundation (FN3.442.0.79), the March of Dimes Birth Defects Foundation (1-727), the National Institutes of Health (CA27713-02) and the Roche Research Foundation.

REFERENCES

Bachvarova R, De Leon V (1980). Polyadenylated RNA of mouse ova and loss of maternal RNA in early development. Dev Biol 74:1-8.

Baranska W, Koprowski H (1970). Fusion of unfertilized mouse eggs with somatic cells. J Exp Zool 174:1-14.

Barlow PW, Sherman MI (1972). The biochemistry of differentiation of mouse trophoblast: studies on polyploidy. J Embroyl Exp Morphol 27:447-465.

Braude PR, Pelham H, Flach G, Lobatto R (1979). Post-transcriptional control in the early mouse embryo. Nature 282: 102-105.

Bromhall JD (1975). Nuclear transplantation in the rabbit egg. Nature 258:719-721.

Bürki K and Ullrich A (1981). Transplantation of the human insulin gene into fertilized mouse eggs. Submitted to Proc Natl Acad Sci USA.

Celis JE, Graessmann A, Loyter A (1980). Transfer of cell constituents into eukaryotic cells, Vol. 31, NATO Advanced Study Institutes Series. New York: Plenum Press.

Chapman VM, West JD, Adler DA (1977). Genetics of early mammalian embryogenesis. In Sherman MI (ed):"Concepts in mammalian embryogenesis", Cambridge, Massachusetts and London, England: The MIT Press, p 94-135.

Gardner RL (1978). The relationship between cell lineage and differentiation in the early mouse embryo. In Gehring W (ed): "Results and Problems in Cell Differentiation Vol IX, Heidelberg: Springer Verlag, p 205-241.

Gardner RL and Rossant J (1976). Determination during embryogenesis. In "Embryogenesis in Mammals", Ciba Foundation Symposium 40 (new series), Amsterdam: Elsevier, Excerpta Medica, p 5-18.

Gordon JW, Scangos GA, Plotkin DJ, Barbosa JA, Ruddle FH (1980). Genetic transformation of mouse embryos by microinjection of purified DNA. Proc Natl Acad Sci USA 77: 7380-7384.

Graham CF (1969). The fusion of cells with one- and two-cell mouse embryos. In Defendi V (ed): "Heterospecific Genome Interaction", Symposium Monograph Vol IX, Philadelphia: Wistar Institute Press, p 13-35.

Hoppe PC, Illmensee K (1977). Microsurgically produced homozygous-diploid uniparental mice. Proc Natl Acad Sci USA 74:5657-5661.

Illmensee K, Hoppe PC (1981a). Nuclear transplantation in *Mus musculus*: developmental potential of nuclei from preimplantation embryos. Cell 23:9-18.

Illmensee K, Hoppe PC (1981b). The potential of transplanted nuclei during mammalian differentiation. In Sauer HW (ed): "Progress in Developmental Biology", Stuttgart New York: Gustav Fischer Verlag, p 67-74.

Illmensee K (1981). Experimental Manipulation of the Mammalian embryo: Biological and genetic consequences.

In "Mammalian Genetics and Cancer": The Jackson Laboratory Fiftieth Anniversary Symposium, New York: Alan R. Liss, p 105-120.

Illmensee K, Bürki K, Hoppe PC, Ullrich A (1981). Nuclear and gene transplantation in the mouse. In Brown DD, Fox CF (eds): "Developmental Biology Using Purified Genes", ICN-UCLA Symposia on Molecular and Cellular Biology, Vol 23, New York: Academic Press, p 607-619.

Jackson BW, Grund C, Schmid E, Bürki K, Franke WW, Illmensee K (1980). Formation of cytoskeletal elements during mouse embryogenesis: I. Intermediate filaments of the cytokeratin type and desmosomes in preimplantation embryos. Differentiation 17: 161-179.

Jaenisch R, Mintz B (1974). Simian Virus 40 DNA sequences in DNA of healthy adult mice derived from preimplantation blastocysts injected with viral DNA. Proc Nat Acad Sci USA 71:1250-1254.

Johnson MH, Pratt HPM, Handyside AH (1981). The generation and recognition of positional information in the preimplantation mouse embryo. In Glasser SR, Bullock DW (eds): "Cellular and Molecular Aspects of Implantation", New York and London: Plenum Press, p 55-74.

Lin TP, Florence J and Oh JO (1973). Cell fusion induced by a virus within the zona pellucida of mouse eggs. Nature 242:47-49.

Markert CL, Petters RM (1977). Homozygous mouse embryos produced by microsurgery. J Exp Zool 201:295-302.

Modlinski JA (1978). Transfer of embryonic nuclei to fertilized mouse eggs and development of tetraploid blastocysts. Nature 273:466-467.

Modlinski JA (1981). The fate of inner cell mass and trophectoderm nuclei transplanted to fertilized mouse eggs. Nature 273:466-467.

Petzoldt U, Illmensee GR, Bürki K, Hoppe PC, Illmensee K (1981). Protein synthesis in microsurgically produced androgenetic and gynogenetic mouse embryos. Mol Gen Genet, in press.

Petzoldt U, Hoppe PC, Illmensee K (1980). Protein synthesis in enucleated fertilized and unfertilized mouse eggs. Wilhelm Roux's Archives 189:215-219.

Piko L, Clegg KB (1981). RNA synthesis in the one-cell mouse embryo. IX. Congress Int Soc Devl Biol, Basel, Abstracts, P236.

Schultz GA, Clough JR, Braude PR, Pelham HRB, Johnson MH (1981). A reexamination of messenger RNA populations in the preimplantation mouse embryo. In Glasser SR, Bullock DW (eds): "Cellular and Molecular Aspects of Implantation", New York: Plenum Press, p 137-154.

Schultz RM, Letourneau GE, Wassarman PM (1978). Meiotic maturation of mouse oocytes in vitro: Protein synthesis in nucleate and anucleate oocyte fragments. J Cell Sci 30: 251-264.

Sherman MI (1979). Developmental biochemistry of preimplantation mammalian embryos. Ann Rev Biochem 48:443-470.

Takagi N (1978). Preferential inactivation of the paternally derived X chromosome in mice. In Russell LB (ed):"Genetic mosaics and chimeras in mammals". New York and London: Plenum Press, p 341-360.

Ullrich A, Dull DJ, Gray A, Brosius J and Sures I (1980). Genetic variation in the human insulin gene. Science 209: 612-614.

Van Blerkom J (1981). Intrinsic and extrinsic patterns of molecular differentiation during oogenesis, embryogenesis and organogenesis in mammals. In Glasser SR, Bullock DW (eds): "Aspects of Implantation", New York: Plenum Publishing Corporation, p 155-176.

Van Blerkom J, Barton SC and Johnson MH (1976). Molecular differentiation in the preimplantation mouse embryo. Nature 259:319-321.

Wagner EF, Stewart TA, Mintz B (1981). The human β-globin gene and a functional viral thymidine kinase gene in developing mice. Proc Natl Acad Sci USA 78:5016-5020.

TRANSPLANTATION OF THE HUMAN INSULIN GENE INTO FERTILIZED
MOUSE EGGS

Kurt Bürki[*] and Axel Ullrich[+]

[*]Department of Animal Biology
University of Geneva
CH-1224 Geneva, Switzerland

[+]Genentech, South San Francisco,
Ca 94080, USA

By means of recombinant DNA technology distinct portions of eukaryotic genomes can be selected and cloned. Various genes including their flanking regions have been isolated and characterized in detail. Contrary to expectation comparative sequence analysis has not directly revealed DNA regions which are responsible for the control of gene expression. To define the control region of a particular gene functional assays are applied where either the unaltered gene or stepwise deleted modifications of the gene are tested. One such functional assay is the introduction of cloned genes into cultured cells via DNA mediated gene transfer (Wigler et al., 1979). Cultured cell lines so far widely used (e.g. fibroblasts) rarely correspond to the cell type where the gene is active in the organism. Cell lines which mimic differentiation processes *in vitro* usually are karyotypically abnormal. Results obtained with these cells might not reflect normal gene expression. Ideally, genes to be assayed should be introduced into cells of early developmental stages, so that they are submitted to all cellular differentiation processes during ontogeny of the organism. The highly successful transfer of cloned genes into cultured cells by direct microinjection (Anderson et al., 1980; Capecchi, 1980) may therefore be adapted to early embryos. In mammals, mouse embryos provide suitable recipients for gene transfer because of their well characterized genetics and the ease with which they can be collected from pregnant females, manipulated and cultured *in vitro* and then be retransferred into the uteri of pseudopregnant foster females for further development. Thus the successful integration and possible expression of foreign genetic material can conveniently be studied at various stages of embryonic, fetal and adult life.

Using this rational Gordon et al. (1980) injected recombinant DNA molecules containing segments of herpes simplex virus, simian virus 40 and pBR322 directly into one of the two pronuclei of fertilized mouse eggs. Two of the newborn mice derived from plasmid-injected eggs were found to contain donor DNA-specific sequences, though in a rearranged form. Recently the introduction of the human β-globin gene and a functional viral thymidine kinase gene into developing mice was reported (Wagner et al., 1981).

In our initial experiments we attempted to introduce the human preproinsulin gene into the developing mouse fetus by injecting this cloned DNA into eggs at the pronuclear stage (Bürki and Ullrich, 1981; Illmensee et al., 1981).

The plasmid used is comprised of a 12.5 kb region of the human genome including the preproinsulin gene (Ullrich et al., 1980) and the 4.3 kb bacterial plasmid pBR322 (Bolivar et al., 1977) (Fig. 1). About 1-2 pl of plasmid solution containing 30.000 - 50.000 copies were microinjected either into the cytoplasm or directly into one of the two pronuclei of C57BL/6J mouse eggs 10-15 hours after fertilization (Fig. 2). 379 manipulated eggs were cultured *in vitro* for four days, of which 282 developed to the morula or blastocyst stage and were transferred into the uteri of pseudopregnant ICR foster females. At days 16-19 of pregnancy the females were sacrificed and their uterine contents dissected. A total of 60 fetuses and corresponding placentas were recovered and their DNA isolated (Wigler et al., 1979).

Undigested DNA was electrophoresed in order to examine whether the injected plasmid was associated with high molecular weight DNA. No fast migrating unintegrated plasmids were detected.

High molecular weight DNA was digested using several restriction enzymes, separated on agarose gels and transferred onto nitrocellulose paper (Southern, 1975). The preparations were probed either with a 310 base pair internal fragment containing most of the human preproinsulin gene (Fig. 1) or with pBR322 DNA.

The restriction endonuclease Pvu II cleaves the injected plasmid close to either side of the human insulin gene sequences, generating a 1.6 kb fragment (Fig. 1). A fragment of the same size is also generated when human placental DNA is digested with Pvu II (Fig. 3). By hybridization with the 310 bp probe a 1.6 kb band was detected in the DNA-digest

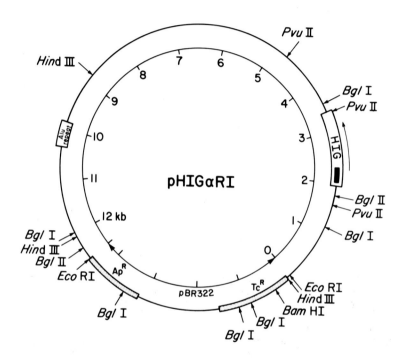

Fig. 1. Restriction endonuclease map of pHIGαRI recombinant DNA plasmid. Not all Pvu II and Bgl I restriction endonuclease cleavage sites are shown. The human insulin gene is shown (HIG) with an arrow designating the direction of transcription (5' to 3'). The solid bar in the human insulin gene region represents that portion of the gene used as hybridization probe, referred to in the text as the 310 bp probe. Tc^R, tetracycline resistance gene; Ap^R, ampicillin resistance gene. Sequences numbered 0 to 12 kb flank the insulin gene within the human genome (Ullrich et al., 1980); pBR322 designates a bacterial plasmid (Bolivar et al., 1977).

of two of the 60 experimentally manipulated mouse fetuses as well as in the DNA-digest of their corresponding placenta, indicating that the human insulin gene had been retained intact in the fetal and placental tissues (Fig. 3).

The intensities of the hybridization bands of fetal or placental DNA are comparable to the hybridization band obtained with equal amounts of human placental tissue. Therefore the positive mouse tissues carry injected DNA roughly equivalent to 1-2 copies per cell. Cleaving the plasmid with Bgl II endonuclease generates a 10 kb fragment containing the insulin gene and about 8.7 kb of the 3' region flanking human DNA sequences (Fig. 1). Hybridization analysis indicates that the 10 kb portion of the injected plasmid had been maintained intact in the two positive fetuses as well as in their corresponding placentas.

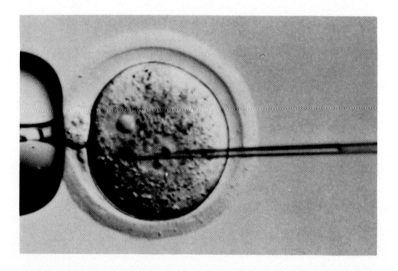

Figure 2. After having attached a fertilized C57BL/6 egg to the holding pipette (right) in the appropriate position, the injection pipette (left) is introduced, preferably into one of the two pronuclei. A volume of 1-2 pl of DNA solution injected into each egg contains about 30 000 copies of a recombinant plasmid composed of a 12.5 kb fragment of human DNA including the entire insulin gene and the 4.3 kb bacterial vector pBR322. (For further details, see Bürki and Ullrich, 1981).

When cleaved with EcoRl the plasmid is split into the 12.5 kb human portion and the bacterial pBR322 portion (Fig. 1). Hybridization analysis using the 310 bp probe revealed that a 12.5 kb fragment containing the human insulin gene was still present in the two positive placentas whereas the DNA of the corresponding fetuses showed hybridization bands which were larger in size by 1.5 kb that the originally injected human DNA. Differences were also detected in the mobilities of the endogenous mouse fetal and placental insulin gene fragments present in all experimental

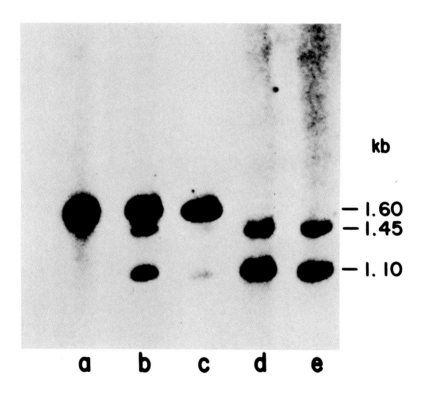

Figure 3. Pvu II restriction endonuclease cleavage in conjunction with Southern hybridization. DNA from (a) human placenta, (b) mouse fetus 46, (c) mouse placenta 46, (d) control mouse fetus, and (e) control mouse placenta, electrophoresed on a 1% agarose gel and hybridized with the 310 bp human insulin gene probe.

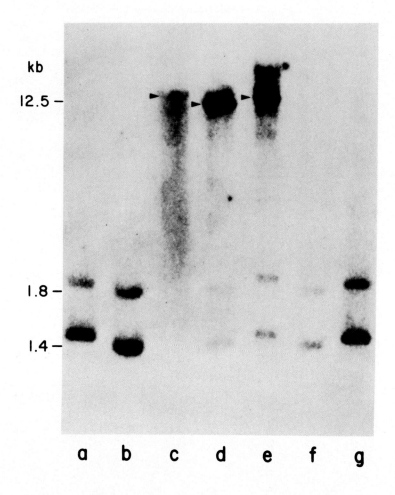

Figure 4. Eco RI restriction endonuclease cleavage in conjunction with Southern hybridization. (a) mouse fetus and (b) mouse placenta derived from non-injected eggs; (c) human placenta; (d) mouse placenta and (e) mouse fetus derived from injection 46; (f) mouse placenta and (g) mouse fetus arising from injection 45. Arrowheads indicate center of human gene fragment bands. A radiolabelled 310 bp human insulin gene fragment was used as hybridization probe.

samples and in uninjected controls (Fig. 4). Further experiments are in progress to examine the molecular basis of these differences observed in the endogenous mouse insulin genes as well as in the injected human insulin gene.

To determine whether chromosomal integration had occured within the pBR322 region of the injected plasmid, EcoRl blots of fetal and placental DNA were probed with pBR322 DNA. The digests of the two placental DNAs and of one of the fetal DNAs showed two hybridization bands each. The blot of the second fetus exhibited four hybridization bands for further details see Bürki and Ullrich, 1981). The simple hybridization patterns obtained with the pBR322 probe indicate that in three of the positive tissues one plasmid had become integrated into the mouse genome by opening in its pBR322 region. In one of the positive fetuses probably two plasmids had integrated, both by opening in their pBR322 region.

Experiments are under way to extend our observations to the adult stage. We are currently investigating whether the injected cloned human DNA can remain stably integrated in the DNA of adult tissues, is tissue-specifically expressed and eventually transmitted via the germ line to the next generation.

Acknowledgements: We thank A. Gray and E. Wong for technical assistance and R. Swanson for support. We are also grateful to K. Illmensee, I. Sures, G.R. Illmensee and S. Pfeffer for their contributions to this work. Inbred mice were kindly donated by the Füllinsdorf Institute of Biomedical Research, Switzerland. This work was also financed by grants awarded to K. Illmensee from the Swiss National Science Foundation (FN 3.442.0.79), the March of Dimes Birth Defects Foundation (1-727) and the National Institute of Health (CA27713-02).

REFERENCES

Anderson WF, Killos L, Sanders-Haigh L, Kretschmer PJ and Diacumakos EG (1980). Replication and expression of thymidine kinase and human globin genes microinjected into mouse fibroblasts. Proc Natl Acad Sci USA 77:5399-5403.

Bolivar F, Rodriguez RL, Greene PJ, Betlach MC, Heyneker HL and Boyer HW (1977). Construction and characterization of new cloning vehicles. II. A multipurpose cloning system. Gene 2:95-113.

Bürki K and Ullrich A (1981). Transplantation of the human insulin gene into fertilized mouse eggs. Submitted to Proc Natl Acad Sci USA.

Capecchi MR (1980). High efficiency transformation by direct microinjection of DNA into cultured mammalian cells. Cell 22:479-488.

Gordon JW, Scangos GA, Plotkin DJ, Barbosa JA and Ruddle FH (1980). Genetic transformation of mouse embryos by microinjection of purified DNA. Proc Natl Acad Sci USA 77: 7380-7384.

Illmensee K, Bürki K, Hoppe PC and Ullrich A (1981). Nuclear and gene transplantation in the mouse. In Brown DD, Fox CF (eds): "Developmental Biology Using Purified Genes," ICN-UCLA Symposia on Molecular and Cellular Biology, Vol 23, New York: Academic Press, p 607-619.

Southern EM (1975). Detection of specific sequences among DNA fragments separated by gel electrophoresis. J Mol Biol 98:503-517.

Ullrich A, Dull DJ, Gray A, Brosius J and Sures I (1980). Genetic variation in the human insulin gene. Science 209: 612-614.

Wagner EF, Stewart TA and Mintz B (1981). The human β-globin gene and a functional viral thymidine kinase gene in developing mice. Proc Natl Acad Sci USA 78:5016-5020.

Wigler M, Sweet R, Sim GK, Wold B, Pellicer A, Lacy E, Maniatis T, Silverstein S and Axel R (1979). Transformation of mammalian cells with genes from procaryotes and eucaryotes. Cell 16:777-785.

GERM LINE TRANSMISSION IN TRANSGENIC MICE

Jon W. Gordon and Frank H. Ruddle
Yale University
Department of Biology
260 Whitney Avenue
New Haven, Connecticut USA 06511

ABSTRACT

Our laboratory has recently demonstrated the successful transfer of genetic material into the genomes of newborn mice by injection of that material into pronuclei of fertilized eggs. Our initial results indicated two patterns of processing of the injected DNA: one in which the material was not integrated into the host genome, and another in which the injected genes became associated with high molecular weight DNA. We now present evidence that these patterns are maintained through further development to adulthood. We provide definitive evidence of the covalent association of injected DNA with host sequences, and transmission of such linked sequences in a Mendelian distribution to 2 succeeding generations of progeny. The observation that injected gene sequences can become integrated and stably transmitted through the germ line should contribute significantly to further studies of this gene transfer system.

INTRODUCTION

The successful introduction of exogenous DNA into cultured mammalian cells (Bachetti and Graham, 1977; Maitland and McDougall, 1977; Wigler et al., 1977; Minson et al., 1978) has led to the development of a novel gene transfer system which has yielded important information about gene regulation in higher eukaryotes. One difficulty with this system, however, is that cultured cells are not capable of organismal development and differentiation. Our laboratory has recently demonstrated that DNA sequences cloned by recombinant DNA technology can be microinjected into the pronuclei of fertilized mouse oocytes and subsequently located in the DNA of newborn mice by Southern blot hybridization (Gordon et al., 1980). This system allows the study of transferred gene sequences in the context of normal embryonic development. Since development is a continual process which includes maturation to adulthood, reproduction, and senescence, it is important to examine the fate of transferred genes beyond the point of birth. In the present study we have followed this injected material through further stages of mouse development.

Two recombinant plasmids have been employed for microinjection. The first, designated pST6 (Gordon et al., 1980), was composed of the HindIII C fragment of SV40 virus and the Herpes virus thymidine kinase (tk) genes cloned in pBR322; the second, pIf (kindly provided by C. Weissmann, Zurich, Switzerland), contained human leukocyte interferon cDNA also cloned in pBR322 (Nagata et al., 1980). A simplified diagram of each plasmid with its relevant restriction sites is shown in Figure 1. Between 1,000 and 35,000 copies of each plasmid were injected into each zygote. All microinjections were carried out as described previously (Gordon et al., 1980). For convenience, we suggest the term "transgenic" for animals into which exogenous genes have been transferred. This term will be employed to describe the mice throughout the ensuing discussion.

Plasmid sequences present in newborn and adult mice were evaluated by Southern blot hybridization (Southern, 1975; Blin and Stafford, 1976; Wahl et al., 1979), with minor modifications described elsewhere (Gordon et al., 1980). In the case of adults, DNA was extracted from spleens. Whether or not the donor material was integrated into the host genome was assessed using 2 criteria: restriction enzyme analysis, and progeny testing. Two classes of transgenic mice were found; those in which donor sequences did not appear integrated, and those with integrated sequences. The application of restriction analysis and progeny testing to these two classes of mice will be considered separately.

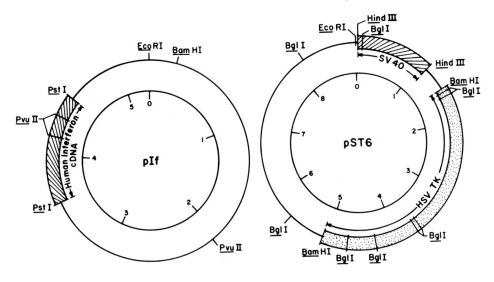

Figure 1. Simplified diagrams of recombinant plasmids used for microinjection of mouse embryos. The *Eco*RI site of pBR322 is marked at 0 kb for reference.

RESTRICTION ANALYSIS

Sequences Retained Without Integration

We have now identified 6 adult mice injected at the pronuclear stage with pST6 whose spleen DNA hybridized to the pST6 probe, but whose restriction patterns indicated failure of the donor material to integrate into the host genome. When the DNA was undigested, the sequences homologous to the plasmid migrated significant distances into 1% agarose gels, despite the fact that ethidium bromide staining showed that the host DNA did not enter the gel. Digests with the restriction enzymes *Xba*I and *Xho*I, which are known not to cut pST6, did not alter the mobility of the bands when compared with undigested DNA. This result showed that host sequences carrying restriction sites for these two enzymes had not become associated with the plasmid DNA. In five of the 6 adult mice demonstrating these patterns, retained sequences hybridized only to the pBR322

portion of pST6. These animals were thus similar to the
first transgenic mouse ever reported, which has been
described in detail elsewhere (Gordon et al., 1980). This
observation raised the possibility that sequences could not
be retained without integration unless extensively modified
such that the tk and SV40 inserts were removed. We were
also compelled to consider the possibility that these
results were not a direct consequence of the microinjection
procedure. However, a sixth mouse, #5.09 was found whose
spleen DNA did contain unintegrated pST6 sequences which
were modified only slightly. Figure 2a shows a *Bam*HI
digest of DNA from #5.09 compared with the same digest
of a positive control.

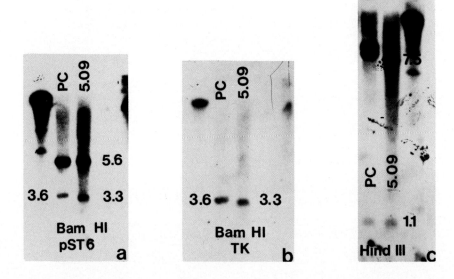

Figure 2. *Bam*HI digest of pST6 added to DNA from uninjected mice at a ratio of 1: 10^6 by weight (PC) and of
#5.09 and probed with either pST6 (a), or tk DNA (b). The
patterns are similar, and tk DNA is present in #5.09.
A *Hind*III digest (c) also shows the presence of the
SV40 insert in #5.09

A 5.6 kilobase (kb) band comigrates with the 5.6 kb band of the positive control, the latter of which comprises the pBR322 and SV40 inserts of the plasmid. A second 3.3 kb band is also produced by BamHI. This band is smaller than the 3.6 kb tk insert of the control. Digestion with BamHI thus produces a pattern almost identical to that of plasmid DNA added to genomic DNA from uninjected mice. Probing this filter with tk DNA alone (Fig. 2b) shows that the 3.3 kb band in #5.09 hybridizes to tk DNA, while the 5.6 kb band does not. This result demonstrates the presence of tk material in #5.09, and suggests that the plasmid material in this animal was unaltered save for a small modification of the tk region. Further support for this interpretation is provided by a HindIII digest of the sample (Fig. 2c). In this instance, a 1.1 kb band which comigrates with the SV40 insert of the control is produced. The remaining 7.5 kb band is slightly smaller than the 7.8 kb band of the control. The larger control fragment comprises the pBR322 and tk inserts of the plasmid in this digest. That the larger HindIII band of #5.09 is smaller than the corresponding control fragment is consistent with the results from BamHI digestion, which suggested that the tk region was reduced in size. Probing the HindIII digested sample with SV40 DNA showed that the 1.1 kb band in #5.09 hybridized to SV40, while the 7.5 kb band did not (data not shown). These results thus demonstrate that SV40 and tk sequences can be retained for many weeks in apparently unintegrated form. These findings strengthen the notion that hybridizing sequences in all six of these mice were derived from the injected pST6 molecule. This conclusion is further supported by failure to detect hybridization in DNA from 57 spleens and 109 newborns that were obtained from uninjected control mice.

Integrated Sequences

Two mice (#73 and #9.02) injected with either pST6 or pST9 (pST9 is identical to pST6 except that the orientation of the SV40 insert is reversed), and one mouse #If-4) injected with pIf, retained plasmid sequences whose restriction patters were consistent with integration. When undigested, the DNA of all three mice gave single bands of high molecular weight on Southern blots. This result suggests an association of the plasmid sequences with high molecular weight DNA of the host genome. When cut with the restriction enzymes XbaI and XhoI, which do not have sites in either recombinant plasmid, DNA from #9.02 and #73 again gave single high molecular weight bands. The mobility of these bands could not be distinguished from each other or from that of the band produced by undigested DNA. When digested with XbaI, DNA from #If-4 yielded a single band of 13.5 kilobases (kb), a much larger size than the original 5.2 kb plasmid. These patterns are again consistent with integration into the host genome at a single site. However, the data do not conclusively demonstrate covalent association of the plasmid with the host DNA. Studies of DNA mediated gene transfer with carrier DNA in cultured cells have shown that large, extragenomic "transgenomes" can be produced which are apparently composed entirely of donor sequences (Scangos et al., 1981). Such transgenomes also produce single high molecular weight bands when undigested or digested with enzymes which do not cut the donor sequence. Thus, such bands do not by themselves show attachment of donor DNA to host sequences. This ambiguity would be resolved if the XbaI and XhoI digests produced single bands of differing mobility. This pattern would indicate integration because the random spacing of sites for these enzymes around the plasmid would result in fragments of different size when the donor material was cut out of the genome. On the other hand, restriction of transgenomes with enzymes which do not recognize donor sequences would always produce a single band of the same size. Efforts to resolve a mobility difference between XbaI and XhoI fragments in #73 and #9.02 were unsuccessful, however. Fragments produced by these digests were of such large size that their mobilities could not be distinguished.

Double digests on the DNA of #9.02 and #73 circumvented this problem. The first digest, with BamHI, was followed by digests with XbaI or XhoI. The rationale for this experiment was that BamHI, which is known to cut within the plasmid sequence, would generate a hybrid fragment defined by a restriction site within the donor sequence and one within the flanking genomic material. Such a hybrid fragment would be expected to be smaller than those produced by XbaI or XhoI, which cut on either side of the entire integrated fragment. Because the hybrid BamHI fragment would be relatively small, alterations in its mobility would be more easily detected. If an XbaI or XhoI site were situated between the genomic BamHI site and the point of integration of the plasmid, a second digest with one of these enzymes would cleave a piece of DNA from the genomic portion of the BamHI hybrid fragment. The consequent reduction in size of this relatively small fragment would then be readily detected. The results of these double digests are shown in Figure 3.

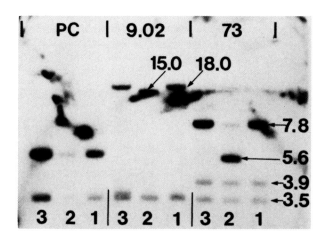

Figure 3. DNA from mouse #73 and #9.02 digested with BamHI (lane 1), BamHI + XbaI (lane 2), and BamHI + XhoI (lane 3). Although the XbaI digest did not proceed to completion in #73, almost all of the 7.8 kb BamHI band was converted to 5.6 kb by XbaI. The 18 kb BamHI band in 9.02 was altered to 15 kb by XbaI. As expected, double digests of the positive control (PC) did not alter the BamHI pattern. Failure of XhoI to alter the mobilities of the BamHI fragments indicates that the genomic portions of these fragments did not contain XhoI sites.

The 7.8 kb BamHI band in #73 was converted to 5.6 kb by XbaI. Similarly, the 18 kb band in #9.02 was reduced to 15 kb by XbaI. These alterations in mobility indicate the acquisition of XbaI sites, a result consistent with integration.

Similar results permitted the same conclusion regarding the state of pIf sequences in the spleen DNA of animal

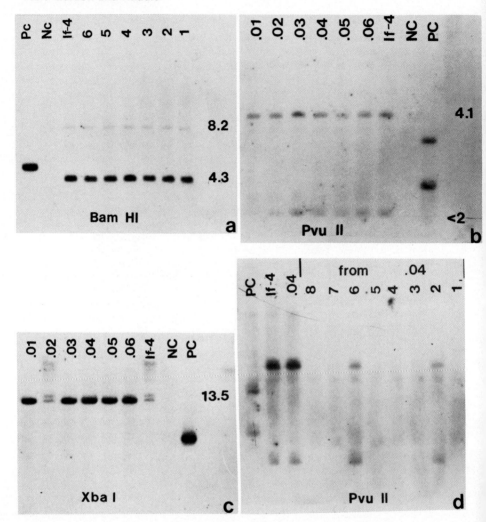

Figure 4. Spleen DNA from mouse #If-4 and six progeny digested with *Bam*HI (a), *Pvu*II (b), and *Xba*I (c), and probed with pIf. These digests demonstrate germ line transmission of pIf sequences. (d) *Pvu*II digest of spleen DNA from #If-4, its offspring #.04, and 8 progeny of #.04. Of these 8 second generation progeny, #2 and #6 show inheritence of pIf sequences. PC indicates the positive control, NC a negative control.

#If-4. A partial digest with $XbaI$ yielded several high molecular weight bands after hybridization with the pIf probe (Figure 4c). The smallest of these bands, 13.5 kb, was the only band produced by a complete digest with $XbaI$. The pattern obtained by partial digestion indicates linkage of the plasmid to DNA sequences containing multiple $XbaI$ sites. The 13.5 kb band was generated by cutting at the closest $XbaI$ sites to the point of attachment of the plasmid, whereas the larger fragments were produced when one or both of these closest sites was missed but more distant sites were cleaved. These results thus demonstrate for #73, #9.02, and #If-4 that plasmid DNA had become ligated to host genomic sequences, but the data do not conclusively demonstrate integration into a host chromosome. An additional test for this possibility is breeding of the transformed adult. Integration into a single chromosomal homolog should result in Mendelian transmission of the plasmid sequence as a heterozygous marker. Since #73 and #9.02 were sacrificed at birth, mouse #If-4 was the only animal subjected to progeny testing whose restriction pattern indicated integration.

BREEDING ANALYSIS

Sequences Retained Without Integration

Progeny testing of the six adult mice whose plasmid sequences did not satisfy tests for integration failed to show transmission of sequences to offspring. Table 1 summarizes the breeding data from these animals. Single litters were evaluated from five mice which retained only pBR322 derived material. As shown in Table 1, these litters were negative. Mouse #5.09 was more thoroughly tested because of the presence of diagnostic SV40 and tk material within its spleen DNA. Hybridization of DNA extracted from 42 offspring have shown no evidence of transmission of pST6 sequences. These data are consistent with restriction analysis which indicated that the donor sequences had not become integrated. The possibility remains, however, that the sequences were integrated but not transmitted because of mosaicism within the mice. Investigation of this possibility is currently under way.

Integrated Sequences

By contrast with the six mice described above, breeding tests on transgenic mouse #If-4 gave results consistent with chromosomal integration of the injected plasmid. This mouse was crossed to an uninjected male. Nine of 15 progeny from three litters thus far tested have inherited the pIf derived sequences (Table 1). The number of positive offspring in each litter were 6 of 6, 0 of 4, and 3 of 5. Consistent with the notion that the sequences were inherited is the observation that the restriction patterns of the offsprings' DNAs were indistinguishable from those of the parent. The restriction pattern of such a sequence would not be expected to differ between parent and offspring. As shown in Figure 4, digests with BamHI, PvuII, and XbaI all gave identical patterns in parent and offspring (Figure 4; a, b, c). Particularly persuasive is the digest with XbaI. A partial digest of one offspring's DNA and of #If-4 gave the same multiple bands. This result shows that not only are the closest XbaI sites in parent and offspring located at similar distances from the plasmid sequence, but more distant sites are also similarly or identically spaced. Subsequent complete XbaI digests of #If-4 and offspring #2 resulted in a single band of the same size as the other five offspring (data not shown). These results provide strong evidence that the pIf sequences were integrated into a host chromosome.

The introduction of foreign DNA in a mouse chromosome without disrupting the meiotic process presents the possibility of producing large colonies of mice carrying transferred sequences. This capability is essential for many important studies of gene transfer into mice. The production of such a colony, however, requires that the transferred material remain stable in the genome over several generations. We tested the stability of the pIf derived sequences in the If-4 line by breeding one of its offspring to an uninjected male mouse to produce F_2 progeny. Whole animals were sacrificed. Their DNA was extracted, digested with PvuII, and subjected to filter hybridization using pIf as probe. Two of the first eight offspring produced by one of the F_1 mice showed clear homology to the probe with a restriction pattern indistinguishable from the F_1 parent or from the original transformed mouse, If-4 (Figure 4d). This second generation of germ line transmission constitutes evidence for the stability of the transferred material.

TABLE 1

Mouse	Sex	Plasmid	# Copies Injected	Restriction Pattern Integration	Restriction Pattern No Integration	Litters # Positives/ # Offspring	Totals
35.25	Male	pST6	35,000		X	0/9	0/9
35.27	Female	pST6	35,000		X	0/8	0/8
35.28	Female	pST6	35,000		X	0/8	0/8
35.30	Male	pST6	35,000		X	0/8	0/8
35.32	Female	pST6	35,000		X	0/8	0/8
5.09	Male	pST6	5,000		X	0/6, 0/8, 0/5, 0/6 0/4, 0/5, 0/8	0/42
If-4	Female	pIf	10,000	X		0/6, 0/4, 3/5	9/15
If-4, Offspring #4	Female	---	---	X		2/8	2/8

DISCUSSION

Cloned gene sequences microinjected into mouse pronuclei exhibit restriction patterns indicative of two major pathways of processing: one in which donor material fails to integrate, and another in which it becomes associated with high molecular weight DNA. Progeny tests of transgenic mice give results consistent with each restriction pattern. When restriction analysis indicates the absence of integration, donor material is not transmitted, while association with host sequences correlates with transmission of donor DNA in ratios consistent with Mendelian inheritence.

The unexpected finding that microinjected plasmids can be maintained in mice for the number of cell generations required for development from the fertilized egg to adulthood raises questions regarding the nature of the mouse gene transfer system. It is not presently known whether donor material is in free circular form or is produced from an integrated fragment whose transcriptional regulation is aberrant. In the latter case, failure to detect DNA which fulfilled criteria of integration might be due to mosaicism in the animal. The presence of numerous transcripts in a few cells could result in detection only of the free material. Such an integrated segment would be detected if transmitted through the germ line, since all of the cells of the progeny would carry the sequence. Failure to observe germ line transmission in these cases thus mitigates against the possibility of integration, but as stated above, mosaicism in the parent could result in a negative germ line. If these sequences exist autonomously, then their persistence over the course of development implies efficient self replication. In this case, it is important to determine if such self replication is an anomaly resulting from the abnormal addition of genetic material to eggs or whether it results from a nascent processing mechanism in mammalian cells which had not heretofore been recognized.

The integration of plasmid sequences and their transmission to offspring will allow many interesting studies in the near future. Mice can be backcrossed to produce homozygotes for the transferred sequences; this situation may allow the study of crossover events within a DNA segment whose sequence is well defined, and facilitate mapping studies by both Mendelian and somatic cell genetic approaches. Sequences present in small organs can be studied by pooling tissue from many animals. Breeding tests can also be used to determine if genes transferred into mice are integrated randomly or reproducibly into a specific site. This issue is of considerable importance if future attempts at gene replacement are to be made.

The data presented here, as well as those accumulating from several other laboratories, indicate promise for the technique of pronuclear injection as a method of studying gene action during mammalian development. Our initial report that such injections could succeed in transferring genes into developing mice has recently been confirmed in other laboratories (Bürke and Ullrich, 1981; Wagner E *et.al.*, 1981; Constantini and Lacy, 1981; Wagner T *et al.*, 1981). The successful transfer of human insulin into fetal mice by pronuclear injection has been reported (Bürke and Ullrich, 1981); and subsequently, the retention of the human β-globin gene in the DNA of fetal mice was described (Wagner E *et al.*, 1981). Supporting evidence for germ line transmission of transferred genes has also been gathered (Constantini and Lacy, 1981; Wagner T *et al.*,1981). Expression of genes injected into the pronucleus has been observed at late fetal states and in adult mice (Wagner E *et al.*, 1981; Wagner T *et al.*,1981), and we are currently examining the If-4 line of transgenic mice for human interferon expression. These early results indicate that pronuclear injection is likely to have broad application to the study of mammalian developmental genetics.

REFERENCES

Bachetti S, Graham FL (1977). Transfer of the gene for thymidine kinase to thymidine kinase deficient human cells by means of purified *Herpes simplex* viral DNA. Proc Natl Acad Sci USA 74:1590.
Blin N, Stafford DW (1976). A general method for isolation of high molecular weight DNA from eukaryotes. Nucleic Acids Res 3:2303.
Burke K, Ullrich A (1981). Abstract presented to the 10th ICN-UCLA Symposium on Molecular and Cellular Biology, in press.
Constantini F, Lacy E (1981). Introduction of a rabbit β-globin gene into the mouse germ line. Nature, in press.

Gordon JW, Scangos GA, Plotkin DJ, Barbosa JA, Ruddle FH, (1980). Genetic transformation of mouse embryos by microinjection of purified DNA. Proc Natl Acad Sci USA 77:7380.

Maitland H, McDougall J (1977). Biochemical transformation of mouse cells by fragments of *Herpes simplex* virus DNA. Cell 11:233.

Minson AC, Wildy P, Buchan A, Darby G (1978). Introduction of the *Herpes simplex* virus thymidine kinase gene into mouse cells using virus DNA of transformed cell DNA. Cell 13:581

Nagata S, Taira H, Hall A, Johnsrud L, Streuli M, Ecsodi J, Boll W, Cantell K, Weissmann C (1980). Synthesis in *E. coli* of a polypeptide with human interferon activity. Nature 284:316.

Scangos GA, Huttner KM, Juricek DK, Ruddle FH (1981). DNA-mediated gene transfer in mammalian cells: Molecular analysis of unstable transformants and their progression to stability. Mol Cell Biol 1:111.

Southern E (1975). Detection of specific sequences among DNA fragments separated by gel electrophoresis. J Mol Biol 98:503.

Wagner EF, Stewart TA, Mintz B (1981). The human β-globin gene and a functional viral thymidine kinase gene in developing mice. Proc Natl Acad Sci USA 78:5016.

Wagner TE, Hoppe PC, Jollick JD, Scholl DR, Hodinka RL, Gault JB (1981). Microinjection of a rabbit β-globin gene into zygotes and its subsequent expression in the adult mice and their offspring. Proc Natl Acad Sci USA, in press.

Wahl GM, Stern M, Stark GR (1979). Efficient transfer of large DNA fragments from agarose gels to diazobenzyloxymethyl-paper and rapid hybridization using dextran sulfate. Proc Natl Acad Sci USA 76:3683.

Wigler M, Silverstein S, Lee L-S, Pellicer A, Cheng T, Axel R (1977). Transfer of purified *Herpes* virus thymidine kinase gene to cultured mouse cells. Cell 11:223.

PROTEINS INVOLVED IN THE VECTORIAL TRANSLOCATION OF NASCENT PEPTIDES ACROSS MEMBRANES

DAVID I. MEYER

European Molecular Biology Laboratory
D-6900 Heidelberg, Germany

INTRODUCTION

Eucaryotic cells are "compartmentalized" into specific organelles each of which performs functions unique to itself. These functions are mediated in large part by proteins which reside in the particular organelle. A question of central importance in cell biology concerns itself with how these organelle-specific proteins, most of which are **synthesized** in the cytoplasmic compartment, are transported from their site of synthesis to their final destination within the cell.

A good case in point are proteins which are either secreted or reside at the cell surface as integral proteins of the plasma membrane. These molecules are synthesized on ribosomes bound to the rough endoplasmic reticulum (rough E.R.) and are translocated across or into the membrane of the rough E.R. cotranslationally (Palade, 1975; Blobel and Dobberstein, 1975; Wickner, 1980). Subsequently, these proteins undergo specific modifications in either the E.R. or the Golgi (Warren, 1981), are sorted out from E.R. or Golgi proteins, are concentrated and are transported on to the cell surface where they either reside or are secreted out of the cell (Palade, 1975).

Based on the fact that most proteins synthesized in the cytoplasmic compartment are neither secretory nor membrane proteins, one can propose that a mechanism must exist whereby nascent secretory and membrane proteins can be discriminated already at the level of the rough E.R. Blobel and Dobberstein (1975) demonstrated quite elegantly that rough microsomes (the centrifugally-derived equivalent of rough E.R.) are capable of recognizing, translocating and processing nascent secretory proteins in vitro. Moreover, this finding implied that the cellular machinery which performs the entire first step in intracellular protein transport, i.e., vectorial translocation of nascent proteins, is easily obtainable and available for further study. In other words, the sequential dissection of rough microsomes, in conjunction with an in vitro assay for vectorial translocation, should enable one to elucidate the initial stages of an important cellular pathway.

Described in this report are the results of a dissection of the rough microsomal membrane. An E.R. membrane component is characterized which is essential to the process of vectorial translocation. It provides a handle with which further components necessary for this function can be identified.

RESULTS AND DISCUSSION

As two types of membrane proteins exist, i.e., integral and peripheral (Singer and Nicholson, 1972), a treatment which removes the peripheral proteins as well as the cytoplasmically-disposed portions of the integral ones was used (Meyer and Dobberstein, 1980a). KCl (500 mM) in conjunction with 1 µg/ml elastase enabled one to effect the dissection scheme described in figure 1. By stripping rough microsomes with high salt and elastase two fractions were obtained: An inactive rough microsome vesicle fraction (RM_i) and a mixture of proteins referred to as elastase extract (E.E.). When the E.E. was added back to the RM_i vectorial translocation activity was restored.

Polyacrylamide gel electrophoresis indicated that E.E. was a complex mixture of peptides (fig. 2, lane A) yet it was hoped that a single component would be the peptide responsible for restoring activity to RM_i. Fortunately the active factor has a high isoelectric point and thus lent itself well to separation on ion-exchange resins (fig. 2, lane B). Further fractionation using gel filtration yielded a species of MW = 60,000 which possessed the required ability to reconstitute an active rough microsome when added back to RM_i (Meyer and Dobberstein, 1980b).

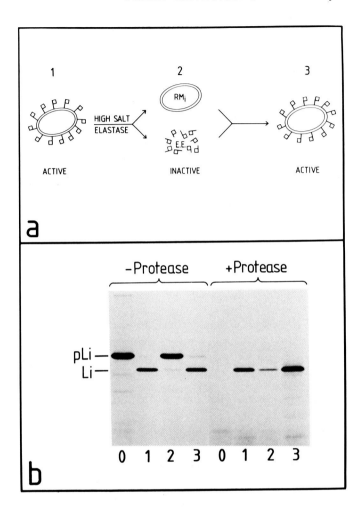

Figure 1. Dissection of translocation/processing activity from rough microsomes. a. Schematic representation of 1) the intact rough microsomes; 2) the removal of peripheral proteins by elastase and high salt treatment yielding inactive rough microsomes (RM_i) and a soluble elastase extract (E.E.); and 3) the recombination of the two fractions restoring the translocation/processing activity. b. Assay of translocation/processing activity using an m-RNA-dependent cell-free protein synthesizing system primed with m-RNA specific for IgG light chain (Li). pLi denotes the 22-amino-acid-longer

light chain precursor (unprocessed). Translocation of nascent chains across the microsomal membrane is confirmed by the resistance of the mature form to proteolytic attack. Shown is light chain synthesized: In the absence of membranes (lane 1); in the presence of intact microsomes (lane 2); in the presence of RM_i (lane 3); and in the presence of RM_i + E.E. (lane 4). (From Meyer et al. 1981.)

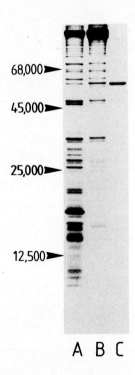

Figure 2. Purification of activity from E.E. Crude elastase extract (lane A) was purified by ion-exchange chromatography on CM-Sephadex. This activity-containing fraction from ion-exchange (lane B) was further fractionated by gel filtration on Sephadex G-150. The protein composition of the most active fraction obtained is shown in lane C. (From Meyer and Dobberstein, 1981b.)

As the 60,000 M.W. component was isolated from a proteolytic digest, it was assumed that this molecule represented a fragment of a larger membrane-associated protein. In order to answer this question as well as to definitively prove that this was indeed the active factor, an antibody was produced against the purified active fragment (Meyer et al., 1982). The data shown in figure 3 characterize the proteins immunoprecipitated by the antibody. From radiolabelled crude E.E. (lane 3), only the 60,000 M.W. fragment was precipitated (Lane 4). When E.E. purified over CM-Sephadex (lane 5) was immunoprecipitated, the fragment was recognized especially well by the antibody (lane 6). In order to determine if a membrane protein would be recognized by the antibody, intact

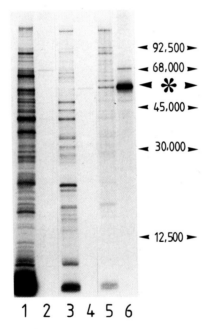

Figure 3. Immunoprecipitation of microsomal proteins using antibody prepared against the 60,000 M.W. fragment. Shown are autoradiograms of proteins labeled with ^{125}I. Lanes 2, 4 and 6 represent the proteins precipitated by antibody from the fractions depicted in the lanes to their left. Lane 1: detergent-solubilized rough microsomal proteins; lane 3: crude elastase extract; lane 5: ion exchange-purified elastase extract. Asterisk denotes fragment against which the antibody was produced. (From Meyer et al., 1982.)

rough microsomes were solubilized in detergent, radiolabeled, and immunoprecipitated (lane 1). In this case there was no peptide at M.W. = 60,000 precipitated, rather a protein of M.W. = 72,000. It can be implied from these data that the native, intact active component has a molecular weight of 72,000 which upon cleavage with elastase and treatment with high salt yields an active fragment of M.W. = 60,000.

To prove that the antibody was directed against the active fragment, anti-fragment IgG was immobilized on protein A-Sepharose beads and the beads were added to active E.E. Following incubation and removal of the beads by centrifugation, the E.E. was tested for activity. If the 60,000 M.W. fragment recognized by the antibody (as shown in fig. 3) was the active factor, it should have been removed from E.E. by this procedure. The results shown in figure 4 confirms this, as E.E. pretreated with anti-fragment loses its ability to restore activity to RM_i.

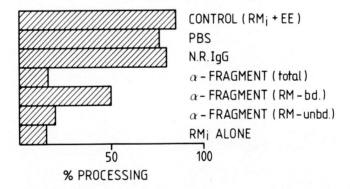

Figure 4. Inhibition of reconstitution by antibody. Elastase extracts (E.E.) were absorbed to protein A-Sepharose-antibody complexes and then tested for their ability to reconstitute translocation and processing when combined with inactive rough microsomes (RM_i). Sample 1: RM_i + unabsorbed E.E.; sample 2: control with protein A Sepharose preincubated with PBS; sample 3: control with protein A preincubated with normal (preimmune) rabbit IgG; sample 4: E.E. preabsorbed to protein A Sepharose preincubated with anti-60,000 M.W. fragment IgG; sample 5 represents RM_i without added E.E. (From Meyer et al., 1982.)

Further evidence demonstrating that the fragment is derived from a larger membrane-associated protein of M.W. = 72,000 comes from studies in which immunoprecipitation was carried out on an appropriate cell line metabolically labeled with ^{35}S-methionine. As can be seen in figure 5, when cells are pulsed (lane 1) and chased for either short (lane 2) or longer (lane 3) periods, or if continuously labeled (lane 4), the anti-fragment precipitates a molecule of M.W. = 72,000. If elastase is added prior to or following immunoprecipitation, the 60,000 M.W. fragment is generated (lane 5).

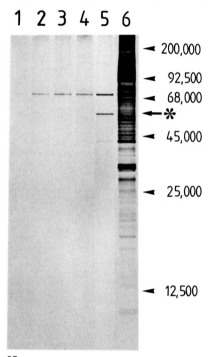

Figure 5. Immunoprecipitation by anti-fragment of pulsed and pulse-chased MDCK cells. Cells were pulsed for 40 min (lane 1) and chased for either 60 min (lane 2) or 3 hours (lane 3); or continuously labeled for 4 h (lane 4). Lane 5 represents continuously labeled samples prepared in the presence of exogenously added elastase at 1 µg/ml. Lane 6 represents the protein composition of detergent-solubilized membranes prior to immunoprecipitation. Asterisk indicates the 60,000 M.W. fragment. (From Meyer et al., 1982.)

It is clear that a great deal remains to be elucidated regarding the function of the molecule which has been characterized in these studies. The role it plays in vectorial translocation has yet to be determined. Recently other workers have described proteins which are also involved in vectorial translocation (Kreibich et al., 1978; Walter and Blobel, 1980), but which at this point seem antigenically as well as structurally unrelated to the protein described here. What is evident is that such an approach has already been successfully employed to "pull" one component out of rough microsomes, and in conjunction with the antibody described, it is hoped that the relationship of this specific molecule to others required for translocation can be determined. In this way ultimately all of the components which mediate this complex process, common to all eucaryotic cells, can be identified and characterized.

ACKNOWLEDGEMENTS

I would like to thank Elke Krause for expert technical assistance, John Stanger for photography and Wendy Moses for typing this manuscript. The work presented here was supported in part by the Deutsche Forschungsgemeinschaft.

REFERENCES

Blobel G, Dobberstein B (1975). Transfer of proteins across membranes. II. Reconstitution of functional rough microsomes from heterologous components. J Cell Biol 67: 852-862.

Kreibich G, Ulrich BL, Sabatini DD (1978). Proteins of rough microsomal membranes related to ribosome binding. I. Identification of ribophorins I and II, membrane proteins characteristic of rough microsomes. J Cell Biol 77: 464-487.

Meyer DI, Dobberstein B (1980a). A membrane component essential for vectorial translocation of nascent proteins across the endoplasmic reticulum: Requirements for its extraction and reassociation with the membrane. J Cell Biol 87: 498-502.

Meyer DI, Dobberstein B (1980b). Identification and characterization of a membrane component essential for the translocation of nascent proteins across the membrane of the endoplasmic reticulum. J Cell Biol 87: 503-508.

Meyer DI, Kvist S, Dobberstein B (1981). Assembly of membrane proteins. In Giebisch, G (ed.): "Membranes in Growth and Development", in press.

Meyer DI, Louvard D, Dobberstein B (1982). Characterization of molecules involved in protein translocation using a specific antibody. J Cell Biol 92 in press.

Palade GE (1975). Intracellular aspects of the process of protein synthesis. Science 189: 347-358.

Singer SJ, Nicholson GL (1972). The fluid mosaic model of the structure of cell membranes. Science 175: 720-731.

Walter P, Blobel G (1980). Purification of a membrane-associated protein complex required for protein translocation across the endoplasmic reticulum. Proc Natl Acad Sci USA 77: 7112-7116.

Warren G (1981) Membrane proteins: Structure and assembly. Comprehensive Biochemistry 1:215-257.

Wickner W (1980). Assembly of proteins into membranes. Science 210: 861-868.

SUSCEPTIBILITY TO MEROCYANINE 540-MEDIATED PHOTOSENSITI-
ZATION AS A DIFFERENTIATION MARKER IN MURINE HEMATOPOIETIC
STEM CELLS

Fritz Sieber, Richard C. Meagher and Jerry L. Spivak

The Johns Hopkins University School of Medicine, 924 Traylor Research Building, Baltimore, Maryland 21205

INTRODUCTION

In vivo and in vitro clonal assays have contributed substantially to our understanding of blood cell development. However, in order to gain insight into molecular processes underlying cellular differentiation, these functional assays should be complemented with suitable molecular markers. Most molecular markers that have been used so far (e.g., hemoglobin or glycophorin) focus on the later phases of differentiation. In this communication, we report on a plasma membrane property (sensitivity to merocyanine 540) which appears to be developmentally regulated and to undergo characteristic changes during the early phases of hematopoietic cell differentiation.

Merocyanine 540 (MC 540) is a negatively charged, fluorescent dye which is incorporated into plasma membranes where it can serve as a reporter molecule for transmembrane potential [Waggoner, 1979; Cohen et al, 1974]. In aqueous media, MC 540 has two excitation maxima, one at 510 nm and a second one at 535 nm which is indicative of a monomer-dimer equilibrium [West, Pearce, 1965; Tasaki et al, 1976]. The emission spectrum peaks at 540 nm. In a hydrophobic environment, both spectra are redshifted to 565 and 585 nm respectively [Easton et al, 1978]. Photoexcitation of membrane-bound dye with green or white light causes increased dye uptake, a breakdown of the normal permeability barrier and eventually cell death [Valinsky et al, 1978; Schlegel et al, 1980a]. Serum contains one or several compounds that bind MC 540 with moderate affinity. Serum can, therefore, successfully compete with low affinity cellular binding sites for dye molecules and thus allows the selective staining of cells with high affinity binding sites for MC 540 such as electrically excitable cells and leukemic cells [Easton et al, 1978; Valinsky et al, 1978; Schlegel et al, 1980a]. Valinsky and collaborators [1978] have also reported that MC-540-mediated photosensitization completely abolishes the ability of normal adult mouse bone marrow cells to form colonies in the spleens of lethally irradiated host animals and to form granulocyte/macrophage colonies in culture. This suggests that pluripotent hematopoietic stem cells ('colony forming unit-spleen' or 'CFU-S') and progenitor cells committed to the granulocyte/macrophage pathway ('colony forming unit-granulocyte/macrophage' or CFU-GM') share a common membrane structure with leukemic and electrically excitable cells. Our study confirms and expands these observations. By use of sublethal combinations of dye concentration, serum concentration and illumination, we were able to determine characteristic rates of photosensitization for four classes of murine hematopoietic progenitor cells.

EXPERIMENTAL PROCEDURES

Staining and Photosensitization of Marrow Cells

Marrow cells were flushed with a small amount of alpha-medium from the femurs of female $B6D2F_1$ mice (6-10 weeks old) into clear polystyrene tubes and diluted to approximately 4×10^6 cells/ml in alpha-medium supplemented with 10% fetal calf serum. Twenty-five μl of a 1 mg/ml stock solution of MC 540 in 50% ethanol was added per 1 ml of cell suspension and the tube was placed into the center of a circular (40 cm diameter) array of eight fluorescent light bulbs and illuminated for the times indicated. Photosensitization was terminated by washing the cells three times by centrifugation and resuspension in alpha-medium supplemented with 5% fetal calf serum. During all subsequent operations, cells were shielded from light. Trypan blue was used to quench the nonspecific absorption of dye by dead cells.

Bioassay of Progenitor Cells

CFU-S were assayed by injecting 10^5 nucleated bone marrow cells via tail vein into recipient animals that had received a lethal dose (950 rad) of total body irradiation [Till, McCulloch, 1961]. Spleens were excised seven days later, fixed in Carnoy's solution, and examined for macroscopic colonies. The assay of committed progenitor cells was based on their capacity to form morphologically recognizable colonies in culture [Iscove et al, 1974; Iscove, Sieber, 1975]: 10^5 nucleated cells were placed into 35 mm plastic petri dishes containing 1 ml of alpha-medium supplemented with 0.8% methylcellulose (4,000 cps), 30% (for erythroid progenitors) or 15% (for granulocyte/macrophage progenitors) fetal calf serum, 10^{-4} M alpha thioglycerol, and a saturating dose of erythropoietin or granulocyte/macrophage colony stimulating factor (WEHI-3-conditioned medium [Williamson et al, 1978]). Colonies formed by late erythroid progenitors ('colony forming unit-erythroid' or 'CFU-E') were scored after two days, colonies formed by CFU-GM after seven days, and colonies formed by early erythroid progenitors ('burst forming unit-erythroid' or 'BFU-E') after ten days in culture. All colonies were identified in situ with an inverted microscope (magnification 40-100X) according to established criteria.

RESULTS AND DISCUSSION

When mouse bone marrow cells were incubated with MC 540 in the presence of 10% fetal calf serum and then examined with a fluorescence microscope, 2%-4% of the cells were visibly fluorescent. This concentration of stained cells correlated well with current prevalence estimates of progenitor cells in adult mouse bone marrow. There was some variability with respect to the intensity of the fluorescence emitted by individual cells. Flow cytometric analysis of MC 540-stained cells also revealed distinct subpopulations. When the staining was carried out in the absence of serum, all cells, including mature erythrocytes, were brightly fluorescent. Prolonged illumination of stained cells with green light emitted by the mercury vapor pressure lamp of the microscope resulted in cell lysis.

Marrow cells which were incubated with MC 540 and simultaneously exposed to a strong source of white light gradually lost their ability to form spleen colonies in lethally irradiated host animals and to form colonies in culture (Fig. 1). Photosensitization was more rapid when lower serum concentrations were used (Table 1), but the relative susceptibility of individual classes of progenitor cells remained unchanged; late erythroid progenitors (CFU-E) were always the most sensitive, granulocyte/macrophage progenitors (CFU-GM) the least sensitive cells. The same rank order of sensitivity has also been found in normal human bone marrow [Sieber et al, 1981b].

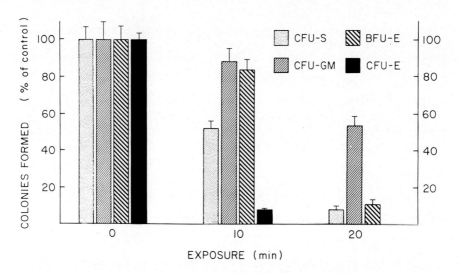

Fig. 1. Representative example of MC 540-mediated photosensitization of mouse hematopoietic progenitor cells.

Suspensions which were not exposed to light (100% at time 0) contained 31 ± 2.1 CFU-S, 37 ± 2.5 BFU-E, 401 ± 39 CFU-GM and 572 ± 20 CFU-E per 10^5 nucleated cells (data expressed as mean colony count of six spleens or four replicate culture dishes respectively $\pm$ SEM). Size and morphological appearance of the reduced number of in vitro colonies grown under conditions of partial photolysis were normal. Control cell suspensions, either stored in medium and in the dark or incubated with ethanol or medium and exposed to light, produced the same number of in vitro colonies as those incubated with MC 540 but shielded from light. (From: Sieber et al, 1981a)

Figure 1 indicates that a structural and/or functional cell surface property of immature blood cells undergoes repeated and, in some instances, quite dramatic changes as the progenitor cells differentiate along a particular pathway: Commitment of the pluripotent stem cell (as defined by the spleen colony assay) to the erythroid or granulocyte/macrophage lineage is accompanied by a moderate decrease in susceptibility to MC 540-mediated photosensitization. This decrease is followed by a marked increase in sensitivity when erythroid progenitors reach the stage of a CFU-E. Terminal differentiation of all progenitors appears to involve another decrease in sensitivity since mature circulating blood cells were found to be refractory to MC 540 [Valinsky et al, 1978]. At least two species, mouse and human, follow a very similar pattern in their developmental regulation of MC 540 sensitivity.

Several laboratories have presented evidence that colony formation by hematopoietic progenitor cells is in part controlled by regulatory cells such as T-lymphocytes and macrophages [Lord, Schofield, 1973; Nathan et al, 1978; Rinehart et al, 1978; Sharkis et al, 1980; Schreier, Iscove, 1980; Reid et al, 1981]. The reduction of colony formation after photosensitization with MC 540 could thus have been due to an effect on regulatory cells rather than a direct effect on colony forming cells. An unequivocal distinction between these two mechanisms is not possible as long as the regulatory cells have not been sufficiently characterized and the isolation of homogeneous populations of progenitor and regulatory cells has not been achieved. Differences in cell culture conditions may account for some of the conflicting reports in the current literature on the influence of regulatory cells on in vitro colony formation. Our culture system was designed to minimize the influence of regulatory cells. The overall concentration of regulatory cells could be varied over two orders of magnitude without noticeably affecting the plating efficiency of erythroid progenitor cells [Iscove, Sieber, 1975]. To further minimize the effects of regulatory cells, the culture medium was supplemented with a saturating amount of GM-CSF derived from a monocytic cell line and a high concentration of a batch of fetal calf serum that had been selected for its ability to optimally support erythroid burst growth (probably due to a

Table 1. Effect of serum concentration on MC 540-mediated photosensitization of hematopoietic progenitor cells. Recovery of progenitors (% of control)

Illumination	10% FCS			4% FCS			0% FCS		
	BFU-E	CFU-GM	CFU-E	BFU-E	CFU-GM	CFU-E	BFU-E	CFU-GM	CFU-E
0 min	100.0	100.0	100.0	100.0	100.0	100.0	100.0	100.0	100.0
10 min	40.5	78.0	6.9	5.5	12.4	0.0	0.0	0.0	0.0
20 min	2.4	25.5	0.0	0.0	0.0	0.0	0.0	0.0	0.0

Aliquots of the same cell preparation were stained with a constant amount (25 µg/ml) of MC 540 in the presence of varying concentrations of fetal calf serum (FCS). Photosensitization was accelerated at lower serum concentrations but the relative sensitivity of the various classes of progenitor cells remained unchanged. Very high concentrations of serum ($\geq$ 30%; data not shown) prevented MC 540-mediated photosensitization. All data are based on mean colony numbers of four replicate culture dishes. (From: Sieber et al, 1981a)

Table 2. Colony formation by mixed cultures of MC 540-treated and untreated mouse bone marrow cells.

	BFU-E	CFU-E	CFU-GM
10^5 untreated cells	37.0 + 1.6	444.0 + 27.2	455.8 + 77.0
10^5 cells treated for 10 min	23.3 + 4.2	16.5 + 5.8	299.5 + 9.4
10^5 cells treated for 20 min	1.5 + 0.6	1.5 + 3.0	51.5 + 7.6
5×10^4 untreated cells and 5×10^4 cells treated for 10 min	32.8 + 2.0*	232.5 + 26.0*	336.0 + 12.2*
(expected)	(30.1 + 2.3)	(230.3 + 13.9)	(377.6 + 38.8)
5×10^4 untreated cells and 5×10^4 cells treated for 20 min	22.0 + 5.0*	208.5 + 34.4*	226.8 + 24.4*
(expected)	(19.3 + 0.9)	(222.8 + 13.7)	(253.6 + 38.7)

Cells which had been stained with MC 540 (25 µg/ml) in 10% serum were cocultured with an unstained aliquot of the same suspension. The colonies produced by the mixed culture were not significantly (*; $P > 0.05$, 2-tailed Student's t-test) different from the sum of colonies produced by treated and untreated cells in separate cultures. Data are expressed as mean colony numbers of four replicate culture dishes + SD. (From: Sieber et al, 1981a) high content of burst promoting activity). In situations where the concentration of putative regulatory cells has been shown to affect the formation of colonies by hematopoietic progenitor cells, the number of colonies formed was frequently a complex, nonlinear function of the number of regulatory cells plated, and apparently exhaustive depletions of regulatory cells sometimes failed to completely abolish colony formation by progenitor cells. Complete suppression of colony formation was, however, readily achieved by MC 540-mediated photosensitization (Fig. 1, Table 1). Cocultures of photosensitized and untreated cells (Table 2) containing at least 50% of the normal concentration of regulatory cells produced colony numbers that were equal to the sum of colonies produced by treated and untreated cells in separate cultures. This finding is most readily explained by a direct effect of MC 540 on colony-forming progenitor cells. Additional support in favor of a direct effect on progenitor cells is offered by the observation that MC 540 stains about 60% of the total number of spleen cells in neonatal mice (at this stage of development, the spleen is a major erythropoietic organ) but few, if any, thymocytes or circulating lymphocytes and monocytes/macrophages [Schlegel et al, 1980a; Valinsky et al, 1978].

The exact nature of the MC 540 binding site, its physiological relevance and the mechanism of photosensitization remain unclear. Fluid-like membrane domains [Williamson et al, 1981] which are rich in acidic phospholipids [Easton et al, 1978] have been implicated as binding sites for MC 540, and it has been hypothesized that these domains may play a role in the selective removal of plasma membrane proteins during development [Schlegel et al, 1980b].

Sensitivity to MC 540-mediated photosensitization and density of MC 540 surface binding sites hold promise as useful molecular markers for blood cell differentiation and should prove complementary to lectins, antibodies, and other membrane probes in attempts to isolate homogeneous populations of live hematopoietic progenitor cells.

ACKNOWLEDGMENTS

We thank Mrs. E. Connor for her expert technical assistance and Mrs. A. E. Fields for typing the manuscript.

This research was supported by grant AM 16702 from the National Institute of Arthritis Metabolism and Digestive Diseases, training grant HL 07143 from the National Heart, Lung and Blood Institute, a Hubert E. and Anne E. Rogers Scholarship awarded to F.S. and Research Career Development Award HL 00480 from the National Heart, Lung and Blood Institute to J.L.S. Figure 1 and Tables 1 and 2 were reproduced with permission from Springer Verlag.

REFERENCES

Cohen LB, Salzberg BM, Davila HV, Ross WN, Landown D, Waggoner AS, Wang C-H (1974). Changes in axon fluorescence during activity: molecular probes of membrane potentials. J. Membr Biol 19:1.

Easton TG, Valinsky JE, Reich E (1978). Merocyanine 540 as a fluorescent probe of membranes: staining of electrically excitable cells. Cell 13:45.

Iscove NN, Sieber F, Winterhalter KH (1974). Erythroid colony formation in cultures of mouse and human bone marrow: analysis of the requirement for erythropoietin by gel filtration and affinity chromatography on agarose-Concanavalin A. J Cell Physiol 83:309.

Iscove NN, Sieber F (1975). Erythroid progenitors in mouse bone marrow detected by macroscopic colony formation in culture. Exp Hematol 3:32.

Lord BI, Schofield R (1973). The influence of thymus cells in hemopoiesis: stimulation of hemopoietic stem cells in a syngeneic, in vivo, situation. Blood 42:395.

Nathan DG, Chess L, Hillman DG, Clarke B, Breard J, Merler E, Housman DE (1978). Human burst forming unit: T cell requirement for proliferation in vitro. J Exp Med 147:324.

Reid CDL, Baptista LC, Chanarin I (1981). Erythroid colony growth in vitro from human peripheral blood null cells: evidence for regulation by T-lymphocytes and monocytes. Br J Haemat 48:155.

Rinehart JJ, Zanjani ED, Nomdedeu B, Gormus BJ, Kaplan ME (1978). Cell-cell interaction in erythropoiesis, role of human monocytes. J Clin Invest 62:979.

Schlegel RA, Phelps BM, Waggoner A, Terada L, Williamson P (1980a). Binding of merocyanine 540 to normal and leukemic erythroid cells. Cell 20:321.

Schlegel RA, Phelps BM, Lofer GP, Massey W, Williamson P (1980b). Cell surface differentiation and membrane phase state in normal and leukemic erythroid cells. J Cell Biol 87:95a.

Sharkis SJ, Spivak JL, Ahmed A, Misiti J, Stuart RK, Wiktor-Jedrzejczak W, Sell KW, Sensenbrenner LL (1980). Regulation of hematopoiesis: helper and suppressor influences of the thymus. Blood 55:524.

Sieber F, Meagher RC, Spivak JL (1981a). Differential sensitivity of mouse hematopoietic stem cells to merocyanine 540. Differentiation 19:65.

Sieber F, Stuart RK, Sensenbrenner LL, Sharkis SJ (1981b). Susceptibility to merocyanine 540-mediated photolysis as a differentiation marker in human hematopoietic stem cells. Blood, in press.

Tasaki I, Warashina A, Pratt M (1976). Studies on light emission, absorption and energy transfer in nerve membranes labeled with fluorescent probes. Biophys Chem 4:1.

Till JE, McCulloch EA (1961). A direct measurement of the radiation sensitivity of normal mouse bone marrow cells. Rad Res 14:213.

Valinsky JE, Easton TG, Reich E (1978). Merocyanine 540 as a fluorescent probe of membranes: selective staining of leukemic and immature hemopoietic cells. Cell 13:487.

West W, Pearce S (1965). The dimeric state of cyanine dyes. J Phys Chem 69:1894.

Williamson PL, Massey WA, Phelps BM, Schlegel RA (1981). Membrane phase state and the rearrangement of hematopoietic cell surface receptors. Mol Cell Biol 1:128.

HUMAN LYMPHOCYTE MEMBRANE PROTEINS IN ACTIVATION BY PHYTOHEMAGGLUTININ

P. Lerch, H. Gmünder and W. Lesslauer

Dept. of Biochemistry, University of Bern

CH-3012   Bern, Switzerland

The cell membrane is a critical site in the intercellular regulatory network of differentiation, growth and inhibition signals in lymphocyte populations. T-lymphocytes have antigen-specific receptors, but recognize antigen only if coincident with syngeneic histocompatibility antigens [Doherty and Zinkernagel, 1975]. The chemical structure of the receptor(s) involved in restricted antigen recognition is unknown. Immunoglobulin heavy chain variable region genes may or may not take part in the complete structural receptor gene [Kurosawa et al, 1981]. The present experiments map the surface proteins of the human T-lymphocyte and try to elucidate a functional significance of membrane protein associations in cell activation.

Human peripheral blood lymphocytes were isolated with Ficoll-Paque density gradients, surface-iodinated by lactoperoxydase and lysed in non-ionic detergent (typically 0.5% Triton X-100 or X-114). The cell lysate was sedimented (400g, 10 min) and the partitioning of surface-labeled proteins in the Triton-soluble supernatant fraction and in the insoluble fraction was studied by autoradiography of two-dimensional gels [O'Farrell, 1975]. The major labeled proteins of the supernatant fraction are represented in Fig. 1. The location of the histocompatibility antigens HLA-AB and β_2-microglobulin in the gel system was determined by immunoprecipitation with an anti-β_2-microglobulin antiserum

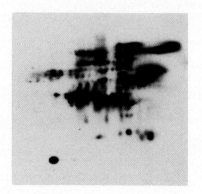

Fig. 1 Solubilised surface proteins (lymphocytes ^{125}I-labeled by lactoperoxydase, lysed in 0.5% Triton X-100 at 4° C, 15 min; supernatant fraction; 2 d-gel with ph-range of about 4.0-7.0 and molecular weight scale as in Fig. 6; autoradiograph enhanced by fluorescent screen)

as shown in Fig. 2. Surface immunoglobulin was identified by immunoprecipitation and by its absence in T-cells purified with nylon wool columns.

Two microheterogeneic proteins of apparent molecular weight of about 45'000 (p45) and 30'000 (p30) partitioned quantitatively into the low speed sediment of the Triton-lysate (Fig. 3). Within the limits of sensitivity of the autoradiography these two proteins are entirely absent from the Triton-soluble fraction.

The two proteins p45 and p30 are membrane proteins and not artefactually labeled intracellular proteins, because (a) other known intracellular proteins (eg. actin) are not labeled, (b) p45 and p30 are enriched in one surface membrane fraction

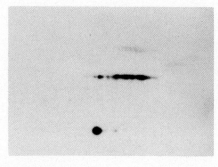

Fig. 2 Immunoprecipitate from Triton-lysate of surface-iodinated cells with anti-β_2-microglobulin-antiserum and protein A/Sepharose (2 d-gel, autoradiograph as in Fig. 1)

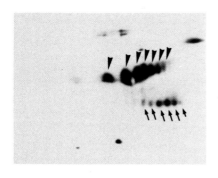

Fig. 3 Triton-insoluble surface-labeled proteins p45 [◂] and p30 [◂] from the low speed sediment of the cell lysate (2 d-gel, autoradiograph as in Fig. 1)

isolated after homogenisation of the cells on sucrose density gradients in the absence of detergent (see Fig. 4), and (c) they are sensitive to neuraminidase treatment of intact cells. The isolated surface membrane fraction enriched in p45 and p30 (Fig. 4) is also rich in glycolipids, but no other surface labeled protein is present in significant amount. The relative content in p45 compared to total protein was estimated (in the order of 1%) from the relative intensities of spots obtained with this membrane fraction prepared from unlabeled cells, but where all proteins were unspecifically tritiated by reductive methylation (see Fig. 5).

A relation of p45 and p30 with components of known immunologic specificities of the cell surface could not yet be assessed, because both proteins were solubilised only under denaturing conditions, and therefore could not be immunoprecipitated. From the molecular weights a relation with histocompatibility antigens might be considered. Both p45 and p30 are present in enriched T-cell preparations which renders a relation of p30 with HLA-D/Dr unlikely, since human Ia antigens were demonstrated exclusively on activated T-cells [Ko et al, 1979].

The physical reasons for the sedimentation of p45 and p30 at low speeds were studied. It was observed that one fraction could be isolated from detergent-lysates of cells on discontinuous sucrose

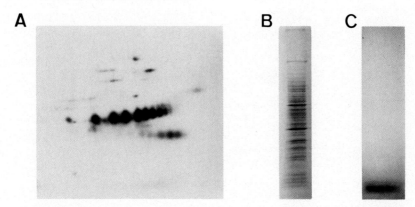

Fig. 4 Isolated lymphocyte surface membrane fraction (cells surface-iodinated, homogenised and separated on discontinuous sucrose density gradient; the least dense fraction of surface mebranes recovered from 22% sucrose; gel and autoradiograph as in Fig. 1 [A]; one-dim. SDS-gels stained with Coomassie [B] and PAS [C]).

density gradients in detergent, which contained p45 in relatively pure form and separated from the supernatant and nuclear pellet fractions (Fig. 6).

It is concluded tentatively that the sedimentation of the two Triton-insoluble proteins at low speed is not due to a (tight) linkage with the nuclei. Rather they appear to be present in heavily aggregated form. Their association is no artefact due to unspecific protein aggregation in non-ionic detergent [Helenius et al, 1979], since p45/p30 are also enriched in one surface membrane fraction isolated from cell homogenates in the absence of detergents. This membrane fraction is free of other surface-labeled proteins; it may reflect a specialised region in the cell membrane rich in p45, p30 and glycolipids (Fig. 4). This same membrane fraction contains also a number of intracellular proteins which may be contaminants or may be at least in part essential to maintain certain molecular configurations in the membrane. The lifetimes of aggregates of p45 and p30 in the cell membrane cannot be determined from the present experiments; they might be shortlived.

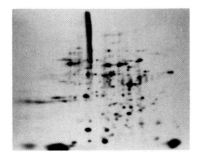

Fig. 5 Isolated surface membrane fraction (unlabeled cells homogenised, same membrane fraction recovered as in Fig. 4, reductive methylation of proteins after membrane isolation using ^{3}H-borohydride; gel as in Fig. 1, autoradiography enhanced by PPO)

The presumed clustering of p45 and p30 in the membrane makes them interesting candidates for a receptor function in cell activation [see eg. Ravid et al, 1978]. In a first approach to learn about a functional significance of the two proteins, the fate of p45 and p30 in the course of mitogen-activation was studied. Peripheral blood lymphocytes were incubated at 4° C for one hour in medium with 5 µg/ml phytohemagglutinin. The non-adherent mitogen was washed off the cells, which were then transferred into fresh mitogen-free medium and kept in culture for up to 40 hours. Aliquots of cells were removed and surface-labeled after various time intervals. This unusual protocol of mitogen activation which does not take into account the two-step activation schedule of mitogen/secondary growth factor [Andersson et al, 1979] was chosen in order

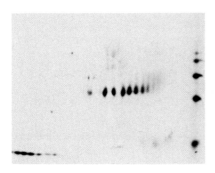

Fig. 6 Purified p45 fraction (cells surface-labeled by reductive methylation using ^{3}H-borohydride, lysed in Triton and separated on sucrose density gradient in Triton; gel and autoradiograph as in Fig. 5; molecular weight markers 22'/43'/68'/94'/116'000)

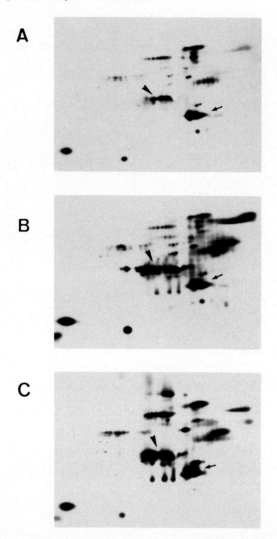

Fig. 7 Surface membrane proteins in the Triton-soluble supernatant fraction of cells labeled 0 hr [A], 16 hrs [B] and 40 hrs [C] after a one hour phytohemagglutinin exposure at 5 µg/ml and 4° C (gel and autoradiograph as in Fig. 1; exposure adjusted to make the phytohemagglutinin spots in A,B, and C of comparable intensities). HLA-AB [►], phytohemagglutinin [→].

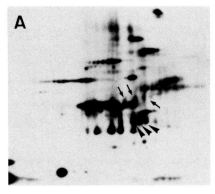
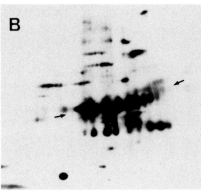

Fig. 8 Surface membrane proteins of lymphocytes labeled 40 hrs after a one hour leucoagglutinin-exposure at 5 μg/ml and 4° C (gels and autoradiographs as in Fig. 1)

A. Triton-soluble supernatant fraction

B. Triton-insoluble fraction in the 400g/10 min sediment

leucoagglutinin [➤]
p45 [➚]

to obtain reproducible conditions for surface-labeling without interference from cell clumping. By ^{3}H-thymidine incorporation the chosen protocol leads depending on the individual donor to partial cell activation.

By surface-iodination after the one hour exposure to phytohemagglutinin, the mitogen was found to cover up large parts of the accessible cell surface (Fig. 7). Immediately after the contact with phytohemagglutinin, the mitogen was the most prominent spot and the endogenous membrane proteins were only faintly visible in the autoradiographs despite extensive washing of the cells after the mitogen exposure. This result supports the observations of Dillner et al [1980] who reported that a large fraction of the different surface proteins is adsorbed from T-lymphocyte detergent lysates by leucoagglutinin/Sepharose affinity chromatography. With increasing time in culture after the mitogen exposure, the endogenous membrane proteins slowly reappeared in the labeling patterns (Fig. 7). However, even after 40 hrs in culture phytohemagglutinin remained one of the most strongly labeled spots. From the present data there appears to exist a defined time sequence in the re-emergence of the endogenous membrane proteins in the labeling pattern, the histocompatibility antigens HLA-AB being among the first to become labeled again strongly.

The long persistence of phytohemagglutinin at the cell surface was unexpected, if it is compared to the time scale eg. of capping. It was considered that only the not or less mitogenic isomeric tetramers of phytohemagglutinin might remain at the cell surface. However, essentially the same results were obtained, when the purified mitogenic tetramer leucoagglutinin was used in the initial exposure of the cells to the activating agent.

In preliminary experiments the fate of p45 was found to be affected by the one hour exposure to leucoagglutinin. While p45 partitioned quantitatively into the low speed sediment of Triton-lysates of unstimulated cells, some p45 was found in the Triton-soluble supernatant fraction after the incubation with leucoagglutinin (Fig. 7). In addition the surface-labeling of p45 immediately after mitogen exposure appears less affected than that of the other proteins of the Triton-soluble fraction. Further experiments with a more refined protocol of mitogen activation [Andersson et al, 1979] will have to establish, whether these observations reflect changes in the aggregation state of one or both of the Triton-insoluble proteins which are functionally significant.

This work is supported by the Swiss National Science Foundation.

Andersson J, Grönvik KO, Larsson EL, Coutinho A (1979). Studies on T-lymphocyte activation. I. Requirements for the mitogen dependent production of T-cell growth factors. Eur J Immunol 9: 581

Dillner ML, Axelson B, Hammarström S, Hellström U, Perlmann P (1980). Interaction of lectins with human T-lymphocytes. Mitogenic properties, inhibitory effects, binding to the cell membrane and to isolated surface glycopeptides. Eur J Immunol 10:434

Doherty PC, Zinkernagel RM (1975). A biological role for the major histocompatibility antigens. Lancet June 28:1406

Helenius A, McCaslin DR, Fries E, Tanford C (1979). Properties of detergents. Meth Enzymol 56G:734

Ko HS, Fu SM, Winchester RJ, Yu DTY, Kunkel HG (1979). Ia determinants on stimulated human T lymphocytes. J Exp Med 150:246

Kurosawa Y, von Boehmer H, Haas W, Sakano H, Traunecker H, Tonegawa S (1981). Identification of D segments of immunoglobulin heavy chain genes and their rearrangement in T-lymphocytes. Nature 290:565

O'Farrell PH (1975). High-resolution two-dimensional electrophoresis of proteins. J Biol Chem 250:4007

Ravid A, Novogrodsky A, Wilchek M (1978). Grafting of triggering sites onto lymphocytes. Requirement of multivalency in the stimulation of dinitrophenyl-modified thymocytes by anti-dinitrophenyl-antibody. Eur J Immunol 8:289

BASEMENT MEMBRANE GLYCOPROTEINS IN THE EXTRACELLULAR MATRIX
OF A TERATOCARCINOMA-DERIVED DIFFERENTIATED CELL LINE

Ilmo Leivo[1,2], and Jorma Wartiovaara[1]

Departments of Electron Microscopy[1] and
Pathology[2]
University of Helsinki, Helsinki, Finland

SUMMARY

Glycoproteins synthesized and deposited into extracellular matrix in cultures of a mouse teratocarcinoma-derived differentiated cell line (PYS-2) were studied by metabolic labeling and immunochemical methods, and by immunofluorescence and electron microscopy. PYS-2 cells synthesized two major high-molecular weight glycoproteins, laminin and type IV collagen, which were deposited in matrix form with no apparent processing. The lamellar subcellular matrix was composed of a loose network of fine fibrils and dense grains. Straight 60-80 nm fibrils attached the matrix to the plasma membranes and were decorated with ferritin-coupled antibodies for laminin. The cells did not synthesize fibronectin, but the matrix bound fibronectin from the culture medium. The PYS-2 matrix provides possibilities for in vitro studies on basal lamina components in early differentiation.

INTRODUCTION

Basement membranes contain type IV collagen (Bornstein and Sage, 1980; Timpl and Martin, 1982), non-collagenous glycoproteins including laminin (Timpl et al., 1979; Foidart et al., 1980) and fibronectin (Stenman and Vaheri, 1978) and proteoglycans containing heparan sulfate and other glycosaminoglycans (Hassell et al., 1980; Kanwar et al., 1981). The continuous mouse teratocarcinoma-derived endodermal cell lines (the PYS lines) secrete abundant basement membrane-like extracellular material (Pierce et al., 1962). The PYS cells resemble parietal endoderm cells of the early mouse embryo which are the first cells in the developing embryo that secrete large amounts of extracellular matrix material, deposited as the Reichert's membrane. Recently, synthesis and deposition into extracellular matrix of laminin-like glycoproteins has been reported in teratocarcinoma-derived cell lines (Chung et al., 1979; Hogan, 1980; Howe and Solter, 1980; Sakashita and Ruoslahti, 1980; Wartiovaara et al., 1980). In this study, PYS-2 cells were found to secrete laminin and type IV collagen, which were the major glycoprotein components in the extracellular matrix composed of fine fibrils and dense grains. Laminin antigenicity was found in straight fine fibrils which attach the matrix network to the plasma membranes.

MATERIALS AND METHODS

Cell Cultures and Preparations of Cell-Free Matrices

Differentiated teratocarcinoma-derived PYS-2 cells, originating from Dr. J. Lehman (Lehman et al., 1974) were cultured as monolayers in Eagle's Minimum Essential Medium (MEM) with 10% fetal calf serum depleted of fibronectin in a gelatin-Sepharose column, and with sodium ascorbate (30 ug/ml and β-aminopropionitrile fumarate (50 ug/ml). For preparation of cell-free matrices cell cultures were treated three times for 10 min at $0^{\circ}C$ with 0.5% sodium desoxycholate in 10 mM Tris-HCl, pH 8.0, containing 1 mM phenylmethylsulphonyl fluoride ($PhCH_3SO_2F$), rinsed in 2 mM Tris-HCl, pH 8.0, containing 1 mM $PhCH_3SO_2F$, and the cell-free material was incubated in DNase I (Worthington, Freehold, NJ) 50 ug/ml for 15 min at room temperature (Leivo et al., 1982).

Microscopical Studies

For immunofluorescence of extracellular antigens cells and matrices were fixed with 3.5% paraformaldehyde for 30 min. Purified antibodies against laminin and type IV collagen from mouse EHS tumor matrix, and against bovine type I collagen and type III procollagen were used (for refs. see Leivo et al., 1982). Antiserum against human plasma fibronectin was previously described. The specificities of the antisera, lack of cross-reactions and immunohistological control tests including blocking experiments were described previously.

For scanning electron microscopy (SEM) cell-free matrices were fixed with 2.5% glutaraldehyde, critical point dried, gold-coated with JFC-1100 ion sputter and studied in a JSM-35 scanning electron microscope or JEM CX-100 Temscan electron microscope. For transmission electron microscopy (TEM) intact cell layers on plastic dishes were fixed with 2.5% glutaraldehyde, prestained with 2% tannic acid, processed routinely and studied in JEOL 100B transmission electron microscope.

For indirect immunoferritin studies cell cultures were fixed with a combination of paraformaldehyde (1% w/v) and glutaraldehyde (0.05% v/v), washed with 0.1 M lysine-HCl, and reacted with normal swine serum, the primary rabbit antibody and ferritin-coupled anti-rabbit IgG (Cappel Laboratories, Cochranville, PA). After extensive washes the cultures were fixed in 2.5% glutaraldehyde and processed for TEM. In control samples, the primary antibody was substituted with normal rabbit serum in low dilution.

Analysis of Matrix Proteins

^{3}H-proline (5 uCi/ml) was added to cell cultures daily. Ammonium sulfate (176 mg/ml) precipitates of cell-free culture media, collected overnight at +4°C in the presence of 1 mM EDTA, 0.9 mM N-ethylmaleimide and 0.2 mM $PhCH_3SO_2F$, were analyzed by gel electrophoresis. Labeled culture media were immunoprecipitated in a high ionic strength buffer and the specificity was controlled with nonimmune rabbit serum or antibodies to type I collagen or type III procollagen. Polyacrylamide gel electrophoresis was performed using 5-10% gradient gels and 5% linear gels, and fluorography or Coomassie Blue protein stain (Leivo et al., 1982).

RESULTS

Secreted Matrix Proteins

In polyacrylamide gel electrophoresis of the secreted labeled proteins of PYS-2 cultures three major doublet bands (Fig. 1, lane 1) were seen. Collagenase treatment removed the doublet bands with the highest electrophoretic mobility (lane 2), which were precipitated with type IV collagen antibodies (lane 3) and resemble pro-α1(IV) chains (upper band) and pro-α2(IV) chains (lower band) with molecular weights of 185 000 and 170 000. No other collagenous polypeptides were detected in the culture media. The polypeptides not degraded by collagenase were precipitated by antibodies to laminin (lane 4). The laminin chains comigrated with those purified from the mouse EHS tumor indicating apparent molecular weights of about 400 000 and 210 000.

Deposited Matrix Proteins

PYS-2 cell cultures were labeled with ^{3}H-proline for five days, the matrices were isolated and proteins were examined by electrophoresis under reducing conditions (Fig. 1). Nearly identical band patterns were detected by fluorography (lane 6) and proteins staining (lane 7). The major bands comigrated with the chains of authentic laminin (lane 9). Two faster moving bands were collagenase-sensitive and showed the mobility of pro-α1(IV) and pro-α2(IV) chains. Possible other proteins in the matrix were represented by additional minor polypeptide bands (Fig. 1, lanes 6 and 7), presumably not related to laminin and type IV collagen.

Morphology of the Matrix

Trypsinized PYS-2 cells attached 30-60 min after seeding and an extracellular amorphous plaque of laminin was observed under each attached cell (Fig. 2A). No extracellular type IV collagen, interstitial collagens or fibronectin were seen. Type IV collagen was first deposited three hours after seeding of the cells and was localized similarly under the cells.

After treatment of the cell layers with desoxycholate and DNase an abundant lamellar extracellular matrix was found on the substratum. Immunofluorescence for laminin and type IV collagen visualized these proteins evenly distributed throughout the matrix lamellae (Fig. 2B). No fluorescence for type I collagen or type III procollagen was seen. In cultures grown in the presence of fibronectin-free fetal calf serum, PYS-2 cells attached and spread

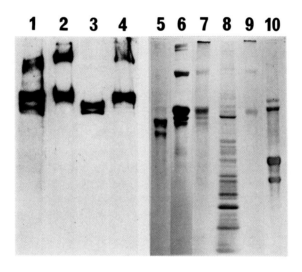

Fig. 1. SDS-polyacrylamide gel electrophoresis under reducing conditions of proteins in PYS-2 cultures. 5-10% gradient gel (lanes 1-4) and 5% linear gel (lanes 5-10). The cells were labeled for 24 h with ^{3}H-proline (10 uCi/ml in serum free medium) and proteins detected by autoradiography (lanes 1-4, 6) or with Coomassie Blue (lanes 5, 7-10). Medium proteins precipitated with ammonium sulfate (176 mg/ml), (lane 1), and medium proteins digested with collagenase (30 IU/ml medium) for 60 min at 37°C prior to precipitation (lane 2). Immunoprecipitation of medium proteins with antibodies to type IV collagen (lane 3) or laminin (lane 4). Acid-soluble type IV collagen as reference (lane 5). Proteins of cell-free matrix from a labeled culture visualized by autoradiography (lane 6); and protein staining (lane 7). Proteins of intact cell layer (lane 8). Other references included laminin (lane 9), and type I collagen (lane 10). (From: Leivo et al., 1982).

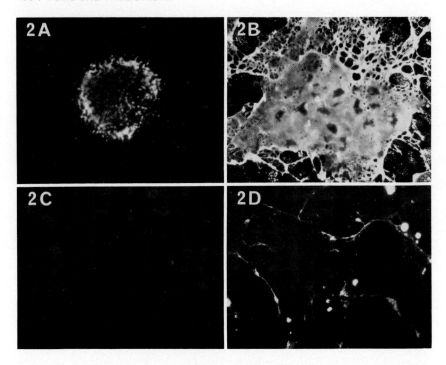

Fig. 2A. Intact attached PYS-2 cell 60 min after seeding on the dish stained for immunofluorescence of extracellular laminin. A multipunctuate matrix deposition is seen under the cell. x 1500.
Fig. 2B. Cell-free matrix preparation. Immunofluorescence for laminin shows a lamellar matrix under the extracted cell cluster and interconnected strands at the periphery. x 600.
Fig. 2C. Intact culture grown in medium depleted from serum fibronectin and stained for intra- and extracellular fibronectin. No fluorescence is seen. x 600.
Fig. 2D. Cell-free matrix preparation of culture grown in the presence of fetal calf serum containing fibronectin, and stained for fibronectin. Strands of fluorescence line matrix periphery which presumably had been exposed to serum during cultivation. x 600. (From: Leivo et al., 1982).

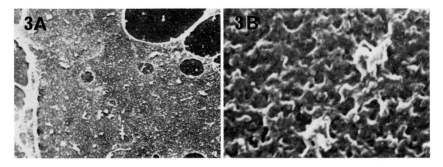

Fig. 3A. Low-power SEM micrograph of matrix preparation demonstrating an even lamellar deposit with only minor amounts of globular or filamentous material. x 1200.
Fig. 3B. High-power SEM micrograph of matrix with a non-fibrillar surface containing low rugosities and small pits. x 50 000.

normally and no fluorescence for fibronectin was seen within or outside of the cells (Fig. 2C). If the cells were grown in the presence of fetal calf serum containing fibronectin, strands of fibronectin fluorescence were found at the edges of matrix islets (Fig. 2D).

In scanning electron microscopy (SEM) at low magnification the matrix preparation appeared as an even lamellar deposit with only minor amounts of overlying fibrillar or spherical material (Fig. 3A). At high magnification with SEM (Fig. 3B) the surface of the matrix showed low rugosities and small pits but thin fibrils seen in sectioned material were not evident on the surface.

In transmission electron microscopy (TEM) of intact cell cultures the matrix was visualized as a loose network of thin interconnected fibrils, up to 4 nm in diameter (Fig. 4A). Numerous electron-dense grains, between 8-12 nm in diameter, were distributed along the fibrils often at their crossing points. The grains were often separated by a distance of 30-40 nm and sometimes formed lattice-like structures with the grains interconnected by thin fibrils. The fibril meshwork was attached to the plasma membrane by thin straight fibrils 60-80 nm long (Fig. 4A).

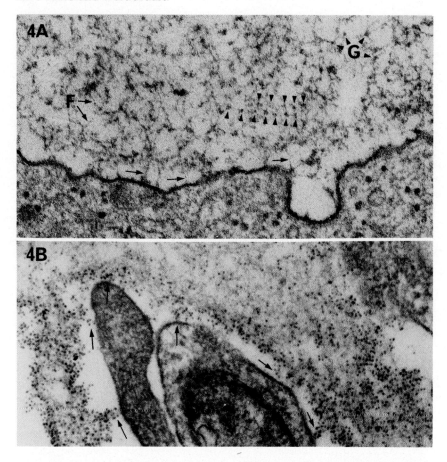

Fig. 4A. TEM micrograph of a horizontally sectioned PYS-2 culture. Note a loose network of interconnected up to 4 nm fibrils (F), and dense 8-12 nm grains (G) along the fibrils often at crossing points. Distance between grains measures 30-40 nm and the grains occasionally form lattice-like structures (arrowheads). Thin fibrils (arrows) connect the matrix network to the plasma membrane. Tannic acid stain. x 80 000. (From: Leivo et al., 1982).
Fig. 4B. TEM micrograph of an indirect immunoferritin staining for laminin. Note numerous ferritin particles binding to the matrix and decorating the fine fibrils which attach the matrix network to the plasma membranes (arrows). No tannic acid stain. x 65 000.

In indirect immunostaining for laminin, numerous ferritin particles were seen associated with the PYS-2 matrix (Fig. 4B). Special attention was focused to the attachment fibrils in areas where the plane of section was perpendicular to the plasma membrane. Ferritin particles decorated the matrix ends of such fibrils and were commonly seen along the straight portions of the fibrils uptil a close proximity (<20 nm) of the plasma membrane (Fig. 4B arrows). The density of ferritin particles within the matrix decreased as a function of distance from a free matrix surface accessible to the immunoreagents suggesting their limited penetration within the matrix. In control samples, only occasional ferritin particles were found within the matrix, even in areas where a free surface of the matrix was observed, indicating a high level of specificity.

DISCUSSION

Two basement membrane components, laminin and type IV collagen were the major high molecular weight glycoproteins in the medium and in the extracellular matrix of cultured PYS-2 cells. The two polypeptides precipitated from the culture media by antibodies to type IV collagen comigrated with similar polypeptides in the matrix preparation corresponding to the pro-α1(IV) and pro-α2(IV) chains (Bornstein and Sage, 1980; Tryggvason et al., 1980). Moreover, the matrix contained polypeptides which comigrated with similar polypeptides precipitated from the culture media by antibodies to laminin. These data suggest that no apparent proteolytic cleavage had occurred during deposition of these proteins into the matrix structure. Only minor amounts of other high-molecular weight proteins were seen by autoradiography or protein staining in matrix preparations solubilized by reduction.

The lack of fibronectin production by the PYS-2 cells is notable as fibronectin is secreted and deposited by various other cells forming extracellular matrices and it promotes cell adhesion (Vaheri et al., 1980). Exposed areas of the matrix were, however, capable of binding exogenous fibronectin from fetal calf serum presumably due to interaction with matrix collagen (Hayman and Ruoslahti, 1979) or glycosaminoglycans (Yamada et al., 1980). Immunofluorescence for keratin and vimentin, and electrophoretic and scanning electron microscopic studies suggested that the PYS-2 matrix preparation was not detectably contaminated by cytoskeletal structures.

The matrix consisted of a loose network of fine fibrils and attached grains. The interrupted surface of the matrix preparations in high-power SEM may represent canalicular openings of this network. The fibrils of the matrix network may contain both type IV collagen and laminin since both proteins have a suitable fibrillar structure also containing globular domains (Engel et al., 1981; Timpl et al., 1981) which may contribute to the grains seen in the PYS-2 matrix. The thin straight fibrils found between the matrix and the plasma membrane resemble fibrils which attached epithelial cells to basal laminae (Cohn et al., 1977; Kanwar and Farquhar, 1979).

Our ferritin tracer data indicate that these fibrils contain laminin antigenicity and they may represent one of the arms of the laminin molecule which in rotary shadowing electron microscopy appears as a four-armed rigid cross-shaped structure (Engel et al., 1981). Similar organization of the matrix attachment may be present in basement membranes in vivo where laminin was found in the laminae rarae (Foidart et al., 1980; Madri et al., 1980) but localization to the fibrillar level was not accomplished, perhaps due to the compact assembly of basement membranes in vivo. The present results conform well with the reported role of laminin in promoting epithelial cell adhesion (Terranova et al., 1980; Vlodavsky and Gospodarowicz, 1981).

The authors would like to thank Dr. Rupert Timpl for his help and interest during these studies.

We also thank Ms. Elina Waris, Ms. Eija Lihavainen, and Mr. Simo Lehtinen for their expert technical assistance. This study was supported by the Finnish Cancer Foundation, The Finnish Medical Research Council and the Sigrid Jusélius Foundation, Helsinki.

REFERENCES

Bornstein P, Sage H (1980). Structurally distinct collagen types. Ann Rev Biochem 49:957.

Chung AE, Jaffe R, Freeman IL, Vergnes J-P, Braginski JE, Carlin B (1979). Properties of a basement membrane-related glycoprotein synthesized in culture by a mouse embryonal carcinoma-derived cell line. Cell 16:277.

Cohn RH, Banerjee SD, Bernfield MR (1977). Basal lamina of embryonic salivary epithelia. Nature of glycosaminoglycan and organization of extracellular materials. J Cell Biol 73:464.

Engel J, Odermatt E, Engel A, Madri JA, Furthmayr H, Rohde H, Timpl R (1981). Shapes, domain organizations and flexibility of two multifunctional proteins of the extracellular matrix: laminin and fibronectin. J Mol Biol 150:97.

Foidart JM, Bere EW Jr, Yaar M, Rennard SI, Gullino M, Martin GR, Katz SI (1980). Distribution and immunoelectron microscopic localization of laminin, a noncollagenous basement membrane glycoprotein. Lab Invest 42:336.

Hassell JR, Gehron Robey P, Barrach H-J, Wilczek J, Rennard SI, Martin GR (1980). Isolation of a heparan sulfate-containing proteoglycan from basement membrane. Proc Natl Acad Sci USA 77:4494.

Hayman EG, Ruoslahti E (1979). Distribution of fetal bovine serum fibronectin and endogenous rat cell fibronectin in extracellular matrix. J Cell Biol 83:255.

Hogan BLM (1980). High molecular weight extracellular proteins synthesized by endoderm cells derived from mouse teratocarcinoma cells and normal extraembryonic membranes. Develop Biol 76:275.

Howe CC, Solter D (1980). Identification of non-collagen glycopeptides of a basement membrane synthesized by the mouse parietal entoderm and endoderm cell line. Develop Biol 77:480.

Kanwar YS, Farquhar MG (1979). Anionic sites in the glomerular basement membrane. J Cell Biol 81:137.

Kanwar YS, Hascall VC, Farquhar MG (1981). Partial characterization of newly synthesized proteoglycans isolated from the glomerular basement membrane. J Cell Biol 90:527.

Lehman JM, Speers WC, Swartzendruber DE, Pierce GB (1974). Neoplastic differentiation: characteristics of cell lines derived from a murine teratocarcinoma. J Cell Physiol 84:13.

Leivo I, Alitalo K, Risteli L, Vaheri A, Timpl R, Wartiovaara J (1982). Basal lamina glycoproteins laminin and type IV collagen are assembled into a fine-fibered matrix in cultures of a teratocarcinoma-derived endodermal cell line. Exp Cell Res 137:15.

Madri JA, Roll FJ, Furthmayr H, Foidart J-M (1980). Ultrastructural localization of fibronectin and laminin in the basement membranes of the murine kidney. J Cell Biol 86:682.

Pierce GB Jr, Midgley AR Jr, Sri Ram J, Feldman JD (1962). Parietal yolk sac carcinoma: Clue to the histogenesis of Reichert's membrane of the mouse embryo. Am J Pathol 41:549.

Sakashita S, Ruoslahti E (1980). Laminin-like glycoproteins in extracellular matrix of endodermal cells. Arch Biochem Biophys 205:283.

Stenman S, Vaheri A (1978). Distribution of a major connective tissue protein, fibronectin, in normal human tissue. J Exp Med 147:1054.

Terranova VP, Rohrbach DH, Martin GR (1980). Role of laminin in the attachment of PAM 212 (epithelial) cells to basement membrane collagen. Cell 22:719.

Timpl R, Rohde H, Gehron Robey P, Rennard SI, Foidart J-M, Martin GR (1979). Laminin - a glycoprotein from basement membranes. J Biol Chem 254:9933.

Timpl R, Wiedemann H, van Delden V, Furthmayr H, Kühn K (1981). A network model for the organization of type IV collagen molecules in basement membranes. Eur J Biochem (in the press).

Timpl R, Martin GR (1982). Components of basement membranes. In Furthmayr H (ed.): "Immunochemistry of the Extracellular Matrix", Vol II Applications. CRC Press Inc., Boca Raton, Florida (in the press).

Tryggvason K, Gehron Robey P, Martin GR (1980). Biosynthesis of type IV procollagens. Biochemistry 19:1284.

Vaheri A, Keski-Oja J, Vartio T, Alitalo K, Hedman K, Kurkinen M (1980). Structure and functions of fibronectin. In Prockop DJ, Champe PC (eds): "Gene families of collagen and other proteins", Elsevier/North-Holland, New York, p. 161.

Vlodavsky I, Gospodarowicz D (1981). Respective roles of laminin and fibronectin in adhesion of human carcinoma and sarcoma cells. Nature 289:304.

Wartiovaara J, Leivo I, Vaheri A (1980). Matrix glycoproteins in early mouse development and in differentiation of teratocarcinoma cells. In Subtelny S (ed.): "The cell surface: Mediator of Developmental processes", Academic Press, New York, p. 305.

Yamada KM, Kennedy DW, Kimata K, Pratt RM (1980). Characterization of fibronectin interactions with glycosaminoglycans and identification of active proteolytic fragments. J Biol Chem 255:6055.

CELL-CELL CONTACT, CYCLIC AMP, AND GENE EXPRESSION DURING
DIFFERENTIATION OF THE CELLULAR SLIME MOLD DICTYOSTELIUM
DISCOIDEUM

Daphne D. Blumberg, Stephen Chung, Scott M.
Landfear, Harvey F. Lodish
Department of Biology, Massachusetts Institute
of Technology, Cambridge, MA 02139

The predominant feature of the developmental cycle of
Dictyostelium discoideum is the aggregation of unicellular,
free-living amoebae into a multicellular organism (Loomis
1975). Differentiation of amoebae within the newly-formed
aggregates generates the three distinct cell types found in
the mature fruiting body: spore cells, stalk cells, and
basal discs. Dictyostelium exhibits many features of development seen in more complex eukaryotic organisms: specific
cell-cell contacts are found; a homogenous cell population
differentiates into discrete cell types; and there is specific cell migration and pattern formation. These morphogenetic changes are accompanied by major changes in the pattern
of gene expression. As we shall discuss here, the transcription of these genes is controlled by many of the same
factors that affect gene expression in other differentiating
systems, in particular, by cell-cell contact and by an
extracellular hormone, cyclic AMP.
    Starvation for amino acids triggers the developmental
cycle and during the first 8 hrs the single celled amoebae
undergo an elaborate program of cAMP-mediated chemotactic
cell signalling, resulting in the formation of aggregates of
about 10^5 cells. Concommitantly, there are major alterations to the cell surface (reviewed in Loomis, 1975, 1979).
The cells acquire a system of EDTA resistant cell-cell contacts which transform the loose aggregates into tightly contacted multicellular mounds (Beug et al. 1973). A number of
lectins, glycoproteins and other antigenetic determinants
have been identified which either are a part of the EDTA
resistant cell-cell contact system or which are involved in
some manner so as to allow the developmentally regulated
appearance of these tight contacts (Müller and Gerisch,

1978; Geltosky et al. 1979; Rosen et al. 1973. Once the multicellular state has been achieved by the formation of these contacts a tip appears on the mound of cells and the program of prespore-prestalk cell differentiation commences.
The pattern of developmental gene expression: Changes in the mRNA population.

Analysis of changes in the mRNA population indicates that there is only one stage when expression of a large number of new genes is initiated - that of cell-cell aggregation (Alton and Lodish 1977; Blumberg and Lodish 1980a, b; 1981). As is summarized in Table 1, growing Dictyostelium cells contain around 4500 discrete species of mRNA, about the same number as in growing yeast cells (Blumberg and Lodish 1980a; Hereford and Rosbash 1977). Approximately 600 of these sequences are present in greater than 160 copies per cell; the rest are present, on the average, at 14 copies per cell. These mRNAs represent the transcription products of 19% of the single-copy Dictyostelium genome (Blumberg and Lodish 1980a).

The number of mRNA species does not change significantly by 6 hr of differentiation (preaggregation stage)(Blumberg and Lodish 1980b). Cross-hybridization studies showed, moreover, that the population of mRNAs present in growing and preaggregating cells were extreemly similar. While other types of studies show that expression of some genes is induced in the preaggregation stage, the number of such genes must be very few (Alton and Lodish 1977a; Williams and Lloyd 1979; Williams et al. 1979).

By contrast, the polysomes of post-aggregation cells contain 7000 discrete mRNA species, the transcription products of 30% of the single-copy genome (Blumberg and Lodish 1980b). The majority of the 4500 mRNA species present before aggregation remain in the cells throughout development, although the average abundance of these mRNAs decreases with time. About one-third of the mass of mRNAs in the post-aggregation stage cells is comprised of the 2000 to 3000 mRNA species which appear only after aggregation. Thus, an additional 11% of the single-copy genome is expressed as polysomal polyadenylated RNA, presumably mRNA, after aggregation. Of these aggregation-stage mRNAs, 100 to 150 sequences are present at 80 copies per cell; the remainder are present at 5 copies per cell.

The population of mRNAs present in culminating cells is indistinguishable, by cross-hydridization, from that in post-aggregation cells. Both contain the same 7500 different species of mRNA. Although synthesis of spore- and stalk-specific proteins occur during the culminmation stage,

Table 1: Complexity of nuclear and cytoplasmic polyadenylated RNA.

Developmental stage	Percentage single copy DNA expressed in total nuclear plus cytoplasmic polyadenylated RNA	Percentage single copy genome expressed in cytoplasmic polyadenylated RNA (mRNA)	Number of genes expressed as mRNA
Vegetative	53.4	19.3	4820
6 hr (preaggregation)	54.0	19.3	4800
13 hr (postaggregation)	82.2	29.8	7420
22 hr (culmination)	76.8	31.0	7750
Mixture: vegetative plus 13 hr cells	79.4	31.0	7750

Legend to Table 1: Percentage single copy DNA expressed is determined from the fraction of the single copy DNA rendered double stranded after hybridization with either total nuclear plus cytoplasmic polyadenylated RNA (column 1) or cytoplasmic polyadenylated RNA alone (column 2). The assumption is made that only one strand of the DNA is transcribed. Therefore the percentage of genomic DNA which is expressed is twice the percentage rendered double stranded in the RNA-driven hybridization reactions. The number of genes expressed as average sized cytoplasmic polyadenylated RNA species (column 3) is calculated assuming that Dictyostelium mRNA has a weight average molecular weight of 400,000 daltons and the size of the single copy portion of the genome is 2×10^{10} daltons. The data are taken from Blumberg and Lodish (1980a, b; 1981).

they are not encoded by a significant fraction of the mRNA sequences present in these cells (Alton and Lodish 1977a; Blumberg and Lodish 1980b; Coloma and Lodish 1981).

The number of induced genes is not large relative to other developmental systems. For example, 1500 new mRNAs are induced during conidia formation in Aspergillus and 3000 to 4000 new mRNA species are present in differentiating myoblasts relative to their embryonal precursor cells (Affara et al. 1977; Timberlake 1980). However, because the Dictyostelium genome is smaller than that of most other eukaryotic cells, these developmentally regulated sequences represent a substantially larger proporation of the genome thereby facilitating the identification, cloning, and further study of their regulation. Indeed, many such genes have been cloned, providing very specific hybridization probes for individual regulated sequences.

The number of developmentally regulated genes

Genetic estimates indicate that the function of only 300-400 additional genes is essential for the normal progression of Dictyostelium through the developmental cycle (Loomis 1978). This is markedly lower than the number of new polyadenylated RNA species which appear on the polyribosomes during differentiation - about 2000 to 3000. These genetic studies focus on the number of genes whose function is essential for normal morphological development under laboratory conditions. There are several reasons why this approach could underestimate the number of developmentally regulated mRNAs. For instance, it would not score genes required for formation and migration of pseudoplasmodia, or for chemotactic or phototactic behavior, or for variability of spore cells. Likewise, it would not score mRNAs derived from two or more identical or nearly-identical genes. Also, during differentiation genes may be activated whose functions are not necessary for spore or stalk formation in the laboratory, but are essential for morphogenesis under certain natural conditions or for differentiation along alternative pathways, such as macrocyst formation. As an example, mutations which destroy the activity of two developmentally regulated enzymes - α-mannosidase and β-glucosidase - do not have any observable effect on morphogenesis (Free et al. 1976; Dimond and Loomis 1976). Thus, during differentiation synthesis of protein and mRNAs not essential for normal morphogenesis may be induced. By analogy with the lac operon a large number of genes may be "activated" at the cell aggregate stage and thereafter transcribed at a low basal level of 4 - 5 copies per cell but only induced to a "functional" level of several hundred copies per cell if and when the need arises.

Appearance of the new post aggregation stage mRNAs is under transcriptional control

The appearance of the 2000 to 3000 new mRNA species in the cytoplasm and on the polysomes is accompanied by a corresponding but larger increase in the complexity of the nuclear polyadenylated RNA species (Blumberg and Lodish 1981). Altogether, 26% of the single copy portion of the genome is newly expressed in total nuclear plus cytoplasmic polyadenylated RNA from late aggregating cells. As indicated in Table I, 53 to 54% of the single copy portion of the genome is expressed in polyadenylated RNA in growing and early aggregating cells. Of this only about one-third or 19% is expressed on polysomes as mRNA. By contrast, during late aggregation and culmination, after the time when tips have formed on the aggregates a substantially greater portion of the single copy genome, 80%, is expressed in the total population of polyadenylated RNA. Again, however, only about one-third or 30% of the single copy genome is expressed as mRNA. Thus, throughout growth and development the complexity of the polyadenylated nuclear RNA is about 1.5 times the complexity of the corresponding mRNA populations. The hn RNA species represent only a very small portion of the mass of the polyadenylated RNA, less than 3%.

Little is known about the biogenesis of mRNA from heterogeneous nuclear RNA in Dictyostelium. In contrast to mammalian cells, about 75% of the radioactivity incorporated into Dictyostelium nuclear polyadenylated RNA species is conserved and transported to the cytoplasm as mRNA; only 25% is degraded in the nucleus (Firtel and Lodish, 1973). The heteronuclear RNA precursors of mRNA species are estimated to be, on the average, only about 20% larger than mRNA molecules (Firtel and Lodish, 1973). Of the individual genes which have so far been cloned and characterized most either lack intervening sequences (actin and discoidin) or contain only very short ones (no more than a few hundred bases) (Bender et al. 1978; Kimmel and Firtel 1980). Thus, the relationship, if any, of these high complexity, low abundance nuclear transcripts to mRNA is not clear. Their presence however does raise an important question: Are the 2000 to 3000 mRNA species which are only detectable on the polysomes late in development present as rapidly degraded polyadenylated heteronuclear RNA species during growth? Alternatively, does the transcription of these mRNAs initiate at the stage of cell-cell aggregate formation?

Several lines of evidence indicate that this later alternative is the most likely. First hybridization of cDNA probes complementary to the late mRNA species to a vast excess of total nuclear plus cytoplasmic RNA from growing cells failed to detect a significant fraction of the late mRNAs (Blumberg and Lodish 1981). Under the conditions used even one copy per 25 cells of any RNA would have been detected. It appears, then that if genes encoding late mRNAs are transcribed in preaggregation-cells, the transcripts are destroyed immediately. It is impossible to obtain enough radioactive RNA in a very short labeling period to test transcriptional control directly. However, nuclei from preaggregation cells synthesize, in vitro, only the "constitutive" class of polyadenylated RNA, while nuclei from 16 hr cells synthesize both aggregation-dependent and "constitutive" mRNA species (Landfear, Lefebvre, Chung, and Lodish,

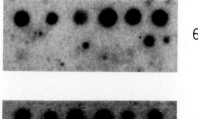
6 hr nuc

15hr nuc

FIGURE 1: Transcription of individual mRNAs measured in isolated Dictyostelium nuclei. Nuclei were prepared from 10^9 pre-aggregation (6 hr nuc) or post-aggregation (15 hr nuc) cells and incubated for 30 minutes at 23˚ in 40 mM Tris-HCl, pH 7.9; 10 mM $MgCl_2$; 250 mM NaCl; 0.16 mM ATP; 0.16 mM GTP; 0.16 CTP; 0.1 mM DTT; 5% glycerol; 10 μCi α[^{32}P]-UTP at 410 Ci/mmole. Polyadenylated RNA was isolated and hybridized to cloned Dictyostelium DNA (0.1 - 0.5 mg per spot) that had been spotted onto nitrocellulose filters (Mangiarotti et al. 1981). The filters were then exposed to Kodak XAR-5 film uisng a Dupont Cronex intensifying screen. For each filter, the top row contains cloned DNA from Dictyostelium genes whose transcripts are present in cells at all stages of development; the bottom row contains cloned DNA from genes whose transcripts are present only in post-aggregation cells (Chung et al. 1981; Mangiarotti et al. 1981).

in preparation).

An example of this analysis is shown in Figure 1, in which [^{32}P] labeled RNA was synthesized in vitro by nuclei from either preaggregation or post-aggregation cells. The polyadenylated RNA was hybridized to a set of immobilized cloned DNAs encoding either "constitutive" mRNAs (top row) or "aggregation-dependent" mRNAs (bottom row). As can be seen, this latter class of mRNAs is synthesized only by aggregation-stage nuclei, while both nuclear preparations synthesize the "constitutive" mRNAs.

All of these observations suggest that transcription of the late mRNA species initiates at the time of cell aggregate formation approximately 8 to 12 hours into the developmental cycle.

Achievement of multicellularity and formation of tight cell-cell contacts are a necessary prerequisite for induction of the 2000 to 3000 late mRNA species.

Analysis of mutants which are blocked at discrete stages of differentiation has indicated that there is a very strong correlation between the ability of cells to form the tight cell-cell contacts which resist the action of EDTA and their ability to express the late mRNA species (Blumberg et al. 1981). This is summarized in Table II.

All aggregation-defective mutants (irrespective of the nature of the mutated gene), fail to induce any late mRNA tested, even though they induce many of the early develop-

Table 2: **Properties of developmentally blocked mutants.**

			1	2	3	4	5	6	7	8
			Very early Development				EDTA Contacts	Terminal Morphology	Level of Aggregation Dependent mRNAs	Aggregation dependent proteins
Strain	Parent	Source	ment	45M	PDE	cAMP-B	tacts			
Agg2	AX3	D. McMahon	+	–	–	–	–	agg⁻	0.4%	–
WL3	AX3	W.F. Loomis	+	+	–	–	–	agg⁻	0.2%	–
JM41	AX3	J. Margolskee	+	+	–	–	ND	agg⁻	0.4%	–
HC72	HC6	M.B. Coukell	+	+	+	–	–	agg⁻	0.2%	–
HC54	HC6	M.B. Coukell	+	+	+	+	–	agg⁻	0.2%	–
HJR-1	NC4	R.A. Lerner	ND	ND	ND	ND	–	ripples, loose aggregates	0.3%	ND
JM84	AX3	J. Margolskee	+	+	+	+	–	ripples, loose aggregates	6 %	–
GM2	AX3	W.F. Loomis	+	+	+	+	+	mounds	39.8%	+
JM35	AX3	J. Margolskee	+	+	+	+	+	mounds with tips	100 %	+
AX3	NC4	W.F. Loomis	+	+	+	+	+	culminates	100 %	+

Footnotes to Table 2: 1) <u>Very early development</u> "+" indicates induction of synthesis of the enzyme N-acetylglucosaminidase. It also indicates a 2- to 4-fold induction of actin synthesis between 0 and 30 min of development as well as a sharp drop in actin synthesis at 90 min of development (Margolskee and Lodish 1980a). 2) <u>45 min</u> "+" indicates induction of the synthesis of two early proteins (the "45 minute" proteins) synthesized between 45 and 90 min of development (Margolskee and Lodish 1980a). 3) <u>PDE</u> Induction of the activity of the cell bound phosphodiesterase. "+" indicates that the mutant cells developed by starvation for 15 hours in a rapidly shaken suspension culture expressed a level of the enzyme at least 4-fold above the background activity observed in growing cells. AX3 wild type cells starved in suspension culture for 15 hours expressed a level 13-fold above the growth phase background. 4) <u>cAMP-b</u> Cell surface cAMP binding activity. "+" indicates that the mutant cells developed by starvation in a rapidly shaken suspension culture for 8 or 15 hours induce a level of binding which is at least four times above the level detected in growing cells. Wild type AX3 cells induced a 6-fold increase under these conditions. 5) <u>EDTA Contacts</u> Formation of EDTA-resistant cell contacts. 6) Agg⁻; a failure to aggregate. 7 and 8) Taken from data in Blumberg et al. 1981. ND: Not done.

mental functions. Mutants (JM84, HJR1) that form loose aggregates of cells but no EDTA-resistant cell contacts also fail to induce significant levels of any of the regulated mRNAs. Thus, chemotaxis to cAMP and the close proximity of cells in aggregates may be necessary for induction of gene expression, but it is not sufficient. All mutants which form morphologically normal cell aggregates with normal EDTA resistant cell-cell contacts (GM2, JM35) make near normal to normal amounts of most late mRNAs even though they fail to differentiate further.

These experiments establish a correlation between formation of EDTA-resistant cell contacts and expression of the entire class of aggregation-stage mRNAs. The point of contact must be past for the late mRNAs to be expressed. Whether this induction is the result of the contact event itself, subsequent coupling of cells through formation of junctions, or interaction with a "hormone" functioning only over very short distances remains to be determined.

Cell contact and cAMP control both the synthesis and stability of the 2000 to 3000 post-aggregation mRNAs

1. Disaggregation results in specific degradation of the late mRNAs

Abundant evidence indicates that continued synthesis of aggregation-specific polypeptides and enzymes is dependent on continued cell-cell interactions. Disaggregation of tightly contacted mounds to single cells results in immediate cessation of accumulation of developmentally regulated enzymes such as UDP glucose pyrophosphorylase and UDP galactose polysaccharide transferase (Newell et al. 1971, 1972; Okamoto and Takeuchi 1976). Reaggregation of the cells into mounds induced a new round of enzyme synthesis. Likewise, expression of 45 predominant polypeptides whose synthesis initiates after aggregation, ceases when the aggregates are dispersed and kept from re-attaching to each other by constant shaking (Landfear and Lodish 1980).

The metabolism of all 2000 to 3000 regulated mRNAs is also coordinately affected by disaggregation (Chung et al. 1981). Synthesis of virtually all the "aggregation-dependent" mRNAs ceases when mounds are disaggregated to single cells and are kept apart by constant shaking. Synthesis of the "conserved" mRNA species, those synthesized both during growth and differentiation, are unaffected (Chung et al. 1981).

Secondly and most surprisingly the 2000 to 3000 postaggregation mRNA species in the cytoplasm are rapidly and specifically degraded following disaggregation (Chung et al. 1981).

These studies have used two types of hybridization experiments to investigate the synthesis and stability of "aggregation-dependent" and "constitutive" mRNAs in disaggregated cells: hybridization of mRNA to a cDNA probe specific for the population of 2000 to 3000 regulated sequences; and hybridization of mRNA to cloned genomic DNAs encoding individual "regulated" or "constitutive" mRNAs (Chung et al. 1981). An example of this latter is shown in Fig. 2.

Cytoplasmic polyadenylated RNA was prepared from cells plated for development for 15 hours, or disaggregated after 15 hours and maintained as a vigorously shaken single cell suspension for 5 hours following disaggregation. The level of several late-specific mRNAs was measured by "Northern" gel analysis (Figure 2; lanes 1, 2, 3, 4). The constitutive RNAs: bands 79, 314b, 314c, 29 and 55a remain at the same level following disaggregation, whereas the late specific RNAs: 253b, 314a, 315b and 55b are rapidly degraded following disaggregation. Both assays indicate that 2.5 hours after cells are disaggregated, the level of late mRA sequences is only 6% that in plated cells, by 5 hrs, it is only 2.5%.

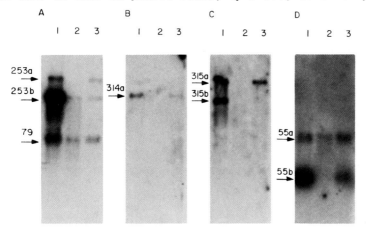

FIGURE 2: Specific degradation of late messages upon disaggregation, and preservation of late messages by cAMP. Two μg of cytoplasmic polyadenylated RNA was loaded on each lane of the gel, electrophoresed, blotted and immobilized on nitrocellulose paper as described in Chung et al. 1981. Filters were hybridized to [^{32}P] labeled DNA of clones 253, 314, 315, 55 and 79 respectively. The RNA encoded by these clones are indicated by the arrows. Lane 1 contains RNA from 15 hr plated cells; lane 2 contains RNA from cells that were plated for 15 hr and then disaggregated for 5 hr; and lane 3 contains RNA from cells plated for 15 hr and then disaggregated for 5 hr in the presence of 100 μM cAMP. Bands 79 and 55a are constitutive mRNAs and bands 253a, 253b, 314a, 315a, 315b and 55b are late, aggregation specific mRNAs.

This indicates that upon disaggregation the late mRNAs are lost from the cytoplasm with a $t_{1/2}$ of 25 to 45 min. This is in marked contrast to the average half life of cytoplasmic polyadenylated RNA. Two methods have been used to determine the stability of Dictyostelium mRNA during growth and differentiation: approach to steady-state labeling of mRNA, and addition of inhibitors of mRNA biosynthesis (Margolskee and Lodish 1980; Mangiarotti et al. 1981). Both methods showed that during growth (8 hr cell doubling) and differentiation, the average half-life of the bulk of the cytoplasmic polyadenylated RNA in the cell is about 4 hrs. These studies, however, did not determine whether the half-life of aggregation-stage mRNAs which represent only 30% of the mass of the late mRNA was different from that of "constitutive" species which represent 70% of the mass. Thus the rapid and specific labilization of these late mRNAs following disaggregation raises the interesting question - what is their half life in normally developing "contacted" mounds? Do they normally turn over with a $t_{1/2}$ of 25 to 45 min? Or are they specifically recognized and selected for degradation upon disaggregation?

Several lines of evidence indicate that this latter interpretation is correct. In normal late developing plated cells under conditions where RNA synthesis is inhibited by actinomycin D and daunomycin both the "aggregation specific" and the developmentally "conserved" mRNAs decay with identical kinetics; an average $t_{1/2}$ of about 4 hrs is obtained (Chung et al. 1981). A potential problem with these experiments, however is that the drugs themselves might act to stabilize the late mRNAs. More recently, it has been established directly that in normal late aggregates both the regulated and conserved mRNAs are indeed long-lived (Mangiarotti et al. 1981). These experiments show that in normal plated cells the kinetics of [^{32}P] labeling of both constitutive and regulated mRNA are linear from 13 to 17 hr of development. The rate of synthesis of those mRNAs is unchanged during this period. The observation that incorporation of radioactivity into specific mRNAs is linear for this prolonged period indicates that both classes of mRNA have half-lives of at least four hours.

How the aggregation-stage mRNA are specifically labilized upon disaggregation is not known. Among several possibilities disaggregation could increase the synthesis or activity of a specific mRNA-degrading enzyme. Such activation of specific mRNA nucleases has been observed in mammalian cells treated with interferon (Nilsen et al. 1980).

We envisage this differential stability of the late mRNAs as providing flexibility to a developing system such as Dictyostelium. The developmental program in Dictyostelium is essentially a means for survival when food sources are depleted. Until very late in the developmental program, when formation of actual spore and stalk cells begins, the committment to differentiation is reversible. Cells disaggregated from tight mounds or slugs will resume normal growth and propagation when a food source is restored. The ability to rapidly labilize the late specific mRNAs upon dispersion of cells from aggregates may represent an early step in the reversal of the developmental program and provide an efficient means for eliminating the molecular vestiges of the program of differentiation. The use of differential mRNA stability as a potential means of reversing the committment to differentiate is quite novel and differs from other systems where changes in the stability of mRNA species have been shown to facilitate the rapid accumulation of high levels of particular mRNAs in terminally differentiating cells (Volloch and Housman 1981; Guyette et al. 1979).

2. cAMP can stimulate the synthesis of aggregation-specific mRNA in disaggregated cells

Cyclic AMP exercises a profound effect on the level of aggregation-specific mRNAs in disaggregated cells. Even five hours after disaggregation, the levels of most translatable mRNAs encoding predominant late polypeptides remain normal providing 10-100 M cAMP is added at the time of disaggregation (Landfear and Lodish, 1980). Likewise the levels of most, but not all, of the 2000 to 3000 late mRNAs are maintained in the disaggregated cells in the presence of cAMP (Chung et al. 1981). Additionally cAMP added to disaggregated cells which have been shaken as a single cell suspension for several hours will restore near normal levels of most of the late mRNAs even if they had already decayed to less than 2% of their original levels (Chung et al. 1981).

These studies have again used the two types of hybridization experiments already described to investigate the levels of the late mRNAs in cells disaggregated in the presence or absence of 100 µM cAMP.

An example of the Northern gel analysis is shown in Figure 2. Cytoplasmic polyadenylated RNA was prepared from late developing cells (15 hours) which had been disaggregated and maintained as a vigorously shaken single cell suspension in the presence or absence of 100 µM cAMP for 5 hours. Figure 2 shows that the levels of most aggregation-specific mRNAs - 253a, 314, 315a, and 55b - are maintained at 20 to 50% of the control value in the cAMP-treated, but not in the untreated, disaggregated cells.

In cells disaggregated in the absence of cAMP, all aggregation-specific mRNAs are labilized, and are present at less than 6% of their normal level after 2.5 hours of vigorous shaking (Chung et al. 1981). Using the bulk cDNA probe to quantitate the levels of expression of the late mRNAs following the addition of cAMP to these cells we estimate that within 3 hrs of addition of cAMP there is a 16 fold increase in the average level of the population of late mRNAs: they are restored from less than 6% of their normal level to within 50 to 100% of their original level (Chung et al. 1981).

While most of the mass of the late mRNAs can be maintained or restored by the addition of 10-100 mM cAMP there are some late mRNAs which do not respond to cAMP at all. For example, aggregation-specific mRNAs, such as 253b and 315b, also disappear from cells 5 hours after disaggregation in the absence of cAMP, but cAMP does not stabilize or induce them to any detectable level (Figure 2 and Chung et al. 1981).

3. cAMP cannot stimulate synthesis of aggregation-dependent mRNAs in cells that have not previously been in contact

The ability of cAMP to stimulate synthesis of many if not most of the late mRNAs raises the question of whether the role of the contact event in late gene expression is not merely a means of bringing the cells into close proximity to each other and therefore to higher levels of cAMP. While this may be the case for maintainance of expression of the aggregation dependent mRNAs cAMP does not appear to be sufficient to initiate the expression of the mRNAs.

Cyclic AMP, in our hands, stimulates the expression of aggregation-stage mRNAs only in single cells which have once formed cell-cell contacts. Cells which have been starved for 15 hrs in a vigorously shaken suspension culture in the same solution as used for differentiation are termed aggregation competent (Gerisch and Hess 1974). These cells possess the cAMP receptor and phosphodiesterase required for cAMP cell signaling and when plated on a solid surface they rapidly form mounds with prominent tips (first fingers) within 7 to 9 hours, several hours faster than plated growing cells induced to differentiation. These suspension-starved, aggregation competent cells fail to induce significant levels of any late mRNA or protein, either in the absence or presence of cAMP added at 4, 8, 13 or 15 hours after initiation of starvation provided they are maintained as single cells (Figure 3; Landfear and Lodish 1980; Chung et al. 1981)'. Expression of at least some aggregation-stage mRNAs by suspension starved cells will occur when they form large agglomerates, a condition that allows cell-cell contact; under these conditions cAMP accelerates the appearance of these mRNAs (Williams et al. 1981).

Thus some aspect of the "contact event" allows most of the late genes to be expressed and to be subsequently regulated through a cAMP mediated interaction. Once the contact event is past cAMP is sufficient to maintain or induce expression of many but not all of these aggregation specific mRNAs. Prior to the "contact event", however most of the late genes will not respond to cAMP.

In related studies, Gross et al. (1981) have shown that the activity of one enzyme, glycogen phosphorylase, can be induced by the addition of very high levels of cAMP (1 mM) in a starved suspension of single wild-type cells. Under similar conditions, however another late enzyme, UDP galactose polysaccharide transferase as well as other late poly-

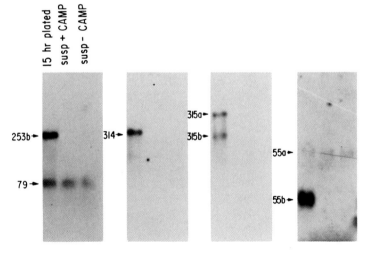

FIGURE 3: Expression of late mRNAs in cells developed by starvation in suspension culture in the presence or absence of cAMP. Two μg of cytoplasmic polyadenylated RNA was loaded on each lane of the gel, electrophoresed, blotted and immobilized on nitrocellulose paper as described in Chung et al. 1981. Filters were hybridied to [^{32}P] labeled DNA of clones 253, 314, 315, 55 and 79 respectively. The RNA encoded by these clones are indicated by the arrows. Lane 1 contains RNA from 15 hr plated cells; lane 2 contains RNA from cells starved in suspension culture for 15 hrs and lane 3 contains RNA from cells starved in suspension culture for 15 hrs and supplemented with 100 μM cAMP for 5 additional hours as described in Chung et al. 1981. Bands 79 and 55a are constitutive mRNAs and bands 253a, 253b, 314a, 315a, 315b and 55b are late, aggregation specific mRNAs.

peptides cannot be induced (Okamoto 1981; Landfear and Lodish 1980). Thus it is important to emphasize that while a great many late gene products are not expressed in aggregation competent suspension starved cells and are not induced in these cells by the addition of cAMP, there are some late gene products which may be exceptions.

The notion that cell-cell contact is an essential prerequisite for activation of developmentally regulated genes is at least superficially at odds with observations by Kay, Gross and their colleagues (Gross et al. 1981). They showed that isolated amoebae of certain mutant strains of Dictyostelium can differentiate as single cells into mature stalk and prespore cells. This differentiation requires that the cells be bound to a special plastic surface, and is dependent on high levels of cAMP, an exact ionic composition and at least for the stalk pathway the interaction of a dialyzable oligosaccharide or lipid-like factor, Dif (reviewed in Gross et al. 1981). It is particularly important to note that it is only in certain mutant strains that these conditions will permit single cells to differentiate. Isolated wild-type cells cannot differentiate into prespore or prestalk cells under similar circumstances. Thus it is very possible that the mutation itself and/or the presence of the adhesive plastic surface may allow the normal requirement of cell-cell contact for morphogenesis and regulated gene expression to be bypassed.

Conclusion

The value of Dictyostelium as a system for investigating several key problems of developmental biology at the molecular level is just now being realized. As one example, formation of specific cell-cell contacts is dependent on a number of developmentally regulated surface proteins. Importantly, formation of these tight cell-cell contacts is an essential prerequisite for both the transcription and the stability of the 2000 to 3000 regulated mRNA species. It should be possible to determine precisely not only how these surface molecules function in cell-cell recognition, but also how interactions at the cell surface are transmitted to the nucleus and cytoplasm in order to affect gene expression. At the level of gene expression, cAMP increases the synthesis, and perhaps the stability of many of the 2000 to 3000 mRNA species which are induced at aggregation; addition of this compound to disaggregated cells (which have ceased transcription specifically, of this class of mRNAs) restores synthesis of these mRNAs. It will be of considerable interest to learn whether, as in prokaryotic cells, cAMP enters the cells and whether cAMP binding proteins interact directly with regulatory sequences in DNA. Intracellular transmitter molecules are believed to mediate the response to extracellular cAMP during chemotaxis, resulting in the pulsatile synthesis and secretion of cAMP. It is also possible that these or other compounds mediate the effects of extracellular cAMP on gene expression. The identification of the nature and mode of action of these intracellular mediators will be of considerable importance.

References

1. Affara N.A., Jacquet M., Jakob J., Gros F (1977). Changes in gene expression during myogenic differentiation. I Regulation of messenger RNA sequences expressed during myotube formation. Cell 12:509.
2. Alton T.H., Lodish H.F. (1977a) Developmental changes in messenger RNAs and protein synthesis in Dictyostelium discoideum. Devel. Biol. 60:180.
3. Bender W., Davidson N., Kindle K.L., Taylor W.C., Silverman M., Firtel R.A. (1978). The structure of M6, a recombinant plasmid containing Dictyostelium discoideum DNA homologous to actin messenger RNA Cell 15:779.
4. Beug H., Katz F.E., Gerisch G. (1973). Dynamics of antigenic membrane sites relating to cell aggregation. J. Cell Biol. 56:647.
5. Blumberg D.D., Lodish H.F. (1980a). Complexity of nuclear and polysomal RNAs in growing Dictyostelium discoideum cells. Devel. Biol. 78:268.
6. Blumberg D.D., Lodish H.F. (1980b). Changes in the mRNA population during differentiation of Dictyostelium discoideum. Devel. Biol. 78:285.
7. Blumberg D.D., Lodish H.F. (1981). Changes in the complexity of nuclear RNA during development of Dictyostelium discoideum. Devel. Biol. 81:80.
8. Blumberg D.D., Margolskee J.P., Barklis E., Chung S.N., Cohen N.S., Lodish H.F. (1981). Specific cell-cell contacts are essential for induction of gene expression during differentiation of Dictyostelium discoideum. Proc. Natl. Acad. Sci. USA. in press.
9. Chung S., Landfear S.M., Blumberg D.D., Cohen N.S., Lodish H.F. (1981). Synthesis and stability of developmentally regulated Dictyostelium mRNAs are affected by cell-cell contact and cAMP. Cell 24:785.
10. Coloma A., Lodish H.F. (1981). Synthesis of spore- and stalk-specific proteins during differentiation of Dictyostelium discoideum. Devel. Biol. 81:238
11. Dimond R., Loomis W.F. (1976). Structure and function of β-glucosidases in Dictyostelium discoideum. J. Biol. Chem. 251:2680.
12. Firtel R.A., Lodish H.F. (1973). A small nuclear precursor of messenger RNA in the cellular slime mold Dictyostelium discoideum. J. Mol. Biol. 79:295.
13. Free S., Schmike R., Loomis W.F. (1976). The structural gene for α-mannosidase-1 in Dictyostelium discoideum. Genetic 84:159.
14. Geltosky J.D., Weseman J., Bakke A., Lerner R.A. (1979). Identification of a cell surface glycoprotein involved in cell aggregation in Dictyostelium discoideum. Cell 18:391.

15. Gross J.D., Town C.D., Brookman J.J., Jermyn K.A., Peacey M.J., Kay R.R. (1981). Cell patterning in Dictyostelium. Philosophial trans. R. Soc. Lond. in press.
16. Guyette W.A., Matusik R.J., Rosen J.M. (1979). Prolactin-mediated transcriptional and post-transcriptional control of Casein gene expression. Cell 17:1013.
17. Hereford L.M., Rosbash M. (1977). Number and distribution of polyadenylated RNA sequences in yeast. Cell 10:453.
18. Kimmel A.R., Firtel R.A. (1980). Intervening sequences in a Dictyostelium gene that encodes a low abundance class mRNA. Nucl. Acids Res. 8:5599.
19. Landfear S.M., Lodish H.F. (1980). A role for cyclic AMP in expression of developmentally regulated genes in Dictyostelium discoideum. Proc. Natl. Acad. Sci. USA 77:1044.
20. Loomis W.F. (1975). "Dictyostelium discoideum: A Developmental System" New York: Academic Press.
21. Loomis, W.F. (1978). The number of developmental genes in Dictyostelium. "Birth Defects: Original Article Series XIV" 497.
22. Loomis W.F. (1979). Biochemistry of aggregation in Dictyostelium. Devel. Biol. 70:1.
23. Mangiarotti G., Lefebvre P., Lodish H.F. (1981). Devel. Biol. in the press.
24. Margolskee J.P., Lodish, H.F. (1980). Half-lives of messenger RNA species during growth and differentiation of Dictyostelium discoideum. Devel. Biol. 74:37.
25. Muller K., Gerisch G (1978). A specific glycoprotein as the target site of adheshion blocking Fab. Nature 274:445.
26. Newell P.C., Longlands M., Sussman M. (1971). Regulation of four functionally regulated enzymes during shifts in the developmental program of Dictyostelium discoideum. J. Mol. Biol. 58:541.
27. Newell P.C., Franke J., Sussman, M. (1972). Control of enzyme synthesis by cellular interaction during development of the cellular slime mold Dictyostelium. J. Mol. Biol. 63:373.
28. Nilsen T.W., Weissman S.G., Baglioni C. (1980). Role of 2',5'-oligo(adenylic acid) polymerase in the degradation of ribonucleic acid linked to double-stranded ribonucleic acid by extracts of interferon-treated cells. Biochem. 19:5574.
29. Okamoto K., Takeuchi I (1976). Changes in activities of two developmentally regulated enzymes induced by disaggregation of the pseudoplasmodia of Dictyostelium discoideum. Biochem. Biophys. Res. Commun. 72:739.
30. Okamoto K. (1981). Differentiation of Dictyostelium discoideum cells in suspension culture. J. Gen. Microbiol. in press.
31. Rosen S.D., Kafka J.A., Simpson D.L., Barondes S.H. (1973). Developmentally regulated, carbohydrate-binding protein in Dictyostelium discoideum. Proc. Natl. Acad. Sci. USA 70:2554.
32. Timberlake W.E. (1980). Developmental gene regulation in Aspergillus nidulans. Devel. Biol. 78:479.
33. Volloch V, Housman D. (1981). Relative stability of globin and nonglobin mRNA in terminally differentiating friend cells. Cell in press.
34. Williams J.G., Lloyd M.M. (1979). Changes in the abundance of polyadenylated RNA during slime mould development measured using cloned molecular hybridization probes. J. Mol. Biol. 129:19
35. Williams J.G., Lloyd M.M., Devine, J.M. (1979). Characterization and transcriptional analysis of a cloned sequence derived from a major developmentally regulated mRNA of Dictyostelium discoideum. Cell 17:903.
36. Williams J.G., Tsang A.S., Mahbubani H. (1981). A change in the rate of transcription of a eukaryotic gene in response to cyclic AMP. Proc. Natl. Acad. Sci. USA 77:7171.

ADHESION OF DICTYOSTELIUM DISCOIDEUM CELLS TO SUGAR DERIVATIZED POLYACRYLAMIDE GELS

Salvatore Bozzaro§ and Saul Roseman

Dept. of Biology and McCollum-Pratt Inst.
The Johns Hopkins University
Baltimore, Md 21218 USA

Immobilized ligands that interact with cell surface receptors are potentially important tools in attempts to understand the behavior of intact cells. We have used this approach to obtain direct evidence for interactions between cell surface and carbohydrates on apposing surfaces.

The rationale for this study was furnished by the suggestion (Roseman,1970) that complex carbohydrates mediate cell-cell recognition. Therefore, cells should possess receptors capable of interacting with these carbohydrates. The same receptors, eventually, may recognize immobilized analogs of the macromolecules.

It must be emphasized that this approach involves two assumptions. The first is that the immobilized mono- and disaccharides can indeed be recognized by membrane receptors usually interacting with larger and more complex carbohydrates. The second is that, following recognition, a strong binding of the receptor to the immobilized ligand occurs, such that the cells will be hold in place during the subsequent washing procedure.

---

§ Present address: Max Planck Institut für Biochemie, 8033 Martinsried bei München, BRD.
This work was supported by a Fellowship of the Cystic Fibrosis Foundation (USA) to S.B.

PROPERTIES OF DERIVATIZED POLYACRYLAMIDE GELS

Derivatized polyacrylamide gels were prepared by the method of Schnaar et al.(1978) with some minor modifications. 6-Aminohexyl O- or S-glycosides were treated with acryloyl chloride and the resulting derivatives were copolymerized with acrylamide and bisacrylamide(Fig.1).

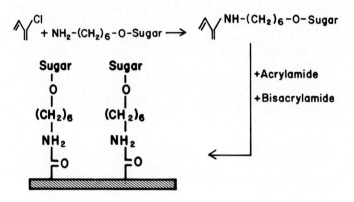

Fig.1:Synthesis of derivatized polyacrylamide gels.

Since the sugar is linked by an amide bond to the matrix, the resulting gels are stable over a wide pH range. Moreover, the ionic charges are very low: in the gels used in these experiments the anionic charges did not exceed 0.6 nmoles per gel; the cationic charges were virtually absent.

The gels were cast in small, flat squares(0.64 cm^2 x0.25mm) to allow visual examination of cell adhesion and rapid quantitation of binding. They were not toxic to the cells so far tested.

ADHESION OF DICTYOSTELIUM DISCOIDEUM TO SUGARS

In the present studies the ability of D.discoideum cells to adhere specifically to a variety of immobilized mono- and disaccharides was investigated. Cells at different developmental stages, from

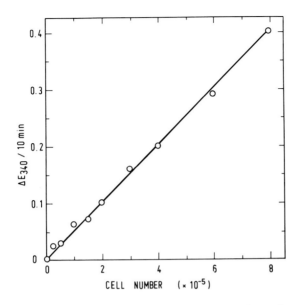

Fig.2: Determination of cell number by the alanine transaminase assay.

D. Discoideum cells were adjusted to 1×10^7 cells per ml by counting in a hemocytometer. An aliquot was diluted 1:10 with 0.2% Triton X-100 in 0.1 M Tricine buffer, pH 7.6. Aliquots of the lysate were transferred to 1-ml plastic cuvettes, the volume adjusted to 0,8 ml with lysis buffer and alanine transaminase assayed by adding 0.1 ml of a mix containing: 4.4 mg L-alanine, 0.05 mg α-ketoglutarate, 0.03 mg lactic dehydrogenase, 0.08 mg NADH in Tricine buffer. The oxidation of the NADH was monitored by measuring the decrease of absorbance at 340 nm in a Beckman spectrophotometer.

growth phase to late aggregation stage, were incubated on the surface of the gels. The gels were washed and the number of adhering cells was determined by the alanine transaminase assay (Fig.2).

Growth phase cells were found to adhere to glucose, maltose, cellobiose, N-acetylglucosamine and mannose (fig.3,A). Aggregation-competent cells fai

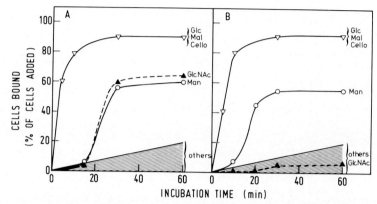

Fig.3: Adhesion of D.discoideum growth-phase(A) or aggregation competent cells(B) to sugar-derivatized gels.

Derivatized gels were equilibrated in 0.017 M Sörensen phosphate buffer pH 6.2 supplemented with 0.06 M NaCl, blotted dry and placed in a circular array in 60-mm Falcon plastic dishes. $2,5 \times 10^5$ cells just washed free of bacteria(A) or previously shaken in suspension for 6 hours to reach aggregation competence(B) were incubated on the gels at 23°C for the time indicated. At the end of the incubation, 10 ml of phosphate-NaCl buffer were gently pipetted at the center of each dish and the dishes swirled at 125 rpm for 1 min on a gyrotory shaker. Thereafter, the buffer was aspirated, the gels were transferred to Linbro wells containing 1 ml of lysis buffer and the number of bound cells was determined by measuring the alanine transaminase released after cell lysis. The abbreviations used are(all D-sugars): Glc,glucose;Mal,maltose;Cello,cellobiose;GlcNAc,N-acetylglucosamine;others:galactose,lactose,aminohexanol.

led to bind to N-acetylglucosamine, while binding to the other sugars was essentially unaltered (Fig.3,B).

Further experiments showed that the ability of the cells to recognize N-acetylglucosamine clearly declined to negligible levels between early and la te aggregation.

In earlier work, Weigel et al.(1979) showed that chicken and rat hepatocytes bound respectively to N-acetylglucosamine and galactose with a typical threshold response to small changes in the concentration of glycoside incorporated in the gel.

Similar results were obtained with D.discoideum cells. As shown in Fig.4, below a concentration of 0.2 μmoles of glucose immobilized in the gel, cell binding was no longer detected. At values above this concentration binding increased rapidly and reached a maximum at a concentration of 0.6 μmoles.

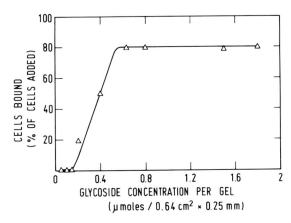

Fig.4: Response of D.discoideum cells to varying concentrations of glucose linked to the gels.
Gels containing different concentrations of glucose were prepared by varying the concentration of acryloylated derivatives during the copolymerization procedure (Fig.1, second reaction). Cells were incubated on gels as described in Fig.3. The incubation time was 30 min.

Preliminary experiments with N-acetylglucosamine and mannose gave similar binding response curves, though the critical concentration values differed greatly for each sugar.

An interesting result with the last two sugars was the finding that even at concentrations of gly coside well above the critical one required for optimal binding, cell adhesion was consistently lower than to glucose gels. This result raises the question whether the ability of D.discoideum cells to adhere to N-acetylglucosamine and mannose, in contrast to binding to glucose, is restricted to a cell subpopulation only. Experiments are in progress to answer this question.

Binding of D.discoideum cells to glucose-derivatized gels was inhibited by free glucose, glucosides and by disaccharides containing glucose at the non reducing terminus. N-acetylglucosamine, mannose, galactose and their derivatives were non inhibitory at the concentrations tested (Table 1).

Table 1: Inhibition of D.discoideum cell binding to glucose gels by carbohydrates.

Cells were incubated with the sugar solution just before transferring onto the gels. Incubation time was 20 min. Cell binding was measured as described in Fig.3.

Carbohydrates tested	Inhibition at 25 mM	Concentration for 50% inhibition
I-Strong inhibitors	%	mM
Glucose		
Methylβ-D-glucoside		
Phenylβ-D-glucoside		
Phenylα-D-glucoside		
Maltose	90-100	1-3
Cellobiose		
Trehalose		
Gentiobiose		
Sucrose		
II-No inhibitors		
N-acetylglucosamine		
Mannose		
L-glucose		
Galactose		
Lactose	0-10	>25
α-D-melibiose		
Methylα-D-galactoside		
Methylα-D-mannoside		
2-Deoxy-glucose		

An interesting result of the inhibition studies was the absence of anomeric specificity of the cell surface receptors. The same lack of anomeric specificity was already observed in the binding to the gels. In fact, cells bound equally well to glucose, maltose and cellobiose gels(Fig.3). Glucose, however, is immobilized as β-glucoside, cellobiose is β-linked at the terminal glycosidic bond, while maltose is α-linked at this position.

The absence of anomeric specificity may prove useful in attempts to characterize the cell surface receptor and distinguish it from other glucose-binding molecules. Glycosides and glycosyltransferases f.e. show a high degree of anomeric specificity.

Inhibition studies with N-acetylglucosamine and mannose were complicated by the fact that maximal cell binding to these sugars required about 40 min of incubation(Fig.3). Higher concentrations of sugar inhibitors were however required in order to block binding for this prolonged period, with apparently deleterious effects on the cells.

For shorter incubation times, a sugar specific inhibition could be observed but quantitative data were difficult to reproduce, because of the variability of the lag period from one experiment to the other.

Is D.DISCOIDEUM BINDING TO SUGARS MEDIATED BY THREE RECEPTORS?

Despite the lack of unequivocal quantitative data in inhibition studies with N-acetylglucosamine and mannose gels, the following lines of evidence suggest that three different receptors are involved in D.discoideum cell adhesion to respectively glucose, N-acetylglucosamine and mannose gels:
(a) the binding kinetics discussed above clearly show relevant differences in both the rate and the extent of adhesion to each of the three sugars;
(b) the inhibition studies show that neither N-acetylglucosamine nor mannose interfere with binding to glucose gels;
(c) binding to N-acetylglucosamine only appears to

be developmentally regulated between the growth phase and the aggregation phase.

Finally, other experiments, not discussed here, show that protein synthesis is differentially required for binding to each of the three sugars to occur.

CONCLUSION

We have shown in these studies that D.discoideum is able to recognize immobilized sugars apparently by means of three different receptors for respectively glucose, N-acetylglucosamine and mannose.

The biological function of these receptors remains to be established. The possibility that they are involved in cell-cell recognition is likely. In fact, glucosyl-, mannosyl-, N-acetylglucosaminyl-moieties are apparently available on the cell surface of D.discoideum cells(Weeks,1975;Reitherman et al,1975). Interactions of these groups with the receptors on apposing surfaces may occur.

Recently, evidence has been presented that in D.discoideum a glucose-binding receptor may be involved in the EDTA-sensitive, "contact sites B"-mediated cell cohesion(Marin et al,1980). A glucose-binding receptor has also been involved in bacterial phagocytosis(Vogel et al,1980).

Moreover, all contact sites so far identified in slime mold possess carbohydrates moieties(Müller et al,1979; Steineman et al,1980; Bozzaro et al, 1981). In some cases, the integrity of the carbohydrates moiety has been shown to be required for the activity of the contact sites(Müller et al,1979;Bozzaro et al,1981;Bozzaro et al,submitted). Nevertheless, its real function is still unknown. The model described in this paper may prove useful in solving this problem.

An unexpected result of our experiments was the lack of cell binding to galactose and lactose, even when aggregation competent cells were tested.

This result is surprising in view of the fact that two galactose-binding proteins, discoidin I and II, have been shown to be present on the cell surface of aggregation-competent cells (for a re-

cent review see Barondes,1981).
Many possible explanations for these negative results are possible. We would like to stress one of these possibilities, namely that cells may interact differently with immobilized ligands than when the same ligands are presented to the cells in solution. In this case, the mere existence of a given receptor on the cell surface may not be sufficient for adhesion to immobilized putative ligands to take place.
Other factors, such as f.e. the conformation and localization of the receptor in the overall architecture of the cell membrane, its mobility, its interaction with neighbouring molecules may play a decisive role. If this happens to be true, then immobilized ligands appear to be a suitable model for cell recognition studies and may eventually be used as a discriminatory test to identify membrane receptors involved in these processes.

REFERENCES

Barondes SH (1981).Lectins: their multiple endogeneous cellular functions. Ann Rev Biochem 50:707

Bozzaro S, Tsugita A, Janku M, Monok G, Opatz K, Gerisch G (1981). Characterization of a purified cell surface glycoprotein as a contact site in Polysphondylium pallidum. Exp Cell Res 134:181

Bozzaro S, Bernstein R, Roseman S. A plasma membrane factor which inhibits cell adhesion in Polysphondylium pallidum. Submitted

Marin FT, Goyette-Boulay M, Rothman FG (1980). Regulation of development in Dictyostelium discoideum.III.Carbohydrate specific intercellular interactions in early development. Dev Biol 80:131

Müller K, Gerisch G,Fromme I, Mayer M, Tsugita A (1979). A membrane glycoprotein of aggregating Dictyostelium cells with the properties of contact sites. Eur J Biochem 99:419

Reitherman RW, Rosen SD, Frazier WA, Barondes SH (1975) Cell surface species-specific high affinity receptors for discoidin: developmental regulation in Dictyostelium discoideum. Proc Natl Acad Sci USA 72:3541

Roseman S (1970). The synthesis of complex carbohy

drates by multiglycosyltransferase systems and their potential function in intercellular adhesion. Chem Phys Lipids 5:270

Schnaar RL, Weigel PH, Kuhlenschmidt MS, Lee YC, Roseman S (1978). Adhesion of chicken hepatocytes to polyacrylamide gels derivatized with N-acetylglucosamine. J Biol Chem 253:7940

Steineman C, Parish RW (1980). Evidence that a developmentally regulated glycoprotein is target of adhesion blocking Fab in reaggregating Dictyostelium. Nature 286:621

Vogel G, Thilo L, Schwartz H, Steinhart R (1980). Mechanism of phagocytosis in Dictyostelium discoideum: phagocytosis is mediated by different recognition sites as disclosed by mutants with altered phagocytic properties. J Cell Biol 86:456

Weeks G (1975). Studies of the cell surface of Dictyostelium discoideum during differentiation. The binding of ^{125}I-Concanavalin A to the cell surface. J Biol Chem 250:6706

Weigel PH, Schnaar RL, Kuhlenschmidt MS, Lee YC, Roseman S (1979) Adhesion of hepatocytes to immobilized sugars. A threshold phenomenon. J Biol Chem 254:10830

THE MOLECULAR BASIS OF SPECIES SPECIFIC CELL-CELL RECOGNITION IN MARINE SPONGES, AND A STUDY ON ORGANOGENESIS DURING METAMORPHOSIS

G.N. Misevic and M.M. Burger

Department of Biochemistry
Biocenter, University of Basel
Klingelbergstrasse 70, CH-4056 Basel

In multicellular organisms cell-cell recognition is a process which plays a crucial role a) during reproduction (sperm and egg recognition), b) during embryonal development (tissue and organ formation), c) in the immune response, and d) in some pathological processes such as organ specific tumor metastasizing. A study about the contribution of the plasma membrane as a carrier of molecules which mediate cell-cell recognition in the processes mentioned above cannot build on much previous knowledge, especially so if molecular explanations are searched for. Even in the best known cases such as the action of cytotoxic lymphocytes, sperm and egg recognition, and the reaggregation of dissociated cells from embryonal tissues, little is known about the molecular mechanism of recognition. One of the systems which is beginning to produce results on the molecular level is that of the reaggregation and reconstruction of marine sponges.

Marine sponges can be easily dissociated into a suspension of single cells, with little cell damage, and without the use of enzymes. In some species these dissociated cells are able to undergo species-specific aggregation, leading eventually to the reconstitution of a functional adult sponge; this can be carried through easily and with only seawater as medium. This simple multicellular organism is particularly appropriate for the study of cellular recognition.

Components necessary for the initial step of species-specific aggregation in marine sponges

The reconstitution of a functional sponge from a single cell suspension is a multistep process (Burger et al; 1980). The initial step, species-specific aggregation into randomly organized groups of cells, was first demonstrated by Wilson in 1907 and Galtsoff in 1925. They showed that two different species of dissociated sponge cells will sort out into clumps, each containing one cell species only. To what degree this sorting out occurs immediately or as a two step process, where cells first adhere randomly and then sort out into homospecific clumps, depends on the two species of sponges used.

During the last two decades, efforts have been made to isolate and characterize molecules mediating such species-specific aggregation in sponges. It was found that for the aggregation of the marine sponge Microciona prolifera a minimum of three components are necessary: a) a proteoglycan-like molecule, called aggregation factor, which can be removed from the cell surface by washing with Ca^{++} and Mg^{++} free seawater (Humphreys, 1963), b) a cell surface receptor for this aggregation factor, the baseplate (Weinbaum and Burger, 1973), and c) Ca^{++} ions, Fig.1.

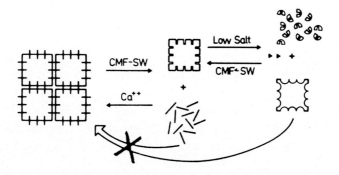

Fig.1. Requirement for the three components for the species-specific reaggregation of sponge cells. Sponge cells dissociated in Ca^{++}/Mg^{++} free seawater (CMF-SW) release aggregation factor (AF). Both AF und Ca^{++} must be added for reaggregation. Treatment of AF-free cells with

low salt releases a receptor or baseplate (BP) which interacts with the AF. Hypotonically treated cells lacking BP do not aggregate in response to AF, but can be restored to do so by preincubation of cells with baseplate.

The Microciona prolifera aggregation factor (MAF) has been characterized as a proteoglycan-like component (Mr = 2.1×10^7) containing 50% protein and 50% carbohydrate by weight (Henkart et al; 1973, Humphreys et al; 1977). Cauldwell et al. demonstrated in 1973 that MAF has 1000-1500 Ca^{++} binding sites. MAF is a fibrous complex molecule appearing on electron micrographs as a "sunburst" with 15-16 arms extending like rays from a circular backbone (Humphreys et al., 1977).

MAF has two functional sites

To investigate the molecular mechanism of MAF promoted species-specific aggregation, the binding of ^{125}I labelled MAF was studied in the following systems: a) homo and heterotypic chemically dissociated cells (cells dissociated through 100 nm nylon mesh and depleted of MAF with CMF-seawater, Humphreys, 1963), b) homo and heterotypic mechanically dissociated cells (cells dissociated in Ca^{++} and Mg^{++} containing seawater and with a lot of MAF on their surface, and c) MAF-substituted agarose beads.

The binding of ^{125}I MAF to chemically dissociated cells showed that the MAF-receptor interaction is species-specific (see Tables 1 and 2), saturable, and that it is not directly Ca^{++} dependent (Jumblatt et al; 1980). In spite of the fact that MAF does bind to the cells in the absence of Ca^{++}, this will not result in aggregation. It is only when Ca^{++} is added back to 10 mM final concentration (seawater concentration) that the cells will aggregate. A model system (Weinbaum and Burger, 1973) was used to investigate whether the Ca^{++}-promoted association between MAF molecules could provide sufficient adhesive forces for cellular aggregation. To investigate this MAF was covalently coupled to Sepharose beads. Under the standard rotary aggregation conditions

(Humphreys, 1963, and Burger and Jumblatt, 1977) the beads coated with MAF aggregated only when the Ca^{++} concentration was equal to or higher than 10 mM. Aggregation was also enhanced by the addition of exogenous MAF. No aggregation was observed in a control experiment with BSA-coated beads, either in the presence of 10 mM Ca^{++} or with addition of exogenous MAF (Jumblatt et al; 1980).

Jumblatt et al., (1980) showed that the binding of ^{125}I MAF to MAF-coated beads would start only in the presence of 10 mM, or higher, Ca^{++} concentration. This binding appeared to be non-saturable, which was in agreement with the already known fact that MAF has 1000-1500 Ca^{++} binding sites and that it can be precipitated by addition of Ca^{++} over 10 mM (Cauldwell et al; 1973, and Henkart et al., 1973). These results indicated that besides the species-specific Ca^{++}-independent MAF cell-receptor interaction sites, Ca^{++} dependent MAF-MAF sites are also present on the MAF molecule. This hypothesis was also fully confirmed by the functional dissociation of the two active sites. Treatment of MAF with 10 mM EDTA for 20 min., heating for 10 min at 50°C or incubation with 5 mM Na periodate for 3 h drastically abolished the MAF aggregation activity, both for cells and for MAF coated beads. None of these treatments, however, altered the cell-receptor interaction site of MAF (Jumblatt et al., 1980, and Table 1).

The existence of both Ca^{++}-independent cell-receptor binding sites and MAF-MAF Ca^{++}-dependent interaction sites also agrees with the fact that binding of ^{125}I MAF to chemically dissociated Microciona cells in the absence of Ca^{++} is saturable; this is opposed to the nonsaturable binding in presence of 10 mM Ca^{++} (Jumblatt et al., 1980). Finally, the binding of ^{125}I MAF to mechanically dissociated Microciona cells with their surface covered with MAF was also nonsaturable in the presence of 10 mM Ca^{++}, and almost no binding was observed in this case in the absence of Ca^{++} (Misevic, unpublished observation).

Table 1. Effect of MAF Pretreatment on Cell or Bead Aggregation Compared to the Binding to Cells and MAF-Conjugated Beads

	pretreatment of ^{125}I MAF[a]	cell or bead aggregation after incubn. with ^{125}I MAF	^{125}I MAF bound[b] (cpm)
Microciona cells	none	+++	20 481
	50°C, 10 min	+	18 024
	10mM EDTA, 30 min	+	31 473
	5mM periodate, 3 h	-	15 358
	baseplate, 2.2 µg/ml	+	5 650
Cliona cells	none	-	2 162
	50°C, 10 min	-	1 324
	10mM EDTA, 30 min	-	1 642
	5mM peridate, 3 h	-	1 510
MAF-conjugated beads	none	+++	25 919
	50°C, 10 min	+	8 381
	10mM EDTA, 30 min	+	9 014
	5mM periodate, 3 h	+	5 501
	baseplate, 2.2 µg/ml	+	5 965
BSA-conjugated beads	none	-	2 537

a) Following the pretreatment, the treated Microciona ^{125}I MAF was dialyzed against 1000 volumes of CaCMFT for 4 h to remove residual EDTA or periodate and was then used for binding assays. Where indicated, the ^{125}I MAF was incubated with baseplate (2.2 µg of protein per ml) for 10 minutes prior to addition of cells or conjugated agarose beads to initiate ^{125}I MAF binding. b) ^{125}I MAF, pretreated as indicated, was incubated at 1 unit/ml with glutaraldehyde-fixed Microciona or Cliona cells (10^7 cells/ml) or with coupled agarose beads 10^6 beads/ml). After 20 minutes of incubation, cell- or bead-associated radioactivity was determined. Data from Jumblatt et al. (1980).

We have shown that the MAF cell-receptor interaction sites carry species-specificity, but it is only together with intact MAF-MAF sites that MAF can promote cell aggregation. We therefore asked whether there is species-specificity in the MAF-MAF type of interaction.

As regards the specificity of the Ca^{++}-dependent MAF-MAF site we have tested the binding of ^{125}I MAF to homo and heterotypic mechanically dissociated cells, in the presence of Ca^{++} ions. When Microciona and Halichondria cells are compared, the binding of ^{125}I MAF appears to be six times higher towards the homotypic species, although binding to Halichondria mechanically dissociated cells in the presence of Ca^{++} was twice as high as to Halichondria chemically dissociated cells tested in the absence of Ca^{++} (Misevic, unpublished observation). The increase of unspecific binding in the presence of Ca^{++} is expected since secondary binding of MAF to already non-specifically bound MAF will occur via Ca^{++} dependent MAF-MAF interactions. One also has to keep in mind that during binding assays in the presence of Ca^{++}, the mechanically dissociated cells are aggregating and thereby unspecifically trapping various molecules (such as MAF) in the aggregates.

In order to assess further to what degree such species-specificity between AF-AF interaction sites exists we are currently preparing AF's from different species, which will be coupled as solid surfaces and to agarose beads; binding of homo or heterotypic ^{125}I AF's will then be tested again on such solid substrates. Such an approach may explain the differences in the degree of the initial aggregation specificity when mechanically dissociated sponge species are used.

We showed recently that several positively charged polymers could substitute for Ca^{++} in the MAF-MAF type interaction (Burkart and Burger, J. Supramol. Structure, in press). The efficiency was not the same for all polycations, and the effect seems to depend on the spacing between charges. It is possible therefore that some or all of the specificity in the AF-AF interaction is carried by the regular spacing of charges in the homotypic AF's, providing a good fitting of sites bridged by Ca^{++}. Such a model does not require a high selectivity for the single site, since a multitude of weak interactions based on evenly spaced charges can provide sufficient force as well as increase the specificity.

Isolation of a Cell-Specific Binding Site

In order to investigate the mechanism and chemical nature of the MAF-cell surface receptor interaction, it is necessary to isolate the smallest functional fragment from the large MAF complex. In a first approach we tried to dissociate MAF into subunits. Since Humphreys had shown in 1977 that after four weeks in the presence of EDTA, MAF will give rise to two types of subcomponents, the circular backbone and the arms or arm fragments, we searched for agents and conditions which could rupture the complex more efficiently. This effort resulted in a simultaneous treatment with urea, EDTA and heat. MAF dissociated in such a way produced two subcomponents after 4 h already, and with reasonable yields (Jumblatt et al., 1978 and Burger et al., 1980).

Binding of the isolated MAF subcomponents (UEP1, Mr = 15×10^6 and UEP2, Mr 1.5×10^4; Urea, EDTA, treatment leading to peak 1 and 2) to homo and heterotypic chemically dissociated cells showed that both still contain the species-specific cell-receptor binding site (Table 2, see also Burger et al., 1980 and Misevic et al., J. Biol. Chem. in press). Chemical analysis revealed that the weight ratio of protein to neutral hexoses and to uronic acid was the same as in the intact MAF, namely 50:40:10; Since the polyacrylamide-SDS-electrophoresis pattern of the UEP1 and 2 was rather complex and their size large, we continued to search for the smallest possible, functionally still active cell receptor binding fragment by means of protease digestion of dissociated MAF.

After trypsin digestion four proteolytically generated

fragments of apparent molecular weights: 1.2×10^5, 7×10^4, 2.7×10^4, 5×10^3 were separated on a Sephacryl S-200 column, (Fig. 2 A) and found to be fairly pure on polyacrylamide SDS electrophoresis (Fig. 2 B). Each one of the four isolated fragments showed still some species-specific binding to chemically dissociated cells supporting the notion that the sites were functionally still active (Table 2). In this case the ratio of protein to carbohydrate was the same in all of the four fragments, but somewhat lower when compared with the ratio in MAF or urea-EDTA subcomponents.

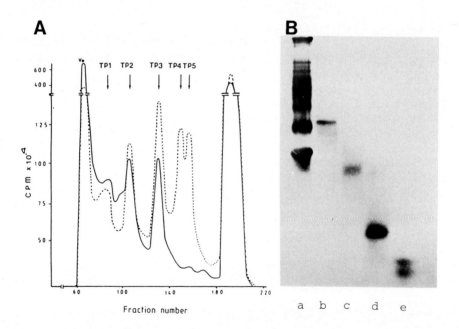

Figure 2 A: Gel filtration on Sephacryl S-200 of trypsin treated dissociated ^{125}I - MAF.
0.05 mg/ml of ^{125}I MAF was treated with trypsin (final concentration 2 mg/ml) for a period of 2 h (————) and 24 h (------). At the end of the incubation time 0.5 mg of MAF was applied to a Sephacryl S-200 column. The column was eluted with 0.25 M NH_4HCO_3 and 1 ml fractions were collec-

ted, counted in a Packard gamma spectrometer and analyzed for protein, neutral hexoses and uronic acid (From Misevic et al; in press, J. Biol. Chem.).
B) SDS electrophoresis on a 7.5 - 20% linear gradient polyacrylamide gel, and autoradiography of ^{125}I labelled MAF tryptic fragments. a) 5.0 µg void volume fraction 22 x 10^4 cpm/µg; b) 2.0 µg TP1 7.2 x 10^4 cpm/µg; c) 2.0 µg TP2, 7.2 x 10^4 cpm/µg; d) 2.9 µg TP3, 6.3 x 10^4 cpm/µg; e) 1.5 µg TP4, 14.6 x 10^4 cpm/µg. After staining and drying the gel was autoradiographed on a NS-2T Kodak X-ray film (From Misevic et al; in press, J. Biol. Chem.).

Table 2. Binding Specificity of ^{125}I -MAF

MAF Fragments	Number of Molecules Applied per Assay x 10^{-12}	% bound to		
		Microciona prolifera	Cliona celata	Mycale fusca
MAF	0.1	33	2.2	2.1
UEP1	0.3	59	2.7	2.3
UEP2	12.7	54	0.6	1.6
TP1	11.4	31	8.1	7.9
TP2	17.8	29	7.3	6.8
TP3	16.1	31	7.8	6.7
TP4	20.0	45	12	11.1

Glutaraldehyde fixed CMF Microciona prolifera, Cliona celata and Mycale fusca cells (10^7 cells ml $^{-1}$) were incubated in CMFT-SW (Ca^{++}, Mg^{++} free seawater buffered with 20 mM TRIS pH 7.4) with serial dilutions of the radioiodinated MAF or fragments thereof under standard assay conditions for 20 minutes at 22°C. After incubation cells were layered on and centrifuged through 0.1 % BSA, 10 % sucrose in CMFT-SW. Supernatants were aspirated and the pellet was counted in a Packard gamma spectrometer (for details see Jumblatt et al., 1980). UEP = urea, EDTA treated MAF fractionated into peak 1 (UEP1) and peak 2 (UEP2). TP = trypsin treated MAF fractionated on Sephacryl S-200 into peaks TP1 to TP4.

We have also tested whether our fragments retain the active MAF-MAF interaction site: the ^{125}I labelled fragments did not bind to MAF coated beads, either in the presence or absence of Ca^{++}, nor were the fragments capable of inhibiting Ca^{++}-dependent binding of MAF to the MAF beads (Misevic et al; J. Biol. Chem., in press). These results confirm the phenomenon described above, i.e. the irreversible inactivation of the MAF-MAF site by urea-EDTA-heat treatment.

Further support for functional identity of the fragment with the active sites of the intact MAF would come from competition of the two for the same receptor on the cell surface. We therefore monitored the extent to which the fragments would inhibit the MAF-promoted aggregation or the binding of MAF to chemically dissociated cells in the absence of Ca^{++}. Since each one of the six fragments showed inhibitory activity for aggregation and for binding the similarity of the native site is preserved among all six fragments. (Misevic et al; J. Biol. Chem., in press).

A Scatchard analysis of the specific binding of fragments to homotypic chemically dissociated cells revealed a direct relationship between fragment size and the binding affinity, as well as an inverse relationship to the maximal number of binding sites present on the cell surface (Fig. 3, Misevic et al; J. Biol. Chem., in press). The inhibition of MAF-promoted cell aggregation by fragments also decreased with their size when the same number of fragment molecules were used. All of these results support the previous concept that high polyvalent interaction between MAF and the cell surface does permit low affinities of single sites and still provides high avidity and specificity,(Burger and Jumblatt, 1977; Burger, 1979).

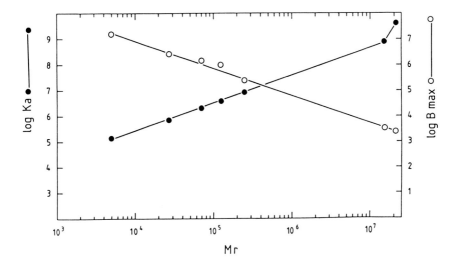

Figure 3. Binding of ^{125}I MAF and ^{125}I MAF fragments as a function of their molecular weight. Glutaraldehyde fixed CMF <u>Microciona prolifera</u> cells (10^7 cells/ml) were incubated in CMFT-SW with serial dilutions of the radioiodinated MAF or fragments under standard assay conditions for 20 minutes at 22°C. Starting amounts per assay were the same as those given in Table 2. After incubation cells were layered on and centrifuged through 0.1% BSA, 10% sucrose in CMFT-SW. Supernatants were aspirated and the pellet counted on a Packard gamma spectrometer (for details see Jumblatt et al., 1980). Binding data were analyzed with a Scatchard plot and K_a's and B_{max}'s were calculated.

This hypothesis about the high valency of the cell-receptor binding site of MAF is also supported by the finding that the fragments occur most likely as repeating units in the intact MAF. A minimal number of theoretical binding sites per MAF complex is 4'200 and this calculation is based on the assumption that TP4 has one site only. Not all 4'200 sites will however be likely to make baseplate or receptor contact but the actual number of contact sites must be considerably less. Judging from the amount of each individual fragment recovered, their molecular weight, and the molecular weight of MAF, the minimal repeating number of fragments is TP1=20, TP2=48, TP3=206, TP4=560 times (Misevic unpublished observation).

Turner and Burger showed in 1973 that aggregation can be specifically inhibited when glucuronic acid is preincubated with the cells. Moreover, β-glucuronidase treatment of MAF decreased its aggregation efficiency. Jumblatt <u>et al</u>. found however (1980) that glucuronic acid failed to inhibit the direct binding of MAF to the chemically dissociated cells. In view of our present results, if glucuronic acid is involved in the MAF-receptor interaction through a lectin-like interaction, as earlier proposed, it would have most likely a relatively weak affinity for the cellular site, and a very high concentration would be required to block the binding of intact MAF, in order to overcome the polyvalence i.e. the avidity. On the other hand glucuronic acid could be necessary for the MAF-MAF type of interaction and thus when added to cells it would specifically inhibit cell aggregation. Still a third hypothesis has to be considered, i.e. that glucuronic acid may play an important role in both types of interaction, MAF-receptor and MAF-MAF. In a first approach to solve this problem inhibition of MAF-interaction with MAF attached to beads should be assessed with glucuronic acid. Further insight should come from an investigation of the chemical basis of the minimal site of interaction between MAF and baseplate receptor, as well as that between MAF and MAF; this should then be followed by further analysis and modification of the smallest protease generated fragments of MAF.

Our present minimal model of species-specific aggregation in the marine sponge <u>Microciona prolifera</u> is shown in Fig. 4. It would involve two processes: a) the species-specific binding of MAF to the cell receptor, where the high valency of the cell binding site will not only guarantee specificity, but will also explain the need for the large size of MAF, b) the Ca^{++} dependent interaction between MAF molecules through MAF-MAF sites finally leading to a linking bridge between adjacent cells. The specificity of this latter interaction would depend on the presence of a multitude of appropriately spaced interaction sites, whereas the selectivity for each individual site would be low.

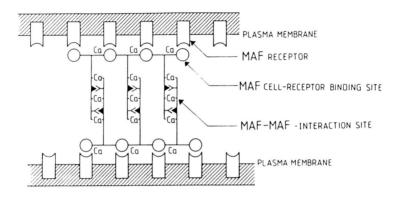

Fig.4. Minimal model of species-specific aggregation in the marine sponge Microciona prolifera.

Organogenesis during metamorphosis

The viviparous sponge Microciona prolifera releases flagellated larvae, about 300 μm in diameter, during late spring. After a short period of free swimming, a larva will adhere to a rock or a shellfish and undergo metamorphosis, resulting in a minute sponge. Histological studies of larvae showed that flagellated epithelial cells surround the core cells which contain most of the adult cell types; these are however not yet organized as tissues. During the process of metamorphosis, which lasts about 4-6 days, the adult tissues, like flagellated chambers, exhalent canals, oscula, and an epithelial layer of pinacocytes are formed.

One of the most interesting events during metamorphosis is the formation of **the chief organ of a sponge**, the flagellated chambers which take in food and pump water through the canal system. In most sponge species it is thought that the **choanocytes**, cells forming the chambers, are derived from the superficial layer of flagellated larval cells which have migrated into the interior of the core cell mass. (Hadzi, 1963; Borojevic, 1966, Ivanov, 1971, Bergquist, 1978). This so-called "inversion of layers" was used to separate sponges from other primitive metazoans, forming the subkingdom of Enantiozoa. We have investigated chamber formation during the larval metamorphosis of the marine sponge <u>Microciona prolifera</u>. Three different approaches were used: a) time lapse cinematography, b) detecting of <u>de novo</u> cell formation via cell division by means of ^{3}H thymidine incorporation, followed by autoradiographic histologic studies, and c) monitoring the fate of the surface labelled flagellated epithelial cells of the larvae during the process of metamorphosis.

Time lapse **cinematography** and histology showed that after larval attachment to the culture vessel wall, spreading proceeds rapidly, together with disappearance of the flagellated cells. Surprisingly, there was no evidence that superficial cells or cell clusters were moving into the core cell mass. The first cell clusters that are transformed into the chambers, built by choanocytes, were visible 24-48 hours after spreading (Burkart, Misevic and Burger, manuscript in preparation).

In order to see whether choanocytes are formed <u>de novo</u> larvae were incubated with ^{3}H thymidine followed by radiography of histological sections of a developing spongelet. Analysis of the sections showed that the grain density over the choanocytes was **seven** times higher than over the other cells. This result **suggests** that flagellated cells of larvae do not simply migrate into the spongelet to become choanocytes without division but that the choanocytes probably derive from stem cells during metamorphosis. Further evidence for the idea that sponge metamorphosis is more than a simple regrouping of already differentiated cells comes from experiments where blockers of DNA and mRNA synthesis were used. We found that the process of metamorphosis and chamber formation could be completely stopped by each **of the two types of blockers.**

Additional and more direct evidence against the hypothesis of "inversion of layers" is the experiment where the fate of the larval epithelial cells labelled with ^{125}I, using the IODO-GEN method was followed. The labelling procedure did not influence the time course of larval settlement or metamorphosis. Autoradiographic studies showed that the label remains on the outermost flagellated cells until the onset of spreading. In the early phase of spreading we detected that a larva loses 40% of the total trichloroacetic acid precipitable counts. At the moment the chambers are formed essentially no label could be detected in the choanocytes. The remaining iodine-marked material could be seen in small areas of mesohyl.

Our observations lead us to conclude that in the metamorphosis of the marine sponge Microciona prolifera, the choanocytes that eventually build the flagellated chambers do not arise from the flagellated epithelial cells, but are formed de novo from stem cells (Fig. 5).

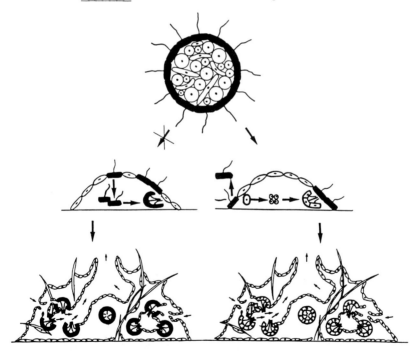

Fig. 5. Formation of the chamber organ. Two possible

pathways. Left, chambers are formed from the flagellated epithelium. Right, chambers are formed de novo from stem cells.

This work was supported by the Swiss National Foundation for Scientific Research, Grant 3.513-0.79 as well as the Ministry of the City and Canton of Basle.

REFERENCES

Bergquist PR (1978). "Sponges". University of California Press, Berkeley and Los Angeles.
Borojevic R (1966). Etude expérimentale de la différentiation des cellules de l'éponge au cours de son développement. Develop. Biol. 14:130.
Burger MM (1979). Early events of encounter at the cell surface. In Nicholls JG (ed). "The Role of Intercellular Signals: Navigation, Encounter, Outcome". Berlin: Dahlem Konferenzen, p 119.
Burger MM and Jumblatt JE (1977). Membrane involvement in cell-cell interactions: A two-component model system for cellular recognition that does not require live cells. In Lash JW and Burger MM (eds). "Cell and Tissue Interactions". Ràven Press, New York, p 155.
Burger MM, Misevic GN, Jumblatt J and Mir-Léchaire FJ (1980). Macromolecules mediating cell-cell recognition. In Salanki J (ed) Adv. Physiol. Sci. Vol. 3. "Physiology of Non-excitable Cells", Budapest: Akademia Kiado and Pergamon Press, Oxford, London, New York, Paris p 233.
Burkart W and Burger MM. The contribution of the calcium-dependent interaction of aggregation factor molecules to recognition: A system providing additional specificity forces. J. Supramol. Struct. Cell Biochem. in press.
Cauldwell CB, Henkart P and Humphreys T (1973). Physical properties of sponge aggregation factor. A unique proteoglycan complex. Biochemistry 12:3051.
Galtsoff PS (1925). Regeneration after dissociation: An experimental study on sponges. J. Exp. Zool. 42:223.
Hadzi J (1963). "The Evolution of the Metazoa". Pergamon

Press, Oxford, London, New York, Paris, p 499.

Henkart P, Humphreys S and Humphreys T (1973). Characterization of sponge aggregation factor: A unique proteoglycan complex. Biochemistry 12:3045.

Humphreys T (1963). Chemical dissolution and in vitro reconstruction of sponge cell adhesion. I. Isolation and functional demonstration of the components involved. Develop. Biol. 8:27.

Humphreys S, Humphreys T and Sano Y (1977). Organization and polysaccharides of sponge aggregation factor. J. Supramol. Struct. 7:330.

Ivanov AV (1971). Embryology of sponges (Porifera) and their position in the animal kingdom (in Russian). J. Biol. Gen. URSS. 32:557.

Jumblatt JE, Schlup V and Burger MM (1978). Segments of sponge aggregation factor which bind specifically to homotypic cells. Biol. Bull. 155:448.

Jumblatt JE, Schlup V and Burger MM (1980). Cell-cell recognition: Specific binding of Microciona sponge aggregation factor to homotypic cells and the role of calcium ions. Biochemistry 19:1038.

Jumblatt JE, Weinbaum G, Turner R, Ballmer K and Burger MM (1976). Cell surface components mediating the reaggregation of sponge cells. In Bradshaw RA, Frazier WA, Merrell RC, Gottlieb DI and Hogue-Angeletti RA (eds): "Surface Membrane Receptors". Plenum Publishing Corporation, New York, p 73.

Misevic GN, Jumblatt JE and Burger MM. Cell binding fragments from a sponge proteoglycan-like aggregation factor. J. Biol. Chem. in press.

Turner RS and Burger MM (1973). Involvement of a carbohydrate group in the active site for surface guided reassociation of animal cells. Nature 244:509.

Weinbaum G and Burger MM (1973). A two component system for surface guided reassociation of animal cells. Nature 244:510.

Wilson HV (1907). On some phenomena of coalescence and regeneration in sponges. J. Exp. Zool. 5:245.

AGGREGATION FACTORS OF SEA URCHIN EMBRYONIC CELLS

M.L. Vittorelli, V. Matranga, M. Cervello, H. Noll.

Institute of Comparative Anatomy

University of Palermo. Italy.

Dissociation and reaggregation of sea urchin embryonic cells has received much attention since 1962 when it was shown (Giudice 1962) that sea urchin embryos can be dissociated into single cells, that under appropriate conditions, can reaggregate into cell clumps which later develop into living larvae. In spite of many studies on the subject, little is known on the mechanism that brings about the correct differentiation of the embryo after mixing togheter the cells. Nevertheless, because of their remarkable ability to reaggregate and differentiate, sea urchin embryonic cells are uniquely suitable for the analysis of those cell interactions that might shed light on the largely unknown mechanism of embryogenesis.

Cell interactions modulate several metabolic activities of these cells. At the blastula stage, for instance, cell contacts are needed for DNA synthesis to occour (Sconzo et al. 1970; De Petrocellis and Vittorelli 1975). The block of DNA synthesis in dissociated cells can be released by a mild treatment with trypsin (Vittorelli et al. 1973), which removes from the membrane a high molecular weight glycoprotein (Matranga et al. 1978). The resumption of DNA synthesis induced by trypsin however is not paralleled by a resuption of histone synthesis (Di Liegro et al. 1978) while, in embryos, these two synthesis are coupled. It is therefore likely that trypsin treatment activate DNA synthesis by a mechanism somehow different from the one operating in the intact embryo.

The molecular components mediating cell contacts and recognition are however still unidentified.

A new breakthrough to the approach of this problem has been opened by our discovery (Noll et al. 1979) that a treatment with diluted butanol in sea water extracts from the surface of dissociated cells a factor(s), whose absence prevents the dissociated cells from reaggregating. The addition of the "butanol extract" to the extracted cells, which had lost their

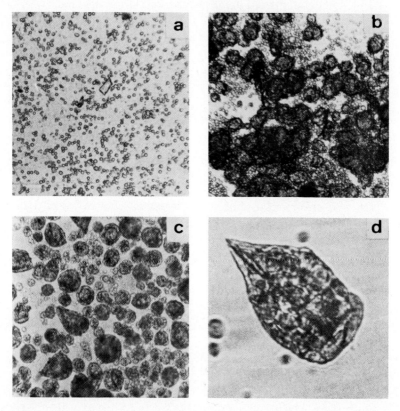

Figure 1.
Reaggregation of butanol extracted cells promoted by butanol extract. 2.5% butanol extracted Paracentrotus lividus cells were reconstituted with the following amounts of extract (µg proteins per .1 ml): (a) none; (b) 18 µg; (c) 21 µg; (d) 5 µg. Photographs were taken after: 7 hrs (a-b); 48 hrs (c); 70hrs (d). Approximate magnification: x 70(a-c); x 300(d).

ability to reaggregate, promotes their aggregation and the differentiation of aggregates into living larvae. Similar results are obtained also with butanol extracts prepared from purified plasma membranes of embryonic cells (Fig. 1). As can be seen in figure 1d, reconstituted larvae often show a skeleton, a digestive tract and an overall shape similar to that of normal plutei. The addition of the butanol extract to dissociated unextracted cells makes aggregation more complete and increases the dimentions of aggregates (Fig. 2).

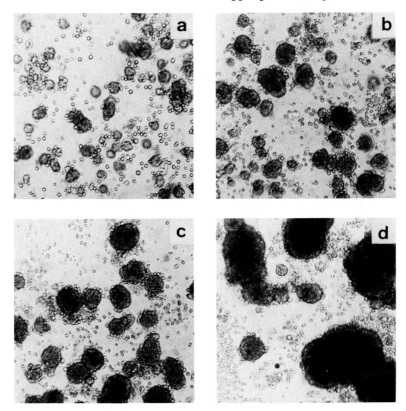

Figure 2.
Reaggregation of unextracted Arbacia lixula blastula cells stimulated by butanol extract. The following amounts of extract from blastla cells were added (µg of proteins per .1 ml culture): none (a); .1 µg (b); .5 µg (c); 4 µg (d). Photographs were taken after 7 hrs. Approximate magnification: x70.

When Fab fragments against plasma membranes are added to dissociated cells, they prevent cell aggregation (Fig. 3). If however Fabs are first mixed with the butanol extract, they loose their inhibitory effect (Fig. 3c). As shown by figure 3e also Fabs against butanol extract are very active on preventing cell aggregation. Moreover both kinds of Fabs can induce cell dissociation in living embryos (Fig. 4). The butanol extract appears therefore to contain one or more aggregation factors.

Fabs against butanol extract prepared from embryos at the blastula stage are active on preventing cell aggregation

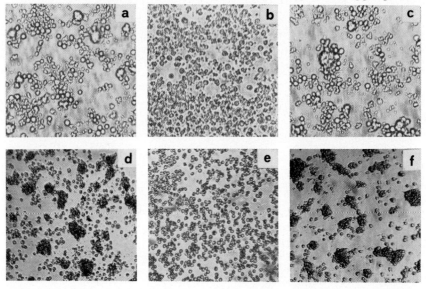

Figure 3.
Effects of Fab fragments on cell aggregation. Paracentrotus lividus (P.l.) cells (a,b,c); Arbacia lixula (A.l.) cells (d, e,f). Additions in .1 ml: (a,d) none; (b) 30 µg Fabs against P.l. blastula plasma membranes; (c) same as (b) but Fabs premixed with 2 µg butanol extract from P.l. blastula plasma membranes; (e) 116 µg Fabs against butanol extract from A.l. blastula plasma membranes; (f) 150 µg Fabs against P.l. blastula plasma membranes. Pictures were taken after 4 hrs. Approximate magnification: x70.

and on causing extensive cell dissociation of embryos also at the pluteus stage (Fig. 4e), and viceversa (Fig. 4b). The aggregation factor extracted by butanol seems therefore to be needed for cell contacts at least until this stage.

As indicated by the work of Mc Clay new surface antigens involved in cell adhesion can be detected in sea urchin embryos after gastrulation (Mc Clay et al. 1978); appearently

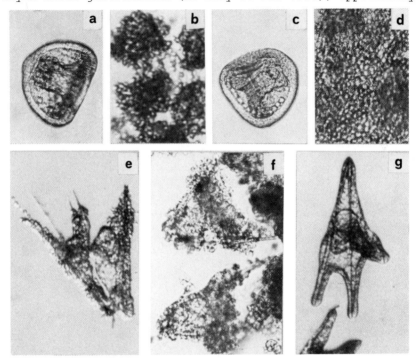

Figure 4.
Effects of Fab fragments on embryos. Paracentrotus lividus (P.l.) blastulae (a-d) and plutei (e,g); Arbacia lixula (A.l.) plutei (f). Additions in .1 ml: (a) none; (b) 57 µg Fabs against butanol extract of P.l. plutei; (c) 185 µg Fabs against butanol extract of A.l. blastulae; (d) 26 µg Fabs against butanol extract of P.l. blastulae; (e) 194 µg Fabs against plasma membranes of P.l. blastulae; (f,g) 186 µg Fabs against butanol extract of A.l. blastulae. Pictures were taken 7.5 (a-d) or 5 (e-g) hrs after Fabs addition. Approx. magnification:x150.

these new aggregation sites are not strong enough to mantain complete cell adhesion when other aggregation sites are blocked.

When Fab fragments against butanol extract prepared from embryos of the sea urchin species Paracentrotus lividus were tested on embryos or dissociated cells of the other mediterranean species Arbacia lixula, they proved unable both to dissociate embryos (Fig. 4c and 4g) and to prevent cell aggregation (Fig. 3f). These results show that butanol extracts contain a species specific aggregation factor.

For a more detailed report of data on the action of Fabs on embryos and dissociated cells, see Noll et al. (1981). In the same paper it is also reported that in spite of the species specific action of Fabs, butanol extracts are active also on promoting reaggregation of etherologus cells.

We can immagine two different explanations for these seeming conflictuary results: a) while being species specific the aggregation factors share enough similarities to be able to interact with plasma membranes of both species; b) the butanol extract contains more than one aggregation factor, and not all the factors are species specific. Purification of the active components would allow to solve this question.

The butanol extract is a rather etherogeneus material containing almost the same amount of proteins and carbohydrates as detected respectively by the Lowry and the Anthrone methods. When analyzed by SDS polyacrylamide gel electrophoresis (Fig. 6a) it resolves in several components, most of which are PAS positive. Table 1 shows how various treatments influence the

Table 1
Effects of various treatments on
the biological activity of the butanol extract

Treatments:	Biological activity:
1 hr 60°C	unchanged
3' 100°C	lost
trypsin digestion	lost
periodate oxidation	partially lost
dialysis	unchanged
lyophilization	unchanged
freezing and towing	unchanged
$(NH_4)_2SO_4$ precipitation	lost

biological activity of the extract. While digestion with trypsin completely unactivate the extract, the oxidation with sodium periodate causes a significant but not complete reduction of its biological activity.

Attemps to purify the active factor(s) were so far only partially succesful. The extract is totally recovered in the void volume of G-150 and G-200 Sephadex columns and does not bind to carboximethyl cellulose. A minor fraction of the extract, which binds to ConA, can be separated from the bulk of the extract by affinity chromatography, but it does not show biological activity on promoting cell aggregation.

A better fractionation is obtained by affinity chromatography on phenyl-sepharose. As shown by figure 5A with this tecnique we can fractionate the extract into three large components. Fraction a, which does not bind to phenyl-sepharose,

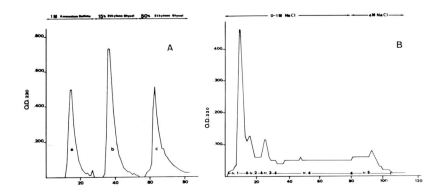

Figure 5.
Fractionation of butanol extract.
A) Phenyl-sepharose affinity chromatography. Step-wise fractionation with the following solutions: 1 M ammonium sulfate (fraction a); 15% ethylen glycol (fraction b); 50% ethylen glycol (fraction c). All solutions were in .01 M phosphate buffer pH 6.8. B) DEAE-sephadex chromatography. Fraction c, obtained from phenyl-sepharose affinity column was fractionated as follows: 0-1 M NaCl continuos gradient in 10 mM Tris pH 8.1; 4 M NaCl pH 5.7.

is completely devoid of biological activity; fraction b has only a very mild stimulatory effect on cell aggregation; fraction c, on the contrary, has a biological activity from two to three times higher than that of total butanol extract.

The electrophoretic pattern of fraction c on SDS polyacrylamide gel is however still very complex (Fig. 6c). This material can be further fractionated by chromatography on DEAE sephadex. We previously attempted to utilize this tecnique for a direct fractionation of entire butanol extract. The

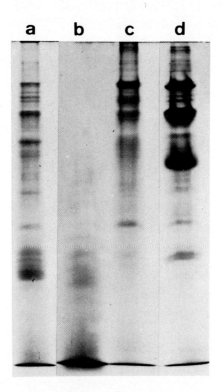

Figure 6.
10% SDS polyacrylamide gel electrophoresis.
a) total butanol extract from Paracentrotus lividus blastula cells; b) fraction b of phenyl-sepharose affinity chromatography; c) fraction c of phenyl-sepharose affinity chromatography; d) fraction 1 of DEAE sephadex chromatography. Gels were performed as described by Laemli.

results we obtained were rather similar to those we are going to describe, but the fractionation patterns were not easily reproducible. DEAE sephadex chromatography applied to material partially purified by phenyl-sepharose chromatography gives more reproducible patterns (Fig. 5B). Biological activity is found partially in a large fraction of material eluted at low ionic strenght (Fig. 5B: fraction 1); and partially in a small fraction (Fig. 5B: fraction 5) with high specific activity. This last fraction is eluted from the column after a 0-1 M NaCl continuous gradient in 10 mM Tris buffer pH 8.1, by the addition of 4 M NaCl pH 5.7. Biological and electrophoretical analysis of this active fraction are still in progress. Since fraction 1 contains several components and the amount of material recovered in fraction 5 is very small, up to now we can not be sure whether these two fractions contain the same aggregation factor or different ones.

Obviously more work will be required to obtain complete purification of the sea urchin embryos aggregation factor(s), solubilized by butanol.

While we proceed by the tecniques which have been described, new tecniques (i.e. immunological methods) are also tested to improve fractionation. Once the factor will be isolated in fact, we should be able to solve several problems.

It will be possible not only to identify its receptors on plasma membranes (if any), but also to verify if indeed the species specific Paracentrotus lividus factor can bind to Arbacia lixula plasma membranes, and viceversa.

Moreover the mechanism by which intercellular contacts can modulate the rate of DNA synthesis and other metabolic activities of the cell could be reinvestigated.

Cell behavior must always be adapted to the enviroment and, in a pluricellular organism, living conditions are greatly modified by the number and kind of surrounding cells. Little is known however about the mechanisms by which a cell translates informations concerning the number of its contacts with cells of different morphogenetic areas into a signal which appropriately modify its metabolic behavior. Isolation of the molecular components involved in cell contacts, is a first important step in this direction.

Butanol extraction has proved successfull in allowing solubilization and purification of contact proteins in several systems (Huesgen and Gerisch 1975; Hausman and Moscona 1976; Muller et al. 1979). In our model system however, the aggregation molecules can be extracted from living cells, that, after reconstitution with isolated factors, can be tested for their aggregation and differentiation properties.

REFERENCES.

De Petrocellis B and Vittorelli ML (1975). Role of cell interactions in development and differentiation of the sea urchin Paracentrotus lividus. Exp Cell Res 94:392.

Di Liegro I, Cestelli A, Ciaccio M and Cognetti G (1978). Block of histone synthesis in isolated sea urchin cells actively synthesizing DNA. Dev Biol 67:266.

Giudice G (1962). Reconstitution of whole larvae from disaggregated cells of sea urchin embryos. Dev Biol 5:402.

Hausman RE and Moscona AA (1976). Isolation of retina-specific cell-aggregating factor from membranes of embryonic neural retina. Proc Natl Acad Sci USA 73:3594.

Huesgen A and Gerisch G (1975). Solubilized contact sites A from cell membranes of Dictyostelium discoideum. FEBS Letters 56:46.

McClay DR (1979). Surface antigens involved in interactions of embryonic sea urchin cells. Current Topics in Dev Biol 13:199.

Matranga V, Giarrusso C, Vasile V and Vittorelli ML (1978). Trypsin treatment which elicits DNA synthesis, removes a high molecular weight glycoprotein from the plasma membrane of sea urchin embryonic cells. Cell Biol Inter Rep 2:147.

Muller K, Gerisch G, Fromme I, Mayer M and Tsugita A (1979). A membrane glycoprotein of aggregating Dictyostelium cells with the properties of contact sites. Eur J Biochem 99:419.

Noll H, Matranga V, Cascino D and Vittorelli ML (1979). Reconstitution of membranes and embryonic development in dissociated blastula cells of the sea urchin by reinsertion of aggregation-promoting membrane proteins extracted with butanol. Proc Natl Acad Sci USA 76:288.

Noll H, Matranga V, Palma P, Cutrono F and Vittorelli ML (1981). Species specific dissociation into single cells of live sea urchin embryos by Fab against membrane components of Paracentrotus lividus and Arbacia lixula. Dev Biol in press.

Sconzo G, Pirrone AM, Mutolo V and Giudice G (1970). Synthesis of ribosomal RNA in disaggregated cells of sea urchin embryos. Biochim Biophys Acta 199:441.

Vittorelli ML, Cannizzaro G and Giudice G (1973). Trypsin treatment of cells dissociated from sea urchin embryos elicits DNA synthesis. Cell Differ 2:279.

THE CEPHALOPOD EGG, A SUITABLE MATERIAL FOR CELL AND TISSUE INTERACTION STUDIES.

Hans Jürg MARTHY
CNRS / Laboratoire Arago
F - 66650 BANYULS-SUR-MER

Morphological and physiological studies on cell and tissue interactions are easier to perform and to interpret in systems in which the cell material is arranged in a "simple architecture", such as the embryonic tissues of Invertebrates (Garrod, Nicole, 1981). Such a remarkably simple architecture is found in the egg and in particular in the pre-organogenetic blastoderm of the Cephalopods. Earlier and current research on Cephalopod development (Marthy 1978/79) has shown the usefulness of this material as a model of cell and tissue interaction phenomena.

Of the various Cephalopod eggs readily available in the Western Mediterranean (Mangold 1963, Boletzky 1974) only the eggs of the squid *Loligo vulgaris* (Fig. 1) are discussed here. For precise observation or experimental work, the gelatineous coatings and the chorion have to be removed with watchmakers forceps. On a translucent egg free of the chorion (Fig. 2) a cellular area can be distinguished from the egg cortex. For details of the spatial organisation of the egg and the blastoderm, an efficient micro-surgery technique is used : after the thin vitellin membrane covering the egg surface is removed with forceps, the egg is placed with the animal pole uppermost, in a small depression made in the agar bottom in a Petri-dish filled with sterile sea water. In this position the blastoderm can be partially or totally removed from the underlying yolk complex and it can be segregated into its components (Fig. 3). Operated eggs are then fixed with 1 % OsO_4 in sea water and prepared for Scanning Electron Microscopy. We gratefully acknowledge the hospitality and the help of Professor A. Haget, Mrs. A. Ressouche and M. Rivière for the SEM work at the Universities of Bordeaux and Montpellier, respectively.

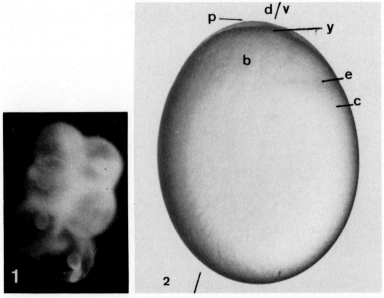

1.- Fragment of an egg string of *Loligo vulgaris* ; about 2 cm
2.- Lateral view of an egg in the pre-organogenetic stage III-IV. b: blastoderm; c: egg cortex; e: edge of the ectoderm; p: polar body at the animal pole; d/v: future dorsal/ventral side of the embryo. Size : 2,2 x 1,6 mm.

The tissue architecture of the pre-organogenetic blastoderm :

After fertilisation and formation of the polar bodies, the blastodisc of the telolecithal egg of *Loligo vulgaris* undergoes cleavage in a centrifugal sequence of events. This results in a flattened blastoderm of a number of cells of different size (Fig. 4), laid out as a monolayer on the yolk syncytium. This can be considered as the "blastula stage" (stage I of Naef(1928)) Disregarding the functional unity of the egg at this stage, its morphology gives clear evidence that the cleavage process definitively splits the egg into an embryo-forming cell material and a large yolk complex. The surface of the latter provides a basis for cell adhesion and attachment and delivers nutritive material to the blastoderm.

The next developmental step is the formation of the germ layers i.e. the "gastrulation process". For the egg of *Loligo peale* this process has recently been studied and discussed by Singley (1977). No fundamental differences in the behaviour of the cells in the *L.v.*blastoderm appear. Figure 5 illustrates a "mid-gastrula"-stage. The inner germ layer forms as a ring of

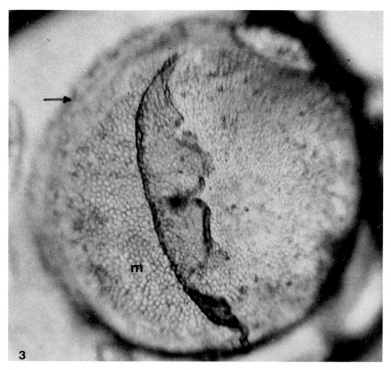

3.-Blastoderm at stage III-IV, freshly prepared. Mes-entoderm (m) partially visible, ectoderm partially removed. Arrow : some ectodermal cell of the blastoderm border. Diameter : 1 mm

cells that become covered by the central population of cells (see also Fig. 4). The gastrulation process is drawn out about one cell division cycle with the parallel occurence of a slow constriction of the ring-shaped inner germ layer.

Germ layers are formed when the covering cells have reached the cortical egg surface peripheral to the outer border of the cell ring (Fig.6). The basic "gastrula" is complete towards stage III, when the mes-entodermal cell complex is entirely covered by an ectodermal cell sheet. It is important to note that the process of germ layer segregation splits the blastoderm into two well defined components within the extent of the original cleavage territory. The typical architecture of the blastoderm at stage III can be termed the "basic spatial organisation of the germ" (Marthy 1976).

Once this spatial organisation is established, the germ layers continue to proliferate independently, but in a well

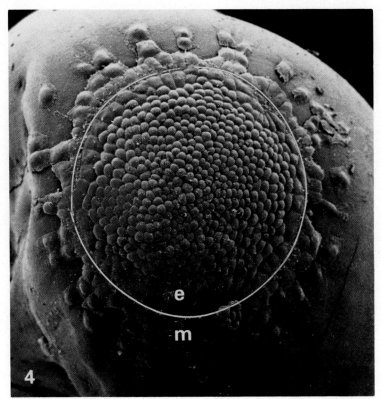

4.- Blastoderm at stage I. About 600 nuclei. White circle shows approximately the future "segregation line" between a ectodermal (e) and a mes-entodermal (m) cell population. 120 x.

defined positional relationship and in close contact. Since the proliferation rate of the ectoderm is higher than that of the mes-entoderm, the distance between the edge of the ectoderm and the outermost mes-entodermal cells increases constantly. Figures 7 and 8 illustrate the morphological situation. The mes-entoderm appears as a compact ring of cells. Numereous intercellular projections are observed and the peripheral cells in particular adhere to the yolk syncytium.

Essentially the basic spatial organisation of the blastoderm is maintained throughout the proliferation phase until organogenesis becomes evident around stage VI. As early as stage IV-V the differentiation of cilia in the ectoderm is observed. This indicates that the proliferation phase is not only a "cellulation phase" of the egg surface, but also a phase of early specific

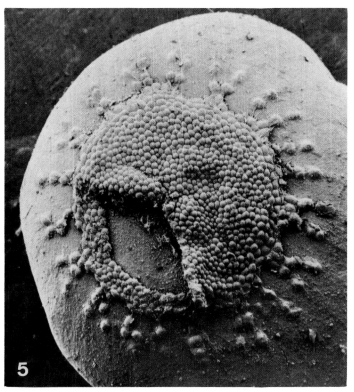

5.- Blastoderm at stage II-III ("mid-gastrula"). Ectoderm partially removed. 100 x.

cell differentiation. As for the mes-entoderm, one observes that the complex spreads in a network-like monolayer of large cells. Figure 9 illustrates the situation immediately before the first organ rudiment appears. The cells of the mes-entoderm are connected to one another by numereous "cytoplasmatic bridges". There are also various plasmatic projections between cells of both germ layers and between these and the yolk syncytium. Whereas in the central part of the mes-entoderm each cell is fixated in its position by the neighbouring cells (Fig. 10), those at the periphery tend to detach (Fig. 11). Detached cells then migrate into the area of the yolk syncytium being covered by the ectoderm (Fig. 12). These cells ultimately differentiate into muscle cells of the external yolk sac.

Discussion

In presenting the general aspects of the pre-organogenetic

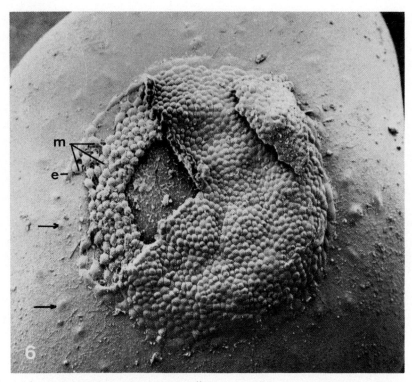

6.- Blastoderm at stage III ("gastrula"). Ectoderm partially removed. e : ectodermal cell ; m : mes-entodermal cells. Arrow : nuclei of the yolk syncytium. 100 x.

blastoderm of the egg of *Loligo vulgaris* the SEM pictures speak for themselves. In fact, at any moment prior to organogenesis, the arrangment of the embryo-forming cell material, laid out on a natural substrate, is clear and accessible to microsurgery. In the context of cell and tissue interaction studies, the blastoderm at the proliferation phase (Fig.6,7) appears particularly interesting : If earlier conclusions (Marthy 1978/79) are correct, the embryo at this developmental stage includes two primary cell and tissue interaction systems. One comprises the contact of the blastoderm with the yolk complex (it determines the nutrition and the physical condition of the cells), the other is the combination of all cell relations within the blastoderm (this is essential for the specialisation of the cells). In the latter interaction system, the "germ layer relation" is considered to be a classical induction - reaction system, in which the mes-entoderm acts as an organizer towards the reacting ectoderm (Marthy 1978).

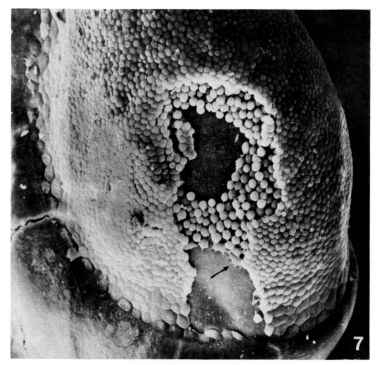

7. - Blastoderm at stage III-IV. Ectoderm partially removed. Arrow : Outermost mesodermal cell attached to the yolk syncytium. 120 x.

According to this model, which may basically "explain" the mode of determination of the cell material into specialised cells and organs, it is likely, that the establishment and the maintenance of cellular interconnections for communication and adhesion are of primary morphogenetic significance. It is also evident, that the morphogenetic significance of cell and tissue interactions can not be drawn from a morphological analysis alone. So far, it can be stated, however : a) it is probable that the transfer and exchange of developmental information between cells of the same germ layer occurs mainly by way of "cytoplasmic bridges" (see fig. 9, 10). In the blastoderm of *Loligo pealei* such bridges have been described by Arnold (1974) and more detailed work was recently been published by Cartwright and Arnold (1980). As to the mes-entoderm we assume that the intercellular bridges guarantee its cytoplasmic unity and its mechanical stability.

8. – Complete mes-entoderm complex (m) at stage III-IV. About 500 cells. Dotted line : bilateral axis ; e : some ectodermal cells from the border of the blastoderm. 130 x.

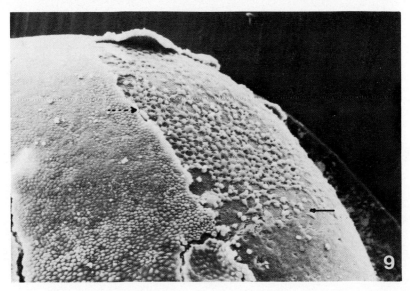

9.– Blastoderm at stage V-VI (close to organogenesis). Ectoderm partially removed. Mes-entoderm complex loosend. Dotted arrow : detail in figure 10. Arrow : detail in figures 11 and 12. 100 x.

Cephalopod Eggs / 231

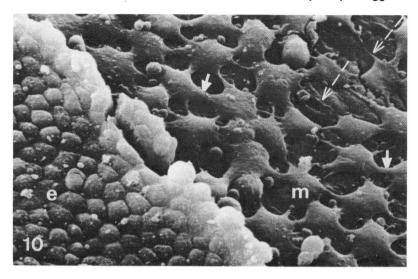

10 - Detail of figure 9. Compact ectodermal sheet (e) in contrast to the scattered mes-entoderm (m). Arrows : cell connecting "plasma bridges". Dotted arrows : filopodia extended to the substrat. 500 x.

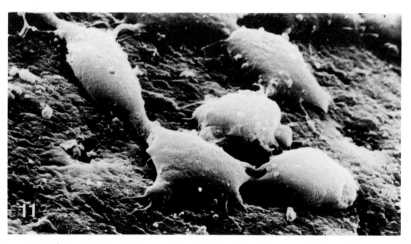

11.- Detail of the figure 9. Cells in the peripheral area of the mes-entoderm. 1400 x.

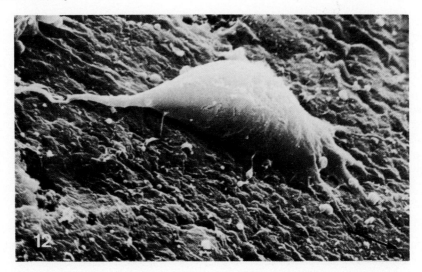

12. – Detail of figure 9. Cell in the peripheral area of the mes-entoderm. By amoeboid movement the cells migrate on the yolk syncytium. Arrow : migration direction. 2300 x.

b) no intercellular bridges are found between cells of different germ layers. The functioning of the induction – reaction system therefore only is conceivable on the basis of membraneous cell contacts and of transmembraneous diffusion. c) the intercellular spaces in the mes-entoderm are thought to be preferential pathways for nutritive substances released by the yolk complex. d) the specific role of the adhesion sites of the cells on their substrate is evident.

In summary it appears that : 1) in the same developmental system adhesive and communicating interactions of cells exist among cells as well as between cells and their natural substrate, and 2) that this developmental system is easily accessible to future structural and functional studies on these cells and their interactions.

References

Arnold, J.M. (1974). Intercellular bridges in somatic cells : cytoplasmic continuity of blastoderm cells of *Loligo pealei*. Diff 2 : 335.
Boletzky, S. von (1974). Elevage des Cephalopodes en aquarium. Vie Milieu 24 : 309

Cartwright, J., Arnold, J.M. (1980). Intercellular bridges in the embryo of the atlantic squid *Loligo pealei*. I. Cytoplasmic continuity and tissue differentiation. JEEM 57 : 3.
Garrod, D.R., Nicol, A. (1981) . Cell behaviour and molecular mechanisms of cell - cell adhesion. Biol Rev Cambridge 56 : 199.
Mangold, K. (1963). Biologie des Céphalopodes benthiques et nectoniques de la mer catalane. Vie Milieu suppl 13 : 1.
Marthy, H.J. (1976). Les déterminismes dans la morphogenèse. Contribution à l'embryologie expérimentale des Céphalopodes. Thèse d'Etat 12426 : 1.
Marthy, H.J. (1978). Recherches sur le rôle morphogénétique du mes-endoderme lors de l'embryogenèse de *Loligo vulgaris*. C r hebd Séanc Acad Sci Paris 287 : 1345.
Marthy, H.J. (1978/79). Embryologie expérimentale chez les Céphalopodes. Vie Milieu 28/29 : 121
Naef, A. (1928). Die Cephalopoden. Fauna Flora del Golfo di Napoli 35 (V-IX) : 1.
Singley, C.T. (1977). An analysis of gastrulation in *Loligo pealei*. Dissertation. University of Hawaii.

ADHESION AND POLARITY OF AMPHIBIAN EMBRYO BLASTOMERES.

Peter B. Armstrong, Marie M. Roberson, Richard
Nuccitelli and Douglas Kline
Department of Zoology, University of California,
Davis, CA 95616, U.S.A. The present address for
M.M.R. is Department of Psychiatry, School of
Medicine, University of California at San Diego,
La Jolla, CA 92093, U.S.A.

Cell-cell adhesion is essential to the ability of individual microscopic cells to construct coherent macroscopic tissues. In addition to this static role in maintaining tissue structure, adhesion plays a role in dynamic processes such as the interaction of cells of the immune system (Martz, 1980), phagocyte-target interaction and control of motility of actively locomoting cells (Steinberg, 1970). Two extreme models can be proposed to account for adhesion. In one, the individual cells are envisioned as being embedded in a voluminous extracellular matrix without direct cell-cell contact. In this model, the mechanical properties of the tissue would be governed largely by the properties of the matrix. In the other, cell adhesion is viewed as being established directly between the surfaces of adjacent cells, without intervening layers of extraneous material. Although the adhesion of the cells of real tissues usually is effected by a combination of both processes, direct cell contact is a predominating element of the adhesion of epithelial cells whereas the extracellular matrix plays an important role in the properties of mesenchymal tissues.

In the review that follows, emphasis will be placed on epithelial-type adhesion. Special attention will be paid to an identification of surface macromolecules of possible importance in adhesion. In this regard, two ideas have received considerable attention. The first is the suggestion that adhesion involves receptor-ligand interactions between molecules on the surfaces of contiguous cells (Weiss, 1947). The second, based on the observation that glycoproteins and glycolipids are abundant in the plasma membrane, proposes that surface carbohydrate moieties are involved in adhesion (Roseman, 1974). In a recent combination of these notions, it has been suggested that binding of cell surface lectins to their oligosaccharide ligands contributes to the bonds linking cohering cells (Barondes and Rosen, 1976). Lectins are sugar-binding proteins of non-immune origin and lacking enzymatic activity toward the sugar ligand. Usually a lectin shows marked specificity, binding strongly to only a restricted number of sugars. This, or any other receptor-ligand, model for adhesion presumes that a significant fraction of receptor-ligand bonds are trans rather than cis.

The cell type used in our study is the superficial blastomere of the cleavage-stage frog embryo. These cells have a polarized organization of the plasma membrane with the apical surface (that part lying at the surface of the embryo) differing from the basolateral surface (that part that contacts adjacent blastomeres). Of particular interest, the apical surface is nonadhesive to other cells and to most artificial surfaces whereas the basolateral surface is adhesive to cells and to glass and plastic surfaces (Holtfreter, 1943; Roberson et al., 1980). These adhesive differences are maintained by individual disaggregated cells in culture. In disaggregated cells, the two membrane domains are reliably recognizable by pigmentation differences. The cytoplasm underlying the basolateral membrane is packed with yolk granules and appears white; that underlying the apical surface is black due to its content of cortical melanin pigment granules. The differences in adhesiveness of the two surface domains was exploited by searching for differences in the composition of apical and basolateral surfaces that might plausibly account for the differences in adhesiveness and, as a consequence, shed light on the fundamental mechanisms of adhesion in this system.

Endogenous tissue lectins are usually detected by their ability to agglutinate erythrocytes. A standard protocol for analysis includes elucidation of the oligosaccharide specificity by determination of the array and potency of the sugars that inhibit hemagglutination. The specificity of some hemagglutinins is such that monosaccharides are ineffectual inhibitors compared with more lengthy oligosaccharide chains. Applying this scheme, we demonstrated in homogenates of amphibian embryo blastomeres the presence of an endogenous hemagglutinin that recognizes oligomannosyl ligands (Table 1: Roberson and Armstrong, 1980). Since the

Table 1. Inhibition of hemagglutinating activity of crude extracts

Inhibitor	Conc. required to reduce hemagglutination by 50%, µg/ml
Mannan	18
Invertase	28
Thyroglobulin	55
Ovalbumin	750
Fetuin	>3000
Transferrin	>3000
Ovomucoid	>3000
Submaxillary gland mucin	>3000
Periodate-oxidized invertase	1000
Periodate-oxidized thyroglobulin	1000
Unit A from thyroglobulin	1*

*Based on a phenol/sulfuric acid assay with mannose as the standard.

hemagglutinating activity of cell homogenates is sensitive to proteases, the hemagglutinin appears to be a protein. Thus the amphibian embryo blastomere appears to contain an endogenous lectin with oligomannosyl specificity. At least a portion of this hemagglutinating activity is present at the cell surface because red cells bind to the surfaces of dissociated cells and this binding is inhibited by the same spectrum of oligomannosyl-containing glycoproteins that inhibit the hemagglutinating activity present in cell homogenates. At the stage of embryonic development used, the blastomeres are large, permitting determination of the regional localization of erythrocyte binding. This is restricted almost entirely to the nonpigmented (adhesive) regions of dissociated superficial cells, indicating that the endogenous, cell surface lectin is confined to the adhesive basolateral regions of the cell surface and is absent from the nonadhesive apical surface (Fig. 1).

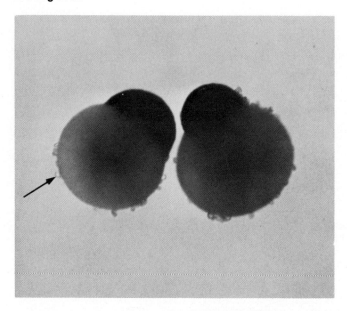

Fig. 1. Binding of formalinized sheep erythrocytes to the surfaces of dissociated superficial blastomeres of *Rana pipiens*. The size difference between the erythrocytes (one is indicated by the arrow) and the blastomeres is striking. Binding is restricted almost completely to the basolateral surface.

Dissociated blastomeres were probed for the presence of oligosaccharide chains with terminal mannose residues with the plant lectin concanavalin A, which binds mannose and glucose. Both FITC-labeled con A (used with epifluorescence illumination) and tritiated con A (used with autoradiography of sectioned cells) were employed. Both probes indicated the presence of abundant con A receptors at the cell surface which were present only on basolateral membrane. If any were present at the apical surface, the density was too low to be detected by either procedure (Fig. 2; Roberson and Armstrong, 1979).

These observations have led to the formulation of the hypothesis that the bonds linking the blastomeres of the cleavage-stage amphibian embryo include those between a cell-surface oligomannosyl lectin and oligomannosyl chains of cell surface glycoconjugates. Presumably the restriction

Adhesion and Polarity in Blastomeres / 239

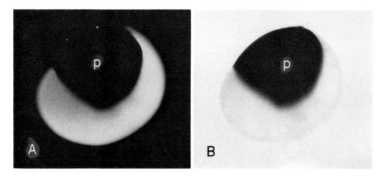

Fig. 2. EDTA-dissociated, FITC-con A-treated superficial blastomere of *Bombina orientalis*. (A) Fluorescence photomicrograph: the light area fluoresced brightly, the dark area was non-fluorescent. (B) Phase-contrast photomicrograph of the same cell. The pigmented portion of the cell surface (p) did not bind con A.

of both lectin and oligomannosyl chains to the basolateral surface is an element in the nonadhesive character of the apical surface. This hypothesis clearly requires further testing. One test will involve the ability of high concentrations of oligomannosyl chains prepared from blastomeres to inhibit aggregation of dissociated cells. The inhibitory ability of the glycoproteins listed in Table 1 will also be tested. Preliminary attempts (Roberson and Armstrong, 1980) have been frustrated by difficulties in developing a method for quantitating rates of aggregation of dissociated cells. The blastomeres are too large and too fragile for rotation-mediated aggregation.

Perhaps as interesting as the use for study of adhesive mechanisms is the potential that the superficial blastomere has for study of the phenomenon of membrane polarity. The cells are large, the pigmentation difference serves as a reliable marker for the apical surface of dissociated cells, and a variety of differences between the two domains of the plasma membrane have been described (Table 2). The two membrane domains have different origins: the apical domain is derived from the plasma membrane of the uncleaved, fertilized zygote whereas the basolateral membrane is inserted during the deepening of the cleavage furrows during cytokinesis of the successive cleavage divisions

Table 2. Differences in properties of apical and basolateral membrane of the superficial blastomeres of cleavage stage amphibian embryos

Character	Localization	Reference
Intramembrane particles (IMPs)	More abundant in luminal than in basolateral membrane	Bluemink (1978)
Glycocalyx	Apical surface covered by fuzzy glycocalyx. Recognizable glycocalyx absent from basolateral surface	Armstrong (unpublished observations), Bluemink (1978)
Concanavalin A receptors	Restricted to basolateral surface	Roberson and Armstrong (1979)
Cell surface endogenous lectin	Restricted to basolateral surface	Roberson and Armstrong (1980)
Transmembrane ion current	Inward positive current at apical membrane; outward current along basolateral membrane	Kline and Nuccitelli (unpublished data)
Specific transmembrane electrical resistance	Apical membrane = $74 k\Omega cm^2$ Basolateral membrane = $1-2 k\Omega cm^2$	De Laat and Bluemink (1974)
Adhesiveness	Apical surface is non-adhesive. Basolateral surface is adhesive to other cells and to artificial surfaces	Holtfreter (1943) Roberson et al. (1980)

(De Laat and Bluemink, 1974; Selman and Waddington, 1955). Polarization of the plasma membrane into apical and basolateral domains apparently is characteristic of the cells of all simple epithelia (epithelia consisting of a single layer of cells). Differences are apparent at the morphological level (e.g., placement of intercellular junctions, microvilli (Cereijido et al., 1978; Mooseker and Tilney, 1975) and the glycocalyx (Ito, 1965; Pisam and Ripoche, 1976)), at the biochemical level (e.g., restriction of a variety of enzymes (Ernst and Mills, 1977; Kyte, 1976; Louvard, 1980; Quinton et al., 1973; Saccomani et al., 1979; Wisher and Evans, 1975, 1977; Ziomek et al., 1980) and concanavalin A receptors (Pisam and Ripoche, 1976; Ziomek and Johnson, 1980) to either apical or basolateral domain), and at the functional level (e.g., restriction of adhesiveness to the basolateral domain (Buck, 1973; Di Pasquale and Bell, 1974; Holtfreter, 1943; Roberson et al., 1980; Vasiliev and Gelfand, 1978), selective release of cytopathic virus from the basolateral surface (Rodriguez Boulan and Sabatini, 1978; Roth et al., 1979) and differences in electrical resistance of apical and basolateral membrane (Bluemink, 1978)).

A problem of fundamental importance is elucidation of the mechanisms responsible for the stabilization of this partitioning of elements into one or the other domain of the continuous, and presumably fluid (Singer and Nicolson, 1972), plasma membrane. A variety of mechanisms may contribute to stability (Table 3).

Table 3. Mechanisms that may contribute to the maintenance of discrete apical and basolateral plasma membrane domains in simple epithelial cells.

No.	Mechanism	References
1	The juxtaluminal junctional complex forms a barrier to exchange of membrane components.	Hoi Sang et al. (1980), Roth et al. (1979), Pisam and Ripoche (1976)
2	Basolateral membrane elements are immobilized by trans bonds established with complementary macromolecules of adjacent cells or the basement membrane.	Weigel (1980)
3	Membrane constituents that stray from their proper domain are selectively internalized; the concentration of unique elements in each domain is maintained by selective insertion from cytoplasmic stores.	
4	The lipid phase of either apical or basolateral membrane is paracrystalline, preventing mixing of membrane elements.	
5	Peripheral membrane proteins at the cell surface cross-link intramembrane proteins into immobile patches that constitute the apical and basolateral surfaces.	
6	Cytoskeletal elements of the cortex immobilize integral membrane proteins in basolateral and apical domains.	Nicolson et al. (1977), Roberson and Armstrong (1978)
7	Transcellular electrical currents electrophorese membrane elements into apical or basolateral domains.	Jaffe (1977)

The present evidence allows a preliminary evaluation of several of these mechanisms vis-a-vis the amphibian blastomere. The observation that the microfilament-depolymerizing agent cytochalasin B causes a rapid loss of the polarized distribution of con A receptors (Roberson and Armstrong, 1979) argues that microfilaments may be involved in the maintenance of polarity, possibly by serving as a cytoskeletal scaffolding to which membrane elements are tethered (Geiger, 1980). Unfortunately cytochalasin B is pleiotropic so use of other cytochalasins that lack some of its side effects (e.g., cytochalasin D) is essential. The observation that membrane polarity is stable in dissociated cells (Roberson and Armstrong, 1979) indicates that mechanism 2 does not contribute to stability and that intact junctional complexes (mechanism 1) are, likewise, not involved. If, however, the intramembranous elements of the zonula occludens junction persist as a continuous belt separating apical and basolateral domains of dissociated blastomeres, they may still play a role in the stabalization of polarity. Freeze-fracture studies will be required to investigate this possibility further. The only data relevant to possibilities #4 and 5 are the observations that con A can induce patching of con A receptors on the basolateral surface. This indicates that receptors are not immobile, as would be expected if possibilities 4 or 5 were valid. As yet, we lack comparable data for the apical domain. Since this domain does contain receptors for lectins other than con A (Roberson and Armstrong, 1979), it should be possible to perform analogous studies using these lectins. If ligand-induced receptor redistribution occurs in the apical domain also, then this would suggest that hypotheses 4 and 5 are inappropriate to this system.

Hypothesis #7 proposes that endogenous transcellular electrical currents play a role in stabilizing membrane polarity by producing a lateral electrophoresis of membrane components. This proposal suggests that the distribution of movable elements in the plasma membrane would be a function of the direction of current flow and the electrostatic charge on that portion of the molecules exposed at the cell surface. This hypothesis is currently under investigation using the vibrating probe microelectrode (Jaffe and Nuccitelli, 1974) to determine transcellular currents in isolated blastomeres. An average positive current of 0.2 ± 0.1 $\mu A/cm^2$ measured 60 μm from the cell surface enters the apical end while an average current of 0.7 ± 0.2 $\mu A/cm^2$ exits the basolateral end (Fig. 3). This is a current pattern that would be expected to electrophorese laterally mobile, negatively charged moieties into the

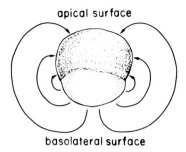

Fig. 3. Pattern of positive electrical current flow through isolated blastomere from the 8-cell *Xenopus laevis* embryo.

basolateral membrane. Once the ionic basis for the transcellular currents has been established, the effects on polarity of the appropriate ion transport inhibitors will be determined. The hypothesis predicts that polarity should be destabilized if the transcellular currents are blocked. Finally, the hypothesis can be tested by determining the effects on polarity of externally imposed electrical fields. If the hypothesis is valid, it should be possible to override the action of the endogenous currents by this method and, for example, to electrophorese con A receptors into the apical surface. These studies will be patterned on those of Poo and Robinson (Poo and Robinson, 1977; Hinkle et al., 1981) who demonstrated that external electrostatic fields imposed on cultured skeletal muscle cells caused a redistribution of surface con A and acetylcholine receptors. It is anticipated that these studies will allow us to provide the evidence necessary to evaluate the validity of the self-imposed lateral electrophoresis hypothesis. The amphibian blastomere is a favorable object for these studies because it is large enough for the vibrating probe analysis.

In summary, studies now in progress are designed to evaluate the various hypotheses that can be envisioned as accounting for the stability of the polarized partitioning of membrane components into basolateral or apical membrane domains. It is envisioned that this investigation will have relevance to the general problems of stabilization of membrane polarity in epithelial cells and in non-epithelial cells that also show differentiated domains in the plasma membrane (e.g., skeletal muscle: Bloch, 1979; and sperm: Friend et al., 1974; Myles et al., 1981).

ACKNOWLEDGMENTS

This study was funded by NSF Grants PCM 77-18950, PCM 78-18047 and PCM 80-24181. Figure 1 and Table 1 are from Roberson and Armstrong (1980) and Figure 2 is from Roberson and Armstrong (1979), reprinted by permission of Academic Press, Inc.

REFERENCE LIST

Barondes SH, Rosen SD (1976). Cell surface carbohydrate-binding proteins: role in cell recognition. In Barondes SH (ed): "Neuronal Recognition," New York: Plenum, p 331.
Bloch RJ (1979). Dispersal and reformation of acetylcholine receptor clusters of cultured rat myotubes treated with inhibitors of energy metabolism. J Cell Biol 82: 626.
Bluemink JG (1978). Use of cytochalasins in the study of amphibian development. In Tanenbaum SW (ed): "Cytochalasins - Biochemical and Cell Biological Aspects," Amsterdam: North-Holland, p 113.
Buck RC (1973). Walker 256 tumor implantation in normal and injured peritoneum studied by electron microscopy, scanning electron microscopy, and autoradiography. Cancer Res 33:3181.
Cereijido M, Robbins ES, Dolan WJ, Rotunno CA, Sabatini DD (1978). Polarized monolayers formed by epithelial cells on a permeable and translucent support. J Cell Biol 77: 853.
De Laat SW, Bluemink JG (1974). New membrane formation during cytokinesis in normal and cytochalasin B-treated eggs of *Xenopus laevis* II. Electrophysiological observations. J Cell Biol 60:529.
De Laat SW, Luchtel D, Bluemink JG (1973). The action of cytochalasin B during egg cleavage in *Xenopus laevis*: dependence on cell membrane permeability. Devel Biol 31: 163.

Di Pasquale A, Bell PB (1974). The upper cell surface: its inability to support active cell movement in culture. J Cell Biol 62:198.

Ernst SA, Mills, JW (1977). Basolateral plasma membrane localization of ouabain-sensitive sodium transport sites in the secretory epithelium of the avian salt gland. J Cell Biol 75:74.

Friend DS, Fawcett DW, Rudolf I (1974). Membrane differentiations in freeze-fractured mammalian sperm. J Cell Biol 63:641.

Geiger B (1980). Transmembrane linkage and cell attachment: the role of vinculin. In Schweiger HG (ed): "International Cell Biology 1980-1981," Berlin: Springer-Verlag, p 761.

Hinkle L, McCaig CD, Robinson KR (1981). The direction of growth of differentiating neurones and myoblasts of frog embryos in an applied electrical field. J Physiol (Lond) 314:121.

Hoi Sang U, Saier MH, Ellisman MH (1980). Tight junction formation in the establishment of intramembranous particle polarity in aggregating MDCK cells. Effect of drug treatment. Exp Cell Res 128:223.

Holtfreter J (1943). Properties and functions of the surface coat in amphibian embryos. J Exp Zool 93:251.

Ito S (1965). The enteric surface coat on cat intestinal microvilli. J Cell Biol 27:475.

Jaffe LF (1977). Electrophoresis along cell membranes. Nature 265:600.

Jaffe LF, Nuccitelli R (1974). An ultrasensitive vibrating probe for measuring steady extracellular currents. J Cell Biol 63:614.

Kyte J (1976). Immunoferritin determination of the distribution of ($Na^+ + K^+$) ATPase over the plasma membranes of renal convoluted tubules I. Distal segment. J Cell Biol 68:287.

Louvard D (1980). Apical membrane aminopeptidase appears at site of cell-cell contact in cultured kidney epithelial cells. Proc Nat Acad Sci USA 77:4132.

Martz E (1980). Immune T lymphocyte to tumor cell adhesion: magnesium sufficient, calcium insufficient. J Cell Biol 84:584.

Mooseker MS, Tilney LG (1975). Organization of an actin filament - membrane complex: filament polarity and membrane attachment in the microvilli of intestinal epithelial cells. J Cell Biol 67:725.

Myles DG, Primakoff P, Bellvé AR (1981). Surface domains of the guinea pig sperm defined with monoclonal antibodies. Cell 23:433.

Nicolson GL, Poste G, Ji TH (1977). The dynamics of cell membrane organization. In Poste G, Nicolson GL (eds): "Dynamic Aspects of Cell Surface Organization," Amsterdam: North-Holland, p 1.

Pisam M, Ripoche P (1976). Redistribution of surface macromolecules in dissociated epithelial cells. J Cell Biol 71:907.

Poo MM, Robinson KR (1977). Electrophoresis of concanavalin A receptors along embryonic muscle membrane. Nature 265:602.

Quinton PM, Wright EM, Tormey JMD (1973). Localization of sodium pumps in the choroid plexus epithelium. J Cell Biol 58:724.

Roberson MM, Armstrong PB (1979). Regional segregation of Con A receptors on dissociated amphibian embryo cells. Exp Cell Res 122:23.

Roberson MM, Armstrong PB (1980). Carbohydrate-binding component of amphibian embryo cell surfaces: restriction to surface regions capable of cell adhesion. Proc Nat Acad Sci USA 77:3460.

Roberson M, Armstrong J, Armstrong PB (1980). Adhesive and non-adhesive membrane domains of amphibian embryo cells. J Cell Sci 44:19.

Rodriguez Boulan E, Sabatini DD (1978). Asymmetric budding of viruses in epithelial monolayers: a model system for study of epithelial polarity. Proc Nat Acad Sci USA 75:5071.

Roseman S (1974). Complex carbohydrates and intercellular adhesion. In Lee EYC, Smith EE (eds): "Biology and Chemistry of Eukaryotic Cell Surfaces," New York: Academic Press, p 317.

Roth MG, Fitzpatrick JP, Compans RW (1979). Polarity of influenza and vesicular stomatitis virus maturation in MDCK cells: lack of a requirement for glycosylation of viral glycoproteins. Proc. Nat Acad Sci USA 76:6430.

Saccomani G, Helander HF, Crago S, Chang HH, Dailey DW, Sachs G (1979). Characterization of gastric mucosa membranes X. Immunological studies of gastric $(H^+ + K^+)$ - ATPase. J Cell Biol 83:271.

Selman GG, Waddington CH (1955). The mechanism of cell division in the cleavage of the newt's egg. J Exp Biol 32:700.

Steinberg MS (1970). Does differential adhesion govern self assembly processes in histogenesis? Equilibrium configurations and the emergence of a hierarchy among populations of embryonic cells. J Exp Zool 173:395.

Vasiliev JM, Gelfand IM (1978). Mechanism of non-adhesiveness of endothelial and epithelial surfaces. Nature 274:710.

Weigel PH (1980). Rat hepatocytes bind to synthetic galactoside surfaces via a patch of asialoglycoprotein receptors. J Cell Biol 87:855.

Weiss P (1947). The problem of specificity in growth and development. Yale J Biol Med 19:235.

Wisher MH, Evans WH (1975). Functional polarity of the rat hepatocyte surface membrane. Isolation and characterization of plasma membrane subfractions from the blood-sinusoidal, bile-canalicular and contiguous surfaces of the hepatocyte. Biochem J 146:375.

Wisher MH, Evans WH (1977). Preparation of plasma membrane subfractions from isolated rat hepatocytes. Biochem J 164:415.

Ziomek CA, Johnson MH (1980). Cell surface interaction induces polarization of mouse 8-cell blastomeres at compaction. Cell 21:935.

Ziomek CA, Schulman S, Edidin M (1980). Redistribution of membrane proteins in isolated mouse intestinal epithelial cells. J Cell Biol 86:849.

EPITHELIAL-DERIVED BASAL LAMINA REGULATION OF MESENCHYMAL CELL DIFFERENTIATION.

Harold C. Slavkin, Elaine Cummings, Pablo Bringas and Lawrence S. Honig.
Laboratory for Developmental Biology, Graduate Program in Craniofacial Biology, University of Southern California, Los Angeles, California 90007, USA

ABSTRACT

The mechanisms by which epithelial-mesenchymal interactions result in differentiation are not known. A number of recombinations between vertebrate tissues associated with epidermal organs (e.g. skin, feather, mammary gland, salivary gland, tooth organ) indicate that regional mesenchymal specificity is instructive for determination and differentiation of epithelial phenotypes. In epidermal organs within which mesenchyme becomes determined and differentiates into a unique phenotype, such as during tooth organogenesis and odontoblast differentiation. Does the epithelial-derived basal lamina regulate mesenchymal differentiation into odontoblasts and the expression of dentine extracellular matrix? Experiments were designed to test the hypothesis that murine or avian epithelial-derived basal lamina possess information which is instructive for determined dental mesenchyme to differentiate into odontoblasts. The strategy was to examine homologous and heterologous tissue recombinants between Theiler stage 25 C57BL/6 molar tooth organs and Hamburger-Hamilton equivalent stage 22-26 Japanese Pharoah quail mandibular processes. Trypsin-dissociated molar epithelium and mesenchyme, reconstituted, secreted a basal lamina within 8 hours and mesenchyme differentiated into odontoblasts and formed dentine matrix within 3 days. Isolated trypsin-dissociated mesenchyme <u>did not differentiate in vitro</u>, whereas heterologous recombinants between odontogenic mesenchyme and quail epithelia resulted in odontoblasts and dentine production. Mouse tooth or quail mandibular epithelia served to regulate odontogenic mesenchyme differentiation. EDTA-dissociated mouse molar mesenchyme, in the absence of epithelium but with adherent basal lamina, routinely differentiated into odontoblasts. Control tooth organs routinely formed both dentine and enamel extracellular matrices within 7-10 days in our serumless, chemically-defined organ culture system. Regulation of determined mesenchymal cells to differentiate into functional and highly specialized odontoblasts appears to be mediated by epithelial-derived basal lamina and is not species or organ-specific.

INTRODUCTION

Epithelial-mesenchymal interactions in vertebrate embryos begin during late neurulation. These secondary embryonic induction interactions between tissues have been studied extensively. As development proceeds there is an increase in the relative rates of differentiation within one of the two tissue interactants. Descriptions of these tissue interactions in a number of diverse organ systems, and evidence for both "instructive" as well as "permissive" interactions is well-known to the international developmental biology community (reviewed in Grobstein, 1967; 1975; Saxen et al, 1976; 1977; and Wessells, 1977).

One intriguing aspect of epithelial-mesenchymal interactions is the specificity of these processes. How do these interactions mediate synchronous epidermal organogenesis (e.g. tooth, skin, feather, scale, salivary gland, mammary gland)? When and where are the interactions directive and when are they permissive? For example, if mouse dental mesenchyme can instruct chick epithelium to differentiate into ameloblasts and produce enamel (as purported by

Kollar and Fisher, 1980), such epithelial-mesenchymal interactions between vertebrate tissues represents a "directive" or "instructive" event; Aves do not have teeth. If an epithelial tissue provides permissive requirements for already determined and differentiating mesenchyme, the interactions are termed permissive (Saxen et al, 1976; Wessells, 1977). The purpose of this contribution is to demonstrate that regulation of determined dental papilla mesenchyme to differentiate into odontoblasts, which produce dentine matrix, appears to be mediated by epithelial-derived basal lamina and is not species or organ-specific. Moreover, evidence is presented which does not confirm recent data suggesting that mouse mesenchyme from tooth organs is capable of inducing chick epithelial cells to become functional ameloblasts and produce enamel.

EXPERIMENTAL DESIGN

The methods we have employed to determine instructive versus permissive interactions required for mesenchyme differentiation to become odontoblasts, include heterologous tissue recombinations between different vertebrates. This method depends upon reliable intrinsic marking techniques which enable unequivocal discrimination between tissue types. In order to interpret the sources of basal lamina influences on odontoblast differentiation, we employed either EDTA or trypsin dissociation procedures; EDTA-dissociated molar tooth organs result in isolated dental papilla mesenchyme with adherent basal lamina (Osman and Ruch, 1981), whereas trypsin-dissociation removed the basal lamina. Finally, to exclude confounding influences from the in vitro culture medium, such as serum and embryonic extract-derived factors, we employed a chemically-defined medium without serum or other exogenous factors which has been published elsewhere (Yamada et al, 1980).

Our design utilized heterologous tissue recombinations between Theiler stages 24-25 murine embryonic molar tooth organ-derived tissues and Hamburger-Hamilton equivalent stages 22-26 quail embryonic mandibular-derived tissues. The stable nucleolar marker in the quail cells provided accurate marking for the quail cells as reported originally by Le Douarin (1973).

The identification of mesenchyme cells along the periphery of the dental papilla which differentiate into odontoblasts was made from histological observations of serial sections. Odontoblasts are non-dividing, highly-polarized, elongated secretory cells which produce dentine extracellular matrix deposited in juxtaposition to the adjacent epithelium. Dentine is initially produced 48 hours before enamel production in rodents.

RESULTS AND DISCUSSION

Homologous and Heterologous Tissue Recombinations: Quail/Mouse

To assess the requirement for directive or possibly permissive epithelial instructions for mesenchyme differentiation into functional odontoblasts, we designed studies to evaluate homologous and heterologous tissue recombinations within and across vertebrate species. Essentially two criteria were monitored in these experiments: (1) embryonic mouse dental papilla mesenchyme differentiation into odontoblasts with production of dentine extracellular matrix; and (2) the time required for mesenchyme differentiation into odontoblasts in juxtaposition to either homologous dental epithelia from mouse, or heterologous mandibular epithelia from quail.

The results are presented in Table I. We detected no odontoblasts and no dentine extracellular matrix in trypsin-dissociated dental papilla mesenchyme explants cultured for periods up to 10 days. Homologous recombinants between trypsin-dissociated dental mesenchyme and enamel organ epithelia, on the other hand routinely contained odontoblasts and dentine matrix within 3 days in vitro. The time and location of mesenchyme diffentiation into odontoblasts was the same as that observed within control cap stage molar tooth organs which had been trypsin-treated but not dissociated and recombined. From these observations we assumed that heterologous recombinants between dental mesenchyme and quail mandibular epithelial tissue would require at least 3 days in order to produce odontoblasts and dentine matrix. Quail/mouse recombinants routinely resulted in the mesenchyme differentiation into odontoblasts and dentine formation.

TABLE I.

Homologous and Heterologous Tissue Recombinations: Quail/Mouse

Experimental Groups	Recovered Explants	Odontoblast Differentiation
Cap Stage Molar Tooth Organs	291	+
Homologous Recombinations	17	+
Trypsin-Dissociated Dental Mesenchyme	44	−
Trypsin-Dissociated Dental Epithelium	40	−
Heterologous Recombinations	41	+

Theiler stage 25 embryonic mouse mandibular first molar tooth organs were trypsin-treated and cultured for periods up to 10-days. Trypsin-treated and isolated cap stage enamel organ epithelial samples were cultured to evaluate possible mesenchyme cellular contamination. Hamburger-Hamilton equivalent stages 22-26 quail mandibular epithelium was isolated by trypsin treatment and mechanical dissociation.

BASAL LAMINA REGULATION OF ODONTOBLAST CYTODIFFERENTIATION

In our previous results, observations of trypsin-dissociated mouse or quail epithelial tissues showed the removal of the basal lamina. In all explants of either homologous or heterologous tissue recombinations in which trypsin dissociative methods were used, we noted de novo production of a basal lamina within 24 hours (Brownell, Bessem and Slavkin, 1981). Since EDTA-treatment of cap stage mouse molar tooth organs removed the basal lamina from the basal surfaces of the dental enamel organ epithelia, yet showed that the basal lamina was retained in association with the progenitor cells of the dental papilla, we tested if retained basal lamina might regulate mesenchyme differentiation into odontoblasts.

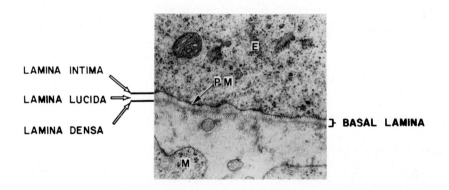

Figure 1. The basal lamina along the undersurface of embryonic mouse molar inner enamel epithelium (E) contains Type IV collagen, laminin, basement membrane proteoglycans and fibronectin (Lesot et al, 1981; Thesleff et al, 1981). EDTA-treatment appears to chelate cations between the epithelial plasma membrane surface (PM) and lamina intima, resulting in isolated dental papilla mesenchyme (M) with adherent basal lamina. Mag. 22,300 X.

Trypsin-treated and EDTA-treated cap stage molar tooth organs as explants in our serumless, chemically-defined medium, produced odontoblasts, dentine matrix, ameloblasts and enamel within 10 days in vitro; odontoblasts formed within 3 days. Trypsin-treated and dissociated dental mesenchyme from cap stage molar tooth organs did not form odontoblasts. EDTA-treated and dissociated dental papilla mesenchyme produced odontoblasts within 3 days. Under these experimental conditions the major trypsin-labile material(s) to be considered are constitutents of the basal lamina.

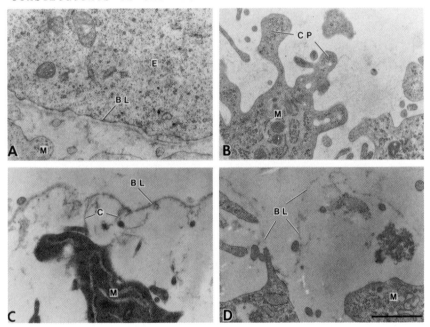

Figure 2. EDTA procedures to isolate dental papilla mesenchyme with adherent basal lamina. (A) Control molar tooth organ prior to EDTA-treatment. E, epithelium; BL, basal lamina; M, mesenchyme cell process. (B) mesenchymal cell surfaces (M, CP) following trysinization. (C) basal lamina (BL) remained adherent to dental papilla mesenchyme (M) following EDTA procedure. C, collagen (D) basal lamina (BL) is evident in association with dental mesenchyme cells after 10 days in vitro. Bar line is 1 um.

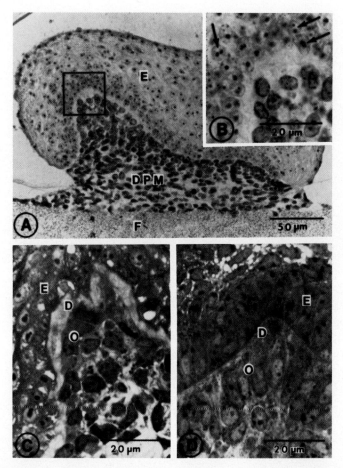

Figure 3. Heterologous tissue recombinants using embryonic mouse and quail tissues. (A) Mouse dental papilla mesenchyme (DPM) recombined with quail mandibular epithelium (E) and cultured on filters (F) for 3 days in serumless, chemically-defined medium. Rectangle shows forming tooth cusp region. Stained with a modified Feulgen-Rossenbeck procedure. (B) Area in rectangle with arrows indicating quail nucleoli. (C) Within 3 days odontoblasts (O) forming dentine extracellular matrix (D) are evident adjacent to quail epithelia (E). (D) Control homologous recombinant, after trypsin-dissociation, shows odontoblasts (O) forming dentine (D) within 3 days in vitro. E, Inner enamel epithelium.

PERMISSIVE VERSUS INSTRUCTIVE INTERACTIONS

A number of curious observations resulted from these investigations. First, a serumless, chemically-defined medium supported a number of complex homologous and heterologous tissue interactions for periods up to 10 days (Yamada et al, 1980; Slavkin, et al, in press). The use of this defined medium reduced a number of confounding variables including the effects of antibiotics, serum and exogenous growth factors often used to supplement organ culture experiments. Second, it appears evident that cap stage dental papilla tissue contains <u>determined</u> mesenchymal cells which will become odontoblasts and produce dentine matrix under suitable environmental conditions; in this context the mesenchyme cells were determined to become a unique phenotype before the experiment was initiated. Thirdly, the presence of a basal lamina in either homologous or heterologous tissue recombinants is associated with odontoblast cytodifferentiation. EDTA-treated dental mesenchyme explants also formed odontoblasts within 3 days <u>in vitro</u>, further supporting the conclusion that epithelial-derived basal lamina constituents regulate determined mesenchyme to differentiate into odontoblasts. None of these observations suggest a directive or instructive interaction according to the definitions previously proposed by Saxen and his colleagues (1976).

A very significant directive or instructive interaction has been reported recently by Kollar and Fisher (1980) in showing that cap stage mouse dental papilla mesenchyme directed adjacent chick pharyngeal epithelial tissue to differentiate into ameloblasts and produce enamel extracellular matrix. In contrast, our heterologous recombinations using cap stage dental papilla mesenchyme and quail mandibular epithelium failed to provide histological or transmission electron microscopic evidence for ameloblast cytodifferentiation. Previous studies from our laboratory of heterologous recombinants using mouse dental papilla mesenchyme and chick limb epithelia also failed to produce ameloblasts (Hata and Slavkin, 1978). Earlier studies by Karcher-Djuricic and Ruch (1973) also failed to show ameloblast differentiation in homologous recombinants between mouse and chick tissues. The disparity in results might represent unique technical differences related to methodology, or the results might be interpreted to suggest that modern birds may not possess the regulatory genes for secretory amelogenesis.

ACKNOWLEDGEMENTS

The authors wish to especially thank Dr. Jean Victor Ruch and his colleagues in Strasbourg for many interesting suggestions and interpretations. Further, we wish to thank Ms. Gwen Airkens for her untiring efforts in preparing this manuscript. Finally, we wish to acknowledge support for this research from the U.S. P.H.S., N.I.H., N.I.D.R. grants DE 02848 and DE 07006.

REFERENCES

Brownell AG, Bessem CC, Slavkin HC (1981). Possible functions of mesenchyme cell-derived fibronectin during formation of basal lamina. Proc Natl Acad Sci 78:3711.

Cummings E, Bringas P Jr, Grodin MS, Slavkin HC (in press). Epithelial-directed mesenchyme differentiation in vitro. Model of murine odontoblast differentiation mediated by quail epithelia. Differentiation.

Grobstein C (1967). Mechanisms of organogenetic tissue interaction. Nat Cancer Inst Monogr 26:279.

Grobstein C (1975). Developmental role of intercellular matrix: retrospective and prospective. In Slavkin HC, Greulich RC (eds): "Extracellular Matrix Influences on Gene Expression," New York: Academic Press, p 9.

Hata, RI, Slavkin HC (1978). De novo induction of a gene product during heterologous epithelial-mesenchymal interactions in vitro. Proc Natl Acad Sci 75:2790.

Karcher-Djuricic V, Ruch JV (1973). Cultures d' associations interspecifiques de constituants mandibulaires d'embryons de souris et de poulet. C R Acad Sci 167:915.

Kollar EJ, Fisher C (1980). Tooth induction in chick epithelium: expression of quiescent genes for enamel synthesis. Science 207:993.

Le Douarin N (1973). A biological cell labelling technique and its use in experimental embryology. Develop Biol 30:217.

Osman M, Ruch JV (1981). Behavior of odontoblasts and basal lamina of trypsin or EDTA-isolated mouse dental papillae in short-term culture. J Dent Res 60:1015.

Saxen L (1975). Transmission and spread of kidney tubule induction. In Slavkin HC, Greulich RC (ed): "Extracellular Matrix Influences on Gene Expression," New York: Academic Press, p 523.

Saxen L, Karkinen-Jaaskelainen M, Lehtonen E, Nordling S, Wartiovaara J (1976). Inductive tissue interactions. In Poste G, Nicolson GL (eds): "The Cell Surface in Animal Embryogenesis and Development," Amsterdam: North-Holland Publishing Co., p 331.

Slavkin HC (1974). Embryonic tooth formation: a tool for developmental biology. Oral Sci Rev 4:1.

Slavkin HC, Trump GN, Brownell A, Sorgente N, (1977). Epithelial-mesenchymal interactions: mesenchymal specificity. In Lash JW, Burger MM (eds): "Cell and Tissue Interactions," New York: Raven Press, p 29.

Slavkin HC, Bringas P Jr, Cummings E, Grodin MS (in press). Initiation of quail and mouse mandibular chondrogenesis and osteogenesis in a serumless, chemically-defined medium. Calcif Tissue Int.

Wessells NK (1977). "Tissue Interactions and Development." California: Benjamin/Cummings p 58, 231.

Yamada M, Bringas P Jr, Grodin M, MacDougall M, Slavkin HC (1980). Developmental comparisons of murine secretory amelogenesis in vivo, as xenografts on the chick chorio-allantoic membrane, and in vitro. Calcif Tissue Int 31:161.

MESENCHYME-DEPENDENT DIFFERENTIATION OF INTESTINAL BRUSH-
BORDER ENZYMES IN THE GIZZARD ENDODERM OF THE CHICK EMBRYO.

K. Haffen, M. Kedinger, P.M. Simon-Assmann and
B. Lacroix
INSERM Unité de Recherches 61
3, avenue Molière
67200 STRASBOURG - Hautepierre   France

Numerous studies have clearly established that interactions between mesenchyme and epithelium are required for normal organogenesis of various organs (for review see Sengel, 1970). However, less numerous are investigations in which the biochemical response of reciprocal epithelial-mesenchymal interactions were analyzed and the data obtained led to opposite conclusions concerning induced morphology and organ specific protein synthesis. For instance, the results concerning recombinants of pancreatic epithelium and salivary mesenchyme which secrete pancreatic enzymes (Rutter and Weber, 1965) or recombinants of mammary epithelium and salivary mesenchyme which synthesize a milk protein (Sakakura et al, 1976) led to the concept that mesenchyme controls morphology whereas the epithelium is the determinant of the cytodifferentiation to be expressed. At variance with this concept, Dhouailly et al (1978) reported examples of heterotopic (chick/chick) and heterospecific (chick/mouse) epidermal-dermal recombinations where morphogenesis and cytodifferentiation, i.e., keratin protein synthesis of chick epidermis were strictly mesenchyme dependent.

During organogenesis of the gut, the morphological aspect of epithelial-mesenchymal interactions is well documented (Le Douarin and Bussonnet, 1966 ; Le Douarin et al, 1968 ; Yasugi and Mizuno, 1974, 1978 ; Gumpel-Pinot et al, 1978 ; Kedinger et al, 1981). Of particular interest in this regard, was the demonstration that in the bird embryo, mesenchyme from the small intestine always dictated the type of differentiation of the endoderm originating from other levels of the digestive tract (Gumpel-Pinot et al, 1978).

The same authors showed that the epithelium from the small intestine was not modified by the different mesenchymes with which it was recombined. These results prompted us to design studies concerned either with the mesenchymal control of the differentiation of intestinal cell cultures (Haffen et al, 1981) or with the biochemical aspect of epithelial-mesenchymal interactions during organogenesis of the small intestine (Kedinger et al, 1981). The present experiments emphasize the instructive action of intestinal mesenchyme on the enzymatic differentiation of gizzard endoderm in comparison with that of allantoic endoderm.

MATERIALS AND METHODS

The experiments were performed on white Leghorn chick embryos. The eggs were incubated at $38 \pm 1°C$ and the developmental stages were referred as days of incubation.

Preparation of Mesenchyme - Endodermal Recombinants and Grafting Procedure

Mesenchyme and endoderm from the small intestine (anlage comprising the jejunum and proximal ileum) and from the gizzard were separated after incubation of the gastrointestinal tract of 5-and 5 1/2-day chick embryos in a 0.03 % solution of collagenase (Millipore : 136 U/mg) for one hour at 37°C, according to Gumpel - Pinot et al (1978). The same procedure was applied to 4-day chick embryonic allantois in order to obtain its endodermal layer. Two kinds of heterotopic reassociations were performed : 5 1/2-day intestinal mesenchyme was associated either with 5-day gizzard endoderm (Im/Ge) or with 4-day allantoic endoderm (Im/Ae). In order to achieve their mutual adhesion, the associations were cultured overnight on agar-solidified medium (Wolff and Haffen, 1952). In order to favor their growth and differentiation, the associations were grafted for 11 or 12 days into the coelomic cavity of 3-day chick embryos above the omphalo-mesenteric vessels. The morphologic or enzymic behavior of these two kinds of heterotopic recombinants were compared: 1) to grafts of 5-day chick gizzard mesenchyme recombined with its own endoderm (Gm/Ge), 2) to gizzard (G) and intestine (I) dissected out from 16-day chick embryos, stage corresponding to that presumably attained by the grafted recombinants.

Scanning Electron Microscopy (SEM)

The grafts were fixed for 1 hour at 4°C with 2 % glutaraldehyde in 0.1 M sodium cacodylate buffer, pH 7.4. They were then dehydrated, dried in a critical point drier (Balzers Union) and coated with gold using a sputter coater (Balzers Union). The specimens were examined in a Philips 501 B scanning electron microscope.

Electron Microscopy

The grafts were fixed in 0.2 M cacodylate buffered 2 % glutaraldehyde (pH 7.4) for 2 hours at 4°C, postfixed in 0.2 M cacodylate buffered 1 % osmium tetroxyde (pH 7.4) for 30 min at 4°C and embedded in araldite. Ultrathin sections were double stained with uranyl acetate and lead citrate and examined with a Philips 300 electron microscope.

Biochemical Analysis

Enzyme activities were measured in the fraction containing the brush border membranes according to the method previously described (Raul et al, 1978 ; Kedinger et al, 1981). Sucrase and maltase activities were determined according to Messer and Dahlqvist (1966), alkaline phosphatase was determined according to Garen and Levinthal (1960). Proteins were assayed by the method of Lowry et al (1951). All enzyme activities were expressed as milliUnits per milligramm of protein. One Unit is defined as the activity that hydrolyzes 1 µmole of substrate/min under the experimental conditions.

Statistical Analysis

Results were expressed as the mean $\pm$ SEM. Student's "t" test was used to analyze the data for statistical significance of differences between means. Differences with a P value of less than 0.05 were considered significant.

RESULTS

Morphological Characteristics of the Heterotopic Recombinants

The heterotypic differentiation of gizzard and allantoic endoderm resulting from their association with small intestinal mesenchyme, previously reported by Gumpel-Pinot et al (1978) and by Yasugi and Mizuno (1974, 1978), is confirmed in the present study.

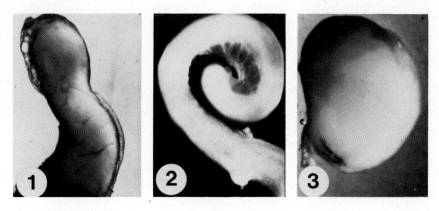

Figs 1,2,3 : Morphological characteristics of the grafted recombinants observed under the dissecting microscope.
Fig. 1 : Im/Ge (x 19) ; Fig. 2 : Im/Ae (x 19) ; Fig. 3 : Gm/Ge (x 10).

As illustrated in figures 1 and 2, both types of recombinants (Im/Ge and Im/Ae) gave rise, as intracoelomic grafts, to well vascularized intestinal segments, whose growth represented generally a tenfold increase of their original size. The gizzard endoderm recombined with its own mesenchyme (Gm/Ge) differentiated always homotypically (Fig. 3). The observation of SEM specimens showed that the Im/Ge recombinants displayed a zig-zag pattern of previllous ridges (Fig. 5) similar to that of the normally developed 16-day embryonic chick small intestine (Fig. 4) already described by Grey (1972). This zig-zag pattern appeared somewhat disturbed in the Im/Ae recombinants (Fig. 6).

At the ultrastructural level, the absorptive cells of Im/Ge recombinants (Fig. 8) have elongated shape, contain numerous mitochondria and exhibit brush borders similar to those present in the enterocytes of normal 16-day embryonic small intestine (Fig. 7). Although, like in the Im/Ge recombinants, intestinalization of the allantoic endoderm occured in the Im/Ae recombinants, it appeared that the brush border of the enterocyte-like cells were composed of short and irregular microvilli (Fig. 9).

Enzymic Characteristics of the Heterotopic Recombinants

They are illustrated in table 1, as well as those of normally developed chick embryonic intestine (I) and gizzard (G) at a stage corresponding presumably to that attained by the grafted recombinants (16 days of incubation). As previously reported by Kedinger et al (1981), the normally developed chick small intestine (I) displays enzyme activities such as sucrase, maltase and alkaline phosphatase. In the normally developed 16-day embryonic gizzard (G) the level of alkaline phosphatase was lower than in the intestine (I) ($p < 0.001$), maltase activity was negligeable and sucrase activity undetectable.

Table 1 : Specific enzyme activities expressed as mU/mg protein (mean $\pm$ SEM) of the normal 16-day small intestine (I), gizzard (G) and of the grafted heterotopic recombinants Im/Ge and Im/Ae. n = number of enzymic determinations.

	Sucrase	Maltase	Alkaline Phosphatase
I (n=30)	133.07 ± 11.35	143.39 ± 17.48	92.35 ± 10.36
G (n=9)	0.22 ± 0.05	3.97 ± 0.24	19.19 ± 2.13
Im/Ge (n=44)	27.33 ± 1.91	25.02 ± 2.29	21.00 ± 1.97
Im/Ae (n=43)	2.74 ± 0.45	6.21 ± 0.53	12.62 ± 0.85

The Im/Ge recombinants displayed typical brush border enzyme activities : sucrase and maltase, but their levels were approximately five times lower ($p < 0.001$) than in the

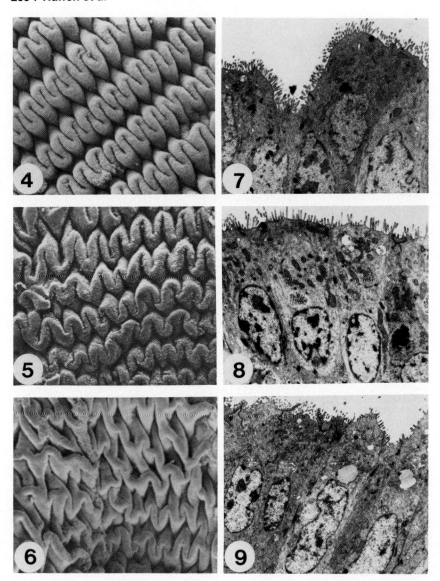

Figs 4,5,6 : SEM observation of the luminal surface : 16-day embryonic intestine. (Fig. 4 ; x 113), grafted Im/Ge (Fig. 5; x 113) and Im/Ae (Fig. 6 ; x 140) recombinants.

Figs 7,8,9 : Electron micrographs of enterocytes : 16-day embryonic intestine (Fig. 7 ; x 4437), grafted Im/Ge (Fig. 8; x 4437) and Im/Ae (Fig. 9 ; x 2936) recombinants.

control intestine (I). In contrast to the disaccharidases, alkaline phosphatase was not enhanced by the mesenchyme since its level was not significantly different from that present in the gizzard (G).

In the Im/Ae recombinants, only traces of sucrase and negligeable amounts of maltase could be detected. Both activities were much lower than those present in the Im/Ge recombinants ($p < 0.001$), sucrase activity being however significantly over that of the gizzard ($p < 0.025$). Concerning alkaline phosphatase, its activity was below that present in the Im/Ge recombinants ($p < 0.001$) and even below that present in the gizzard ($p < 0.05$).

DISCUSSION

In the present paper we presented evidence that 5-day gizzard endoderm can be induced by intestinal mesenchyme to synthesize brush border enzymes (sucrase and maltase) in conformity with the induced-intestinal morphology previously reported by Gumpel-Pinot et al (1978). However, the levels of brush border enzyme activities attained in the recombinants were considerably lower than those present in the normally developed 16-day intestine, despite the similarities in the ultrastructural differentiation of the absorptive cells. Our results suggest a delayed enzymic differentiation which cannot be attributed to the intracoelomic grafting procedure since previous data showed no modifications in the enzymic pattern of grafted 5 1/2-day chick intestine for example (Kedinger et al, 1981). An age-related loss of competence can also be excluded since, as shown in some additional experiments (n=7) more enzyme synthesis occured in the gizzard endoderm associated at 5 1/2 days of incubation (sucrase : 51.70 ± 5.03 mU/mg prot. ; maltase : 75.47 ± 7.61 mU/mg prot.). Anyhow, this enzymic differentiation, although delayed, represents the first documentation about the synthesis of enzymic proteins foreign to the organ from which the epithelial anlage was taken. These results corroborate the findings of Hata and Slavkin (1978) which provided evidence of de novo induction of type II collagen in the mouse tooth mesenchyme as a direct consequence of its recombination with avian limb epithelium.

Concerning the biochemical responsiveness of the allantoic endoderm, it is noteworthy that, taken at 4 days of incubation, the latter was unable to synthesize noticeable amounts of true brush border enzymes when recombined with intestinal mesenchyme, despite an ultrastructural differentiation of intestinal type. Considering this fact, it is interesting to note that during normal development, as soon as microvilli appear at the surface of intestinal cells (11 days of incubation according to Overton and Shoup, 1964 and to Chambers and Grey, 1979), brush border enzymes were present (unpublished data of our laboratory). The absence of correlation between morphology and biochemistry can be related to some observations made previously by Yasugi (1979). According to this author, the allantoic endoderm easily differentiates into an intestinal epithelium when recombined with aged oesophagus or gizzard mesenchymes having lost their region-specific inductive ability. Thus, our biochemical data strongly suggest that the intestinal mesenchyme when associated with allantoic endoderm behaves like a "neutral" one, since it does not alter the natural direction of differentiation of this endoderm. They also demonstrate that unlike the 5-day gizzard, the allantoic endoderm, at least at 4 days, has little if any competence for de novo synthesis of intestinal brush border enzymes. Its self-differentiation capacity represents at this stage a morphological event only, but the question arises whether a biochemical responsiveness could be expressed with younger allantois.

REFERENCES

Chambers C, Grey RD (1979). Development of the structural components of the brush border in absorptive cells of the chick intestine. Cell Tissue Res 204 : 387.

Dhouailly D, Rogers GE, Sengel P (1978). The specification of feather and scale protein synthesis in epidermal-dermal recombinations. Develop Biol 65 : 58.

Garen A, Levinthal C (1960). A fine structure genetic and chemical study of the enzyme alkaline phosphatase of E. Coli I. Purification and characterization of alkaline phosphatase. Biochim Biophys Acta 38 : 470.

Grey RD (1972). Morphogenesis of intestinal villi. I. Scanning electron microscopy of the duodenal epithelium of the developing chick embryo. J Morph 137 : 193.

Gumpel-Pinot M, Yasugi S, Mizuno T (1978). Differenciation d'épithéliums endodermiques associés au mésoderme splanchnique. CR Acad Sci 286 : 117.

Haffen K, Kedinger M, Simon PM, Raul F (1981). Organogenetic potentialities of rat intestinal epithelioid cell cultures. Differentiation 18 : 97.

Hata RI, Slavkin HC (1978). De novo induction of a gene product during heterologous epithelial mesenchymal interactions in vitro. Proc Natl Acad Sci 75 : 2790.

Kedinger M, Simon PM, Grenier JF, Haffen K (1981). Rôle of epithelial-mesenchymal interactions in the ontogenesis of intestinal brush-border enzymes. Develop Biol 86 : 339.

Le Douarin N, Bussonnet C (1966). Détermination précoce et rôle inducteur de l'endoderme pharyngien chez l'embryon de poulet. CR Acad Sci 263 : 1241.

Le Douarin N, Bussonnet C, Chaumont F (1968). Etude des capacités de différenciation et du rôle morphogène de l'endoderme pharyngien chez l'embryon d'oiseau. Ann Embryol Morph 1 : 29.

Lowry OH, Rosebrough NJ, Farr AL, Randall RJ (1951). Protein measurement with the Folin phenol reagent. J Biol Chem 193 : 265.

Messer M, Dahlqvist A (1966). A one-step ultramicromethod for the assay of intestinal disaccharidases. Ann Biochem Biophys 14 : 376.

Overton J, Shoup J (1964). Fine structure of cell surface specializations in the maturing duodenal mucosa of the chick. J Cell Biol 21 : 75.

Raul F, Simon PM, Kedinger M, Grenier JF, Haffen K (1978). Separation and characterization of intestinal brush border enzymes in adult rats and in suckling rats under normal conditions and after hydrocortisone injections. Enzyme 23 : 89.

Rutter WJ, Weber CS (1965). "Developmental and metabolic control mechanisms and neoplasia". Baltimore : Williams and Wilkins, p 195.

Sakakura T, Nishizuka Y, Dawe CJ (1976). Mesenchyme dependent morphogenesis and epithelium-specific cyto-differentiation in mouse mammary gland. Science 194 : 1439.

Sengel P (1970). Study of organogenesis by dissociation and reassociation of embryonic rudiments in vitro. In Thomas JA (ed) : "Organ culture". New York : Academic Press, p 379.

Wolff Et, Haffen K (1952). Sur une méthode de culure d'organes embryonnaires in vitro. Texas Rep Biol Med 10 : 463.

Yasugi S (1979). Chronological changes in the inductive ability of the mesenchyme of the digestive organs in avian embryos. Develop Growth Differ 21 : 343.

Yasugi S, Mizuno T (1974). Heterotypic differentiation of chick allantoic endoderm under the influence of various mesenchymes of the digestive tract. Wilhelm Roux'Arch 174 : 107.

Yasugi S, Mizuno T (1978). Differentiation of the digestive tract epithelium under the influence of heterologous mesenchyme of the digestive tract in the bird embryos. Develop Growth Differ 20 : 261.

Acknowledgments : The authors are particularly grateful to
    Dr. M. GUMPEL-PINOT for helpful discussions. They
    would like to thank E. Alexandre, D. Daviaud and
    F. Gosse for their excellent technical assistance.
    This work was supported by the CNRS, a grant from the
    INSERM (CRL N° 80.70.17) and the Ziring Donation.

ACID PHOSPHATASE OF SCHIZOSACCHAROMYCES POMBE IS INVOLVED IN MORPHOGENETIC AND BEHAVIOURAL DIFFERENTIATION

M.E. Schweingruber, A.M. Schweingruber,
M.E. Schüpbach* and F. Schönholzer
Institute for General Microbiology, Bühlstr. 24
CH- 3012 Bern
*Biological and Pharmaceutical Research Department
F. Hoffmann-La Roche and Co., Ltd. CH-4002 Basel

INTRODUCTION

The fission yeast Schizosaccharomyces pombe can be regarded as a simple system exhibiting behavioural and morphogenetic differentiation. Behavioural differentiation is manifested as an alteration in the pattern of cell agglutination, while morphogenetic differentiation is expressed as change in both cell size and shape. Both types of differentiation can be induced by appropriate growth media. For example, media low in nitrogen induce cell agglutination (Egel, 1971) while rich media yield larger cells than do poor media (Fantes and Nurse, 1977).
The biochemical basis of morphogenetic and behavioural differentiation is expected to be complex. It will certainly include biochemical alterations that affect cell surface structures. The cell surface, defined here as the complex of cell wall and cell membrane, determines both cell form and reactivity with other cells. To gain access to the biochemical structures and mechanisms regulating cell surface differentiation, we are studying the regulation and modulation of acid phosphatase. Previous studies with acid phosphatase of Saccharomyces cerevisiae have led us to the speculation that this enzyme is involved in morphogenetic and behavioural differentiation (Schweingruber and Schweingruber, 1981).
Acid phosphatase of S.pombe is, like the corresponding enzyme of the budding yeast, a cell surface glycoprotein regulated by extracellular orthophosphate (Dibenedetto, 1972). It has been purified to homogeneity and characterized (Dibenedetto

and Cozzani, 1975; Dibenedetto and Teller, 1981).The active protein has a molecular weight of roughly 360'000 and can be dissociated into subunits with a molecular weight of about 90'000. Neutral sugars account for about 66 % of the total molecular weight, mannose and galactose constituting the major components of the carbohydrate moiety.
In this communication we present data indicating that morphogenetic and behavioural differentiation of S.pombe is partly achieved by differential regulation and modulation of acid phosphatase.

RESULTS

Active and Inactive Forms of Acid Phosphatase

We have recently shown that acid phosphatase of S.cerevisiae can exist in an inactive as well as an active form. The inactive form seems to be mainly membrane-bound whereas the major portion of the active form is secreted and associated with the cell wall (Schweingruber and Schweingruber, manuscript submitted). To determine whether an inactive form also exists for S.pombe, we purified the enzyme essentially according to the published method (Dibenedetto and Cozzani, 1975). The last purification step - chromatography on Sepharose CL-6B - separates the applied material into two fractions. These are designated as I and II in Fig. 1. Fraction II represents active acid phosphatase as described by Dibenedetto and Cozzani (1975). It can be further separated into multiple active forms (Schönholzer and Schweingruber, unpublished). Material of peak I is enzymatically inactive, but crossreacts with antibodies raised against active acid phosphatase (data not shown). To establish the structural relationship between the material of peak I and active acid phosphatase, their tryptic peptide patterns were compared. As shown in Fig. 2, the material of peak I reveals many peptides which are identical to those of active acid phosphatase (for details see legend to Fig. 2). Thus, on the basis of immunological crossreactivity and partial identity of the peptide maps we conclude that the material of peak I represents inactive acid phosphatase. We do not yet know anything about the structure of the inactive form except that it is more complex than the active form. Preliminary data indicate, that it is associated with carbohydrates, lipids, phosphate and with strongly UV-absorbing compounds which have so far remained unidentified. The active form barely enters a 4 % polyacrylamide gel under denaturing conditions. A major

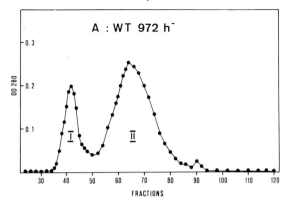

Figure 1: Elution pattern of acid phosphatase protein from wildtype 972.
Starting with 20 g wet cells which were grown in rich low orthophosphate medium (Schweingruber and Schweingruber, 1981) and harvested at the end of the logarithmic growth phase, acid phosphatase was prepurified and concentrated essentially as described by Dibenedetto and Cozzani (1975). The partially purified enzyme was applied in a volume of 2 ml, to a Sepharose CL-6B column (1.5 cm x 90 cm) and eluted with 0.1 M sodium acetate buffer pH 4.0. Flow rate was adjusted to 3.6 ml/h. Fractions of 1.2 ml were collected and measured for absorbance at 280 nm. Peak I designates inactive and peak II represents active acid phosphatase. For tryptic peptide maping the fractions 38-45 (I) and 58-74 (II) were pooled, dialyzed and lyophilized.

fraction of the inactive form can only be isolated when TritonX-100 is included in the extraction buffer (Schönholzer et al. unpublished). This indicates that, as in the case of S.cerevisiae, the inactive form of S.pombe is at least partially membrane-bound.

Mutants Altered in the Stucture of Acid Phosphatase

To examine the possible role of acid phosphatase in behavioural and morphogenetic differentiation, we isolated 71 mutants which were deficient in acid phosphatase activity and examined them for alterations in such properties as cell morphology and agglutination. The results are summarized in Table 1. The mu-

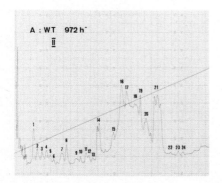 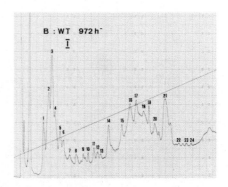

Figure 2: Tryptic peptide maps of active and inactive acid phosphatase form wildtype 972.
Acid phosphatase protein was digested with trypsin and the resulting peptides were separated on a C18 reverse phase column as described in detail elsewhere (Schweingruber and Schweingruber, manuscript submitted) by HPLC. The graphs represent the elution profiles of peptides recorded at 210nm. The oblique line marks the phosphoric acid-acetonitrile elution gradient (10% - 60% acetonitrile). To facilitate comparison of the peptide pattern, the corresponding peptides or peptide groups have been designated with a number. A and B: Tryptic peptide patterns of active (II) and inactive (I) acid phosphatase. The two proteins differ only in the yield of peptides 1-7. Slight differences, as observed for example for peptides 14 and 21 are due to run to run variations.

tants can be roughly divided into 3 classes. Apart from their lack of acid phosphatase activity, mutants of the largest class, I, do not substantially differ from wildtype. Mutants of class II exhibit aberrant cell morphology. The sole mutant assigned to class III, differs from wildtype not only in morphology, but also in agglutination behaviour. As indicated by the results of tetrad analyses (see legend to Table 1), aberrant cell morphology and agglutination behaviour cosegregate with the deficiency in phosphatase activity and can therefore not be accounted for by second-site mutations. It has to be stressed that the classification of the mutants is arbitrary. Many mutants of class I do differ slightly from wildtype in morphology and agglutination behaviour but the effects are rather slight, while most mutants of class II exhibit, in

Table 1: Number and properties of phol-mutants lacking acid-phosphatase activity

Class	Number of mutants	Morphology	Agglutination
I	47	normal	normal
II	23	altered	normal
III	1	altered	altered

The mutants listed in the Table were selected on the basis of their lack of acid phosphatase activity in colonies on plates. The isolation procedure has been described elsewhere (Schweingruber et al., manuscript submitted). Morphology was examined under the microscope (see Fig. 3), agglutination was detected macroscopically (flocking cultures) and microscopically (see Fig. 3). The one mutant of class III and 3 strains of class II were crossed to the wildtype, and progeny cells of the spores of tetrades were examined for acid phosphatase activity, cell morphology and agglutination. In all tetrads analyzed (7 for mutant of class III, and 25 for mutants of class II) there was a 2:2 segregation for wildtype and acid phosphatase deficient phenotype. Aberrant cell morphology and agglutination behaviour cosegregated with deficient phosphatase activity.

addition to altered cell morphology, slight alterations in agglutination behaviour. Many of the mutant strains of class II and III are genetically unstable. This is indicated by the relatively high frequency of sectored colonies after straining for phosphatase activity. In many of the strains sporulation is also impaired. Photographs of wildtype cells and of cells from a representative mutant of class II and III are shown in Fig. 3. All the mutants were crossed to 3 tester phol-mutants, which had been shown in preliminary experiments to belong to the same locus of at least to be closely linked. 50-500 meiotic progeny clones were examined for phosphatase activity. In no instance did we find colonies containing phosphatase activity. This indicates that all mutants given in Table 1 map in the same genetic locus, called phol. Apparently this locus controls the activity of acid phosphatase as well as being important for cell morphology and agglutination.

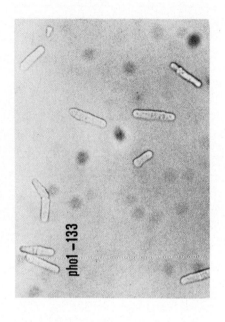
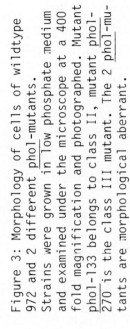
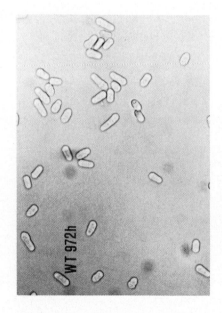
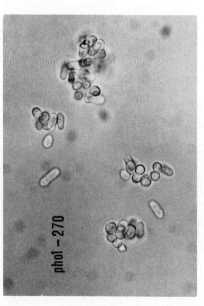

Figure 3: Morphology of cells of wildtype 972 and 2 different phol-mutants. Strains were grown in low phosphate medium and examined under the microscope at a 400 fold magnification and photographed. Mutant phol-133 belongs to class II, mutant phol-270 is the class III mutant. The 2 phol-mutants are morphological aberrant.

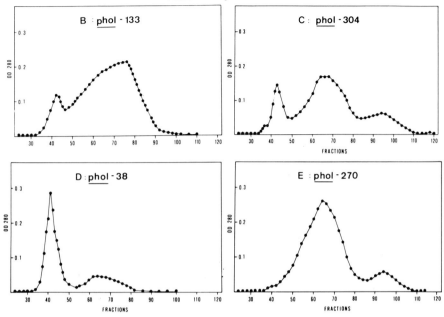

Figure 4: Elution patterns of acid phosphatase protein from 4 different phol-mutants.
Cells were grown and harvested and acid phosphatase was purified as described in legend to Fig. 1. Acid phosphatase of the phol-mutants is enzymatically inactive and recoveries could not be followed. Mutant phol-38 belongs to class I, mutants phol-133 and phol-304 are from class II and mutant phol-270 is the class III mutant. For tryptic peptide mapping the following fractions were pooled, dialyzed and lyophilized: phol-304: 40-46, phol-270: 50-76. For comparison with the elution pattern of the wildtype see Fig. 1.

To investigate the nature of the phol-locus, we purified mutationally inactivated acid phosphatase from 4 different phol-mutants. The protein elution patterns from the Sepahrose CL-6B column are given in Fig. 4. The absence of the inactive form (peak I) in mutant phol-270 is obvious. Mutant phol-38 exhibits an elution pattern almost reciprocal to that of the wildtype with low amounts of peak II and relatively large amounts of peak I material. For mutant phol-304, the protein recovery seems to be slightly reduced and for mutant phol-133 the shape

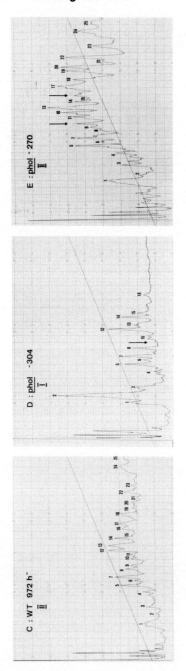

Figure 5: Tryptic peptide maps of acid phosphatase from wildtype 972 and two mutant strains. Peptide mapping was performed as described in legend of Fig. 2. The peptide maps cannot be compared with those of Fig. 2, since they were generated on a different column brand. C, D and E: Tryptic peptide patterns of active acid phosphatase from wildtype, the corresponding mutant protein of phol-270 and of inactive form of phol-304. The map of the phol-270 protein differs from the wildtype map at the two positions indicated by an arrow. The inactive form of phol-304 seems to lack almost completely the peptides eluting in the hydrophobic region of the gradient. In addition a peptide change is observed at the position indicated by an arrow. The small differences in yield observed for some peptides are, according to our experience, not significant. We have verified the indicated peptide changes by mixing digests of wildtype and mutant proteins (data not shown).

of elution peak II is unusual.
The purified proteins of mutant pho1-270 and the inactive form of pho1-304 were digested with trypsin and the resulting peptides were mapped by high pressure liquid chromatography and compared with the map of the wildtype protein (Fig. 5). For both mutants we detected many peptides that were identical to wildtype peptides. This demonstrates that the isolated mutant proteins really represent inactivated acid phosphatase and not a copurified contaminant. We find a few alterations in the peptide map of the active and inactive acid phosphatase (for details see legend to Fig. 5). This indicates that the acid phosphatase of both mutants is structurally altered, and that the active form as well as the inactive form of acid phosphatase is under the genetic control of the pho1-locus.

Acid Phosphatase from Cells Grown under Different Physiological Conditions.

There is evidence in S.cerevisiae that the yeast cell has not only the capacity to regulate the amount of acid phosphatase but also the ability to modulate the properties of the enzyme (Schweingruber and Schweingruber, 1981). If acid phosphatase is involved in morphogenesis and cell agglutination as suggested by the phenotype of some pho1-mutants it follows that the modulation and regulation of acid phosphatase must constitute a part of the mechanisms responsible for the morphogenetic and behavioural differentiation of the yeast cell. To examine this question in S.pombe we purified and compared the enzyme from cells grown in a minimal medium supporting only slow growth and in a rich medium allowing fast growth. The cells cultured in the minimal medium are smaller than those grown in the rich medium (Fantes and Nurse, 1977). In addition to these cell size differences we also observed differences in agglutination behaviour after cells had been transferred from either minimal or rich medium to a medium inducing agglutination. We do not yet have a good quantitative assay for agglutination and can therefore not express differences in clumping behaviour in quantitative terms. Amounts and some properties of purified acid phosphatase isolated from the cells grown in the two different media are given in Table 2. It is apparent that acid phosphatase from the two cell populations differs mainly in the ratio of inactive to active form. This ratio is shifted towards the side of the active form in cells grown in the minimal medium by both decreasing the amount of inactive and increasing the amount of

Table 2: Amount of acid phosphatase isolated from cells grown in a rich low orthophosphate medium (YPD-P) and a minimal medium (EMM-P).

	Inactive form	Active form	
YPD-P	7.2	28.8	(335)
EMM-P	2.2	307	(309)

Cells were grown in a rich low orthophosphate medium (YPD-P, legend Fig. 1) and in a minimal medium containing limiting amounts of phosphate (EMM-P, Bostock, 1970). Acid phosphatase from logarithmic phase cells was isolated according to the method mentioned in the legend to Fig. 1. The recovery of the active enzyme was 30% for the cells from the YPD-P medium and 42% for the EMM-P cells. The numbers in the Table refer to the amount of isolated inactive and active acid phosphatase given as μg protein recovered from 1g wet cells. The numbers in parenthesis give the specific activities (U/mg protein, Schweingruber and Schweingruber, 1981) of the active enzyme. Active acid phosphatase from the two enzyme preparations differed in the ratio of carbohydrates to protein. The percentage of protein (w/w) was 2.4 for cells grown in YPD-P and 15.8 for cells grown in the minimal medium.

active acid phosphatase. The active forms seem also to differ in their degree of glycosylation. These results indicate that differential glycosylation and differential expression of active and inactive acid phosphatase are at least partly responsible for the control of morphogenesis and cell-cell interaction.

DISCUSSION

We have shown in this paper that acid phosphatase of S.pombe exists in at least two forms, an active form and an inactive form. Preliminary results indicate that the inactive form is membrane-bound, as it is in S.cerevisiae. Mutations in the genetic locus pho1 cause inactivation of acid phosphatase and concomitantly lead in many cases to alterations in cell morphology and additionally, in one case, to alterations in cell agglutination. To examine the nature of the pho1-locus, acid phosphatase of 4 different mutants was purified and subjected to peptide mapping. The results revealed that the mutants were altered in the structure of acid phosphatase and in the ratio between active and inactive forms. The most likely interpretation of the data is that the pho1-locus codes for the protein moiety of the acid phosphatase. These results imply that acid phosphatase is involved in the control of morphogenesis and cell agglutination. They both corroborate and extend the results of Ballou and coworkers (Ballou et al., 1980) who recently showed that mutants of S.cerevisiae defective in the carbohydrate moiety of cell surface mannoproteins are clumpy and altered in cell morphology. The analysis of the mutants has so far revealed nothing about the molecular mechanism by which acid phosphatase controls morphogenesis and cell-cell contact. The finding that out of 23 mutants aberrant in cell morphology only one is also abnormal in agglutination could possibly mean that the structural domain of the acid phosphatase molecule responsible for agglutination is small relative to the domain controlling morphogenesis. Of course other interpretations are possible. A comparison of acid phosphatase from cells differing in morphology and agglutination capacity suggests that glycosylation and the balance between inactive and active forms may be important factors determining morphogenetic and behavioural differentiation.

The mechanisms controlling cell size are part of the mechanisms controlling morphogenetic differentiation. Fantes and Nurse (1977) showed that cell size control is also related to cell cycle control. They found that the cell has a mechanism for monitoring cell size and communicating this to the cell division cycle. The fact that some pho1-mutants are altered in cell size suggests that acid phosphatase might be part of this cell sizing system.

A pho1-mutant not discussed in this report exhibits in addition to altered cell morphology slow growth (Schweingruber et al., manuscript submitted). This indicates that acid phosphatase is not only involved in differentiation but also in growth control.

ACKNOWLEDGEMENT

The authors thank Dr. M. Cunningham for critical reading of the manuscript and Mrs Hutmacher for typing the manuscript. This work was supported by the Swiss National Science Foundation.

REFERENCES

Ballou L, Cohen RE and Ballou CE (1980). Saccharomyces cerevisiae mutants that make mannoproteins with a truncated carbohydrate outer chain. J. Biol. Chem., 255: 5986.

Bostock CI (1970). DNA synthesis in the fission yeast Schizosaccharomyces pombe. Exp. Cell Res. 60: 16.

Dibenedetto G (1972). Acid phosphatase in Schizosaccharomyces pombe. Biochem. Biophys. Acta 286: 363.

Dibenedetto G and Cozzani I (1975). Nonspecific acid phosphatase from Schizosaccharomyces pombe. Biochemistry 14: 2847.

Dibenedetto G and Teller DC (1981). Molecular properties and active form of nonspecific acid phosphatase from Schizosaccharomyces pombe. J. Biol. Chem. 256: 3926.

Egel R (1971). Physiological aspects of conjugation in fission yeast. Planta 98: 89.

Fantes P and Nurse P (1977). Control of cell size at division in fission yeast by a growth-modulated size control over nuclear division. Exp. Cell Res. 107: 377.

Schweingruber ME and Schweingruber AM (1981). Modulation of a cell surface glycoprotein in yeast: Acid phosphatase. Differentiation 19: 68.

PATTERN FORMATION IN THE DEVELOPMENT OF *DICTYOSTELIUM DISCOIDEUM*

Ikuo Takeuchi, Masao Tasaka, Masakazu Oyama,
Akitsugu Yamamoto and Aiko Amagai
Department of Botany, Faculty of Science
Kyoto University
Kyoto 606, Japan

INTRODUCTION

The development of multicellular organisms is characterized by differentiation of various cell types not only in certain topological arrangements but also in certain proportions, thus resulting in formation of specific patterns.

In the development of the cellular slime molds, aggregation of amoebae is soon followed by differentiation of two types of presumptive cells, prestalk and prespore cells, along the antero-posterior axis. Fig. 1 shows the sequence of this differentiation in *D. discoideum* (Takeuchi et al., 1978). Prespore cells as identified by the staining with FITC conjugated immunoglobulin produced against spores of *D. mucoroides* (Takeuchi, 1963), first appear in the cell mound which is about to form a tip and are located around the basal region of the tip. As the tip elongates, prespore cells increase in number but are left behind the tip. On the other hand, the tip is always composed of unstained prestalk cells. In this manner, the formation of the normal prestalk-prespore pattern is completed before the standing slug stage. This indicates that this organism is a superb model system to study the problem of pattern formation, in that only two types of cells are arranged in a linear order.

A similar prestalk-prespore pattern is also configurated in a cell aggregate obtained in submerged conditions (Forman and Garrod, 1977; Sternfeld and Bonner, 1977; Takeuchi et al., 1977). When cultured in a roller tube, vegetative cells form a spheroidal aggregate, in which differentiated prestalk and prespore cells occupy each hemisphere, despite the fact that the aggregate has no apparent polarity.

How such a prestalk-prespore pattern forms during the slime mold development is not known, but two different mechanisms are conceivable. One is that the pattern is formed by each cell's differentiating according to its position within the aggregate; a cell which happens to come to the anterior of a slug somehow obtains an information on its position and is directed to become a prestalk cell. Another is that the pattern is formed by sorting out of cells which have beforehand differentiated independently of their positions; cells differentiate first into either prestalk or prespore cells and then they are sorted out to form the pattern.

On the other hand, there is ample evidence that cell sorting actually occurs during formation of a slug (Bonner, 1959; Takeuchi, 1969; Bonner et al., 1971; Leach et al., 1973; Maeda and Maeda, 1974; Feinberg et al., 1979). For example, Leach et al., (1973) showed that cells grown in a glucose-containing medium (referred to as G(+) cells) are mixed with

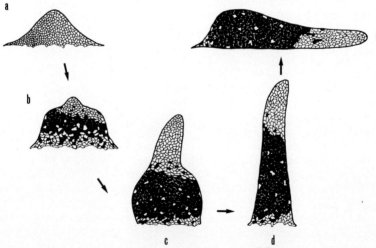

Fig. 1. Formation of prestalk-prespore pattern during aggregation and slug formation in D. discoideum NC4. Sections of cell aggregates were stained with FITC conjugated immunoglobulin produced against spores of D. mucoroides (Takeuchi, 1963). Black cells, stained prespore cells; white cells, unstained cells. From Takeuchi et al. (1978).

those grown in a non-glucose medium (referred to as G(-) cells), G(-) cells tend to occupy the anterior while G(+) cells the posterior portion of a slug. Despite many such evidence, the role of sorting out in the formation of the prestalk-prespore pattern has not been crucially investigated.

In this paper, we will present evidence to show (1) the role of cell sorting in the pattern formation, (2) the relationship between the proportion regulation and the pattern and (3) the topology of cell type conversion in the prespore isolate of a slug.

ROLE OF CELL SORTING IN PATTERN FORMATION

First, we asked the question whether sorting out precedes differentiation or is a consequence of it (Tasaka and Takeuchi, 1981). To examin this, we used an experimental system of G(+) and G(-) cells. We labeled G(-) cells of strain AX2 with [^{3}H]thymidine during the growth period and mixed with unlabeled G(+) cells at 1:2 ratio. The mixed cells were either placed on non-nutrient agar to allow them to undergo normal morphogenesis or cultured in roller tubes to make them form submerged round aggregates. Cell aggregates were fixed and sectioned at various stages of development. A section was first stained with the FITC conjugated antispore immunoglobulin to locate prespore cells and the same section was processed for autoradiography to locate G(-) cells.

During the normal development, prespore cells first appeared in an aggregate which was about to form a papilla (Hayashi and Takeuchi, 1976), as shown in Fig. 2a. The autoradiograph of the same section as Fig. 2a showed that this was also the first time we could see non-random distribution of G(+) and G(-) cells (Fig. 2b). Thus, differentiation of prespore cells and sorting out between G(+) and G(-) cells appeared to occur simultaneously during the normal morphogenesis. It is worth remarking here that most G(-) cells in the G(+) region corresponded to unstained, i.e., probably prestalk cells.

The results we obtained with submerged aggregates were somewhat different. Fig. 3a shows a section of an aggregate cultured in a roller tube for 15 h, in which stained prespore cells appeared for the first time in a certain portion.

However, as shown in the autoradiograph (Fig. 3b), labeled G(-) cells were still randomly distributed. This is the same in an 18 h aggregate where stained prespore cells now increased in number, but G(-) cells still showed a uniform distribution. In a 24 h aggregate, the sorting out between G(+) and G(-) cells was for the first time recognized (Fig. 4b) and their distribution coincided with that of stained and unstained cells, which now gave a typical prestalk-prespore pattern (Fig. 4a).

The above results showed that in a roller tube culture, the prespore differentiation preceded the cell sorting. On the other hand, the cell sorting coincided in time with the formation of the prestalk-prespore pattern. This was true during both the normal development and submerged culture.

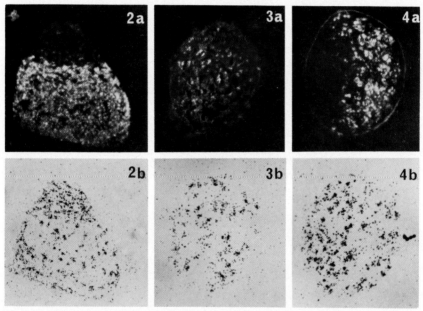

Fig. 2-4. Distributions of prespore (a) and labeled G(-) cells (b) within cell aggregates. The mixture of labeled G(-) and unlabeled G(+) cells (at 1:2 ratio) were cultured either on agar or in roller tubes (see text). Sections of the aggregates were stained with the FITC conjugated anti-spore globulin (a) and then processed for autoradiography (b). (2) An aggregate initiating papilla formation. (3 and 4) Aggregates fixed after 15 h and 24 h of roller tube culture, respectively. From Tasaka and Takeuchi (1981).

From these findings, we like to conclude that cells differentiate first into either prestalk or prespore cells and then the differentiated cells are sorted out to form the pattern. A similar idea has been advanced by Forman and Garrod (1977) based on a different ground. We now came to the same conclusion on the basis of the above findings.

If the above conclusion is right, the sorting out we observed in the above experiment must not have occurred between G(+) and G(-) cells, but between prestalk and prespore cells. That this was actually the case is indicated by the following fact. When more G(-) cells were mixed with less G(+) cells, more G(-) cells were distributed in the prespore region. This is probably because excess G(-) cells were forced to differentiate into prespore cells due to the proportion regulation. Accordingly, the sorting pattern between G(+) and G(-) cells became obscured, but the prestalk-prespore pattern was as clear as ever. This indicates that sorting occurred between prestalk and prespore cells. That sorting appeared to occur between G(+) and G(-) cells is merely due to the fact that they have predisposition to become prespore and prestalk cells respectively when mixed.

Turning back to normal development, there seems to be no reason to suppose that the mechanism of pattern formation is different in the aerial and submerged aggregates. In the normal morphogenesis, we presume that both prestalk and prespore cells differentiate independently of their position, in the upper part of a cell aggregate which is about to form a tip. However, the differentiated prestalk cells will soon be sorted out to form a tip, while prespore cells are left behind. During differentiation and subsequent sorting out of cells, a typical prestalk-prespore pattern will be established. Accordingly, the observed difference in the timing of sorting out between the aerial and submerged aggregates seems only due to the fact that in the aerial culture conditions, sorting out is accelerated by morphogenetic movement of the aggregate.

This indicates that in addition to differential adhesion, cell movement plays an important role in cell sorting. This idea has been supported by findings of Matsukuma and Durston (1979) and Sternfeld and David (1981). We have also shown that the anterior isolate of a slug has higher motive force than the posterior isolate, and that this difference in fact represents the difference between prestalk and prespore cells (Inouye and Takeuchi, 1980). According to our model of slug migration based on a dynamic equilibrium, cells with higher motive forces should become equilibrated at the anterior positions, while those with lower motive forces at the posterior ositions of a slug. Therefore, the difference in motive force is responsible for the sorting out, at least in part, (Inouye and Takeuchi, 1979).

CELL PROPORTION AND PATTERN FORMATION

The following two examples indicate that the regulation of the proportion of prespore cells within a cell aggregate is independent of the normal prestalk-prespore pattern.

Liquid Shake Culture

Recently, Okamoto (1981) developed a shaking culture system in which dissociated aggregative cells form small cell clumps and differentiate into prespore cells. When cells dissociated at the early aggregation stage were shaken in 20 mM KH_2-phosphate buffer (pH 6.0) containing 5 % glucose, 2 % albumin, 2 mM EDTA and 1 mM cAMP, they formed small clumps of 100-300 cells. At the beginning of the culture, no prespore cells were found, but during the incubation, cells stained with the antispore immunoglobulin increased in number and within 6 h, reached the maximum which amounted to 70-80 % of total cells, a value comparable to the normal prespore ratio found in an ordinary slug. We made sections of such a clump and found that prespore cells were scattered and showed no particular pattern of distribution (Fig. 5), indicating that cells can differentiate without having a definite pattern (Oyama, Okamoto and Takeuchi, unpublished). It is most likely that under the culture conditions, cell sorting was somehow

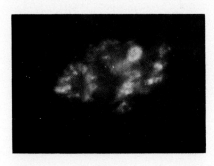

Fig. 5. A section of a cell clump formed in a liquid shake culture. Dissociated aggregative cells were cultured as described in the text and fixed after 18 h. Sections were stained with the FITC conjugated antispore globulin.

inhibited, but not the regulation of prespore proportion.

Here, two additional remarks should be made. One is that after a period of culture, about 20 % of cells became strongly stained with neutral red. This indicates that cells in clumps differentiated into prestalk cells as well as prespore cells. Secondly, the cell clumps were entirely devoid of a surface sheath, indicating that the sheath is not required for prespore or prestalk differentiation.

To examine whether the cells in clumps are able to regulate the proportion of prespore cells, without having the normal prestalk-prespore pattern, we dissociated migrating slugs and fractionated prestalk cells according to the method of Tsang and Bradbury (1981). The prestalk fraction contained about 10 % prespore cells (Tsang and Bradbury, 1981), but during the incubation in the above-mentioned culture medium, the proportion of prespore cells increased to a level which is close to the normal ratio in an ordinary slug (Table 1). In contrast, unfractionated dissociated cells kept the prespore ratio during the culture. This indicates that the normal two-zoned prestalk-prespore pattern is not required for regulation of proportion.

Table 1. Regulation of prestalk fraction in liquid shake culture

Culture time (h)	Percentage of prespore cells	
	Unfractionated slug cells	Prestalk fraction
0	77.2	12.9
18	75.2	55.5

Dissociated slug cells and prestalk fraction prepared therefrom (Tsang and Bradbury, 1981) were cultured as described in the text and stained with the FITC conjugated antispore globulin after being dissociated.

A Mutant

The above point is also illustrated by a temperature sensitive aggregateless mutant isolated by Ishida in our laboratory. The mutant cells aggregated normally at 21°C, but not at 27°C. When the cells were allowed to aggregate at 21°C and then shifted up to 27°C, they formed slugs, but finally almost all the cells differentiated into spores. We examined the distribution of prespore cells in migrating slugs and found that it was normal at 21°C (Fig. 6a) while the pattern was considerably deranged at 27°C, during 3 or 6 h of migration (Fig. 6b and c). Nevertheless, the proportion of prespore cells in slugs was found to remain normal during the period, either at 21°C or 27°C (Amagai, Ishida, Takeuchi, unpublished). This indicates that at 27°C, the mutant becomes defective in sorting out, but not in the regulation of proportion and again shows that the normal prestalk-prespore pattern is not a prerequisite to the proportion regulation.

REGULATION OF PRESPORE ISOLATES

We have examined whether the conversion of cell type (from prespores to prestalks) in the posterior isolate of a slug occurs dependent or independent of the slug axis. Observations of this sort is usually obscured by the fact that the posterior isolates stop migration and undergo morphological changes to start culmination. To avoid this, we used the method devised by Inouye and Takeuchi (1980). We let a slug migrate into a capillary of the slug diameter, made of 3 blocks of agar. When one-third of the slug went into the second agar block, this block was removed and the third block was connected to the first one. In this manner, the posterior two-thirds isolate left in the first block continued to migrate without a moment's stop. It is interesting to note

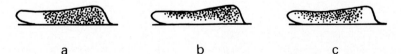

a   b   c

Fig. 6. Sections of slugs of a mutant migrating at 21°C (a) or 27°C (b and c). The mutant slugs were fixed after 6 h of migration. Sections were made perpendicular to the agar surface and stained with the FITC conjugated antispore globulin. Black dots represent stained prespore cells.

that it kept moving forward even before it began to regenerate its own tip after 1 h of isolation. Cells had previously been stained with neutral red to identify not only anterior (prestalk) cells but also anterior-like cells present in the prespore region (Sternfeld and David, 1981). To discriminate between newly converted prestalk cells (due to proportion regulation) and pre-existing anterior-like cells, we added Nile blue sulphate to the agar, so that during migration of the isolate, the latter picked up the stain to become violet while the former were stained blue. After letting it migrate for a period, we made secondary isolation in a similar way.

Fig. 7 shows the time courses of appearance of blue cells after the primary and secondary isolations of prespore region. A larger number of blue cells appeared in the secondary isolation, probably because cell type conversion was suppressed, due to the presence of the anterior-like cells, confirming

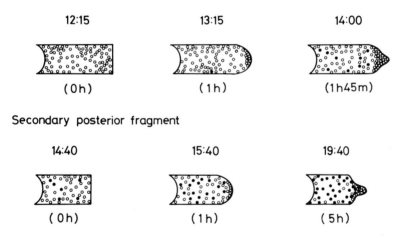

Fig. 7. Appearance of prestalk cells in the prespore isolates of a slug. Prespore isolate (two-thirs of the total length) was cut out in an agar capillary as described in the text. After 1 h and 45 min, the secondary isolation was made in a similar manner. The isolate continued to migrate after either isolation. Open circles, pre-existing anterior-like cells stained violet; closed circles, newly formed prestalk cells stained blue (see text). The numbers above and below (in parenthesis) each figure represent the real times of the day and the times after each isolation, respectively.

the result of Sternfeld and David (1981). In either isolation, however, blue cells, *i.e.*, cells converted due to regulation, did not appear in an antero-posterior sequence, but almost independently of their positions. The reason why it does not appear so during the normal process of regulation (Bonner *et al.*, 1955) is probably because cell sorting becomes more predominant due to the absence of migratory movement.

Unlike the posterior isolate, the evidence we now have with the anterior isolate indicates that cell type conversion is dependent upon the cell positions:when prestalk isolates were fixed after a period of regulation and stained with the antispore immunoglobulin, a single prespore cell was rarely found in the prestalk region (Takeuchi *et al.*, 1977). Whether this is because a gradient is secondarily set up in the prestalk region after formation of the pattern, or because prestalk cells are sorted out according to their predispositions so that more anterior cells are less liable to become converted, is not known at present.

CONCLUSION

Using glucose and non-glucose grown amoebae, we have shown that prespore differentiation precedes sorting out and that the sorting out coincides in time with the formation of the prestalk-prespore pattern. On the basis of these findings, it was concluded that cells differentiate first into either prestalk or prespore cells and the differentiated cells are then sorted out to form the pattern. We have also presented evidence that prestalk and prespore cells appear in a variety of configuration other than the normal two-zoned pattern. This strongly indicates that cells within an aggregate differentiate independent of their positions, and therefore seems to be inconsistent with the idea of positional information. Furthermore, the fact that the proportion regulation proceeds even in disorganized aggregates indicates that this process is also independent of a specific cell patterning. That the same principle is applicable to redifferentiation was shown in the prespore isolates where prestalk cells appeared independently of their positions, although redifferentiation in the prestalk isolates seems to be position-dependent.

Bonner JT, Chiquoine AD, Kolderie MQ (1955). A histochemical study of differentiation in the cellular slime molds. J Exp Zool 130:133.

Bonner JT (1959). Evidence for the sorting out of cells in the development of the cellular slime molds. Proc Natl Acad Sci USA 45:379.

Bonner JT, Sieja TW, Hall EM (1971). Further evidence for the sorting out of cells in the differentiation of the cellular slime mold *Dictyostelium discoideum*. J Embryol Exp Morphol 25:457.

Feinberg AP, Springer WR, Barondes SH (1979). Segregation of pre-stalk and pre-spore cells of *Dictyostelium discoideum*: Observations consistent with selective cell cohesion. Proc Natl Acad Sci USA 76:3977.

Forman D, Garrod DR (1977). Pattern formation in *Dictyostelium discoideum* II Differentiation and pattern formation in non-polar aggregates. J Embryol Exp Morphol 40:229.

Hayashi M, Takeuchi I (1976). Quantitative studies on cell differentiation during morphogenesis of the cellular slime mold *Dictyostelium discoideum*. Dev Biol 50:302.

Inouye K, Takeuchi I (1979). Analytical studies on migrating movement of the pseudoplasmodium of *Dictyostelium discoideum*. Protoplasma 99:289.

Inouye K, Takeuchi I (1980). Motive force of the migrating pseudoplasmodium of the cellular slime mould *Dictyostelium discoideum*. J Cell Sci 41:53.

Leach CK, Ashworth JM, Garrod DR (1973). Cell sorting out during the differentiation of mixtures of metabolically distinct populations of *Dictyostelium discoideum*. J Embryol Exp Morphol 29:647.

Maeda Y, Maeda M (1974). Heterogeneity of the cell population of the cellular slime mold *Dictyostelium discoideum* before aggregation, and its relation to the subsequent locations of the cells. Exp Cell Res 84:88.

Matsukuma S, Durston AJ (1979). Chemotactic cell sorting in *Dictyostelium discoideum*. J Embryol Exp Morphol 50:243.

Okamoto K (1981). Differentiation of *Dictyostelium discoideum* cells in suspension culture. J Gen Microbiol (in press).

Sternfeld J, Bonner JT (1977). Cell differentiation in *Dictyostelium* under submerged conditions. Proc Natl Acad Sci USA 74:268.

Sternfeld J, David CN (1981). Cell sorting during pattern formation in *Dictyostelium*. Differentiation (in press).

Takeuchi I (1963). Immunochemical and immunohistochemical studies on the development of the cellular slime mold

Dictyostelium mucoroides. Dev Biol 8:1.
Takeuchi I (1969). Establishment of polar organization during slime mold development. In Cowdry EV, Seno PS (eds): "Nucleic Acid Metabolism Cell Differentiation and Cancer Growth," Oxford: Pergamon Press, p 297.
Takeuchi I, Hayashi M, Tasaka M (1977). Cell differentiation and pattern formation in *Dictyostelium.* In Cappuccinelli P, Ashworth JM (eds): "Development and Differentiation in the Cellular Slime Moulds," Amsterdam: Elsevier/North-Holland, p 1.
Takeuchi I, Okamoto K, Tasaka M, Takemoto S (1978). Regulation of cell differentiation in slime mold development. Bot Mag Tokyo Special Issue 1:47.
Tasaka M, Takeuchi I (1981). Role of cell sorting in pattern formation in *Dictyostelium discoideum.* Differentiation 18:191.
Tsang A, Bradbury JM (1981). Separation and properties of prestalk and prespore cells of *Dictyostelium discoideum.* Exp Cell Res 132:433.

INTRASPECIFIC TISSUE INCOMPATIBILITIES IN THE METAGENETICAL
Podocoryne carnea M. SARS (*Cnidaria, Hydrozoa*)*

Pierre Tardent and Max Bührer

Zoological Institute, University of Zurich-Irchel
Winterthurerstrasse 190, 8057 Zürich, Switzerland

1. INTRODUCTION

The metagenetical cycle of *Podocoryne carnea* M. Sars (*Anthomedusae*) includes two alternating polymorphic generations like those of many other marine *Hydrozoa* (Tardent 1978). The asexually multiplying colonial polyps are attached to the substrate by a network of root-like, tubular stolons (Fig. 1). A particular type of polyps, the gonozoids, produces blastogenetically planktonic medusae (Avset 1961, Frey 1968, Bölsterli 1977, Fig. 5e) which represent the sexual phase of the cycle. The fertilized eggs shed by these gonochoristic medusae develop into planula larvae, which upon settling on an adequate substrate (Müller 1973) transform into a primary polyp. This autozoid forms stolons, out of which arise secondary polyps and later also medusa-producing gonozoids (Fig. 5d). The species, therefore, features three levels of organization, the stolon, the polyp and the medusa, listed here in the presumable order of their evolutionary sequence and complexity (Tardent 1978). The development of the medusa is a true neo-differentiation, since the cellular inventory of the latter is qualitatively different from that of the stolon and polyp.

When outgrowing stolons of *Podocoryne* collide with other stolons belonging to another clone they exhibit, like those

---

*)This investigation was supported by the Swiss National Science Foundation Grant Nr. 3.317.0.78.

of the closely related genus *Hydractinia* (Hauenschild 1954, 1956; Müller 1964, Ivker 1972), unmistakable reactions of incompatibility (Jauch 1977). This is expressed in the failure to anastomose with the opponent as occurs when stolons of the same clone meet. Stolons are therefore capable of distinguishing between "self" and "non-self" and of reacting accordingly.

Based on this finding the present study is dedicated to the following questions: 1. Does what holds for the stolons also apply to the two other levels of organization (polyp, medusa)? 2. Has the structural and biochemical neo-differentiation of the medusa gone so far, that its tissues have estranged to such a degree that they react to the polyp from which they were generated as if they were "non-self"? The experimental approaches from which answers to these questions were expected, followed 2 pathways. One, using the grafting technique, attempted to safeguard the normal cellular structure of the two opponents which were to be confronted homo- and heteroclonally as much as possible. The other intentionally destroyed this integrity by dissociating the tissues and letting the intra- and interclonally as well as homo- and heterotypically produced cell-mixtures reaggregate and regenerate.

2. MATERIAL AND METHODS

2.1. Material

The clones Nr. 2 (♂) and Nr. 6 (♀) of *Podocoryne carnea* were selected for experimental purposes using their different sexual status as a convenient marker. Grown on glass-slides (Braverman 1962) they were kept in artificial sea-water (WIMEX,KREFELD) at 15° C at a 12 hr day/night rhythm. They were fed twice a week with larvae of *Artemia*. The medusae were kept in glass-bowls (20 ml).

2.2. Stolonial confrontations

For testing the reactions of stolons following their intra- and interclonal encounters 2 small fragments of colonies were fastened to a slide at a distance of about 1 cm. Collisions of the outgrowing stolons could be observed in this intermediate region.

2.3 Grafting

Because of their cuticular covering (periderm) stolons could be grafted neither "inter se" nor onto polyps or medusae. Grafting between polyps (autozoids) was performed by implanting distal halves of donor individuals into the tubular bodies of host polyps (Fig. 2), which were always part of a small group of autozoids. The graft had to be pressed for some time against the rim of a hole previously opened in the flank of the receiver.

Transplantations between medusae were limited to the exchange of the manubria. By exerting a gentle pressure to the apex of the umbrella the manubrium protrudes from the subumbrellar cavity. It can be amputated with ophthalmological scissors and can be replaced by a homologous organ from another specimen. The graft had to be pressed gently against the cut curface produced by the removal of the host-manubrium. The same technique was adopted when the amputated host-manubrium was replaced by the distal portion of an autozoid polyp (Fig. 3).

2.4. Separation and recombination of polyp ecto- and endoderm

The ecto- and endodermal layers of autozoid polyps were mechanically separated from each other according to the procedure developed by Achermann (1980). When the isolated, tubular ectodermal epithelia of 2 specimens were to be joined by grafting they were pushed together over a glass-capillary. Recombinations of ecto- and endoderm were brought about by wrapping the latter into the ectodermal tube.

2.5. Dissociation and reaggregation

Dissociation of stolonial, polypoid and medusoid tissues was brought about by their pretreatment either with Ca^+- and Mg^+-free sea water (15'; Humphreys 1963) or heat-shock (20'

at 30° C; Schmid et al. 1981) followed by mechanical means
consisting of sucking the specimens in and out of the narrow
tip of a pipette. The cell-suspensions were filtered through
nylon fabric for removing remaining cell-clusters. Suspensions
from 2 different specimens were then mixed and brought to reaggregate in normal sea-water by centrifugation (3 $\frac{1}{2}$' at 2150
rpm). The aggregates were kept and observed in the depression
of glass-blocks containing 5 ml of millipore-filtered seawater to which 0.1 mgr streptomycine per ml had been added.
As soon as polyps had regenerated from reaggregates (Fig. 5c)
they were fed with *Artemia* larvae.

2.6 Histology

For histological examinations of chimaeras and aggregates
the specimens were fixed for 1 hr in a 2 $\frac{1}{2}$ % glutaraldehyde-
sea water-solution (4° C) and postfixed for 1 hr in 2% OsO_4
dissolved in sea water. After aceton-dehydration and embedding
in Spurr they were sectioned with a Reichert-Ultramicrotome
(0.5) µm and stained with Toluidine blue.

3. RESULTS

3.1 Intra- and interclonal parabiosis of intact tissues

Stolonial confrontations: In all of the 10 intraclonally
opposed stolonial systems the outgrowing tips readily anasto-
mosed (Jauch 1977) with stolons of the same clone standing in
their way, thus indicating that their tissues mutually recognized themselves as belonging to the same clone and reacted
accordingly. Analogous interclonal encounters opposing in 10
separate trials stolons of the clones 2 (♂) and 6 (♀) on the
other hand never lead to anastomosis. The stolonial tips continued their advance by proceeding either over and across or
alongside their opponents (Fig. 1). At the stolonial level
the 2 clones were therefore incompatible with each other. Such
stolonial tests were routinely repeated throught the entire
experimental period (2 years) in order to check whether or not
this situation remained stable.

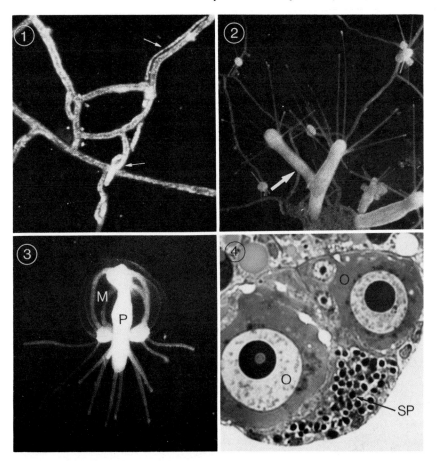

Fig. 1. Heteroclonal stolonial encounter. The arrows indicate the sites where stolons instead of anastomosing cross each other or run parallel to each other (34 x).

Fig. 2. Interclonal insertion (arrow) of a donor polyp into a host-polyp (autozoid) 16 d after grafting (13.5 x).

Fig. 3. Heterotypical and heteroclonal parabiosis between a polyp (P) and a host-medusa (M) the manubrium of which has been replaced by the graft 24 h before (25 x).

Fig. 4. Semi-thin section through a hermaphroditic manubrial gonad of a medusa which has been liberated by a colony reconstituted from a heteroclonal aggregate of polyp cells (cfr. Fig. 5, 850 x). O = oocytes, SP = spermatocytes.

Polyp-polyp-chimaeras: A total of 10 intraclonal and reciprocal transplantations of autozoid polyps (see 2.2) were carried out and observed for 28 days. The distal halves of donor polyps which had been grafted into the body column of receiver specimens were neither rejected nor could we detect any other signs of negative interactions. The grafts became integrated constituent of the host such as to form a double-headed, Y-shaped specimen (Fig. 2) of which both "heads" could be fed.

All the 14 interclonally produced polyp-polyp chimaeras exhibited unmistakable signs of incompatibility: After the grafts had taken well and had ingested food, both graft and host started resorbing their tentacles. After an average of 6 postoperational days the 11 grafts fell off and desintegrated while the host polyps gradually were resorbed by their colonies: In 3 instances the graft was resorbed together with its host. In 8 out of 14 cases this "killing effect" was spreading in the host colony leading also to the resorption of polyps which had been neighbours of the individual which had received the "non-self" graft. Although expressed in a different manner than in stolons the heteroclonal incompatibility holds true also for parabiotic polyps.

The exchange of manubria in between medusae: Medusoid manubria were homoclonally (n = 28) and heteroclonally (n = 34) exchanged (see 2.2). Four specimens died for unknown reasons. All other medusae not only survived the operation but accepted the grafted manubrium without suffering any ill-effects even when the graft came from the other clone. This means, that unlike in stolons and polyps, interclonally transplanted parts of medusae were compatible with each other.
23 of the heteroclonally produced chimaeras became gametogenetically active. The ectoderm of their manubria harboured spermatocytes and oocytes side by side which indicates that in these hermaphroditic specimens both the host's manubrial residues and the grafted manubrium had produced gametes in accordance with their obviously most stable status of sexual determination. From such sexual chimaeras planula larvae which originated from "self-fertilized" eggs were obtained. In a number of cases the graft had taken in a somehow eccentric position permitting the host medusa to regenerate a second manubrium close to the grafted one. In the cases of interclonal transplantation one manubrium produced eggs, the other spermatozoa.

The medusa-polyp chimaeras: The transplantations of distal halves of polyps to the site of a previously amputated medusa manubrium initially were unsuccessful because the grafted polyps were too large (Fig. 3). The excess weight hindered the locomotory activities of the host medusae, which sank to the bottom of the jars and died. This handicap was overcome by using smaller donor polyps.

Of the 8 successful attempts in which polyps were homoclonally transplanted to the medusa the polypoid grafts were not rejected but gradually resorbed by the host in between the 2nd and 5th post-operational days. In all cases the medusa subsequently regenerated a manubrium at the site where graft resorption had taken place. This homoclonal behaviour cannot be evaluated as being a reaction based on a true tissular incompatibility as it became evident in the case of 8 interclonal transplantations. Here, following the resorption of the grafted polyps the host medusae not only failed to regenerate the missing manubria but were subjected to a progressing disintegration that led to their death between the 9th and 12th post-operational days. Thus, it is clear that this heteroclonal and heterotypical parabiosis featured a "killing effect" similar to that found in polyp-polyp combinations.

3.2 Intra- and interclonal and allotypical cell-aggregates

The cell-aggregates produced (see 2.3) by intra- and interclonal mixing of elements from 2 different specimens were of spherical shape. Their variable volumes ranged between 0.2 and 0.8 mm^3. The spacial segregation of ectodermal and endodermal cells took place soon after reaggregation had occurred.

Stolon-stolon-combinations: From each clone 5 homoclonal reaggregates of stolonial cells were obtained of which one, throughout 32 days, did not exhibit any morphogenetic activity. The others attached themselves to the glass-surface by means of the periderm which is excreted by the ectodermal cells. The first stolonial outgrowths (Fig. 5 b) appeared 5-14 days after reaggregation and the reconstitution of the first autozoid polyp took 10-25 days. All these aggregates gave rise to healthy growing colonies.

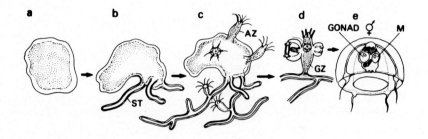

Fig. 5. Morphogenetic performances of a heteroclonal aggregate of polyp cells. a = 1 d after reaggregation, b = stolonial outgrowths (ST), 15 d; c = regenerated autozoid polyps (AZ), 25 d; d = individual gonozoid (GZ) with medusa buds, 150 d; e = hermaphroditic medusa with oocytes and spermatocytes developing on the manubrium (cfr. Fig. 4).

Of the 11 aggregates produced by mixing stolonial cells of both clones 1 failed to grow stolons and polyps while all others developed into normal colonies without any signs of clone-specific segregation of cells. This means that stolonial cells, when heteroclonally confronted in a dissociated and reaggregating state are compatible with each other and collaborate morphogenetically.

Polyp-polyp combinations: The homoclonal reaggregates prepared from polyp cells behaved in principle like the stolonial counter parts (see above). From each clone the fate of 10 such reaggregates was followed. Five from clone 2 disintegrated for unknown reasons before attaching to the substrate by means of the periderm and before having produced any stolonial outgrowth. The others fastened themselves to the substrate (3-9 d), 10 regenerated stolons (5-6 d) and 6 reconstituted polyps, of which finally 6 expanding colonies were obtained.

Of the 11 heteroclonal reaggregates of polyp cells 6 disintegrated within 14 d. From the 5 remaining specimens 3 developed stolons but 1 failed to regenerate polyps while 2 completed this ultimate reconstitutional process. Finally one rapidly growing colony was obtained which after 153 days started producing medusae (Fig. 5). The latter were all hermaphroditic as revealed by histological examination of the manubrial ectoderm (Fig. 4), which contained oocytes and spermatocytes side by side. From such medusae we obtained self-fertilized eggs that developed into viable planula larvae. This fact indicates that like stolonial cells the polypoid cells originating from different clones are compatible with each other and are capable of jointly realizing all 3 levels of organisation, the stolon, the polyp and the medusa.

Medusa-medusa-combinations: A total of 43 homoclonal reaggregates composed of medusa cells were prepared. Five of them decomposed within 9 days without having shown any morphogenetic activities. From the remaining 38 specimens, which failed to produce a periderm covering, 31 developed outer flagellae (cfr. Schmid 1972). Due to the beating of the latter the reaggregates were forced into rotational movements. Of these specimens 23 regenerated medusa-typical, umbrellar tentacles and 22 reconstituted a functional manubrium which eventually could be fed. In 13 cases the development of gametes was observed which appeared predominantly in the peripherial portions of the aggregate. Ten specimens, however, disintegrated after having regenerated various medusoid organs.

Of the 24 heteroclonal reaggregates 4 disintegrated within 9 days. The remaining 20 specimens behaved like the homoclonal analogons. Ten regenerated manubria and 4 of them produced simultaneously spermatozoa and eggs indicating that they harboured cells of both clones. From 2 individuals we obtained self-fertilized eggs. Neither homoclonal nor heteroclonal reaggregates of medusa cells ever regenerated any stolonial or polypoid structures nor did they reconstitute whole medusae. Their reconstitutional abilities were limited to the neoformation of isolated umbrella tentacles and of functional manubria including gonads.

Polyp-medusa-combination: Mixed reaggregates composed of polyp- and medusa cells were made homoclonally (n = 20) and heteroclonally (n = 23). Segregation of polyp cells from medusa cells never occurred. The cells originating from 2 different levels of organisation collaborated in the following manner: Within 10 days medusoid organs such as umbrellar tentacles and manubria were reconstituted (n = 30). Two homoclonal aggregates (clone 2) produced ripe testes and 4 (clone 6) released eggs. Eighteen specimens had produced external flagellae. After about 3 weeks these medusa typical products were gradually resorbed. The ectoderm of 18 aggregates started to produce a stolon-specific periderm cover and in one single case a small polyp was generated but soon resorbed again. Most of the aggregates decomposed after varying length of time without having reconstituted any defined and lasting polypoid or medusoid structures. This may be interpreted as being the expression of an unsolved conflict between the divergent morphogenetic tendencies of polyp- and medusa cells uncapable of segregating but equally uncapable of collaborating morphogenetically.

3.3 Heteroclonal reassembly of separated ecto- and endoderm.

After interclonal incompatibility has been demonstrated to exist in stolons and polyps in the cases of transplantational parabiosis (3.1) and to be absent in all combinations of cellular reaggregates (3.2) it was of some interest to check the corresponding behaviour of the isolated but intact and heteroclonally recombined ectodermal and endodermal layers.

Ectoderm-ectoderm: The ectoderm of 8 polyps of each clone was separated from the underlying mesogloea and endoderm. The tubelike ectoderm of 2 polyps from different clones were joined together over a glass-needle which was removed after the tissues had joined firmly. Two of the grafts separated from each other after 7 and 15 days while the remaining 6 regenerated the missing endoderm according to Achermann's (1980) observations. Within 4 weeks all specimens had produced stolons and polyps which partly arose from the stolons and partly directly from the central cell mass. In 2 instances the regenerated polyps produced medusae buds, which however never terminated their development. Although originating from different clones the 2 joined ectodermal epithelia of polyps not only failed to show any signs of incompatibility but jointly regenerated the missing endoderm and reconstituted stolons and polyps.

Ectoderm-endoderm: The separated ectodermal epithelia and the endodermal cell-masses were interclonally recombined. In less than 24 hrs the ectoderm of all 24 recombinants had completely enclosed the endoderm. Two specimens remained inactive at that stage while the ectoderm of the other synthesized a peridermal envelope. Seventeen specimens regenerated stolons and 2 produced autozoid polyps. The heteroclonal reassembly of polypoid ecto- and endoderm was not followed by dramatic interactions as observed in interclonal polyp parabiosis but lead to a harmonious collaboration between the 2 components.

4. DISCUSSION

In some *Cnidaria* as in *Hydra* (Kolenkine 1967, Artmann 1969, Stidwill 1981) the tissue and cellular incompatibility is limited to interspecific parabiotic confrontations. In *Podocoryne carnea* (Jauch 1977) and the closely related *Hydractinia echinata* (Hauenschild 1954, 1956; Müller 1964, Ivker 1972) however, there are also intraspecific i.e. interclonal reactions which are undramatically expressed at the level of stolons which anastomose with "self" but fail to do so with "non-self". The present study has shown that in parabiotic situations this interclonal incompatibility does apply also to the polyp level of organization but surprisingly enough not to the medusa, the cells of which along with their neo-differentiation (Bölsterli 1977) have either lost the appropriate receptors or the ability to react efficiently against a foreign graft. Another explanation would be that receptors are indeed present but that at the medusoid level informative macromolecules which would identify a specimen as belonging to another clone are missing. We are inclined to favour this last hypothesis because at the present state of knowledge we think that the actual carriers of the clone identifying information are contained in the extracellular matrices of the stolon and the polyp. As possible candidates we propose the extracellular matrices of the stolons and polyps (Chapman 1974): One is the external periderm which encloses the stolons in a chitinous tube and covers the polyps body as a mucuous layer; the other is the inner mesogloea separating the ectoderm from the endoderm. The following facts testify in favour of this hypothesis:
1. An advancing stolonial tip recognizes an opponent as self or non-self as soon as it has entered physical contact with

the obstacle's periderm, but before any cellular contacts are established (Jauch 1977). 2. While heteroclonal parabiotic confrontations of intact stolonial and polypoid tissues, which have retained their periderm and mesogloea, exhibit reactions of incompatibility the same combinations produced by dissociating and reaggregating cells not only failed to show any signs of non-compatibility but harmoniously collaborated morphogenetically. During the dissociation process, both the periderm and the mesogloea are removed. In a reaggregated state these extracellular matrices are reconstituted by both components and must therefore constitute a molecular mosaic agreeable to the cells of both clones. 3. Heteroclonal grafting of isolated ectodermal epithelia and the interclonal recombinations of ecto- and endoderm of polyps remained without negative consequences. In both cases at least the mesogloea had been removed before the components were joined.

The medusa has no periderm cover but a voluminous umbrella mesogloea. The question is, therefore, why does the latter not act as a clone specific signal of identification as with stolons and polyps ? We cannot offer any plausible answer to this question, as nothing is known so far about the possible differences regarding the structural and biochemical properties of the polypoid and the medusoid mesogloea. The latter is in fact a new product and may well differ from that of the polyp and stolon in as much as it could lack any clone specific components. Another still unsolved problem is related to the unilateral and mutual killing effects which have been reported from interspecific chimaeras of the genus *Hydra* (Artmann 1969 ,Noda 1970, Stidwill 1981) and observed here in *Podocoryne* when polyps were heteroclonally grafted "inter se" and polyps were implanted into a medusa of another clone.This phenomenon is most probably due to an invasion of the host and/or graft by cells of the opponent and to severe complications in establishing a harmonious collaboration between the host cells and the invading elements.

REFERENCES

Achermann J (1980). The fate and regeneration capacity of isolated ecto- and endoderm in polyps of *Podocoryne carnea* M. Sars (Hydrozoa, Athecata). In Tardent P, Tardent R (eds): "Developmental and cellular biology of coelenterates", Amsterdam, Elsevier/North-Holland, p. 273.

Artmann G (1969). Unverträglichkeitsreaktionen während und nach der Parabiose zwischen *Hydra attenuata* und *Hydra fusca*. Thesis, Zool. Inst. University of Zurich.
Avset K (1961). The development of the medusa *Podocoryne carnea* M. Sars. Nytt Mag Zool 10:49.
Bölsterli U (1977). An electron microscopic study of early developmental stages, myogenesis, oogenesis and cnidogenesis in the anthomedusa *Podocoryne carnea* M. Sars. J Morph 154: 259.
Braverman MH (1962). Studies on hydroid differentiation. I. *Podocoryne carnea*, culture methods and carbon dioxide induced sexuality. Exp Cell Res 27:301.
Chapman GB (1974). The skeletal system. In Muscatine L, Lenhoff HM (eds): "Coelenterate Biology. Reviews and Perspectives", New York: Academic Press, p 93.
Frey J (1968). Die Entwicklungsleistungen der Medusenknospen und Medusen von *Podocoryne carnea* M. Sars nach Isolation und Dissoziation. Wilhelm Roux' Archives 160:428.
Hauenschild C (1954). Genetische und entwicklungsphysiologische Untersuchungen über Intersexualität und Gewebeverträglichkeit bei *Hydractinia echinata* Flemm. (*Hydrozoa, Bougainvill.*). Wilhelm Roux'Archives 147:1.
Hauenschild C (1956). Ueber die Vererbung einer Gewebeverträglichkeits-Eigenschaft bei dem Hydroidpolypen *Hydractinia echinata*. Z Naturf 11b:132.
Humphreys T (1963). Chemical dissolution and in vitro reconstruction of sponge cell adhesions.I. Develop Biol 8:27.
Ivker FB (1972). A hierarchy of histo-incompatibility in *Hydractinia echinata*. Biol Bull 143:162.
Jauch U (1977) Ultrastrukturelle Untersuchungen des Stolonensystems von *Podocoryne carnea* M Sars (*Cnidaria,Athecata*). Thesis Zool Inst University of Zurich.
Kolenkine X (1967). Caractéristiques antigéniques et histocompatibilité chez trois espèces d'hydres d'eau douce: *Pelmatohydra oligactis, Hydra attenuata* et *Hydra vulgaris*. C R Acad Sci Paris 265:473.
Müller AW (1964). Experimentelle Untersuchungen über Stockentwicklung, Polypendifferenzierung und Sexualchimären bei *Hydractinia echinata*. Wilhelm Roux'Archives 155: 181.
Müller AW (1973). Metamorphose-Induktion bei Planulalarven. I. Wilhelm Roux'Archives 173:107.
Noda K (970). On the incompatibility of two species of *Hydra* (*H.magnipapillata* and *P. robusta*) in mixed culture. J Fac Sci Hokkaido Univ Ser VI 17:440.

Schmid V (1972). Untersuchungen über Differenzierungsvorgänge bei Medusenknospen und Medusen von *Podocoryne carnea* M Sars Wilhelm Roux'Archives 190:143.
Schmid V Stidwill R Bally A Marcum B Tardent P (1981). Heat dissociation and maceration of marine cnidaria. Wilhelm Roux'Archives 190:143.
Stidwill R (1981). Interspezifische Inkompatibilitäten zwischen Arten der Gattung *Hydra (Hydrozoa,Cnidaria)*. Thesis Zool Inst University of Zurich.
Tardent P (1978) "Morphogenese der Tiere: *Coelenterata, Cnidaria.*" Jena: VEB Gustav Fischer Verlag, p.71.

EFFECTS OF RETINOIC ACID ON DEVELOPMENTAL PROPERTIES OF THE FOOT INTEGUMENT IN AVIAN EMBRYO

by DANIELLE DHOUAILLY

*Laboratoire de Zoologie et Biologie animale
Université Scientifique et Médicale de Grenoble
ERA CNRS 621, BP 53 X   38041 Grenoble Cedex,
France.*

Feathers are the characteristic cutaneous appendages of birds. In most species however, as in the chicken, feathers do not cover the entire body; feet are covered with scaly skin. Three types of scales are distinguished on the feet of chickens : large distally overlapping scales (*scuta*), arranged in two longitudinal rows on the anterior face of the tarsometatarsus and one row on the upperface of each toe; smaller proximally overlapping scales (*scutella*), arranged in two longitudinal rows on the posterior face of the shank; smallest non-overlapping scales (*reticula*) covering the remainder of the posterior face of the foot.

In previous years, we have shown that the administration of a simple molecule, retinoic acid, to the chick embryo (from Leghorn breed or F2 crosses of Wyandotte x Rhode Island Red breeds) may cause the development of feathers on the scales (Dhouailly and Hardy, 1978). Further studies (Dhouailly, Hardy and Sengel, 1980) have shown that a strict correspondence may be established between the time sequence of appearance of scales and their sensitivity to the retinoid. In normal embryo, the large anterior tarsometatarsal *scuta* appear first, during the 10th day of incubation; posterior *scutella* later, on the 11th day; and *reticula* last, from 12th day to 14th day. Correspondingly, only those *scuta* and *scutella*, which become discernible as opaque patches, or *reticula*, which appear as elevations, during the 24 h following injection, are affected by the retinoic acid treatment.

In view of the existence of domestic breeds of fowl

with feathered feet, it is clear that retinoic acid (RA) somehow interferes with normal scale morphogenesis and thereby reveals a latent ability of avian foot integument to produce feathers. The mechanisms of action of RA on the chick foot integument remains to be determined. Several questions arise : What changes in integument may be observed in the hours following injection, at the morphological level and at the ultrastructural level ? What is the minimum time required for RA action ? Does the drug act on both skin tissues directly, or only on one of them, and in the latter case, does the drug act on dermis or epidermis ?

In order to answer these questions, the early effects of RA on the integument of the anterior face of tarsometatarsus were studied after a single injection on day 10, which causes the formation of feathers on *scuta* in 57% of treated embryos (Dhouailly, Hardy and Sengel, 1980).

Retinoic acid *(all-trans-β-retinoic acid,* provided by Hoffmann-La-Roche, Basel) was dissolved in dimethylsulfoxide (DMSO), and injected (125 μg/0.03 ml/embryo) into the amniotic cavity of 10-day chick embryos. At increasing intervals after injection, pieces of tarsometatarsal skin were removed from both feet; one piece was fixed and impregnated with osmium tetroxide, then embedded in Epon, for optimal visualization of scale and feather primordia, and then prepared for electron microscopy observations; the other sample was explanted on the chick chorioallantoic membrane (CAM) to test its morphogenetic capacity. To answer the question of RA-target tissue, heterotypic dermal-epidermal recombinants were prepared, in which one of the two tissues was obtained from a RA-treated embryo, the other tissue being obtained from a non-treated embryo. As in the previous experiments, pieces of skin were removed from both feet, one was used for heterotypic recombination, the other one for control homotypic recombination. Skin tissues were dissociated with trypsin and the recombinants were cultured on the chick CAM for 5 days.

MORPHOLOGICAL OBSERVATIONS

The *scutum* primordia appear as opaque placodes **of rectangular shape,** during the 10th day of incubation. At 12 days the normal rectangular scales (at hump stage or definitive scale ridge stage, Sawyer 1972) are formed (Fig. 1).

The immediate effect of retinoic acid on the foot skin is to inhibit the formation of scale primordia by transforming them into feather primordia. In some cases, 24 h after treatment, the tarsometatarsal skin appeared as an homogeneous structure; in other cases after 24 h and in those after 48 h (Fig. 2), several smaller rudiments of more or less circular shape were seen instead of the normal rectangular scale rudiments. Each of these new abnormally shaped placodes may be interpreted as a feather rudiment, which has the property to give rise to a feather. These well-formed feathers, seven days later, were borne by the distal tip of scales (Fig. 3). The inhibition of scale morphogenesis thus appears to be temporary, so that, as soon as the drug is eliminated from the tissues, scales resume their development.

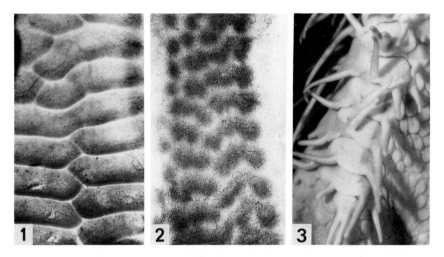

Figs. 1, 2 and 3 : anterior tarsometatarsal chick skin at 12 days of incubation (1) (DMSO-treated-control embryo) and 48 h (2) and 7 days (3) after one injection of retinoic acid at 10 days. (1) and (2) G x 20; (3) G x 13.

Feather morphogenesis is irreversibly under way, in all of the cases (that is 57% of treated embryos) only 48 h after the treatment. Indeed pieces of skin, showing feather primordia and explanted on the CAM 24 h after injection, produced, in more than half of the cases, only scales (Dhouailly, 1980 and Dhouailly, submitted).

ULTRASTRUCTURAL OBSERVATIONS

Several changes of the dermal-epidermal junction (DEJ) may be observed in the 24 h following injection (Dhouailly, submitted). These changes correspond to the destruction of the typical DEJ scale pattern. At 11 days, the scale DEJ is characterized (Demarchez, Mauger and Sengel, 1981) by the presence of a large number of small tubular dermal cell processes which have a characteristic orientation, parallel to the long axis of the scale. In longitudinal sections of the tarsometatarsus of a 24 h-DMSO-treated control embryo, these tubular cell processes appear as small round circles (Fig. 4). After RA-treatment, this privileged orientation of dermal cell processes was disrupted and it was replaced by very close parallel apposition of dermal cell membranes against the basement membrane *lamina densa* of the epidermis (Fig. 5), which is a characteristic feature of feather-forming skin.

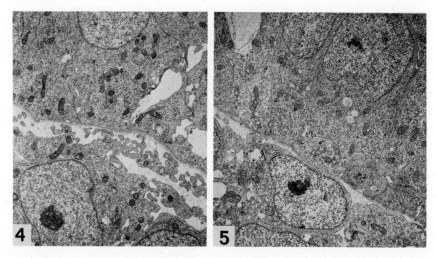

Figs. 4 and 5 : Tarsometatarsal dermal-epidermal junction at 11 days, that is 24 h after the injection of DMSO (control embryo) (4) and retinoic acid (5). G x 5000.

There were other feather-forming features which appeared in the tarsometatarsal skin after RA treatment. These are the formation of so-called anchor filaments, which are filaments at right angle to the basement membrane of the epidermis and which are the first indication of feather morphogenesis (Fig. 6). At the level of insertion of the anchor filaments in the basement membrane, direct cell to cell contacts, between either dermal or epidermal cell processes, were frequently observed (Fig. 7). Therefore it appears that the early effect of retinoic acid is to transform the ultrastructure of the dermal-epidermal junction of the tarsometatarsal skin into a morphological organisation which resembles to that of feather-forming skin. However, this transformation is still reversible (see above) in more than half of the cases.

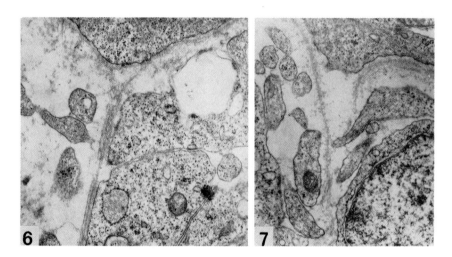

Figs. 6 and 7 : formation of anchor filaments in tarsometatarsal skin 24 hours after one injection of retinoic acid at 10 days. G x 20 000.

DERMAL-EPIDERMAL RECOMBINANTS

Heterotypic dermal-epidermal recombinants were prepared in which one of the two tissues was obtained from embryos treated with RA 24 h or 48 h before recombination; the other tissue was obtained from a normal 10-day or 11-day embryo. The development of explants comprising a RA-treated epidermis and an untreated dermis was in all cases similar to that of its contralateral recombinant of treated epidermis and treated dermis : recombinants involving 24 h-RA-treated epidermis (homotypic and heterotypic) produced feathered scales in less than 20% of cases. Recombinants involving 48 h-RA-treated epidermis (homotypic and heterotypic) differentiated feathered scales in 46% of cases (Fig. 8) : a percentage similar to that of feathered feet of 10-day RA-treated embryos. The reverse combination of normal epidermis and 24 h-RA-treated dermis led to the formation of normal scales, identical to those of control recombinants of untreated epidermis and untreated dermis. The scale morphogenesis was obtained even if the RA-treated dermis originated from a 48 h-treated embryo but the scales had a more or less rounded shape (Fig. 9).

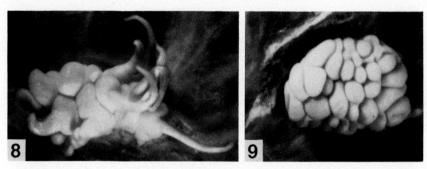

Figs. 8 and 9 : Dermal-epidermal recombinants of 48 h-RA-treated epidermis with normal dermis (8) and of normal epidermis with 48 h-RA treated dermis (9), after 5 days of culture on CAM. G x 20.

These results demonstrate that the drug affects the epidermis only and that the effects observed in the dermis, as the morphology of the dermal cell processes, are mediated through the epidermis which is thus shown to transmit a feather-forming message to the dermis. Furthermore the dermal changes are reversible, provided a normal untreated epidermis is recombined with the treated dermis. This reversion to normality leads to the expression of scale-forming properties by the tarsometatarsal dermis, despite of its treatment. This view is confirmed by the result of the recombination of treated tarsometatarsal dermis with a non-treated epidermis from the featherless midventral apterium of a 10-day embryo, a region which does not form cutaneous appendages normally; recombinants of this kind, indeed, formed normal scales (Dhouailly and Cadi, submitted).

IN CONCLUSION, these experiments demonstrate that retinoic acid acts directly on the **chick foot** epidermis by repressing scale morphogenetic properties and thus leaving latent feather morphogenetic properties free to be expressed. A possible explanation is that RA could act directly or indirectly on the epidermal cell genome. Indeed the promotor role of RA in cell differentiation has been demonstrated in cultures of teratocarcinoma cell lines (Jetten, Jetten and Sherman, 1979). After RA treatment, the tarsometatarsal epidermis becomes able, like a normal feather-forming epidermis, to inhibit the acquisition of scale-forming properties by the chick foot dermis. Furthermore these results, and particularly the formation of scales in recombinants of tarsometatarsal dermis and epidermis from the midventral *apterium*, lead to a better understanding of the morphogenesis of dorsal-tarsometatarsal heterotopic recombinants of Sengel (1958) and Rawles (1963) : contrary to a previous interpretation, tarsometatarsal dermis is endowed with scale-inducing capacity before 12 days of incubation and does not acquire this capacity by contact with tarsometatarsal epidermis only. Scale morphogenesis which starts at 10 days, remains labile until 11 days or 12 days and tarsometatarsal dermis can be induced to participate to feather morphogenesis if it is associated with feather-forming epidermis, from the back of normal embryos or from the foot of RA-treated embryos.

REFERENCES

Demarchez M, Mauger A, Sengel P (1981). The dermal-epidermal junction during the development of cutaneous appendages in the chick embryo. Arch Anat microsc Morphol exp (in press).
Dhouailly D (1980). Action of retinoic acid on chick cutaneous appendages morphogenesis. J Invest Derm 74:455.
Dhouailly D (1982). Sequential study of tarsometatarsal chick skin after retinoic acid treatment. (Submitted).
Dhouailly D, Cadi R (1982). Effects of retinoic acid on chick tarsometatarsal dermo-epidermal interactions (submitted).
Dhouailly D, Hardy MH (1978). Retinoic acid causes the development of feathers in the scale-forming integument of the chick embryo. Wilhelm Roux's Arch 185:195.
Dhouailly D, Hardy MH, Sengel P (1980). Formation of feathers on chick foot scales : a stage-dependent morphogenetic response to retinoic acid. J Embryol exp Morph 58:63.

Jetten AM, Jetten MER, Sherman MI (1979). Stimulation of differentiation of several murine embryonal carcinoma cell lines by retinoic acid. Exp Cell Res 124:381.

Rawles ME (1963). Tissue interactions in scale and feather development as studies on dermal-epidermal recombinations. J Embryol exp Morph 11:765.

Sawyer RH (1972). Avian scale development. I. Histogenesis and morphogenesis of the epidermis and dermis during formation of the scale ridge. J exp Zool 181:365.

Sengel P (1958). Recherches expérimentales sur la différenciation des germes plumaires et du pigment de la peau de l'embryon de poulet en culture *in vitro*. Ann Sci nat Zool 20:431.

I am most grateful to Mrs. J. Lana for the typewriting.

This work was supported by financial help from INSERM (Contrat de recherche libre n° 7915142).

REGULATION OF THE ASSEMBLY OF THE Z-DISC IN MUSCLE CELLS

Elias Lazarides, David L. Gard, Bruce L. Granger, Clare M. O'Connor, Jennifer Breckler, and Spencer I. Danto

Division of Biology, California Institute of Technology, Pasadena, California 91125 USA

Perhaps one of the most striking visual aspects of skeletal and cardiac muscle cells is their striated appearance. It has long been suspected that there is present in these cells a framework responsible for holding the constituent myofibrils in place. The presence of structural elements transverse to a muscle fiber's long axis is suggested by the tendency for the Z-discs, M-lines and other sarcomeric units of neighboring myofibrils to be in axial register across the fiber. Numerous ultrastructural observations suggest linkage of cardiac and skeletal sarcomeric units, and in particular of the Z-disc, transversely to other Z-discs and to the plasma membrane (Bennett and Porter, 1953; Bergman, 1958, 1962; Garamvölgyi, 1962, 1965; Pease and Baker, 1949; Walker et al., 1968). The molecular nature of this integrating matrix was until recently unknown. Recent studies on the biochemical composition of intermediate filaments, their solubility properties and cytoplasmic distribution have allowed us to approach this question from a new point of view. With the development of a new technique for the visualization of avian muscle Z-discs we have been able to define distinct molecular domains within this structure and demonstrate the existence of a cytoskeletal system responsible for laterally integrating myofilaments and perhaps also the various membranous organelles at the peripheries of the Z-discs. The main molecular components underlying this integrating matrix are the intermediate (100 Å) filament subunits, desmin and vimentin. By the use of immunofluorescence with antibodies specific for the various structural components of the Z-disc, we could follow and define during myogenesis in vitro the major morphogenic steps that culminate in the assembly of the various Z-disc domains and the development of this integrating matrix. Below we summarize these observations and describe morphological and biochemical evidence which suggests that the assembly of this matrix may be regulated through developmental changes in the composition and structure of avian skeletal muscle intermediate filaments.

INTERMEDIATE FILAMENTS IN MUSCLE CELLS

The cytoplasm of many higher eukaryotic cells contains a filamentous system which by a variety of morphological, biochemical and immunological criteria is distinct from actin filaments, microtubules and myosin filaments. As their characteristic diameter of 100 Å is between that of the smaller actin filaments (60 Å) and the larger microtubules (250 Å) and myosin filaments (150 Å), this filamentous system has come to be known as the "intermediate filaments" (Ishikawa et al., 1968). Recent biochemical and immunological evidence has shown that this class of filaments is composed of chemically heterogeneous subunits (for a review, see Lazarides, 1980). Their subunit composition defines five major classes of intermediate filaments: 1) keratin (tono) filaments, found in epithelial cells and cells of epithelial origin; 2) neurofilaments found in neurons; 3) glial filaments found in glial cells and cells of glial origin; 4) desmin filaments found predominantly in muscle cells and 5) vimentin filaments which exhibit the widest distribution from all other intermediate filament proteins. Earlier observations indicated that desmin was the major intermediate filament protein in smooth, skeletal and cardiac muscle cells (Lazarides and Hubbard, 1976; Cooke, 1976; Small and Sobieszek, 1977; Izant and Lazarides, 1977) and that in adult skeletal and cardiac muscle this protein was a component of myofibril Z-discs (Lazarides and Hubbard, 1976). In our efforts to understand the role of intermediate filaments in muscle cells, we have concentrated on their relationship to the Z-disc in adult muscle and in the assembly of this structure during myogenesis.

THE EXISTENCE OF AN INSOLUBLE Z-DISC SCAFFOLD IN ADULT SKELETAL MUSCLE

High concentrations of glycerol (50%) are frequently used for the long term storage of muscle fibers. The glycerol disrupts the plasma membrane and other membranous organelles, releasing soluble proteins without affecting the ultrastructure and contractile properties of the fibers. When such glycerol-extracted skeletal muscle is subjected to shearing forces, the fibers fragment along their long axes into individual myofibrils, or, more frequently, bundles of myofibrils of variable lengths. However, when strips of glycerinated muscle are incubated in a solution containing high concentrations of KI, a chaotropic agent capable of solubilizing actin and myosin filaments, a considerable amount of their protein mass is solubilized. We observed that what remains is an insoluble residue that maintains its original three-dimensional structure and that is composed predominantly of Z-discs (Granger and Lazarides, 1978). If these KI extracted muscle fibers are now subjected to shearing forces, a different cleavage pattern is obtained. Cleavage occurs preferentially along planes perpendicular to the long axis of the muscle fiber; what is generated are planar arrays of Z-discs giving rise to a honeycomb-like appearance. Such sheets of Z-discs can be clearly visualized by conventional phase contrast microscopy as well as by electron microscopy, revealing a face-on view of the Z-disc and demonstrating that each Z-disc is physically linked laterally to neighboring Z-discs within a given Z-disc plane.

Presence of the Intermediate Filament Proteins Desmin and Vimentin in Adult Muscle

Myosin, actin, tropomyosin and troponin represent more than 90% of the total myofibrillar protein mass; with the exception of α-actinin, a well-known component of the Z-disc in avian and mammalian muscles, the proteins of the Z-disc are minor components by comparison. However, extraction of myofibrils with a buffer containing 0.6 M KI (i.e., the preparation of Z-disc scaffolds) results in an enrichment in the insoluble components of the Z-disc. Analysis by two-dimensional gel electrophoresis of scaffolds of myofibrils purified from chicken pectoralis muscle has revealed actin, the intermediate filament proteins desmin and vimentin, α-actinin and the 68-70,000 dalton heat shock proteins thermin a and thermin b (Wang et al., 1981) as predominent components (Granger and Lazarides, 1979). Vimentin is not confined to pectoralis myofibrils; examination of high salt insoluble residues from slow muscle (anterior latissimus dorsi) and twitch muscles (posterior latissimus dorsi, peroneus longus, iliotibialis) by the same electrophoretic techniques has revealed the presence of both desmin and vimentin as major components in addition to actin. An intriguing observation is that the ratio of desmin to vimentin is variable in these different muscles; it is approximately 1:1 in PLD, 2:1 in iliotibialis, 2.5:1 in pectoralis, and 7:1 in ALD (M. Price and E. Lazarides, in preparation).

The variable ratio of vimentin to desmin in adult muscle raises an important issue about the functional significance of having both molecules present. Other studies suggest that the presence of vimentin is not obligatory since in the mammalian Purkinje fibers (Eriksson and Thornell, 1979) and in mammalian fast muscles (O'Shea et al., 1981), desmin is the predominant skeletal muscle intermediate filament protein and vimentin is present in very low quantities if not absent. In addition, some mammalian smooth muscle cells express predominantly desmin, others express predominantly vimentin, and still others express both desmin and vimentin (Gabbiani et al., 1981; Frank and Warren, 1981). Most of these studies have relied on immunofluorescence for the detection of desmin or vimentin, so that the ratio of desmin to vimentin is not evident. In muscles that contain both desmin and vimentin, these molecules exhibit indistinguishable solubility properties, and both molecules copurify in a constant ratio through cycles of solubilization at pH 2.1 and precipitation at pH 5.0-7.0 (Granger and Lazarides, 1980). All these results indicate that desmin and vimentin can exhibit wide variations in their ratio in different cell types, and suggest that these two proteins may be equivalent. Since vimentin and desmin are homologous but not identical molecules (Gard et al., 1979; O'Connor et al., 1981; Geisler and Weber, 1981), variability in the ratios of the two proteins in different cell types may have important physiological consequences in mediating changes in the overall structure of intermediate filaments in the different muscle types.

Identification of Synemin as an Intermediate Filament Associated Protein

Scaffolds, similar to those obtained with skeletal muscle, are also obtained with smooth muscle. Cooke (1976) was first to show that extraction of chicken gizzard with solutions of high ionic strength, leaves insoluble a network of intermediate filaments attached to the smooth muscle analogues of Z-discs, the dense bodies. This network can be solubilized at low pH (Small and Sobieszek, 1977; Hubbard and Lazarides, 1979) or with high concentrations of urea (Huiatt et al., 1980), and the constituent proteins purified by cycles of precipitation and resolubilization between pH 5.0-7.0 and 2.1 (Hubbard and Lazarides, 1979). The major component of this material is desmin with lesser quantities of vimentin, actin and a high molecular weight protein (MW 230,000) which we have named synemin (Granger and Lazarides, 1980). The amount of actin that copurifies with desmin is variable and can be removed by ion exchange chromatography in the presence of 6-8 M urea (Huiatt et al., 1980). The ratio of synemin to desmin is approximately 1:50-100 and this protein appears to bind tightly to desmin. Immunoautoradiography with antibodies specific to synemin has shown that this protein is also a component of skeletal myofibrils (Granger and Lazarides, 1980).

DISTINCT MOLECULAR DOMAINS IN THE Z-DISC

The microscopic evidence summarized earlier indicated that the myofibril Z-disc may be composed of two major domains. Our ability to view the Z-disc by light microscopy side-on in myofibrils (Z-line) and face-on in isolated Z-disc sheets allowed us to map within this structure by immunofluorescence the various proteins that remain insoluble with this structure after removal of actin and myosin filaments with high salt. In unextracted myofibrils indirect immunofluorescence showed that in addition to the well known Z-disc protein, α-actinin, this structure contained the high molecular actin binding protein filamin (Bechtel, 1979; Gomer and Lazarides, 1981) and the three intermediate filament proteins, desmin, vimentin and synemin (Granger and Lazarides, 1979, 1980). Immunofluorescence on Z-disc sheets has demonstrated that the Z-disc is composed of two distinct, yet interconnected domains: a central one which contains α-actinin and actin and a peripheral one which contains actin, filamin and the three intermediate filament proteins desmin, vimentin and synemin (Table 1). These two domains were established from the complementary distribution of the antigens. Demonstration that these antigens are indeed located in the Z-line and do not flank this structure was provided by side-on views of whole myofibrils as well as KI-extracted "ghost" myofibrils. These two views of the Z-disc have allowed us to conclude that within this structure these antigens are segregated into two spatially distinct domains.

Table 1

Protein Composition of the Z-Disc in
Adult Chicken Skeletal Muscle

Peripheral Domain		Central Domain	
Protein	Molecular Weight	Protein	Molecular Weight
Actin	43,000	Actin	43,000
Filamin	250,000	α-Actinin	100,000
Desmin	50,000	Intermediate filament proteins	
Vimentin	52,000		
Synemin	230,000		

INTERMEDIATE FILAMENT PROTEINS AS A MECHANICALLY INTEGRATING MATRIX IN MUSCLE CELLS

On the basis of the observations summarized above we have proposed that as major components of the peripheral domain of the Z-disc the intermediate filament proteins, desmin, vimentin and synemin, may mechanically integrate all contractile actions of a muscle fiber by linking individual myofibrils laterally at their Z-discs and by linking Z-discs to the plasma membrane and other membranous organelles. During the biogenesis and assembly of myofibrils, the three intermediate filament proteins may play an important role in the generation of the striated appearance of muscle by promoting or stabilizing the lateral registry of adjacent Z-discs. Longitudinal and cross sections of adult muscle fibers have shown that desmin is also located beneath the plasma membrane (Lazarides and Hubbard, 1976). Furthermore, freshly isolated bundles of myofibrils exhibit occasional desmin-containing material which is located at the periphery of myofibrils, extending from Z-line to Z-line along the long axis of the myofibril. Thus, the intermediate filament matrix appears to integrate myofibrils not only laterally at the level of the Z-disc, but also longitudinally under the plasma membrane and occasionally between myofibrils along the fiber axis.

ASSEMBLY OF THE TWO Z-DISC DOMAINS DURING SKELETAL MYOGENESIS IN VITRO

Immunofluorescence

The demonstration that the Z-disc is composed of two distinct but interconnected molecular domains raises the question of the temporal sequence of events during skeletal myogenesis that culminate in the assembly of these two domains. Skeletal myoblasts grown in tissue culture provide an ideal system for the study of the steps in the assembly of the Z-disc and the regulation of such processes. Myoblasts isolated from embryonic chick muscle will fuse in vitro into multinucleated myotubes which will assemble fully contractile sarcomeres in the absence of innervation. Using antibodies to α-actinin as a probe for the newly assembling central domains of the Z-discs, and antibodies to desmin, vimentin or synemin as probes for intermediate filaments and the peripheral domain of the Z-disc, we could show that these two domains assemble sequentially during myogenesis (Gard and Lazarides, 1980; Granger and Lazarides, 1980). Earlier studies with cardiac muscle cells grown in tissue culture provided the first evidence that desmin existed in the form of cytoplasmic filaments in the less mature cells and in association with the Z-lines in the more mature cells (Lazarides, 1978). Thus it became obvious that at some point during muscle differentiation the subunits of intermediate filaments would alter their cytoplasmic distribution and begin to associate with the Z-disc. This sequence of events was mapped out and studied more accurately in skeletal myotubes. In primary myogenic cultures (consisting predominantly of fibroblasts, myoblasts and early myotubes), immunofluorescence revealed the presence of vimentin in all the cells. However, double immunofluorescence showed the absence of synemin and desmin from fibroblastic cells positive for vimentin. In myoblasts, only a small proportion of these cells stained positively for desmin and synemin, suggesting that synthesis is restricted to mainly postmitotic fusing myoblasts and multinucleate myotubes (Gard and Lazarides, 1980; Granger and Lazarides, 1980). These results indicated that the expression of vimentin and desmin, or vimentin and synemin, is non-coordinate unless upon the onset of fusion, the fibroblastic (myoblastic) form of vimentin is turned off and a muscle specific form of vimentin is turned on. This possibility is unlikely in light of the in vitro translation experiments discussed later on. After seven days in culture, a striking redistribution of desmin and vimentin takes place within the myotubes. Many cells begin to exhibit a transversely striated pattern of desmin- and vimentin-specific fluorescence. Correlation of these striations with phase microscopy and the distribution of α-actinin indicated that the desmin and vimentin containing striations corresponded to the myofibril Z-lines. This transition of the three filament proteins to the Z-line takes place several days after the development of α-actinin-containing Z-line striations and with a concomitant decrease in their association with cytoplasmic filaments. Close examination of cells at the time of transition of the three proteins to the Z-line has revealed two important features of this molecular rearrangement. The first is that all Z-lines throughout the duration of the transition contain simultaneously desmin, vimentin and synemin as revealed by double immunofluorescence. Thus within the resolution of this technique the association of any one of the three proteins within the Z-disc is always accompanied with the association of the other two proteins. The second important feature is that, the association of the three proteins with the Z-disc is an asynchronous process. If cells are examined by immunofluorescence around the onset of transition, they reveal the existence of all three of the filament proteins both in association with Z-lines and in association with cytoplasmic filaments. The number of Z-lines with any of the three proteins in association with them varies in different myotubes depending on the degree of maturation of a given cell at the time of examination. However, if the cells are examined a few days after the onset of transition, then the majority of the desmin, vimentin and synemin fluorescence is found in association with Z-lines (Gard and Lazarides, 1980; Granger and Lazarides, 1980).

The results summarized thus far demonstrate that the two Z-disc domains of chick skeletal myotubes assemble sequentially during differentiation. The central domain assembles first and after a specific point in its maturation has been attained, the peripheral domain begins to assemble by the association of the three filament proteins, desmin, vimentin and synemin. Association of these proteins with the periphery of the Z-disc coincides with the lateral registration of Z-discs and the appearance of well developed striations in the muscle fiber. Thus these three proteins may not only play a passive role in maintaining the striated phenotype of muscle cells, but may also play an active role in bringing Z-discs into lateral registry upon their association with the Z-disc. How is this process of the association of the three filament proteins with the Z-disc regulated? We can envision at least a dual regulation of the process. In the first the periphery of the Z-disc becomes competent to nucleate newly synthesized subunits of intermediate filaments which begin to polymerize from Z-disc to Z-disc. In the second the three filament proteins undergo certain chemical changes which make them competent to bind to the Z-disc. These may include the expression of specific genes for the form of the proteins to associate with the Z-disc, chemical modifications such as phosphorylation, enzymatic processing such as proteolysis, reutilization of preexisting monomers, etc. Most likely both levels of regulation outlined above are operative in some way in the cell. Below we discuss evidence which allows us to exclude certain possibilities and focus on others.

Colcemid Sensitivity

Sensitivity of intermediate filaments to Colcemid was originally shown by Ishikawa et al. (1968) in skeletal myotubes. In both myotubes and non-muscle cells grown in tissue culture, intermediate filaments aggregate into cytoplasmic bundles in cells exposed to Colcemid as shown by electron microscopy. We have observed that Colcemid sensitivity of the filaments can be used as a probe to define changes in the filaments which take place before, during and after their association with the Z-disc. Using this approach we have defined three states of the intermediate filament proteins. The first one is a Colcemid-sensitive state and is observed in young myotubes well before the association of the three proteins with the Z-disc. Exposure of cells to Colcemid results in the aggregation of all three proteins into cytoplasmic filament bundles; double immunofluorescence shows that within these filament bundles the three proteins exhibit an indistinguishable distribution, as is the case prior to their aggregation. The other two are Colcemid-insensitive states. One is observed after the association of the three proteins with the Z-disc; the Z-line associated forms of the three proteins resists this Colcemid-induced rearrangement but the three proteins still maintain indistinguishable distributions. The other insensitive state has been defined with monoclonal antibodies to desmin. Prior to and during the association of desmin with the Z-disc a subclass of the desmin fluorescence is found in association with cytoplasmic filament bundles which are localized in between myofibrils. This subclass of desmin fluorescence is also Colcemid-insensitive and its identification and a description of its properties are discussed further below. These observations have suggested that intermediate filaments undergo a lateral aggregation prior to and during their association with the Z-disc as evidenced by their insensitivity to Colcemid.

Probing Molecular Changes in Desmin During Myogenesis with Monoclonal Antibodies

The availability of a general map of the pathway of assembly of desmin to the Z-disc, allowed us to probe further into any biochemical and immunological changes in desmin using monoclonal antibodies. We focused our analysis on two monoclonals selected from a bank of monoclonal antibodies generated against adult chicken gizzard desmin (Danto and Fischman, 1981) which had been purified by two cycles of solubilization and precipitation (Hubbard and Lazarides, 1979) prior to immunization. The two monoclonals were directed against different sites on the desmin molecule as revealed by their additive binding in competitive binding studies and by their differential detection of desmin peptides in one-dimensional Staphylococcus aureus V-8 protease maps as determined by immunoautoradiography (Danto, Fischman and Lazarides, in preparation). Similarly to the desmin-specific polyclonal (rabbit) antisera described above, the two monoclonal antibodies react with desmin from all three adult muscle types (smooth, skeletal and cardiac), thus establishing once more the antigenic homology of the three desmins. Furthermore both monoclonals react specifically with the Z-line of skeletal myofibrils and in particular the periphery of the Z-disc as determined by immunofluorescence or isolated intact Z-disc sheets. Finally, the two monoclonals react with both the phosphorylated and unphosphorylated forms of desmin as determined by two-dimensional immunoautoradiography, as is the case with the desmin-specific polyclonal antibodies (Danto et al., in preparation). The difference in the reactivity of the two monoclonals was revealed when myotubes were examined by immunofluorescence at different stages of differentiation. One of the monoclonals (D3) revealed an immunofluorescent pattern indistinguishable from that of the polyclonal desmin antibodies at all stages of myogenesis in vitro as determined by double immunofluorescence. The other monoclonal antibody (D76), failed to stain desmin in myotubes at stages before desmin had begun its transition to the Z-disc although double immunofluorescence with the polyclonal antiserum revealed the presence of desmin in these cells. Immunofluorescent staining by the D76 monoclonal antibody was first manifested as small focal patches of desmin longitudinally oriented between the developing myofibrils. Double immunofluorescence with the polyclonal desmin antiserum at this stage of differentiation showed that while the monoclonal antibody revealed exclusively these longitudinally oriented streaks, the polyclonal antibody reveals desmin, in addition to these patches, also along the Z-line and to a lesser extent along cytoplasmic filaments. Ultimately in fully developed myotubes, the D76 monoclonal antibody revealed an immunofluorescent pattern identical to that of the polyclonal antibody; both antibodies revealed the presence of desmin along Z-lines with a concomitant reduction in the longitudinally oriented desmin streaks. One-dimensional SDS immunoautoradiography on whole myotube extracts from different stages of myogenesis using the D76 monoclonal antibody, revealed the presence of desmin in all stages of myogenesis, even in stages where the antibody exhibited no desmin by immunofluorescence. Both the cytoplasmic and the Z-line form of the desmin fluorescence revealed by the D76 monoclonal antibody were insensitive to Colcemid aggregation; thus the longitudinal filament bundles revealed by this antibody show the presence of a third form of desmin that is Colcemid-insensitive and substantiate further the hypothesis that the Colcemid-insensitivity of any of the three intermediate filament proteins reflects some sort of biochemical change in the molecules (most likely a lateral aggregation of desmin filaments) in the process of their association with the Z-disc. The immunoautoradiographic evidence summarized above indicated that the difference in antigenicity detected by the D76 monoclonal antibody was not due simply to phosphorylation since it recognizes both phosphorylated and unphosphorylated (see below) variants of desmin. Detection of the antigen throughout myogenesis by immunoautoradiography argues that the appearance of fluorescence is probably not due to de novo synthesis of a new form of desmin. Finally since the antibody recognizes urea- or acetic acid-denatured desmin in radioimmunoassay as well as proteolytic fragments of desmin by immunoautoradiography, the antigenic site is probably determined solely by the molecule's primary structure. Even though the exact molecular nature of this observation has not yet been determined, the results suggest the existence of immunologically distinct, but morphologically and biochemically indistinguishable subpopulations of desmin molecules which follow different developmental programs during the assembly of the peripheral domain of the Z-disc. This immunologically distinct form of desmin may be due to a specific conformational change in the molecule or the removal of a specific binding protein both resulting in the exposure of a new antigenic site. Whichever the case might be, these results suggest that there are two immunologically distinct forms of desmin during myogenesis and that the two of them assemble at the periphery of the Z-disc sequentially. The distribution of the D76 desmin antigen suggests that this class of desmin molecules might be specifically involved in the process of lateral alignment of myofibrils.

Biosynthesis and Chemical Modification of Desmin and Vimentin During Myogenesis

The immunofluorescence results presented above indicated that vimentin is present throughout myogenesis, in myoblasts and myotubes in the form of cytoplasmic filaments and in mature myotubes and adult myofibrils in association with the Z-disc. However, desmin and synemin are found predominantly in fusing myoblasts and post-fusion multinucleate myotubes in a distribution indistinguishable from that of vimentin. In accordance with the immunofluorescence results, 2D IEF/SDS-PAGE autoradiography have shown that myoblasts labeled within 1 hr after dissociation from 10 day old thigh muscle synthesize negligible amounts of desmin while vimentin constitutes one of the major synthetic products of these cells. Desmin synthesis is detectable within 18 hr after plating as is the synthesis of α-actin and the skeletal muscle forms of tropomyosin. The relative rate of desmin synthesis continues to increase at least 10-fold during the first three days in culture by which time it represents approximately 0.5% of the total protein synthetic activity. During this time there is also a dramatic increase in the relative synthetic rates of other muscle-specific proteins such as α-actin and the tropomyosins. Synthesis of vimentin also increases during myogenesis, though less dramatically than that of desmin. Both desmin and vimentin continue to be synthesized through at least 20 days in culture (Gard and Lazardies, 1980). Two conclusions can be derived from these results: first, they support those results obtained by immunofluorescence that vimentin is synthesized throughout myogenesis in vitro, while synthesis of desmin is restricted to postmitotic fusing myoblasts, and myotubes. In addition, they suggest that the synthesis of desmin and vimentin is not coordinately regulated, which results in a dramatic change in the relative composition of intermediate filaments during myogenesis.

The presence of both vimentin and desmin in postmitotic myoblasts and early myotubes has also been reported by Holtzer and his collaborators using one-dimensional SDS gel electrophoretic analysis of cytoskeletons from these cells as well as immunofluorescence microscopy with antibodies to vimentin and to desmin (Bennett et al., 1978a, b). These authors have also observed that desmin begins to associate with myofibrils at some point of myogenesis with a concomitant decrease in the association of this protein with cytoplasmic filaments (Bennett et al., 1979). Results similar to those summarized above have also been obtained by Devlin and Emerson (1979) who have shown that quail skeletal myoblasts and myotubes (which are free from contaminating non-myogenic cells) express both desmin and vimentin as judged by pulse labeling experiments and in vitro translation experiments of purified mRNA from 3-5-day-old myotubes. A close examination of desmin and vimentin in early myotubes where the two proteins are associated with cytoplasmic filaments and in late myotubes where the two proteins are found in association with Z-discs reveals no apparent differences in electrophoretic mobility, number of isoelectric variants and isoelectric point (Granger and Lazarides, 1979; Gard and Lazarides, 1980). Thus regulation of the transition of the two proteins from cytoplasmic filaments to the Z-disc does not appear to involve any gross chemical changes in the two molecules themselves or the synthesis of any specific variants. These observations suggest further that the Colcemid-insensitivity and the immunological changes observed upon the association of the two proteins (as exemplified by desmin) with the Z-disc are probably not due to a chemical change (new gene product, chemical modification) of the proteins.

Even though the mechanism regulating the transition of desmin and vimentin to the Z-disc remain unknown, we have observed that both proteins are phosphorylated (O'Connor et al., 1979) suggesting that phosphorylation may be involved in this process. Considerable evidence has suggested that phosphorylation of both desmin and vimentin is mediated by the cAMP-dependent protein kinases (O'Connor et al., 1981a, b). All of the phosphorylation of desmin and most of the vimentin can be shown to be cAMP-dependent; however a small fraction of the vimentin phosphorylation is cAMP-independent (O'Connor et al., 1981a). These results demonstrate that desmin and vimentin may be regulated similarly through cAMP-dependent phosphorylation reactions but that vimentin may be capable of being differentially phosphorylated from desmin through a cAMP-independent kinase. The possibility that desmin and vimentin could have a common site of regulation was substantiated by peptide mapping of the phosphorylation sites of these two molecules. Direct comparison of the tryptic phosphopeptides of α-desmin and vimentin by co-electrophoresis revealed that at least one of the major phosphopeptides of vimentin comigrate with the major phosphopeptide of desmin (O'Connor et al., 1981b). These results have suggested the existence of some sequence homology between the phosphorylation sites of these two intermediate filament proteins. Furthermore the existence of both non-homologous (cAMP-dependent and cAMP-independent) and homologous (cAMP-dependent) phosphorylation sites on these two molecules indicates the existence of both common and different potentially regulatory sites between desmin and vimentin.

The physiological relevance of phosphorylation of these two proteins has recently come from experiments on the modulation of the phosphorylation of desmin and vimentin in myogenic cells by hormones and cAMP analogs. Addition of β-adrenergic agonists such as isoproterenol (10^{-6} M) or cAMP analogs (8-Br cAMP, 10^{-3} M) to 7-14 day old cultures of chicken embryonic skeletal myotubes (in these cells, desmin and vimentin are already in association with Z-discs) yields a significant increase in $^{32}PO_4$ incorporation into desmin and vimentin. This response of desmin and vimentin to either 8-BrcAMP or isoproterenol is not observed in younger myogenic cells (1-3 day old cultures; in these cells desmin and vimentin are in the form of cytoplasmic filaments). Desmin appears to be phosphorylated at a number of sites by cAMP-dependent kinases but a small number of sites appears to be especially prone to modulation by the kinases (Gard and

Lazarides, 1982). Since desmin and vimentin appear to become especially sensitive to modulation by hormones and cAMP upon their association with the Z-disc, modulation of cytoplasmic cAMP levels during myogenesis may play an important regulatory role in the association of desmin and vimentin with the Z-disc. In vitro translation of muscle poly(A) has indicated that the primary translation products for desmin and vimentin are the unphosphorylated forms of these two proteins (O'Connor et al., 1981c). From these results we have concluded that desmin and vimentin are synthesized predominantly if not exclusively as one single form and that the more acidic variants are derived from the parent molecules by posttranslational phosphorylation. Thus from the available evidence thus far, the forms of desmin and vimentin which associate with the Z-disc do not appear to be distinct (new gene products) from those that associate with filaments earlier in myogenesis.

ISOLATION OF NATIVE DESMIN AND VIMENTIN FILAMENTS FROM CHICKEN EMBRYONIC SKELETAL MUSCLE

In our attempts to further analyze the biochemical events which culminate in the association of the three intermediate filament proteins, desmin, vimentin and synemin, with the Z-disc, we have sought to find a method of purifying and analyzing intermediate filaments from embryonic skeletal muscle cells under physiological conditions and have used such a method to investigate changes in the structure and composition of the filaments at different stages of embryonic muscle development. We observed that high speed supernatants (145,000 g at R_{max}) prepared from 14 day-old chick embryonic thigh muscles in a low ionic strength buffer (130 mM KCl, 5 mM EGTA, 20 mM Tris-HCl pH 7.5) and in the absence of detergents, are enriched in actin, α and β tubulin and the two intermediate filament proteins, desmin and vimentin, as judged by two-dimensional IEF/SDS-PAGE (Granger and Lazarides, 1979). Gel filtration chromatography of these extracts through a column of Biogel A-5m (range 10^4 -5 x 10^6 MW) and subsequent analysis of the column fractions by one-dimensional SDS-PAGE indicated that the void volume fractions of the column are enriched in vimentin, desmin, synemin and a new high molecular weight protein, which we have named paranemin (MW 280,000) (Breckler and Lazarides, 1982). Examination of the void volume fraction by electron microscopy has revealed the presence of numerous filaments with a diameter of 80-120 Å and variable lengths. Sedimentation experiments and immunofluorescence have shown that paranemin is associated with desmin and vimentin filaments and can therefore be considered as a new intermediate filament associated protein (Breckler and Lazarides, 1982).

DEVELOPMENTAL REGULATION OF THE EXPRESSION OF PARANEMIN

Immunofluorescence on cultured myogenic cells, using antibodies to paranemin, has revealed that this protein has the same spatial distribution as desmin and vimentin in both early myotubes, where it associates with cytoplasmic filaments, and in late myotubes, where it is associated with myofibril Z-lines. However, examination by immunofluorescence of frozen sections of developing embryonic skeletal muscle has revealed a gradual diminution in the anti-paranemin fluorescence, under conditions where desmin is clearly present. Paranemin is barely detectable in adult skeletal and smooth (gizzard) muscle as shown by immunofluorescence on frozen sections. Paranemin is also undetectable by immunoautoradiography on whole extracts of chicken gizzard and skeletal muscle or on whole extracts of isolated skeletal myofibrils (Breckler and Lazarides, 1982). In chick embryonic fibroblasts grown in tissue culture, only a subpopulation of the cells is reactive with antibodies to paranemin even though these cells all contain vimentin. In the reactive cells, vimentin and paranemin exhibit an indistinguishable cytoplasmic filamentous network, which aggregates into filamentous bundles when the cells are exposed to Colcemid (Breckler and Lazarides, 1982). From these observations we have concluded that paranemin is associated with intermediate filaments containing either vimentin alone or vimentin, desmin and synemin. Since this protein is not found in all nonmuscle cell types where vimentin is present and since it appears to be gradually removed from muscle cells after the association of desmin, vimentin and synemin with the Z-disc, the expression of paranemin appears to be developmentally regulated and does not appear to depend on the expression of the other three intermediate filament proteins. Thus this protein may perform a specific function in the structure of vimentin filaments in some non-muscle cells and during early myogenesis and may be required for the association of the other three intermediate filament proteins with the myofibril Z-disc. Its presence in exceedingly low concentrations in adult smooth and skeletal muscle and its gradual reduction during development suggests further that its function is regulatory in the assembly of intermediate filaments at the periphery of the Z-disc, but is not required for the maintenance of the structure of the Z-disc, subsequent to the assembly of the sarcomere. A comparative summary of the intermediate filament proteins found in embryonic and adult chicken skeletal muscle is shown in Tables 1 and 2.

Table 2

Protein Composition of the Z-Disc in
Embryonic Chicken Skeletal Muscle

Peripheral Domain		Central Domain	
Protein	Molecular Weight	Protein	Molecular Weight
Actin	43,000	Actin	43,000
Filamin	250,000	α-Actinin	100,000
Desmin	50,000	Intermediate filament proteins	
Vimentin	52,000		
Synemin	230,000		
Paranemin	280,000		

DEVELOPMENTAL REGULATION OF THE STRUCTURE OF DESMIN AND VIMENTIN FILAMENTS

The results summarized here indicate that the structure and composition of intermediate filaments is developmentally regulated and changes in response to the different cytoplasmic associations of their two major subunit proteins, desmin and vimentin, and that associated proteins such as paranemin may regulate changes in their structure in response to their different cytoplasmic associations. Thus far we can distinguish three major types of developmental regulation of the structure and composition of this class of filaments.

Differential Expression of Vimentin and Desmin

Initial studies suggested that vimentin was exclusively expressed in non-muscle cells, while desmin was expressed exclusively in muscle cells. However it is now evident that the subunit composition of the filaments varies in different muscle types and in different stages of differentiation. Certain types of smooth muscle appear to express predominantly vimentin, others desmin and still others a mixture of the two (Granger and Lazarides, 1980; Gabbiani et al., 1981; Frank and Warren, 1981). Avian skeletal and cardiac muscle appear to express both vimentin and desmin (see above). However it is important to bear in mind that the available detailed examination of the composition of muscle intermediate filaments indicates that the ratio of desmin to vimentin in different muscle types can vary widely in adult muscle tissues (e.g., 100 D: 1 V in chicken gizzard and 2.5 D: 1 V in chicken pectoralis muscle) and changes from low to high during skeletal myogenesis. Furthermore certain types of muscle, in particular mammalian muscles, may express exclusively either desmin or vimentin (see above). What is the functional significance of such a variation in the expression of the two proteins? From the available evidence the expression of the two proteins appears to be non-coordinate and the expression of any one of the two proteins does not depend on the expression of the other. We are inclined to believe that the two proteins are functionally equivalent but that such a differential expression of the two proteins might allow different muscle cells to modulate subtly or grossly the structure (rigidity, length, etc.) of their intermediate filament in response to the different cytoplasmic associations of these filaments as well as the physiology of a given type of muscle. All circumstantial evidence thus far indicates that desmin and vimentin copolymerize into one and the same filament. If we assume that such a copolymerization is general, then the variable ratio of desmin to vimentin in different cell types suggests that copolymerization in vivo might require the permissiveness of a non-uniform intermediate filament structure. How can this be reconciled chemically? Considerable evidence summarized above has indicated that desmin and vimentin share a number of properties including polymerization and solubility properties, and common phosphorylation sites by cAMP-dependent kinases, but they also exhibit a number of chemical differences including isoelectric point, molecular weight and antigenic differences. To reconcile the chemical heterogeneity and similarity of desmin and vimentin we have proposed that these two proteins share conserved regions responsible for their conserved polymerization properties and have other regions of divergent structure that account for the unique functional properties of each subunit (Lazarides, 1980, 1981). Indeed recent amino acid sequence of the carboxyl terminal end of desmin and vimentin has revealed regions of homology and regions of divergence between the two proteins (Geisler and Weber, 1981).

Differential Expression of Synemin and Paranemin

The second level of developmental regulation in the structure and composition of desmin- and vimentin-containing intermediate filaments is in the expression of their associated proteins as exemplified by that of synemin and paranemin. Synemin was originally discovered as a protein that associates with desmin, when this latter molecule was purified from chicken smooth muscle (gizzard) (Granger and Lazarides, 1980). Subsequently synemin was shown by immunoautoradiography and immunofluorescence to be present in skeletal myotubes in association with desmin and vimentin filaments and in isolated adult myofibrils in association with the periphery of the Z-disc (Granger and Lazarides, 1980). More recent studies on the composition of avian erythrocyte intermediate filaments have shown that they are composed of vimentin and synemin. Under the same experimental conditions desmin was not detected in these cells (Granger et al., 1981). These results indicate that synemin can associate with either desmin- or vimentin-containing intermediate filaments or with filaments containing both desmin and vimentin, irrespective of the ratio of the latter two proteins. Thus the expression of synemin does not appear to depend on the expression of either desmin or vimentin. Similar results have been obtained with paranemin. Paranemin is present in some, but not all, non-muscle cells in association with vimentin filaments; it is also present in myogenic cells in association with desmin and vimentin filaments. However paranemin is absent from adult smooth and skeletal muscle (see above). Thus, as is the case with the expression of synemin, the expression of paranemin does not appear to depend on the expression of either desmin, vimentin or synemin. Finally, since both synemin and paranemin are absent from some non-muscle cells which contain vimentin filaments, the molecules are not necessary for the formation of vimentin filaments or for the maintenance of their structure. Rather they may play a specific role in the structure and function of these filaments in certain differentiated cells. These observations provide further evidence for our earlier postulate (Gard et al., 1979; Lazarides, 1980, 1981) that the structure of a desmin-vimentin-containing filament can be widely variable not only in different cell types, but also in the same cell type at different stages of differentiation.

Developmentally Specific Changes in the Structure of Desmin and Vimentin Filaments

The third level of developmental regulation that appears to be operative on the structure and composition of intermediate filaments are specific developmental changes in desmin and vimentin as manifested by changes in their sensitivity to Colcemid, immunological changes detected with monoclonal antibodies and increased phosphorylation mediated by hormones and cAMP. Although the molecular nature of these changes or their functional consequences are as yet undetermined, these changes appear to occur specifically upon the association of these two proteins with the Z-disc during myogenesis. Insensitivity of vimentin and synemin to Colcemid has also been observed in mature avian erythrocytes and may represent a specific change in the structure of this class of intermediate filaments in response to a differentiated function.

In conclusion, the structure and composition of the desmin-vimentin class of intermediate filaments appears to be developmentally regulated in differentiating muscle and non-muscle cells in response to specific functions assumed by the filaments in the differentiated cell. This is exemplified by the association of desmin-vimentin and their associated proteins with the periphery of the Z-disc. Further biochemical elucidation of this process will provide a paradigm on the molecular control of cytoplasmic morphogenesis.

ACKNOWLEDGEMENTS

This work has been supported by grants from the National Institutes of Health (PHS-GM-06965), the National Science Foundation, the Muscular Dystrophy Associations of America and a Biomedical Research Support Grant to the Division of Biology. The work with the monoclonal antibodies summarized here was carried out while S. I. Danto was a visiting graduate student from D. A. Fischman's laboratory at the Department of Anatomy and Cell Biology at the SUNY Downstate Medical Center. A full report of this work will be published elsewhere (S. I. Danto, E. Lazarides and D. A. Fischman, in preparation). Portions of this manuscript have appeared in the Cold Spring Harbor Symposia on Quantitative Biology, volume 46.

REFERENCES

Bechtel PJ (1979). Identification of a high molecular weight actin-binding protein in skeletal muscle. J Biol Chem 254:1755.

Bennett HS, Porter KR (1953). An electron microscope study of sectioned breast muscle of the domestic fowl. Am J Anat 93:61.

Bennett GS, Fellini SA, Croop JM, Otto JJ, Bryan J, Holtzer H (1978a). Differences among 100-Å filament subunits from different cell types. Proc Nat Acad Sci USA 75:4364.

Bennett GS, Fellini SA, Holtzer H (1978b). Immunofluorescent visualization of 100 Å filaments in different cultured chick embryo cell types. Differentiation 12:71.

Bennett GS, Fellini SA, Toyama Y, Holtzer H (1979). Redistribution of intermediate filament subunits during skeletal myogenesis and maturation in vitro. J Cell Biol 82:577.

Bergman RA (1958). An experimental study of the non-fibrillar components in frog striated muscle. Bull Johns Hopkins Hosp 103:267.

Bergman RA (1962). Observations on the morphogenesis of rat skeletal muscle. Bull Johns Hopkins Hosp 110:187.

Breckler J, Lazarides E (1981). Isolation of desmin and vimentin filaments from avian embryonic skeletal muscle and their association with a new high molecular weight protein. J Cell Biol (submitted).

Cooke P (1976). A filamentous cytoskeleton in vertebrate smooth muscle fibers. J Cell Biol 68:539.

Danto SI, Fischman DA (1981). Detection and localization of desmin in adult and embryonic chicken muscle using monoclonal antibodies. Anat Rec 199:63A.

Devlin RB, Emerson CP Jr (1979) Coordinate accumulation of contractile protein mRNAs during myoblast differentiation. Dev Biol 69:202.

Ericksson A, Thornell LE (1979). Intermediate (skeletin) filaments in heart Purkinje fibers. A correlative morphological and biochemical identification with evidence of a cytoskeletal function. J Cell Biol 80:231.

Frank ED, Warren L (1981). Aortic smooth muscle cells contain vimentin instead of desmin. Proc Nat Acad Sci USA 78:3020.

Gabbiani G, Schmid E, Winter S, Chaponnier C, De Chastonay C, Vanderkerchove J, Weber K, Franke WW (1981). Vascular smooth muscle cells differ from other smooth muscle cells: Predominance of vimentin filaments and a specific α-type actin. Proc Nat Acad Sci USA 78:298.

Garamvölgyi N (1962). Interfibrillare Z-verbindungen im Quergestreiften Muskel. Acta Physiol Acad Sci Hung 22:235.

Garamvölgyi N (1965). Inter-Z bridges in the flight muscle of the bee. J Ultrastruct Res 13:435.

Gard DL, Bell PB, Lazarides E (1979). Coexistence of desmin and the fibroblastic intermediate filament subunit in muscle and non-muscle cells: identification and comparative peptide analysis. Proc Nat Acad Sci USA 76:3894.

Gard DL, Lazarides E (1980). The synthesis and distribution of desmin and vimentin during myogenesis in vitro. Cell 19:263.

Gard DL, Lazarides E (1982). Cyclic AMP-modulated phosphorylation of intermediate filament proteins during myogenesis in vitro. Cell (in press).

Geisler N, Weber K (1981). Comparison of the proteins of two immunologically distinct intermediate sized filaments by sequence analysis: Desmin and vimentin. Proc Nat Acad Sci USA 78:4120.

Gomer RH, Lazarides E (1981). The synthesis and deployment of filamin in chicken skeletal muscle. Cell 23:524.

Granger BL, Lazarides E (1978). The existence of an insoluble Z-disc scaffold in chicken skeletal muscle. Cell 15:1253.

Granger BL, Lazarides E (1979). Desmin and vimentin coexist at the periphery of the myofibril Z-disc. Cell 18:1053.

Granger BL, Lazarides E (1980). Synemin: a new high molecular weight protein associated with desmin and vimentin filaments in muscle. Cell 22:727.

Granger BL, Repasky EA, Lazarides E (1982). Synemin and vimentin are components of intermediate filaments in avian erythrocytes. J Cell Biol (in press).

Hubbard BD, Lazarides E (1979). Copurification of actin and desmin from chicken smooth muscle and their copolymerization in vitro to intermediate filaments. J Cell Biol 80:166.

Huiatt TW, Robson RM, Arakawa N, Stromer MH (1980). Desmin from avian smooth muscle. J Biol Chem 255:6981.

Ishikawa H, Bischoff R, Holtzer H (1968). Mitosis and intermediate-sized filaments in developing skeletal muscle. J Cell Biol 38:538.

Izant JG, Lazarides E (1977). Invariance and heterogeneity in the major structural and regulatory proteins of chick muscle cells revealed by two-dimensional gel electrophoresis. Proc Nat Acad Sci USA 74:1450.

Lazarides E (1978). The distribution of desmin (100 Å) filaments in primary cultures of embryonic chick cardiac cells. Exp Cell Res 112:265.

Lazarides E (1980). Intermediate filaments as mechanical integrators of cellular space. Nature (London) 283:249.

Lazarides E (1981). Intermediate filaments—chemical heterogeneity in differentiation. Cell 23:649.

Lazarides E, Hubbard BD (1976). Immunological characterization of the subunit of the 100 Å filaments from muscle cells. Proc Nat Acad Sci USA 73:4344.

O'Connor CM, Balzer DR, Lazarides E (1979). Phosphorylation of subunit proteins of intermediate filaments from chicken muscle and nonmuscle cells. Proc Nat Acad Sci USA 76:819.

O'Connor CM, Gard DL, Lazarides E (1981a). Phosphorylation of intermediate filament protein by cAMP-dependent protein kinases. Cell 23:135.

O'Connor CM, Gard DL, Asai DJ, Lazarides E (1981b). Phosphorylation of the intermediate filament proteins, desmin and vimentin, in muscle cells. In: "Protein Phosphorylation," Vol. 8, Cold Spring Harbor Conferences on Cell Proliferation, Cold Spring Harbor, New York (in press).

O'Connor CM, Asai DJ, Flytzanis CN, Lazarides E (1981c). In vitro translation of the intermediate filament proteins desmin and vimentin. Mol and Cell Biol 1:303.

O'Shea JM, Robson RM, Hartzer HK, Huiatt TW, Rathbun WE, Stromer MH (1981). Purification of desmin from adult mammalian skeletal muscle. Biochem J 195:345.

Pease DC, Baker RF (1949). The fine structure of mammalian skeletal muscle. Am J Anat 84:175.

Small JV, Sobieszek A (1977). Studies on the function and composition of the 10 nm (100 Å) filaments of vertebrate smooth muscle. J Cell Sci 23:243.

Walker SM, Schrodt GR, Bingham M (1968). Electron microscope study of the sarcoplasmic reticulum at the Z-line level in skeletal muscle fibers of fetal and newborn rats. J Cell Biol 39:469.

Wang C, Gomer RH, Lazarides E (1981). Heat shock proteins are methylated in avian and mammalian cells. Proc Nat Acad Sci USA 78:3531.

ORGANISATIONS OF ACTIN AND FIBROBLAST LOCOMOTION

J. V. Small

Institute of Molecular Biology of the
Austrian Academy of Sciences
Billrothstr. 11, 5020 Salzburg, Austria.

INTRODUCTION

Studies carried out over the last ten years or so have firmly established that proteins closely analagous to the contractile proteins of muscle, in particular actin and myosin serve fundamental roles in the movement of non-muscle cells. In consequence, the advanced unterstanding of muscle contraction has helped a great deal in investigations of the structure and biochemistry of non-muscle systems. But at the same time and through intensive investigations carried out in more recent years (see symposium volumes by Goldman et al., 1979: Hatano et al., 1979 and Albrecht-Buehler & Watson, 1981) it has become clear that for non-muscle cells the possible interactions of the contractile proteins extend beyond those that occur in muscle.

Actin is a major protein in vertebrate non-muscle cells, constituting in the cases measured 10% - 20% of the total cellular protein (Fine & Bray, 1971; Pollard & Weihing, 1974; Bray & Thomas, 1976; Fine & Taylor, 1976). Moreover, and in contrast to muscle about 50% of this protein is in filamentous form and 50% in a depolymerised state, the latter being apparently maintained by proteins that modulate the polymer state of actin (see review by Lindberg et al., 1979). The cyclic turnover of actin from monomer to polymer thus appears to constitute an important feature of non-muscle motile processes (see also Tilney, 1975). In addition, other proteins exist that cross-link actin, to itself, to membranes and other structures and there are clearly others that act to regulate these interactions.

A rather complex picture is thus emerging in which various potential mechanisms for movement may be envisaged (see for example Huxley; 1979) and that may, in vivo, act together in concert.

As an outgrowth of studies on the contractile apparatus of smooth muscle and the organisation of the cytoskeleton of cultured cells we have been focusing our attention on gaining more insight into the mode of fibroblast movement. The locomotive activity shown by fibroblasts when grown on planar substrata and characterised in some detail by Abercrombie and others (see reviews by Wessells et al., 1973; Abercrombie et al., 1977;and Vasiliev and Gelfand, 1977) has proved to be a useful system for the study of metazoan cell movement. While the relation between this type of movement, seen in vitro and cell migrations that occur in a tissue remains to be clarified it seems reasonable to assume that the motile machinery employed in each case is essentially the same.

In this report we shall describe some recent progress in defining the detailed organisation and composition of the motile regions of fibroblasts and draw some conclusions about the possible mode of cell movement. At the same time we shall touch on a very basic problem related to the destruction of actin meshworks by the conventional procedures used for electron microscopy and which poses a current barrier to studies of cell locomotion and cytoarchitecture in general.

THE LEADING LAMELLA

The translocation of fibroblastic cells over a substrate comprises cycles of forward growth of a peripheral convex lamellar zone, resulting in the extension or polarisation of the cell in one direction, followed by a recoil of the drawn out trailing cell part towards the advanced lamella regions

Fig. 1. Triton-glutaraldehyde cytoskeletons of chick heart fibroblasts stained in aqueous uranyl acetate. The broad, convex cell front corresponds to the leading edge. Note specialisation of this region in fine "microspike" bundles (ms) and intervening homogeneous zones containing the actin meshworks (am). Farther in the cell in (a) the stress fibre bundles (sf) may be recognised. Magnifications: a) x 2100; b) x 16,500. a) from Small (1981), with permission.

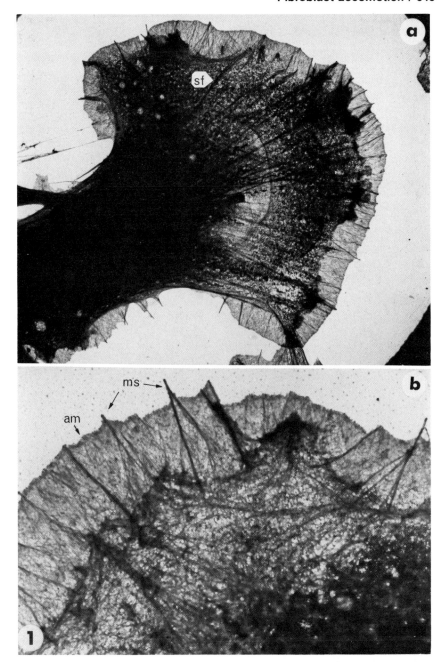

(see e.g. Abercrombie et al., 1977). Thus, the advancing lamella region and in particular the "leading edge" of this region (Abercrombie et al., 1970a) is the site of the primary motile activity. As Abercrombie and coworkers showed, this zone does not advance in a regular fashion but shows phases of standstill, protrusion and withdrawal, "a slight and varying excess of protrusion over withdrawal accounting for the forward translocation" (Abercrombie et al., 1970a). The leading edge of the lamellae may also remain approximately parallel to the substrate or perform upward and backfolding movements to produce so-called "ruffles" (Abercrombie et al., 1970b).

Although earlier electron microscope studies using embedded material had indicated the existence of a meshwork of filaments in the region of the leading edge (Buckley & Porter, 1967; Spooner et al., 1971) the intricate nature of this network has become apparent only more recently with the development of improved procedures to visualise the cytoskeletal components of cultured cells (Small & Celis, 1978; Small et al., 1978, 1980, 1981; Small, 1981; Höglund et al., 1980; Lindberg et al., 1981). These procedures, involving detergent extraction, glutaraldehyde-fixation and negative staining of cells cultured on plastic support films reveal rather clearly the details of the filament arrangements in the thin lamella regions.

Figure 1 shows low power views of chick fibroblasts prepared by detergent extraction. The broad convex band at the cell front, corresponding to the leading edge, is clearly defined as well as some of the large fibre bundles, the stress fibres, that traverse the remaining cytoplasm of the cell. In these examples, and those that follow, extraction was performed with a glutaraldehyde-Triton X-100 mixture (Höglund et al., 1980) which gives slightly improved preservation over that obtained by initial exposure of cells to Triton X-100 alone (Small & Celis, 1978). By cinematographic analysis we have been able to show that such glutaraldehyde-Triton mixtures arrest cell movement within less than 1/2 sec and cause no retraction or other morpho-

Fig. 2. Leading edge region of cell as in Fig. 1a at higher magnification showing the delicate actin meshwork and an actin filament bundle (microspike). Stained with aqueous sodium silicotungstate. Magnification, x 95,000.

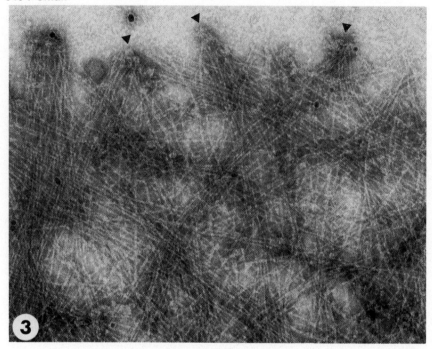

Fig. 3. *The actin meshwork at the very front of the leading edge in the region of the leading membrane. Note the convergence of groups of actin filaments onto foci containing extra material (arrowheads). With weaker extractions the foci are seen to be situated on the inner side of the cell membrane. Sodium silicotungstate stain. Magnification, x 125,000.*

logical changes of the cell edge as may be recognised at the resolution of the light microscope (Small et al., 1981), excepting an apparent collapse of ruffles onto the cell or the substrate. Thus the morphology of these regions as seen in the electron microscope may be related fairly confidently to an arrested stage of the motile process.

The leading edge, defined clearly by a densely packed lamellum of actin filaments shows two morphological states of actin: actin meshworks and thin bundles of actin, or microspikes, that lie within this meshwork and radiate from it (Figs 1b and 2). Without exception the electron microscope images show a continuity between the filaments in the meshworks and those in the bundles, indicating that the two

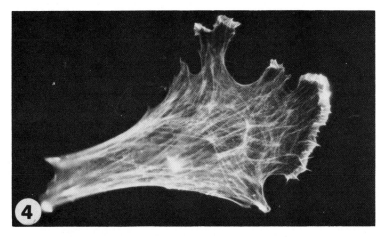

Fig. 4. Staining pattern of anti-gizzard filamin antibodies with a chick fibroblast Triton-glutaraldehyde cytoskeleton. The leading edge meshworks and microspikes are strongly stained as well as the stress fibre bundles. From Small et al., (1981), with permission. Magnification, x 790.

forms are readily interchangeable. Close inspection of the meshworks that lie between the microspike bundles also shows that the actin filaments here converge in groups at the cell front, corresponding to the position of the leading membrane (Fig 3). At the sites of convergence extra material is present, presumably a protein other than actin and that is also concentrated at the tips of the microspikes (Figs 2 & 3; Small et al., 1980; 1981; Höglund et al., 1980). Stereo electron microscopy (Small et al., 1981; Small, 1981) shows that three dimensional order is maintained in the negatively-stained preparations and reveals the complexity of the networks; the filaments form frequent cross-links with each other and extra material is present on the filaments in the region of the cross connections. The microspikes may contain up to 30 - 40 actin filaments in parallel array. Labelling of cells with myosin subfragment-1 shows that the actin filaments in the meshworks and microspikes have the same polarity, directed away from the leading membrane (Small & Celis, 1978; Small et al., 1978).

Studies using antibodies to actin-associated proteins have furnished some additional information about the composition of the leading edge. In addition to reacting with antibodies to actin (see e.g. Lazarides, 1976) the leading edge regions are stained strongly with antibodies to smooth muscle filamin (Heggeness et al., 1977; Fig 4) and ⍺-actinin (Geiger, 1979) as well as to the more recently characterised protein "fimbrin" from the intestinal brush border (Bretscher & Weber, 1980). Interestingly, the protein tropomyosin, while present in the stress fibre bundles is absent from the leading edge (Lazarides, 1976). Further comments on the presence of the above proteins and the organisation of the actin networks will be made when discussing a possible mechanism of cell locomotion. But before this, some attention should be drawn to the problem of preserving actin meshworks for electron microscopy.

DESTRUCTION OF ACTIN MESHWORKS BY OSMIUM TETROXIDE AND DEHYDRATION

Since earlier studies had not shown such an organised network of filaments as revealed by the current methods it was of interest to determine the reason underlying the discrepancy. To answer this question we took the cytoskeletons as prepared by the above methods and subjected them to the additional processing steps commonly employed for conventional electron microscopy, that is leading to plastic-embedded or critical point-dried preparations. The primary steps involved here are those of post-fixation in osmium tetroxide, normally at 1% concentration for 15 min or longer followed by dehydration in organic solvents. These investigations, reported in detail elsewhere (Small, 1981) showed that for the normal exposures employed, OsO_4 post-fixation caused a disordering and kinking of the actin filaments in the leading edge. This was consistent with the effects re-

Fig. 5. Appearance of leading edge actin meshworks after dehydration and critical point drying. a) Triton-glutaraldehyde cytoskeleton, fixed further in glutaraldehyde, then dehydrated in acetone and negatively-stained in 1% uranyl acetate in methanol. To the left of centre is a microspike bundle. Inset shows, at the same magnification, appearance of network in control preparation (like Fig 2) prior to dehydration. b) Preparation as in (a) processed further through a critical point drying step. Again, inset shows how the same network appears at the beginning of the processing steps. Magnifications: x 90,000; b) x 80,000. From Small (1981), with permission.

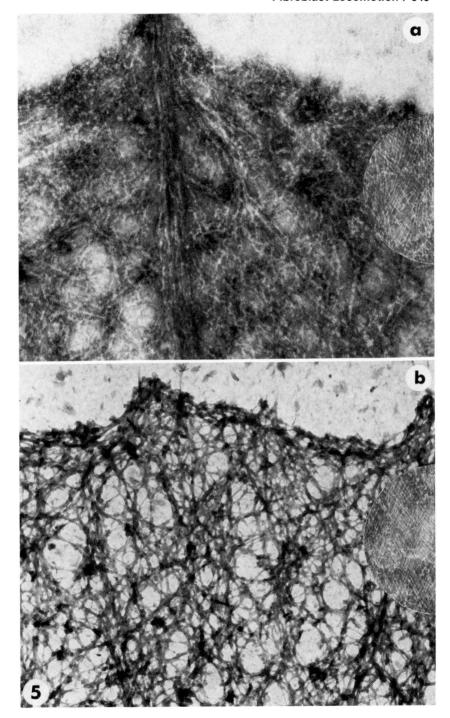

ported for OsO_4 on muscle F-actin in vitro (Maupin-Szamier & Pollard, 1978). In addition it could be shown that dehydration in organic solvents caused, irregardless of the fixation procedure, a rearrangement and aggregation of the filament arrays in the leading edge. Thus, the originally ordered networks (e.g. Fig 2) were converted into an irregular reticulum of fibres (Fig 5a, b) reminiscent of parts of the microtrabecular lattice described by Wolosewick & Porter (1979) and taken by them to represent the "ground substance" of the cytoplasm.

From these results it was clear why so little information about the structure of actin meshworks and their association with other cellular components has been obtained from critical point-dried and embedded material. Further progress in this area will require concerted efforts to overcome these ominous barriers.

THE STRESS FIBRES

Because they may be readily visualised in the light microscope the large fibre bundles or "stress fibres" of cultured cells (see e.g. Figs 1a and 4) were recognised early by histologists (for references see Buckley & Porter, 1967) and have more recently received concentrated attention. In a careful study, Buckley and Porter (1967) established several features of these fibres, including their composition of parallel thin filaments, later identified as actin (Goldman & Knipe, 1973; Lazarides & Weber, 1974) and their relative immobility compared to the motile lamella zones of the cell (see also below). The subsequent application of techniques such as interference reflection- and immunofluorescence microscopy has yielded further information about these fibre bundles.

In line with Buckley and Porter's own suggestion the stress fibres are associated with substrate adhesions, a feature we shall return to below. Attachment of the stress fibres to the membrane at these adhesions appears to be mediated by at least two proteins: α-actinin (Lazarides & Burridge, 1975) and vinculin (Geiger, 1979), although the precise mode of attachment is as yet unclear. Additional proteins, again like those that occur in muscle are present along the length of the stress fib res. These include myosin (Weber & Groeschel-Stewart, 1974), tropomyosin (Lazarides, 1976), filamin (Wang et al., 1975 ;Fig 4)

α-actinin (Lazarides & Burridge, 1975) and myosin light chain kinase (de Lanerolle et al., 1981). With antibodies against tropomyosin, myosin and α-actinin the stress fibre staining has usually been found to be periodic while actin and filamin are distributed evenly. As a result, the stress fibre has been likened to a muscle myofibril with sarcomeric units (Gordon, 1978 and others) an unfortunate and misleading analogy in the light of the continuous distribution of actin. Microinjection experiments have indeed shown that a periodic distribution of α-actinin does exist in the living cell (Feramisco & Blose, 1980) but the claimed alternation of anti-myosin and anti α-actinin staining has more recently been contested (Fujiwara & Pollard, 1980).

But whatever their detailed structural organisation is at the molecular level the stress fibre bundles do show the ability to contract in the presence of ATP (Isenberg et al., 1976), a property that implicates them in the motile process. We might only add the reservation that for cultured cells showing high rates of locomotion, such as those emerging from explants, stress fibres may be completely absent (Abercrombie et al., 1977; Couchman & Rees, 1979) indicating that these fibres, as a morphological entity recognisable in the light microscope, are not absolutely essential for cell locomotion (see also Herman et al., 1981).

ON THE MECHANISM OF FIBROBLAST LOCOMOTION

What would be most useful of course would be a direct correlation between the movement of a cell, recorded in the light microscope prior to its arrest by detergent-glutaraldehyde treatment and the images of the actin meshworks, microspikes and stress fibres in the same cell in the electron microscope. Work to achieve this aim is now only in progress so that no data can yet be presented. Nevertheless, we shall venture to make some speculations on the motile process on the basis of current data.

From studies by light microscopy and particularly those utilising the interference reflection method (Abercrombie & Dunn, 1975; Izzard & Lochner, 1976, 1980; Couchman & Rees, 1979) it has been shown that fibroblasts make two types of contacts with the substrate: 1), relatively large "close contacts" with a cell-substrate separation of about 30 nm and 2), localised "focal contacts" with a cell-substrate separation of 10 - 15 nm. The close contacts are mainly associated with the leading lamella and advance steadily beneath and behind the leading edge as it produces its protrusive and retractive movements (Izzard & Lochner, 1980). The focal contacts correspond to the termini of the stress fibres and once formed remain, for their lifetime, stationary relative to the substrate. Thus, in a given time period, the leading edge and close contacts advance relative to the focal contacts; subsequently the focal contacts disappear and are superceded by new ones formed within or behind the leading edge. Significantly, the formation of focal contacts was found to invariably occur within close contacts in the region of the leading edge and to be preceded by the formation of a microspike or filament bundle within it (Izzard & Lochner, 1980). On the basis of

Fig. 6. A schematic diagram indicating the possible mode of extension of the leading edge via the polymerisation of actin. The essential features of the model are:
1) a nucleation of actin filament growth by a specific nucleating protein situated on the leading membrane,
2) a concomittant cross-linking of the extending filaments by actin-binding proteins (filamin, α-actinin, fimbrin?) to form a semi-rigid lamellum, and 3) traction of this lamellum against the close contact sites or inner cytoskeleton, producing forward extension.
Left part of diagram shows enlarged region of leading edge at early stage of extension. Addition of actin monomers from an actin pool is presumed to occur at the free filament end (see Small et al., 1981). The dark bars depict the actin binding (cross-linking) proteins. With further extension of the leading edge (right part of diagram) the lateral association of nucleation sites together with a bundling of the meshwork filaments would give rise to microspike bundles and signal focal contact and stress fibre formation (see text). Here (right) the cross links have been depicted by solid diamonds and triangles. According to regulation by other factors that break the cross-links and depolymerise actin (Chaponnier et al.,1979, Norberg et al., 1979, Hasegawa et al., 1980, Bamburg et al., 1980, Hinssen, 1981, Harris & Gooch, 1981, Glenney & Weber, 1981) the meshwork filaments may be recycled once the forward position of the cell margin has been established. Modified from Small et al. (1981), with permission.

Fibroblast Locomotion / 353

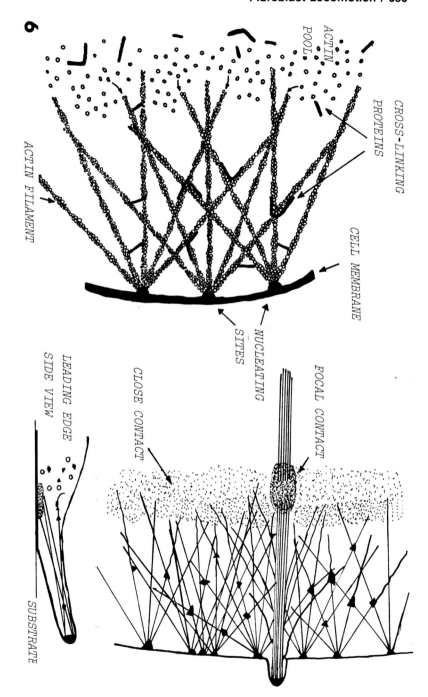

6

these studies the close contacts appear to be the sites against which the leading edge produces traction during its forward growth (Izzard & Lochner, 1980). The focal contacts then provide anchors for the subsequent drawing forward of the cell body effected by the detachment of the trailing cell end and shortening of the axial stress fibres (see e.g. Wessels et al.,1973; Goldman et al.,1976; Chen,1981).

By what means does the leading edge extend itself? From our studies we have suggested that the polymerisation of actin, as in the ejection of the acrosomal process of echinoderm sperm (Tilney,1975), produces the forward growth of the leading edge. This would be consistent with the unipolar organisation of actin filaments in the leading edge relative to the marginal membrane (Small & Celis,1978; Small et al., 1978) as well as the absence of significant amounts of myosin in this region (Heggeness et al.,1977; Herman et al., 1981). A schematic diagram incorporating the proposed features of this model is shown in Fig.6. A similar mechanism, but differing in some details has been proposed by Lindberg et al., (1981).

In addition to the data referred to above we clearly lack information on how the different components, some yet to be discovered, are organised in actin meshworks at the ultrastructural level, how they interact with actin and how these interactions are controlled. Also, can we demonstrate directly the involvement of actin polymerisation in fibroblast locomotion? Unfortunately, space here is too short to review the more recent advances (Albrecht-Buehler & Watson, 1981) in related areas. But significant progress in characterising the types and interactions between actin and actin-associated proteins is being made and we may expect answers to at least some of the basic questions in the near future.

ACKNOWLEDGEMENTS

These studies were supported by grants from the "Austrian Fonds zur Förderung der wissenschaftlichen Forschung" and the "Volkswagen Foundation".
I thank Mr P. Jertschin for technical assistance.

REFERENCES

Abercrombie, M., Heaysman, J.E.M. & Pegrum, S.M. (1970a). The locomotion of fibroblasts in culture. I. Movements of the leading edge. Expl Cell Res. 59:393-398.
Abercrombie, M., Heaysman, J.E.M. & Pegrum, S.M. (1970b) The locomotion of fibroblasts in culture. II. "Ruffling". Expl Cell Res. 60:437-444.
Abercrombie, M. & Dunn, G.A. (1975). Adhesions of fibroblasts to substratum during contact inhibition observed by interference reflection microscopy. Expl Cell Res. 92:57-62.
Abercrombie, M., Dunn, G.A. & Heath, J.P. (1977). The shape and movement of fibroblasts in culture. In "Cell and Tissue Interactions" (ed by J.W. Lash & M.J. Burger) pp 57-70. Raven press, N.Y.
Albrecht-Buehler, G. & Watson, J.D. (1981). Cold Spring Harb. Symp. on the "Organisation of the Cytoplasm". Cold Spring Harbor Laboratory, in press.
Bamburg, J.R., Harris, H.E. & Weeds, A.G. (1980). Partial purification and characterization of an actin depolymerizing factor from brain. FEBS lett. 121:178-182.
Bray, D. & Thomas, C. (1976). Unpolymerised actin in fibroblasts and brain. J.Mol.Biol. 105:527-544.
Bretscher, A. & Weber, K. (1980). Fimbrin, a new microfilament associated protein present in microvilli and other cell surface structures. J.Cell Biol. 76:335-340.
Buckley, I.K. & Porter, K.R. (1967). Cytoplasmic fibrils in living cultured cells. Protoplasma 64:349-380.
Chaponnier, C., Borgia, R., Rungger-Brändle, E., Weil, R. & Gabbiani, G. (1979). An actin destabilising factor is present in human plasma. Experientia 35:1039-1040.
Chen, W.-T. (1981). Mechanism of retraction of the trailing edge during fibroblast movement. J. Cell Biol.90: 187-200.
Couchman, J.R. & Rees, D.A. (1979). The behaviour of fibroblasts migrating from chick heart explants changes in adhesion, locomotion and growth, and in the distribution of actomyosin and fibronectin. J. Cell Sci. 39:149-165.
de Lanerolle, P., Adelstein, R.S., Feramisco, J. & Burridge, K. (1981). Proc. Natl Acad. Sci., in press.
Feramisco, J.R. & Blose, S.H. (1980). Distribution of fluorescently labelled α-actinin in living and fixed fibroblasts. J. Cell Biol. 86:608-615.
Fine, R.E. & Bray, D. (1971). Actin in growing nerve cells. Nature New Biology 234:115-118.

Fine, R.E. & Taylor, L. (1976). Decreased actin and tubulin synthesis in 3T3 cells after transformation by SV40 Virus. Expl Cell Res. 102:162-168.

Fujiwara, K. & Pollard, T.D. (1980). Relative disposition of myosin, alpha-actinin and tropomyosin in stress fibres. J. Cell Biol. 87:222a.

Geiger, B. (1979). A 130K protein from chicken gizzard: its localisation at the termini of microfilament bundles in cultured chicken cells. Cell 18:193-205.

Glenny, J.R. & Weber, K. (1981). Calcium control of microfilaments: uncoupling of the F-actin-severing and -bundling activity of villin by limited proteolysis in vitro. Proc. Natl Acad. Sci. 78:2810-2814.

Goldman, R.D. & Knipe, D.M. (1973). Functions of cytoplasmic fibres in non-muscle cell motility. In "The Mechanism of Muscle Contraction" Cold Spr. Harb. Symp. Quant. Biol. Vol. 37: pp 523-533, Cold Spring Harb. Laboratory.

Goldman, R.D., Schloss, J.A. & Starger, J.M. (1978). Organisational changes of actinlike microfilaments during animal cell movement. In "Cell Motility", Cold Spr. Harb. Conf. on Cell Proliferation", (ed. by R. Goldman, T. Pollard & J. Rosenbaum) pp 217-245. Cold Spr. Harb. Laboratory.

Gordon, W.E. (1978). Immunofluorescent and ultrastructural studies of "sarcomeric" units in stress fibres of cultured non-muscle cells. Expl Cell Res. 117:253-260.

Harris, H.E. & Gooch, I. (1981). An actin depolymerising protein from pig plasma. FEBS Lett. 123:49-53.

Hasegawa, T., Takahashi, S., Hayashi, H. & Hatano, S. (1980). Fragmin: a Ca^{2+}-sensitive regulatory factor on the formation of actin filaments. Biochemistry 19:2677-2683.

Hatano, S., Ishikawa, H. & Sato, H. (1979). Cell Motility: Molecules and Organisation. University Park Press, Baltimore.

Heggeness, M.H., Wang, K. & Singer, S.J. (1977). Intracellular distributions of mechanochemical proteins in cultured fibroblasts. Proc. Natl Acad. Sci. 74:3883-3887.

Herman, I.M., Crisona, N.I. & Pollard, T.D. (1981). Relation between cell activity and the distribution of cytoplasmic actin and myosin. J. Cell Biol. 90:84-91.

Hinssen, H. (1981). An actin-modulating protein from Physarum polycephalum. I. Isolation and purification. Eur. J. Cell Biol. 23:225-233.

Höglund, A.-S., Karlsson, R., Arro, E., Frederiksson, B.-E. & Lindberg, U. (1980). Visualisation of the peripheral weave of microfilaments in glia cells. J. Muscle Res. Cell Motil.1:127-146.

Huxley, H.E. (1979). In "Cell Motility: Molecules and Organisation" (ed by S. Hatano, H. Ishikawa and H. Sato) pp 3-9. University Park Press, Baltimore.

Isenberg, G., Rathke, P.C., Hülsmann, N., Franke, W.W. & Wohlfarth-Bottermann, K.E. (1976). Cytoplasmic actomyosin fibrils in tissue culture cells. Direct proof of contractility by visualisation of ATP-induced contraction in fibrils isolated by laser microbeam dissection. Cell Tiss. Res. 166:427-443.

Izzard, C.S. & Lochner, L.R. (1976). Cell-to-substrate contacts in living fibroblasts: an interference-reflexion study with an evaluation of the technique. J. Cell Sci. 21:129-159.

Izzard, C.S. & Lochner, L.R. (1980). Formation of cell-to-substrate contacts during fibroblast motility: an interference-reflexion study. J. Cell Sci. 42:81-116.

Lazarides, E. & Weber, K. (1974). Actin antibody: the specific visualisation of actin filaments in non-muscle cells. Proc. Natl Acad. Sci. USA 71:2268-2272.

Lazarides, E. & Burridge, K. (1975). α-actinin: immunofluorescent localisation of a muscle structural protein in non-muscle cells. Cell 6:289-298.

Lazarides, E. (1976). Two general classes of cytoplasmic actin filaments in tissue culture cells: the role of tropomyosin. J. Supramol. Struct. 5:531-563.

Lindberg, U., Carlsson, L., Markey, F. & Nyström, L.E. (1979). The unpolymerised form of actin in non-muscle cells. In "Methods and Achievements in Experimental Pathology - The Cytoskeleton in Normal and Pathologic Processes", (ed by G. Gabbiani) Vol.8: pp 143-170. Karger AG, Basel, Switzerland.

Lindberg, U., Höglund, A.S. & Karlsson, R. (1981). On the ultrastructural organisation of the microfilament system and the possible role of profilactin. Biochimie 63:307--323.

Maupin-Szamier, P. & Pollard, T.D. (1978). Actin filament destruction by osmium tetroxide. J. Cell Biol. 77:837-852.

Norberg, N., Thorstensson, R., Utter, G. & Fagreus, A. (1979). F-actin depolymerizing activity of human serum. Eur. J. Biochem. 100:575-583.

Pollard, T.D. & Weihing, R.R. (1974). Actin and myosin and cell movement. CRC Crit. Rev. Biochem. 2:1-65.

Small, J.V. & Celis, J.E. (1978). Filament arrangements in negatively-stained cultured cells: the organisation of

actin. Cytobiologie 16:308-325.
Small, J.V., Isenberg, G. & Celis, J.E. (1978). Polarity of actin at the leading edge of cultured cells. Nature (Lond.) 272:638-639.
Small, J.V., Celis, J.E. & Isenberg, G. (1980). Aspects of cell architecture and locomotion. In "Transfer of Cell Constituents into Eukaryotic Cells" (ed by J.E. Celis, A. Graessmann and A. Loyter) pp 75-111. Plenum, N.Y.
Small, J.V., Rinnerthaler, G. & Hinssen, H. (1981). Organisation of actin msehworks in cultured cells: the leading edge. In "Cold Spring Harb. Symp. on the Organisation of the Cytoplasm" (ed by G. Albrecht-Buehler and J.D. Watson) Cold Spring Harbor Laboratory, in press.
Small, J.V. (1981). Organisation of actin in the leading edge of cultured cells: influence of osmium tetroxide and dehydration on the ultrastructure of actin meshworks. J. Cell Biol., in press.
Spooner, B.S., Yamada, K.M. & Wessels, N.K. (1971). Microfilaments and cell locomotion. J. Cell Biol. 49:595-613.
Tilney, L.G. (1975). The role of actin in non-muscle cell motility. In "Molecules and Cell Movement" (ed by S. Inoué and R.E. Stephens) pp 339-388. Raven Press, N.Y.
Vasiliev, J.M. & Gelfand, I.M. (1977). Mechanism of morphogenesis in cell cultures. Int. Rev. Cytol. 50:159-274.
Wang, K., Ash, J.K. & Singer, S.J. (1975). Filamin, a new high-molecular-weight protein found in smooth muscle and non-muscle cells. Proc. Natl Acad. Sci. 72:4483-4486.
Weber, K. & Groeschel-Stewart, U. (1974). Antibody to myosin: the specific visualisation of myosin-containing filaments in non-muscle cells. Proc. Natl Acad. Sci. USA 71:4561-4564.
Wessels, N., Spooner, B. & Luduena, M. (1973). In "Locomotion of Tissue Cells" Ciba Fdn. Symp. 14:53-82. Elsevier Excerpta Medica, North Holland.
Wolosewick, J.J. & Porter, K.R. (1979). Microtrabecular lattice of the cytoplasmic ground substance: artefact or reality. J. Cell Biol. 82:114-139.

NEURAL CREST CELL DEVELOPMENT

James A. Weston

University of Oregon

Eugene, Oregon 97403

I. NEURAL CREST CELL MIGRATION AND DIFFERENTIATION

The neural crest provides a source of stem cells that differentiate into diverse cell and tissue phenotypes throughout the vertebrate embryo (see Table 1). The migration of neural crest cells and the differentiation of its derivatives have been recently and repeatedly reviewed (Hörstadius, 1950; LeDouarin, 1980a,b; Morriss & Thorogood, 1978; Noden, 1978a; Noden, 1980; Weston, 1970; Weston, 1980; Weston, 1981a,b), and another review is not needed. I will, however, briefly summarize salient aspects of crest development and discuss a few of the many important questions that remain to be answered.

The basic facts concerning neural crest development may be listed as follows:

1. Migration and Localization of Crest Cells.

   a. Crest migratory pathways appear to be characteristic of different axial levels of the embryo (Hörstadius, 1950; LeDouarin, 1980b; Noden, 1978a; Noden, 1980; Weston, 1970; Weston, 1981b).

   b. These pathways seem to be determined, in part, by the structure of the embryonic environment (Pratt et al., 1975; LeDouarin et al., 1978; Löfberg et al., 1980; Noden, 1975; Noden, 1978b; Tosney, 1978; Weston et al., 1978).

   c. The environment undergoes developmental changes that

Table 1
DERIVATIVES OF THE NEURAL CREST

A. Structures arising from cranial neural crest only:

1. Skeletal and connective tissue
   a) Bones and cartilages of the head and face
      Upper and lower jaw
      Dental papilla (odontoblasts)
      Palate
      Cranial vault floor
   b) Connective tissues
      Corneal stromal fibroblasts
      Contributions to dermis and subcutaneous adipose tissue of face, jaw and upper neck
      Lining of forebrain (meninges)
   c) Muscles
      Ciliary muscles (striated)
      Smooth muscle of cranial blood vessels and dermis

2. Endocrine and exocrine tissues
      Mesenchyme of pituitary, thyroid, parathyroid thymus and salivary glands

B. Structures derived from both cranial and trunk neural crest:

1. Neural derivatives
   a) Sensory neurons of cranial and spinal ganglia (peptidergic and ?)
   b) Sympathetic neurons of pre- and paravertebral ganglia (noradrenergic)
   c) Parasympathetic neurons of visceral ganglia and plexuses (cholinergic, serotonergic and ?)
   d) Neurosecretory cells
      Calcitonin-producing (C) cells of the thyroid
      Adrenal medullary cells (adrenergic)
      Possible contribution to neuroactive peptide-secreting cells of pituitary

2. Supportive cells of the peripheral nervous system
      Glia
      Schwann sheath cells
      Satellite cells of ganglionic neurons

3. Pigment (melanin or other pigment granule-containing) cells of skin, hair (feathers) and iris

affect the onset and cessation of crest and crest-derived cell migration (Pratt et al., 1975; Toole & Trelstad, 1971; Toole et al., 1980; Weston & Butler, 1966; Weston et al., 1978), and perhaps the direction as well (Erickson et al., 1980; Weston, 1963, 1970).

d. Environmental factors alone cannot account for all aspects of crest cell morphogenetic behavior, since the crest cells themselves exhibit exceptional migratory abilities distinct from other embryonic tissues (Bronner-Fraser & Cohen, 1980; Erickson et al., 1980).

2. The Crest Cell Migratory Environment.

a. Crest cells migrate in embryonic spaces filled with extracellular matrix (ECM) (Derby, 1978; Löfberg et al., 1980; Pintar, 1978; Pratt et al., 1975; Toole et al., 1980; Tosney, 1981; Weston et al., 1978).

b. The predominant macromolecules present in this matrix include various collagen types (von der Mark, 1980; Hay, 1978; Wartiovaara et al., 1980), laminin in the basal lamina (Wartiovaara et al., 1980; Timpl et al., 1979; Kleinman et al., 1980), the glycoprotein fibronectin (Mayer et al., 1981; Newgreen & Thiery, 1980; Pearlstein et al., 1980; Sieber-Blum et al., 1981; Weston, 1980; Weston, 1981b; Yamada, 1980), and various proteoglycan glycosaminoglycans (GAG) such as hyaluronic acid, chondroitin sulfates and heparin sulfate in the interstitial spaces and at cell surfaces (Comper & Laurent, 1978; Solursh & Morriss, 1977; Pratt et al., 1975; Toole et al., 1980; Weston et al., 1978; Weston, 1981b). These macromolecules interact in characteristic ways to form the structural components (basal lamina and interstitial fibrils) of the matrix (Bernfield, 1980; Hay, 1978; Pearlstein et al., 1980; Yamada et al., 1978; Yamada et al., 1980).

c. The amounts and kinds of GAG are characteristic of different embryonic regions and stages (Derby, 1978; Morriss & Thorogood, 1978; Pintar, 1978; Pratt et al., 1975; Toole et al., 1980; Weston et al., 1978; Weston, 1980; Weston, 1981b). It is not yet known, however, whether there are corresponding regional and temporal changes in the amounts of matrix glycoproteins such as the collagens and fibronectin (see below).

d. Crest cells, as well as the embryonic tissues surrounding the crest migratory spaces, produce characteristic kinds and amounts of ECM macromolecules. For example: (1) Neural crest cells produce a dramatically higher level of hyaluronic acid than other embryonic tissues (Glimelius & Pintar, 1981; Greenberg & Pratt, 1977; Pintar, 1978), whereas somite and ectodermal epithelium produce proportionally more sulfated GAG, and less hyaluronic acid (Pintar, 1978). (2) All the embryonic tissues that are associated with the crest migratory spaces produce collagens in a developmentally-regulated fashion (Cohen & Hay, 1971; Greenberg et al., 1980; Minor, 1973; Newsome, 1976). (3) In contrast, although somitic mesenchyme cells appear to make fibronectin, trunk crest cells lack this macromolecule at their surfaces, and apparently do not make it (Loring et al., 1977; Mayer et al., 1981; Newgreen & Thiery, 1980; Sieber-Blum et al., 1981; Weston, 1980; Yamada, 1980; K. Ranney, unpublished results).

e. The components of the matrix in the crest migratory spaces undergo developmental changes as a consequence of changes in the synthetic and degradative activity of the embryonic cells associated with these spaces. For example, the pattern of hyaluronate synthesis undergoes significant developmental changes in vivo and in vitro (Glimelius & Pintar, 1981; Greenberg & Pratt, 1977; Pintar, 1978; Pratt et al., 1975). This, in turn, is correlated with the onset and extension of crest cell migration (Pratt et al., 1975; Toole & Trelstad, 1971), and the cessation of migration and coalescence of crest cells to form ganglia in the somite (Toole, 1972; Weston et al., 1978; Kinsella & Weston, in preparation).

3. Crest Cell Differentiative Capacity.

a. Some crest cells are developmentally labile, and differentiate in response to developmental cues that they encounter in their migratory environment. Thus, (1) after heterotopic grafting, some crest cells can differentiate according to their location in the host embryo rather than their original source (LeDouarin et al., 1975; LeDouarin et al., 1978; LeLievre et al., 1980; Noden, 1978b). Likewise, (2) some crest or crest-derived cells can alter their differentiation in response to environmental cues in culture (see Weston, 1981a,b; and below).

b. In contrast, although local environmental cues may still be required to elicit cell differentiation, developmental restrictions are imposed on some crest cell subpopulations very early in development. Thus, (1) cranial crest cells give rise to parts of the skull and connective tissue of the head, whereas trunk crest cells are unable to do so (see Hörstadius, 1950; Hall, 1980; Noden, 1975; Noden, 1978a; Newsome, 1976; Weston, 1981b). Likewise, (2) the ability of crest cells to give rise to some neurons is apparently precociously restricted (see Erickson et al., 1980; LeLievre et al., 1980; Weston, 1981a,b; and below). Finally, (3) sensitive and specific assays have recently detected phenotypic heterogeneity in crest cell populations, even at the onset of migration (see Barald & Wessells, 1981; Chun et al., 1980; Ciment & Weston, 1981; Fauquet et al., 1981; Vulliamy et al., 1981; and below).

II. THE REGULATION OF CREST CELL MIGRATORY AND DIFFERENTIATIVE BEHAVIOR

The "declarations of fact" listed above contain important deficiencies. For example, (1) as mentioned above, qualitative and quantitative data are not yet available for matrix components other than some of the GAGs. Specifically, we do not yet know how the amounts and kinds of matrix glycoproteins vary during development in the crest migratory spaces. Moreover, (2) although they must surely play some role, it is not yet known how the components of the extracellular matrix affect crest cell morphogenetic and differentiative behavior. (3) We do not know what special properties of crest cells permit them to migrate so actively and extensively in the embryo, nor how these properties change when migration ceases and (some) crest cells coalesce. Finally, (4) although crest cells clearly can differentiate in response to environmental cues (see above), it has become inreasingly evident that the crest is not always, if it is ever, a homogeneous cell population. It will be important, therefore, to detect developmentally distinct subpopulations, and to document when and under what conditions such subpopulations segregate in the crest cell lineage (see below).

Fortunately, opportunities do exist to gain insights into these important problems. In the rest of this presentation, therefore, I propose to summarize briefly some of our efforts to address a few of the questions I have posed.

1. What Environmental Cues Mediate Crest Cell Differentiation in vivo and in vitro?

   Melanogenesis in vitro. When they are cultured, most crest cells express one phenotype--melanin pigment--even though, presumably, many would have produced a variety of other phenotypes characteristic of the crest lineage if they had been left in the embryo. The extent of melanogenesis in culture can be influenced by the conditions of culture (Glimelius & Weston, 1981a). Such results suggest, therefore, that the culture environment alters the normal pattern of crest cell differentiation. In this case, as yet unknown molecular components or physical conditions in vitro promote melanogenesis at the expense of other phenotypes (see also Glimelius & Weston, 1981b).

   The glial cell alternative. If crest-derived spinal ganglia are explanted in vitro, some cells, responding to unknown developmental stimuli in culture, can also differentiate into melanocytes (Cowell & Weston, 1970; Nichols & Weston, 1977; Nichols et al., 1977). We have suggested that, in this case, crest-derived cells undergo melanogenesis that would normally have differentiated as supportive (glial) cells (Nichols & Weston, 1977; Nichols et al., 1977). We proposed that, on the artificial culture substratum, these supportive cell precursors are "lured" away from the neurons, and are influenced by the culture environment (see above) to become melanocytes (also see Weston, 1981a). To test this, Nichols cultured ganglia in a standard nutrient medium on various substrata. Substrata such as agar or a fibroblast monolayer discouraged the dissociation of the neurons and glial cell precursors, whereas control tissue culture substrata (plastic or collagen-coated glass coverslips) permitted such dissociation. When dissociation was prevented, adventitious pigment cell differentiation did not occur.

   Conversely, when interaction of the crest-derived, non-neuronal cells with neurons was promoted or maintained in such ganglion cultures, glial cell differentiation could be observed (Holton & Weston, 1982a,b). Thus, immunocytochemical methods revealed positive staining for the S100 protein, characteristic of nerve supportive (glial) cells (Moore, 1965), in intact or cultured embryonic spinal ganglia. Such staining appeared primarily, if not exclusively, in cells that were associated with neuron cell bodies or fibers (Holton & Weston, 1982a). Likewise, Holton has shown by immunoprecipitation with antibodies against S100 protein that S100 production by cultured glial cell precursors was promoted even by interaction with paraformaldehyde-killed ganglionic neurons (Holton & Weston, 1982b). These results suggest that interaction with neurons present in developing spinal ganglia mediates the choice between at least two possible phenotypes (pigment and supportive cell). They further suggest that these developmental cues are presented at the neuron cell surface or in the immediate pericellular environment. The molecular basis for these interactions remains to be elucidated.

One neuronal alternative. Culture conditions also affect the differentiation of crest-derived neurons (Bunge et al., 1978; Cohen, 1977; Norr, 1973; Fauquet et al., 1981; LeDouarin, 1980a; Patterson & Chun, 1977; Patterson, 1978; Smith et al., 1977; Smith et al., 1980; Ziller et al., 1979). When crest cells are cultured in close proximity to developing somitic mesodermal or notochordal cells, some of them exhibit formaldehyde-induced fluorescence (FIF). This property is thought to reflect the cells' ability to take up and/or produce, and to store neurotransmitter molecules (catecholamines) characteristic of the adrenergic nervous system (Falck & Owman, 1965). It is one of a "battery" of sensitive markers for sympathetic neuron differentiation (see LeDouarin, 1980a; Patterson, 1978). When applied with caution, it can facilitate quantitation of crest cell differentiation in culture.

The somite-induced "catecholamine-fluorescence" is not expressed if the stimulating tissue (embryonic somite) is separated from the cultured crest by a Millipore filter membrane (Norr, 1973). This suggests that crest cells require close proximity to somite cells or their extracellular products to express this neuronal trait. Since cultured embryonic fibroblasts deposit a nondiffusible "microexudate" on the culture substratum (Culp, 1974; Loring et al., 1982; Schwartz et al., 1979; see above), we could directly compare the differentiation of crest cells grown on such substrata with cultures in the same nutrient medium grown on control (plastic) substrata. When this was done (Loring et al., 1982; see also Sieber-Blum et al., 1981), striking differences were noted. First, some of the population that would differentiate exclusively into pigment cells on plastic culture substrata, produced the FIF characteristic of catecholamine-containing cells when grown on substrata "conditioned" with somite fibroblast-derived microexudate (Fig. 1). Second, "artificial" substrata constituted with some of the individual components of this ECM (e.g., collagen, fibronectin, collagen + fibronectin, hyaluronic acid, or chondroitin sulfates) elicit some characteristic patterns of differentiation in crest cells grown in otherwise identical nutrient conditions (Loring et al., 1982). For example, collagen reduced the proportion of crest cells undergoing differentiation as pigment cells, whereas fibronectin both reduced melanogenesis and promoted the appearance of FIF in crest cell cultures (Fig. 1). Similar conditions for promoting catecholamine-fluorescence have also been reported by other workers (Sieber-Blum & Cohen, 1980; Sieber-Blum et al., 1981). It should be noted, however, that in our studies, differentiation of catecholamine-containing cells appeared to depend on the amount of fibronectin added to the cultured cells, further emphasizing our need to know both the distribution and the amounts of the matrix components normally present in the crest cells' migratory spaces (see above).

Although such analyses are clearly only in their initial stages, it seems reasonable to conclude first, that crest cells in vivo and in vitro encounter and often appear to respond to growth factors and other developmentally relevant macromolecules that constitute part of the substrata on which they reside (see Rutter et al., 1978). The effects of such substrate-bound constituents on crest cell differentiative behavior are largely unknown, but it seems likely that they may effect the choice, as well as the expression of their phenotype. The exact identity and effective dose of the developmental cues involved, their mode of action and the mechanism of detection and response by crest cells all remain to be investigated.

2. Early Detection of Phenotypic Heterogeneity in Neural Crest Cell Populations.

Crest cells from all axial levels can produce sensory and autonomic ganglia (see Table 1). Moreover, after "heterotopic" transplantations, the type of ganglion formed by grafted crest cells corresponds to what normally develops at the site of the graft rather than its source (LeDouarin et al., 1975; Noden, 1978b; Smith et al., 1977). However, in some experiments, cultured neural crest cells (as opposed to intact crest), or cells from crest-derived ganglia, other

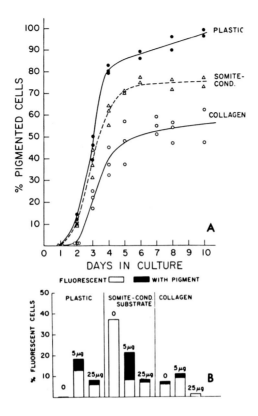

Fig. 1. The effect of various substrata on neural crest cell differentiation in culture.

A. Time course of melanogenesis in dissociated quail neural crest cells cultured on different substrata. On plastic substrata, crest cells differentiated homogeneously into pigment cells. Both collagen and somite microexudate-conditioned substrata reduced the proportion of the population expressing melanin pigment. Most of the unpigmented cells on somite-conditioned substrata, and a few of the unpigmented cells on collagen exhibited FIF (see B).

B. Effect of exogenous fibronectin on the expression of catecholamine-fluorescence in dissociated crest cells cultured on different substrata for seven days. Note altered responses to different amounts of fibronectin on the different substrata. Note, too, that in some cases, the presence of exogenous fibronectin causes a portion of the cell population to express, transiently, both FIF and pigment granules simultaneously (from Loring et al., 1982).

than spinal ganglia, have been grafted ectopically into host embryos. Results from such grafts indicate that these cells apparently do not give rise to sensory neurons even when given appropriate "permissive" environmental cues (Erickson et al., 1980; LeLievre et al., 1980). This may mean that a population of cells in the neural folds, which is depleted or absent from crest cells derived from cultured neural tubes or other crest derived tissue, is specified very early as spinal ganglion neurons (see Weston, 1981a,b; and above). Such a conclusion is certainly consistent with the observations that some neurons differentiate very precociously in nascent ganglia, and can provide the appropriate stimulus for non-neuronal crest-derived cells to differentiate into supportive (glial) cells of the ganglion (see Holton & Weston, 1981a,b; and above).

Finally, the "hybridoma" technology (Milstein & Lennox, 1980) has recently been exploited to produce monoclonal antibodies recognizing as yet unidentified antigenic determinants characteristic of peripheral neurons. Some of these antibodies can mediate the immunocytochemical staining of subpopulations of neural crest-derived cells (Barald & Wessells, 1981; Chun et al., 1981; Ciment & Weston, 1981; Vulliamy et al., 1981). Such results appear to confirm directly the inference that subpopulations of crest cells differentiate precociously (Fig. 2).

3. The Progressive Developmental Restrictions Imposed on Neural Crest-Derived Cells During Normal Development.

There has been much discussion about whether embryonic cell populations in general, and crest cells in particular, are "pluripotent"--that is, whether they are able to differentiate into any of a variety of derivatives characteristic of a particular lineage. The operational notion of pluripotency, however, meaningfully applies only to the genetic determinants that cells carry. Since we know that developmental restrictions do exist, and that the environment can influence the expression of specific phenotypes, the developmentally interesting problem is to establish when, and in what order the particular genetic determinants that specify a cell lineage are chosen to be regulated.

In the case of crest cells, the results that have been discussed above suggest that in fact there may be a characteristic order in which the developmental repertoire of the

Neural Crest Cell Development / 369

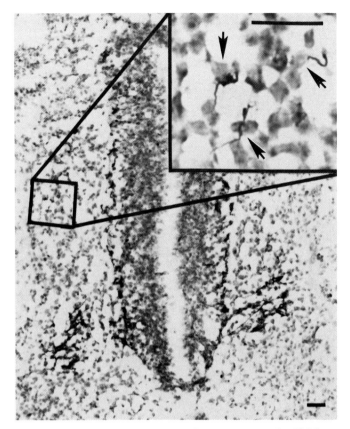

Fig. 2. Evidence for precocious neuronal differentiation in the avian embryo. Oblique, transverse cryostat section through the cervical level of a 2-day (Stg 13) chicken embryo. Section stained immunochemically using a monoclonal antibody raised against 7-day embryonic chicken spinal ganglia, screened for specific binding to cultured 9-day ganglionic neurons, and conjugated to horseradish peroxidase (see Ciment & Weston, 1981). This antibody also binds to a variety of other neuronal derivatives of the neural crest in older embryos and newly-hatched chicks. Note that antibody mediates the staining of fibers in the neural tube, as well as otherwise undistinguished (crest?) cells (arrows) in the mesenchyme lateral to the neural tube. Inset: Higher magnification showing antibody-mediated staining of cells in the mesenchyme. Bars = 25μ. (Preparation kindly provided by Dr. G. Ciment, and expertly photographed by Mr. Harry Howard.)

Fig. 3. Diagrammatic scheme suggesting a possible sequence of binary "choices" leading to the segregation of various crest cell lineages (see Table 1). The circles represent cell populations which may be found in various embryonic locations. Lower case letters within the circles represent the developmental capability <u>remaining</u> in each population. The time that certain restrictions occur is suggested by the large arrow heads indicating the onset of cell migration from the cranial crest (solid arrow) and trunk crest (open arrow). Some ectomesenchymal cells, for example, may have been specified before cranial crest cell migration begins. Likewise, some spinal ganglion neurons may be precociously determined before the onset of trunk crest cell migration (see Weston, 1981a, b).

Abbreviations: cnf, cranial neural folds; nf, trunk neural crest and cranial neural crest after the cranial crest cells (ccc) that will give rise to ectomesenchyme have departed; a, adrenergic neuroblasts (FIF-positive); A, adrenergic neurons; auto, stem cells of the autonomic nervous system (probably also FIF-positive); c, cholinergic neuroblasts; C, cholinergic neurons; g, supportive cell precursors; G, differentiated glial or satellite cells; m, melanoblasts; M, melanocyte; n, sensory neuroblasts; Nv, ventrolateral (early-differentiating) neurons of spinal ganglia; Nm, mediodorsal (late-differentiating) neurons of the spinal ganglia; p, precursor cells of adrenal medulla; P, pheochromocytes of adrenal medulla. Dashed lines denote interactions that may (?) or do occur between neurons and supportive cell precursors (see Holton & Weston, 1981; and text). Small arrow heads indicate terminal differentiations. Adapted from Weston (1981a and 1981b).

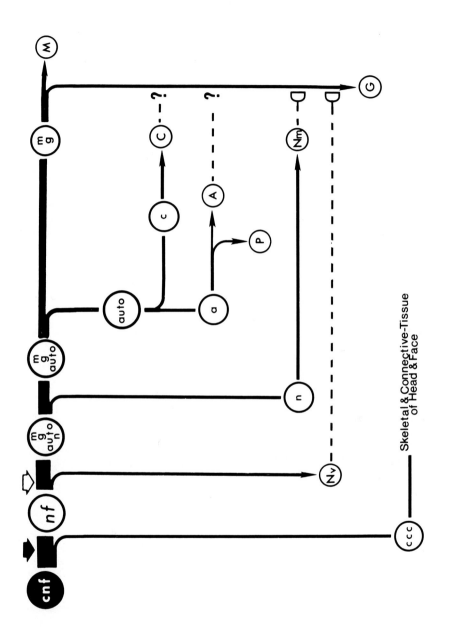

crest lineage is disclosed. In the cranial neural folds, for example, a population of cells is present that gives rise to skeletal and connective tissue structures of the head and face (Hall, 1980; Noden, 1980; Weston, 1981b). If other crest cells ever have this capability, they apparently lose it very early in development (see Hall & Tremaine, 1979; Weston, 1981b). Likewise, a subpopulation of cells that appear to acquire neuronal traits is segregated from the remainder of the crest cell population early in embryogenesis (see Barald & Wessells, 1981; Ciment & Weston, 1981; Smith et al., 1980; Weston, 1981b; and above). Nevertheless, cells still remain in the migratory crest population that can differentiate as neurons of the sensory and the autonomic nervous system in response to specific environmental cues. Finally, cells are left that are neither stromal nor neuronal derivatives of the crest. These appear to respond to specific cues to differentiate either as supportive (glial) cells of the peripheral nervous system or as pigment cells (see Holton & Weston, 1982a,b; Weston, 1981b). Figure 3 suggests one, but by no means the only, possible sequence of restrictions that could generate the known phenotypic diversity of the neural crest lineage (see also LeLievre et al., 1980).

It is of interest that during their morphogenetic phase--the extensive migration through the ECM-filled spaces of the developing embryo, followed in some instances by coalescence of cells into tightly-interacting populations--crest cells sequentially encounter environmental cues that we know can promote specific cellular differentiation. Therefore, the choice of which genetic determinants to regulate may ultimately be determined by the strict developmental control of the pattern of spatial distribution of the peripatetic crest cells.

Attempts to verify the order of segregation of the various members of the crest cell lineage may now be possible. Likewise, it may be possible to establish what causal mechanisms are operating that impose these restrictions. To do so, we may exploit our ability to isolate reasonably large, developmentally homogeneous cell populations (Glimelius & Weston, 1981b; Loring et al., 1981), and to present them with environments in vitro that emulate the now partially defined composition of the ECM of the crest migratory spaces (Loring et al., 1982; Sieber-Blum & Cohen, 1980; Sieber-Blum et al., 1981). Under these conditions, crest cell adhesion to their substrata and to each other can be characterized, and crest cell morphogenetic and differentiative behavior can be analyzed. For the latter task, it seems abundantly clear that the "library" of crest cell phenotypes with their specific, sensitive and reasonably convenient assays will provide unusually fruitful opportunities.

Research in the author's lab has been supported by U.S. P.H.S. Grant # DE-04316, and N.S.F. Grant # PCM-7904577.

III. REFERENCES

Barald KF, Wessells NK (1981). Monoclonal antibodies against neuronal populations of the developing chick. Nature, in press.
Bernfield M (1980). Organization and remodeling of the extracellular matrix in morphogenesis. In Brinkley L, Carlson B, Connally T (eds): "Morphogenesis and Pattern Formation: Implications for Normal and Abnormal Develment." New York: Raven Press, pp. 139-162.
Bronner-Fraser M, Cohen A (1980). Analysis of the neural crest ventral pathway using injected tracer cells. Devel Biol 77:130.
Bunge R, Johnson M, Ross C (1978). Nature and nurture in development of the autonomic neuron. Science 199:1409.
Chun L, Patterson P, Cantor H (1980). Preliminary studies on the use of monoclonal antibodies as probes for the development of sympathetic neurons. J Exp Biol 89:73.
Ciment G, Weston JA (1981). Immunochemical studies on avian peripheral neurogenesis. In McKay R, Raff M, Reichardt L (eds): "Workshop on Monoclonal Antibodies against Neural Antigens." New York: Cold Spring Harbor Labs.
Cohen A (1977). Independent expression of the adrenergic phenotype by neural crest cells in vitro. PNAS 74:2899.
Cohen A, Hay ED (1971). Secretion of collagen by embryonic neuroepithelium at the time of spinal cord-somite interaction. Devel Biol 26:578.
Comper W, Laurent T (1978). Physiological function of connective tissue polysaccharides. Physiol Rev 58:225-315.
Cowell L, Weston JA (1970). An analysis of melanogenesis in cultured chick embryo spinal ganglia. Devel Biol 22:670.
Culp LA (1974). Substrate attached glycoproteins mediating adhesion of normal and virus transformed mouse fibroblasts. J Cell Biol 63:71.
Derby MA (1978). Analysis of glycosaminoglycans within the

extracellular environments encountered by migrating neural crest cells. Devel Biol 66:321.
Erickson CA, Tosney KM, Weston JA (1980). Analysis of migratory behavior of neural crest and fibroblastic cells in embryonic tissues. Devel Biol 77:142.
Falck B, Owman C (1965). A detailed methodological description of the fluorescence method for the cellular demonstration of biogenic monoamines. Acta Univ Lund II:1.
Fauquet M, Smith J, Ziller C, LeDouarin N (1981). Differentiation of autonomic neuron precursors in vitro: Cholinergic and adrenergic traits in cultured neural crest cells. J Neurosci 1:478.
Glimelius B, Weston JA (1981a). Analysis of developmentally homogeneous neural crest cell populations in vitro. II. A tumor-promoter (TPA) delays differentiation and promotes cell proliferation. Devel Biol 82:95.
Glimelius B, Weston JA (1981b). Analysis of developmentally homogeneous neural crest cell populations in vitro. III. Role of culture environment in cluster formation and differentiation. Cell Diff 10:57.
Glimelius B, Pintar J (1981). Changes in GAG production by neural crest cell populations in vitro. Cell Diff 10:173.
Greenberg J, Pratt R (1977). Glycosaminoglycan and glycoprotein synthesis by cranial neural crest cells in vitro. Cell Diff 6:119.
Greenberg J, Foidart J-M, Greene R (1980). Collagen synthesis in cultures of differentiating neural crest cells. Cell Diff 9:153.
Hall BK (1980). Chondrogenesis and osteogenesis of cranial neural crest cells. In Pratt R, Christiansen R (eds): "Current Research Trends in Prenatal Craniofacial Development." New York: Elsevier North-Holland, pp 47-63.
Hall BK, Tremain R (1979). Ability of neural crest cells from the embryonic chick to differentiate into cartilage before their migration away from the neural tube. Anat Rec 194:469.
Hay ED (1977). Interaction between the cell surface and extracellular matrix in corneal development. In Lash JW, Burger MM (eds): "Cell and Tissue Interctions." New York: Raven Press, pp 115-137.
Hay ED (1978). Fine structure of embryonic matrices and their relationship to the cell surface in ruthenium red-fixed tissues. Growth 49:399.
Holton B, Weston JA, (1982a). Analysis of glial cell differentiation in periphral nervous tissue. I. S100 accumu-

lation in quail spinal ganglion cultures. Devel Biol 89: 64-72.
Holton B, Weston JA (1982b). Analysis of glial cell differentiation in peripheral nervous tissue. II. Neurons promote S100 synthesis by purified glial precursor cell populations. Devel Biol 89:72-81.
Hörstadius S (1950). "The Neural Crest." London: Oxford University Press.
Kleinman H et al. (1980). Role of matrix components in adhesion and growth of cells. In Pratt R, Christiansen R (eds): "Current Research Trends in Prenatal Craniofacial Development." New York: Elsevier North-Holland, pp 277-295.
LeDouarin N (1980a). The ontogeny of the neural crest in avian embryo chimaeras. Nature 286:663.
LeDouarin N (1980b). Migration and differentiation of neural crest. Curr Topics Dev Biol 16:31-85.
LeDouarin N, Renaud D, Teillet M-A, LeDouarin G (1975). Cholinergic differentiation of presumptive adrenergic neuroblasts in interspecific chimeras after heterotopic transplantations. Proc Nat Acad Sci (USA) 72:728.
LeDouarin N, Teillet M-A, Ziller C, Smith J (1978). Adrenergic differentiation of cells of the cholinergic ciliary and Remak ganglia in avian embryos following in vivo transplantation. Proc Nat Acad Sci (USA) 75:2030.
LeLievre CS, Schweizer GG, Ziller CM, LeDouarin N (1980). Restrictions of developmental capabilities in neural crest cell derivatives as tested by in vivo transplantation experiments. Devel Biol 77:362.
Löfberg J, Ahlfors K, Fallstrom C (1980). Neural crest cell migration in relation to extracellular matrix organization in the embryonic axolotl trunk. Devel Biol 75:148.
Loring J, Erickson C, Weston J (1977). Surface proteins of neural crest, crest-derived and somite cells in vitro. J Cell Biol 75:71a.
Loring J, Glimelius B, Erickson C, Weston J (1981). Analysis of developmentally homogeneous neural crest cell populations in vitro. I. Formation, morphology and differentiative behavior. Devel Biol 82:86.
Loring J, Glimelius B, Weston J (1982). Extracellular matrix materials influence quail neural crest cell differentiation in vitro. Devel Biol 89.
Mayer BW, Hay ED, Hynes R (1981). Immunocytochemical localization of fibronectin in embryonic chick trunk and Area Vasculosa. Devel Biol 82:267.
Milstein C, Lennox E (1980). The use of monoclonal antibody

techniques in the study of developing cell surfaces. In Friedlander M (ed): "Immunological Approaches to Embryonic Development and Differentiation, II." Curr Top Devel Biol 14:1-32.

Minor RR (1973). Somite chondrogenesis. A structural analysis. J Cell Biol 56:27.

Moore BW (1965). A soluble protein characteristic of the nervous system. Bioch Biophys Res Comm 19:739.

Morriss G, Thorogood P (1978). An approach to cranial neural crest migration and differentiation in mammalian embryos. In Johnson MH (ed): "Development in Mammals, v 3." Amsterdam: Elsevier North-Holland, pp 363-412.

Newgreen D, Thiery J-P (1980). Fibronectin in early avian embryos: Synthesis and distribution along the migration pathways of neural crest cells. Cell Tis Res 211:269.

Newsome DA (1976). In vitro stimulation of cartilage in embryonic chick neural crest cells by products of retinal pigmented epithelium. Devel Biol 49:496.

Nichols D, Weston JA (1977). Melanogenesis in cultures of peripheral nervous tissue. I. The origin and prospective fate of cells giving rise to melanocytes. Devel Biol 60:217.

Nichols D, Kaplan R, Weston JA (1977). Melanogenesis in cultures of peripheral nervous tissue. II. Environmental factors determining the fate of pigment forming cells. Devel Biol 60:226.

Noden D (1975). An analysis of the migratory behavior of avian cephalic neural crest cells. Devel Biol 42:106.

Noden D (1978a). Interactions directing the migration and cytodifferentiation of avian neural crest cells. In Garrod D (ed): "The Specificity of Embryological Interactions." London: Chapman and Hall, pp 4-49.

Noden D (1978b). The control of avian cephalic neural crest cytodifferentiation. Devel Biol 67:296.

Noden D (1980). The migration and cytodifferentiation of cranial neural crest cells. In Pratt R, Christiansen R (eds): "Current Research Trends in Prenatal Craniofacial Development." New York: Elsevier North-Holland, pp 3-25.

Norr S (1973). In vitro analysis of sympathetic neuron differentiation from chick neural crest cells. Devel Biol 34:16.

Patterson P (1978). Environmental determination of autonomic neurotransmitter functions. Ann Rev Neurosci 1:1.

Patterson P, Chun L (1977). The induction of acetylcholine synthesis in primary cultures of dissociated rat sympa-

thetic neurons. Devel Biol 60:473.
Pearlstein E, Gold L, Garcia-Pardo A (1980). Fibronectin: A review of its structure and biological activity. Molec Cell Biochem 29:103-128.
Pintar JE (1978). Distribution and synthesis of glycosaminoglycans during quail neural crest morphogenesis. Devel Biol 67:444.
Pratt R, Larson M, Johnston MC (1975). Migration of cranial neural crest cells in a cell-free hyaluronate-rich matrix. Devel Biol 44:298.
Rutter WJ, Pictet R, Harding J, Chirgwin J, MacDonald R, Przybyla A (1978). An analysis of pancreatic development: Role of mesenchymal factor and other extracellular factors. In Papaconstantinou J, Rutter W (eds): "Molecular Control of Proliferation and Differentiation." New York: Academic Press, pp 205-227.
Schwartz C, Hoffman L, Hellerqvist C, Cunningham L (1979). Scanning electron microscopic visualization of microexudate prepared by the release of cells by urea. Exp Cell Res 118:427.
Sieber-Blum M, Cohen A (1980). Clonal analysis of quail neural crest cells. They are pluripotent and differentiate in vitro in the absence of non-crest cells. Devel Biol 80:96.
Sieber-Blum M, Sieber F, Yamada KM (1981). Cellular fibronectin promotes adrenergic differentiation of quail neural crest cells in vitro. Exp Cell Res 133:285.
Smith J, Cochard P, LeDouarin N (1977). Development of choline acetyltransferase and cholinesterase activities in enteric ganglia derived from presumptive adrenergic and cholinergic levels of the neural crest. Cell Diff 6:199.
Smith J, Fauquet M, Ziller C, LeDouarin N (1980). Acetylcholine synthesis by mesencephalic neural crest cells in the process of migration in vivo. Nature 282:853.
Solursh M, Morriss G (1977). Glycosaminoglycan synthesis in rat embryos during the formation of the primary mesenchyme and neural folds. Devel Biol 57:75.
Timpl R, Rohde H, Gehron-Robey P, Rennard S, Foidart J, Martin G (1979). Laminin--a glycoprotein from basement membrane. J Biol Chem 254:9933.
Toole B (1972). Hyaluronate turnover during chondrogenesis in developing chick limb and axial skeleton. Devel Biol 29:321.
Toole B, Trelstad R (1971). Hyaluronate production and removal during corneal development in the chick. Devel

Biol 26:28.
Toole B, Underhill C, Mikuni-Takagaki Y, Orkin RW (1980). Hyaluronate in morphogenesis. In Pratt R, Christiansen R (eds): "Current Research Trends in Prenatal Craniofacial Development." New York: Elsevier North-Holland, pp 263-275.
Tosney KW (1978). The early migration of neural crest cells in the trunk region of the avian embryo: an electron microscopic study. Devel Biol 62:317.
von der Mark K (1980). Immunological studies on collagen type transition in chondrogenesis. Curr Topics Devel Biol 14:199-226.
Vulliamy T, Rattray S, Mirsky R (1981). Cell-surface antigen distinguishes sensory and autonomic peripheral neurones from central neurones. Nature 291:418.
Wartiovaara J, Leivo I, Vaheri A (1980). Matrix glycoproteins in early mouse development and in differentiation of teratocarcinoma cells. In Subtelny S, Wessells N (eds): "The Cell Surface: Mediator of Developmental Processes." New York: Academic Press, pp 305-324.
Weston J (1963). A radioautographic analysis of the migration and localization of trunk neural crest cells in the chick. Devel Biol 6:279.
Weston J (1970). The migration and differentiation of neural crest cells. Adv Morphogen 8:41-114.
Weston J (1980). Role of the embryonic environment in neural crest morphogenesis. In Pratt R, Christinasen R (eds): "Current Research Trends in Prenatal Craniofacial Development." New York: Elsevier North-Holland, pp 27-45.
Weston J (1981a). Th regulation of normal and abnormal neural crest cell development. In Riccardi V, Mulvihill J (eds): "Neurofibromatosis (von Recklinghausen Disease)." Adv Neurol 29:77-94.
Weston J (1981b). Motile and social behavor of neural crest cells. In Bellairs R, Curtis A, Dunn G (eds): "Cell Behaviour." Cambridge: Cambridge Univ Press, pp 429-469.
Weston J, Butler S (1966). Temporal factors affecting localization of trunk neural crest cells in the chicken embryo. Devel Biol 14:246.
Weston J, Derby M, Pintar J (1978). Changes in the extracellular environment of neural crest cells during their early migration. Zoon 6:103.
Yamada KM (1980). Structure and function of fibronectin in cellular and developmental events. In Pratt R, Christiansen R (eds): "Current Research Trends in Prenatal

Craniofacial Development." New York: Elsevier North-Holland, pp 298-313.

Yamada K, Olden K, Pastan I (1978). Transformation-sensitive cell surface protein: Isolation, characterization and role in cell morphology and adhesion. Ann NY Acad Sci 321:256-277.

Yamada K, Olden K, Hahn L-H (1980). Cell surface protein and cell interactions. In Subtelny S, Wessells N (eds): "The Cell Surface: Mediator of Developmental Processes." New York: Academic Press, pp 43-77.

Ziller C, Smith J, Fauquet M, LeDouarin N (1979). Environmentally directed nerve cell differentiation: In vivo and in vitro studies. Prog Brain Res 51:59.

DEVELOPMENTAL HISTORY OF THE TWO M-LINE PROTEINS MM-CREATINE KINASE AND MYOMESIN DURING MYOGENESIS

H.M. Eppenberger, E.E. Strehler, Th.C. Doetschman,
U.B. Rosenberg, J.C. Perriard, Th. Wallimann
Institute for Cell Biology, Swiss Federal Institute
of Technology (ETH), ETH-Hönggerberg
CH-8093 Zurich, Switzerland

INTRODUCTION

The terminal differentiation of myogenic cells represents a system extremely well suited for the study of regulatory processes going on during muscle development. When myogenic cells become postmitotic and fuse into multinuclear muscle fibers, large amounts of "new" proteins necessary for the specialized functions of muscle tissue are produced. The components of the contractile organelles, many enzymes involved in the metabolic pathways of energy production as well as membrane constituents bringing about the specialized functions of the muscle cell surface are among the most abundantly synthesized proteins during this time period. This in turn means that the genes and/or the messages for these proteins must become very active in a well-coordinated way at that time. At present, however, not much is known about the regulatory mechanisms underlying the "switching on" of this muscle-specific program. Moreover, the sequence of events leading to the highly ordered structure of a myofibril is still unclear. In spite of the fact that the most abundant proteins of the contractile apparatus have been characterized in some detail the factors controlling the appearance and determining the organization of these proteins during myofibrillogenesis are still unknown. In this report the developmental characteristics of two proteins will be described: creatine kinase (CK), the dimeric enzyme catalyzing the regeneration of ATP from ADP and creatine phosphate, and myomesin, a protein located in the M-line region of the myofibrillar sarcomeres.

RESULTS AND DISCUSSION

The Creatine Kinase Isoenzyme Switch

In chicken embryonic myogenic cells as well as in many adult chicken tissues like brain, smooth muscle and heart BB-CK represents the predominant form of creatine kinase (Eppenberger et al. 1964, Morris et al. 1972, Turner et al. 1974, Perriard et al. 1978b). During terminal differentiation of skeletal muscle cells, however, the muscle-specific M-CK subunit gradually replaces the B-subunit. In fact, the large increase in total cellular CK activity found in differentiating skeletal muscle cells in culture and in vivo is mostly due to a drastic accumulation of MM-CK within these cells (Turner et al. 1974, Perriard et al. 1978a). This isoenzyme switch which in vivo leads to an almost complete replacement of BB-CK by MM-CK has been shown to be entirely due to newly synthesized CK-subunits (Caravatti et al. 1979). The levels of total cellular and polysomal mRNA coding for the two different CK-subunits have been determined in cell-free translation systems using sensitive immunoprecipitation methods discriminating between the produced M- and B-subunits (Perriard et al. 1978b, Perriard 1979). From these experiments the conclusion could be drawn that the increased synthesis of M-CK is mostly due to the new appearance of translatable specific mRNA for this protein. Currently experiments are underway to isolate and characterize the M- and B-CK genes. The comparison of the arrangement and structural details of these genes might bring some important information about the possible mechanisms underlying the switching on of muscle-specific genes.

It is assumed that the large increase in CK-activity taking place during terminal differentiation in myogenic cells is related to the enhanced activity of the energy metabolism. Why, however, should a cell switch on a "new" CK-gene in order to ensure the increased need for this protein? On one side, a higher rate of transcription of the M-CK gene, or a higher stability and/or translational activity of the resulting M-CK mRNA might well be a reason for the observed phenomenon. On the other hand, unique properties of M-CK when compared to B-CK could equally well account for its necessity within a muscle cell. MM-CK has indeed been found to differ significantly from BB-CK with respect to its ability to bind to the contractile organelles. While it became clear that most of the

Development of Two M-Line Proteins / 383

MM-CK exists as soluble enzyme in the cytoplasm a small, yet significant, amount of this protein (at least 5 % of the total MM-CK) is firmly and specifically bound to the center of the myofibrillar sarcomeres (Fig. 1) where it makes up for the bulk of electron-density of the M-line (Wallimann et al. 1977, 1978; Strehler et al. 1980). In contrast, BB-CK is never present in the M-line during any stage of myogenesis even though it is present in small amounts in the I-band during the early stages (Wallimann et al. 1977).

FIGURE 1: Indirect immunofluorescence localization of M-line bound MM-CK in differentiated skeletal muscle cells cultured for 5 days. Cells were permeabilized with Triton X-100, washed extensively to remove the bulk of the soluble MM-CK, and then stained with affinity-purified anti-MM-CK IgG (Eppenberger et al. 1981). By superimposing fluorescence and phase contrast images staining can be localized within the middle of the A-band of the sarcomeres. (a) Fluorescence; (b) phase contrast picture of the same culture area. Bar, 10 μm.

In recent experiments we not only have been able to show that the M-line-bound MM-CK is enzymatically active but also that the amount and specific activity of this enzyme fraction might be sufficient to regenerate a considerable amount of the ADP produced during contraction (T. Schlösser and T. Wallimann, in preparation). It is thus quite possible that the presence of MM-CK within a myofibril provides an effective way of ATP regeneration during contraction. The properties enabling MM-CK to bind to the middle of the myofibrillar sarcomeres are still unknown. The fact that different muscle types differ with respect to the amount of M-line MM-CK and that MM-CK may be incorporated into this structure after other components are already present suggests that other properties of a muscle cell, e.g., the myosin type present in the myofibrils, determine whether and at what time MM-CK will be incorporated in the M-line.

Myomesin, a Myofibril-specific Protein of Cross-striated Muscles

The M_r = 165,000 M-protein (Masaki and Takaiti 1974, Trinick and Lowey 1977, Strehler et al. 1979, 1980), now called myomesin (Eppenberger et al. 1981), is another example of a developmentally regulated muscle-specific protein As observed for MM-CK the synthesis of myomesin starts with the onset of terminal myogenic differentiation (Eppenberger et al. 1981). In contrast to CK where a switch from the more ubiquitous embryonic to the muscle-specific form of CK takes place during myogenesis, no embryonic form of myomesin has been detected up to now. Using an antibody against chick myomesin, myomesin has been identified only in the M-line region of the cross-striated muscles of a wide range of vertebrate species indicating the highly conserved nature of its strict tissue specificity (Table 1). Not much is known about the level at which the synthesis of this protein is controlled, this to some extent being due to the difficulties encountered in isolating the mRNA coding from myomesin, a protein which constitutes at most 0.04 % of the total muscle proteins synthesized during that time (Eppenberger et al. 1981).

TABLE 1: Cross-reactivity of rabbit polyclonal antibodies against chicken skeletal muscle myomesin (immunofluorescence).

Species	Skeletal muscle myofibrils	Heart muscle myofibrils	Skeletal muscle cells	Heart muscle cells
Chick	+	+	+	+
Quail	+	ND	+	ND
Rat	+	+	ND	ND
Mouse	+	ND	ND	ND
Hamster	+	+	+	+
Cow	ND	+	ND	ND
Cat	ND	+	ND	ND
Sheep	ND	+	ND	ND
Man	+	+	ND	ND
Xenopus	+	+	ND	ND
Trout	+	+	ND	ND
Crab	-	-	ND	ND
Drosophila melanogaster (flight muscle)	-	ND	-	ND

ND, not determined

Even though myomesin is present in low amounts relative to other myofibrillar proteins, its exclusive localization within the narrow M-line region makes it detectable by the indirect immunofluorescence technique (Fig. 2). Myomesin can already be identified in early postmitotic myoblasts (Eppenberger et al. 1981) or suspended cells (Puri et al. 1980) in a regularly cross-striated pattern. If young myotubes are stained for myomesin by incubation with specific antibodies and then processed for electron microscopy the localization of myomesin in the M-line region of nascent myofibrils can easily be demonstrated (Fig. 3b). Incubation of young myotubes with preimmune IgG shows clearly that an electron-dense M-line (to which MM-CK makes the most essential contribution) is often not yet or only partially developed in the myofibrils at that time (Fig. 3a). Myomesin thus appears to be an attractive candidate for a protein playing a role in the assembly process of cross-striated myofibrils.

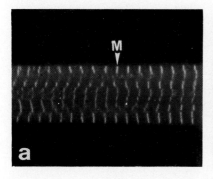 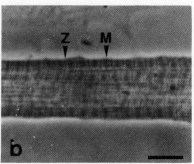

FIGURE 2: Indirect immunofluorescence using affinity-purified anti-myomesin IgG. Localization of myomesin in a myotube of a chicken embryonic breast muscle cell culture 13 days after plating. (a) Fluorescence; (b) phase contrast. Z, Z-line; M, M-line region. Bar, 10 µm.

Although an interaction of myomesin with myosin has been reported to occur in vitro (Mani and Kay 1978), the exact nature and the functional significance of this phenomenon have not yet been established. Experiments designed to study the influence of myomesin on the formation of myosin thick filaments as well as on the aggregation of these filaments in vitro are underway in our laboratory. In addition, we will investigate the functions of myomesin and MM-CK with microinjection techniques employing fluorescently labelled myomesin and other myofibrillar components as well as antibodies against these proteins (Fig. 4). By injecting such proteins into living cultured cells of differing origin and age we hope to learn more about the process of myofibrillogenesis within muscle cells.

ACKNOWLEDGEMENTS

This work was supported by grant 3.187-0.77 from the Swiss National Science Foundation and by a grant to H.M. Eppenberger from the Muscular Dystrophy Association, Inc.

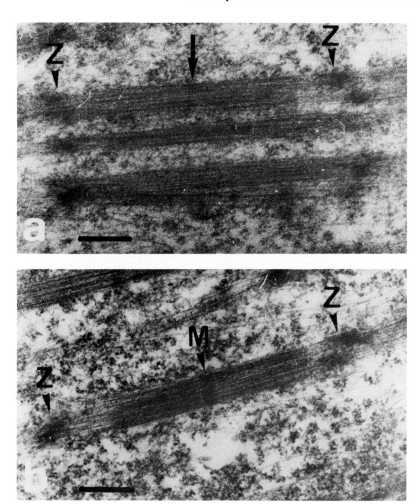

FIGURE 3: Localization of myomesin in nascent myofibrils in small myotubes of chicken breast muscle cells 48 hrs after plating. (a) Incubation with preimmune IgG. Beginning incorporation of electron dense M-line material (MM-CK) into some sarcomeres (arrow). (b) Incubation with affinity-purified anti-myomesin IgG. Heavy antibody decoration of myomesin already present in the M-line region. Z, Z-line; M, M-line region. Bars, 0.4 um.

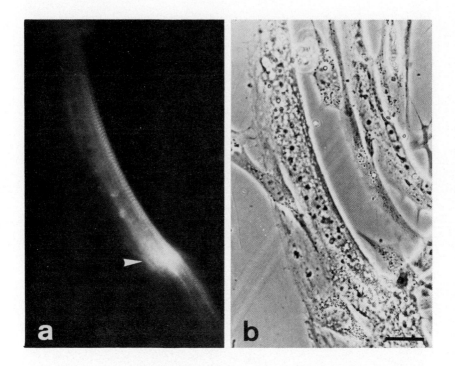

FIGURE 4: Direct immunofluorescence of rhodamine-labelled anti-myomesin IgG microinjected into a living myotube at 3 days of culture. (a) Fluorescence; (b) phase contrast pictures of the myotube 30 min. after injection. Arrowhead shows site of injection. Bar, 20 μm.

REFERENCES

Caravatti M, Perriard J-C, Eppenberger HM (1979). Developmental regulation of creatine kinase isoenzymes in myogenic cell cultures from chicken. Biosynthesis of creatine kinase subunits M and B. J Biol Chem 254:1388.
Eppenberger HM, Eppenberger ME, Richterich R, Aebi H (1964). The ontogeny of creatine kinase isoenzymes. Dev Biol 10:1.
Eppenberger HM, Perriard J-C, Rosenberg UB, Strehler, EE (1981). The M_r 165,000 M-protein myomesin: A specific protein of cross-striated muscle cells. J Cell Biol 89:185.

Mani RS, Kay CM (1978). Interaction studies of the 165,000 dalton protein component of the M-line with the S2 subfragment of myosin. Biochim Biophys Acta 536:134.

Masaki T, Takaiti O (1974). M-protein. J Biochem (Tokyo) 75:367.

Morris GE, Cooke A, Cole RJ (1972). Isoenzymes of creatine phosphokinase during myogenesis in vitro. Exp Cell Res 74:582.

Perriard J-C (1979). Developmental regulation of creatine kinase isoenzymes in myogenic cell cultures from chicken. Levels of mRNA for creatine kinase subunits M and B. J Biol Chem 254:7036.

Perriard J-C, Caravatti M, Perriard ER, Eppenberger HM (1978a). Quantitation of creatine kinase isoenzyme transitions in differentiating chicken embryonic breast muscle and myogenic cell cultures by immunoadsorption. Arch Biochem Biophys 191:90.

Perriard J-C, Perriard ER, Eppenberger HM (1978b). Detection and relative quantitation of mRNA for creatine kinase isoenzymes in RNA from myogenic cell cultures and embryonic chicken tissues. J Biol Chem 253:6529.

Puri EC, Caravatti M, Perriard J-C, Turner DC, Eppenberger HM (1980). Anchorage-independent muscle cell differentiation. Proc Nat Acad Sci USA 77:5297.

Strehler EE, Pelloni G, Heizmann CW, Eppenberger HM (1979). M-protein in chicken cardiac muscle. Exp Cell Res 124:39.

Strehler EE, Pelloni G, Heizmann CW, Eppenberger HM (1980). Biochemical and ultrastructural aspects of M_r 165,000 M-protein in cross-striated chicken muscle. J Cell Biol 86:775.

Trinick J, Lowey S (1977). M-protein from chicken pectoralis muscle: Isolation and characterization. J Mol Biol 113:343.

Turner DC, Maier V, Eppenberger HM (1974). Creatine kinase and aldolase isoenzyme transitions in cultures of chick skeletal muscle cells. Dev Biol 37:63.

Wallimann T, Turner DC, Eppenberger HM (1977). Localization of creatine kinase isoenzymes in myofibrils. I. Chicken skeletal muscle. J Cell Biol 75:297.

Wallimann T, Pelloni G, Turner DC, Eppenberger HM (1978). Monovalent antibodies against MM-creatine kinase remove the M-line from myofibrils. Proc Nat Acad Sci USA 75:4296.

ANALYSIS OF THE EXPRESSION OF MYOSIN AND ACTIN GENES DURING
MOUSE MYOBLAST DIFFERENTIATION USING SPECIFIC RECOMBINANT
PLASMIDS

B. ROBERT, M. CARAVATTI, A. MINTY, A. WEYDERT,
S. ALONSO, A. COHEN, P. DAUBAS, F. GROS & M.
BUCKINGHAM
Pasteur Institute, Dept. Molecular Biology
25, rue du Dr. Roux - 75724 Paris Cedex 15

INTRODUCTION

    Myogenesis constitutes an interesting biological system in which to study the expression of a set of developmentally regulated genes. The different proteins composing the functional myofiber have been well characterized (reviewed by Buckingham, 1977). Recently, most of them have been shown to arise from multigene families and to be expressed in a tissue specific way in different muscle and non muscle cells, or at different stages of development (Garrels & Gibson, 1976 ; Whalen et al., 1978, 1979 ; Montarras et al., 1981). Furthermore, muscle cells are able to undergo differentiation in cell culture and under these conditions, the muscle specific proteins appear together, which would suggest a coordinate regulation controlling the expression of the corresponding genes (e.g. Devlin & Emerson, 1978). A central question concerning such a mechanism is to determine at which level, transcriptional or post-transcriptional, this coordinate accumulation is regulated. A number of indirect approaches including metabolic inhibitors (Yaffé & Dym, 1972) or studies of mRNA metabolism (Buckingham et al., 1976) suggested that muscle specific RNAs were present in myoblasts prior to cell fusion. This, however, was not confirmed by the analysis of the products from in vitro translation of RNA from differentiating myoblasts, which showed a correlation between the appearance of translatable RNAs and proteins (Yablonka & Yaffé, 1977 ; Devlin & Emerson, 1979 ; Daubas et al., 1981). These experiments did not exclude the possibility that the messengers are stored in a non translatable form. The development of recombinant DNA technology makes it possible to approach the regulation of muscle genes directly. We have thus cloned recombinant plasmids containing sequences complementary to actin and myosin RNAs. We describe here the application of these probes to the evaluation of the steady state level of the corresponding mRNAs during terminal differentiation of a mouse muscle cell line.

CHARACTERIZATION OF THE RECOMBINANT PLASMIDS

cDNA was synthesized against polyadenylated RNA from skeletal muscle of 8-10 day old mice and cloned in the Pst1 site of pBR 322 after dG-dC tailing (Minty et al., 1981). Table 1 summarizes the properties of the plasmids which have been identified from the cDNA library.

Plasmid n°	Hybridizes to RNA for	Size of insertion	coding/ non coding	Size of RNA detected
p32	Myosin heavy chains	1150	+/(+)	6900
p161	Fast myosin light chains LC_1 and LC_3	380	-/+	1050 900
p81	Actins	1080	+/-	1600 (α) 2000 (β,γ)
p91	Actins	1350	+/+	1600 (α) 2000 (β,γ)
p91-1	Skeletal muscle actin	200	-/+	1600 (α)

Table 1

Actin and myosin recombinant plasmids isolated from the cDNA library of new born mouse skeletal muscle

Plasmid 32 hybridizes to an mRNA abundant in skeletal muscle, with the size (about 7 kb) expected for the myosin heavy chain (MHC) RNA. It was definitely identified as a MHC plasmid by its cross hybridization with the rat MHC plasmid characterized by Nudel et al. (1980), which they kindly provided. This plasmid hybridizes strongly with RNA from late embryonic, new born or adult muscle RNA, as well as with cardiac muscle RNA ; but poorly with RNA from cell lines which synthesize exclusively the embryonic MHC, like L6 (Whalen et al., 1979). We therefore conclude that it contains a sequence for the adult skeletal or neonatal (Whalen et al., 1981) form of the MHC.

Two different actin plasmids were isolated from the cDNA library. According to its nucleotide sequence (Fig. 1) plasmid 91 contains a DNA insertion complementary to the mRNA for skeletal muscle α-actin (Minty et al., 1981). This insert is 1350 base pairs long, which represents 90 % of the coding sequence of the mRNA, plus 300 base pairs from the 3' non translated region. A fragment of 200 base pairs from this region, which represents exclusively a 3' non coding sequence, has been subcloned in pBR 322 (p91-1). This untranslated region appears to be very specific for the mRNA of α-actin from skeletal muscle. In contrast, the total plasmid 91 hybridizes with the RNAs of all the actin isoforms expressed in different mouse tissues, and throughout evolution.

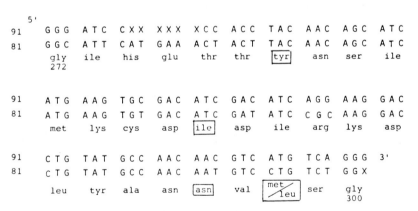

Figure 1 : Partial nucleotide sequence of the actin plasmids Positions 278, 286, 296 indicate that both these plasmids code for muscle actins. Position 298 in p91 (met) is specific for skeletal muscle actin ; position 89 (thr), 298 (leu) and 357 (ser) in p81 are characteristic of an α-cardiac actin. The figures refer to the amino acid sequences (Vandekerckhove & Weber, 1979).

Plasmid 81 has a sequence characteristic of the α-cardiac actin at the positions 89, 298 and 357 (fig. 1) and it shows most hybridization to heart mRNA on a Northern blot. The fact that is was cloned from a skeletal muscle RNA, together with its preferential hybridization to embryonic versus adult skeletal muscle RNA leads us to think that there is expression of cardiac actin in the skeletal muscle at early stages of development.

Plasmid 161 hybridizes with both the RNAs for myosin light chains 1 (LC1) and 3 (LC3) from fast skeletal muscle (Fig. 2). These two light chains are closely related, func-

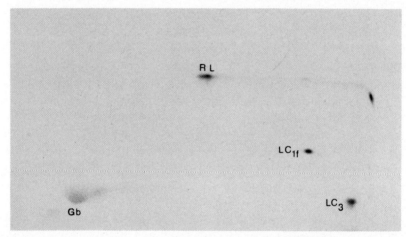

Figure 2 : In vitro translation products of mRNAs which hybridize to plasmid 161.
Mouse skeletal muscle RNA was hybridized to DBM-bound DNA of p161 according to Smith et al. (1979). The hybridized RNA was translated in a nuclease treated reticulocyte lysate (Pelham & Jackson, 1976) and run on a non-equilibrium two dimensional gel (O'Farrell et al., 1977). LC1F and LC3 were characterized by reference to the migration of purified myosin. R.L. and globin are two proteins synthesized from endogenous messengers of the lysate.

tionally as well as structurally, since they share the same C terminal sequence and diverge only at the N terminus (Frank

& Weeds, 1974). However, the nucleotide sequence of the inserted cDNA corresponds exclusively to the 3' untranslated region of the mRNA. Considering the rapid rate of divergence in the non coding sequences of eukaryotic mRNAs even between those coding for highly conserved proteins (illustrated above by the case of the actins), this result was unexpected. The two alkali light chains are found in birds as well as mammals and their occurrence is thus not a recent evolutionary event. We suspect that they were prevented from diverging by a specific genetic phenomenon, possibly involving a common 3' sequence in the DNA. Apart from this cross hybridization the plasmid is highly specific as expected for a non coding fragment and does not hybridize with the myosin light chain RNAs from other tissues or species, except for a faint cross-reaction with the RNA coding for the embryonic form of LC1, LC1 emb.

ANALYSIS OF THE ACCUMULATION OF SPECIFIC RNAs DURING MYOBLAST DIFFERENTIATION

The cloned cDNA probes have been used to investigate the accumulation of the corresponding mRNAs during the differentiation of a mouse myoblast cell line, T 984, derived from a teratocarcinoma (Jakob et al., 1978). Figure 3 shows the kinetics of differentiation. During the first 80 hours after plating, the cells divide rapidly. They are then changed to a medium containing low foetal calf serum concentration (2 %). The cells stop dividing and some cell death is observed. Morphological and biochemical differentiation of muscle fibres is initiated (as monitored here by the accumulation of creatinephosphokinase).

At different times during differentiation, total cellular polyadenylated RNA was extracted from the cultures and analyzed on Northern blot (Alwine et al., 1979) (Fig. 4) The blot was cut into three pieces corresponding to the large, middle and small RNA size classes which were hybridized respectively to the myosin heavy chain, the actin or the myosin light chain probes.

No RNA coding for α-actin, myosin heavy chain or myosin light chains is detected in the proliferating myoblasts. The only forms of actin RNA revealed and those of β and γ actins, which migrate at about 2000 nucleotides, and a minor form, slightly lighter than the α-actin messenger ; none of

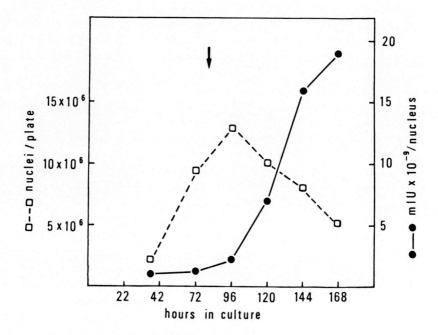

Figure 3 : Kinetics of differentiation of T 984 mouse myoblast cell line.
The number of nuclei per plate was calculated from the DNA content of cell cultures, assuming a figure of 5 pg per haploid genome (Britten & Davidson, 1971). Creatine kinase activity was measured by using the "CK NAC-activated UV-system" (Boehringer, Mannheim) (——●——).

these are revealed with the muscle α-actin specific probe, p 91-1. As differentiation is initiated (96 hrs), myosin heavy chain and α-actin RNAs begin to accumulate in a coordinate fashion. Their amount increases till 144 hrs and then begins to decrease. In parallel with the accumulation of α-actin RNA, the amount of non muscle actin mRNAs decreases rapidly from 96 hrs.

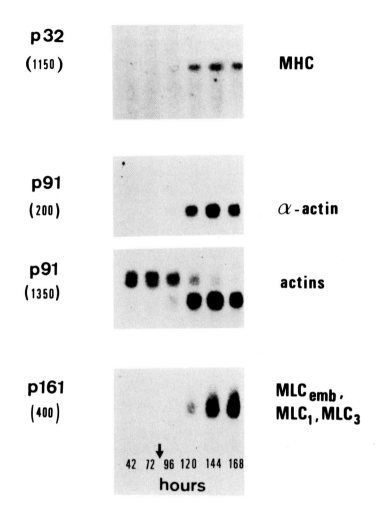

Figure 4 : Northern blot analysis of the RNA from differentiating T 984 myoblasts.
RNA (2 μg of poly (A$^+$) RNA per sample) was denatured and run on a 1.25 % agarose gel as described by Carmichael & McMaster (1980). It was blotted to DBM paper, the blots were hybridized to plasmid DNA (^{32}P-labelled by nick-translation (Minty et al., 1981)) and treated according to Alwine et al. (1979).

The accumulation of light chain RNA appears to be delayed by about 24 hrs relative to that of the actin and myosin heavy chain RNAs. However, the myosin light chain probe does not reveal in T 984 RNA the two discrete bands of about 900 and 1050 nucleotides observed in mouse and rat muscle RNA (fig. 5) but rather a smear around 1000 nucleotides.

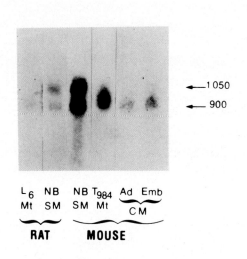

Figure 5 :
RNAs hybridizing to p161 in different cells and tissues.

RNAs (1 µg per sample) from the different sources were analyzed on Northern blot as in Fig. 4. The blots corresponding to skeletal muscle from new born animals (NB SM) and T 984 myotubes (T984 Mt) were exposed for 14 hrs those corresponding to L6 myotubes (L6 Mt) and cardiac muscle (CM) were exposed for 44 hrs.

This is probably related to the phenotypic properties of this cell line, which expresses mainly the embryonic form of LC1, LC1emb together with LC3 and a small amount of the adult form of LC1, LC1F (Robert et al., 1981). The messenger for LC1emb (which is the only form of LC1 synthesized in L6 (Whalen et al., 1978)) shows a faint cross hybridization with plasmid 161 and has a size intermediate between the two RNAs for the adult forms (Fig. 5). (Note that an RNA of the same size is detected in RNA from embryonic and adult heart, which expresses the same LC1 isoform (Whalen et al., 1980a, b)). The smear detected in T 984 RNA is likely to represent mainly hybridization with LC3 mRNA, together with the cross hybridization of LC1emb RNA. The slower kinetics of accumulation of the light chain RNA reflects probably the later synthesis of LC3 in this cell system. It does not exclude the possibility that the accumulation of LC1emb RNA begins at the same time as myosin heavy chain and α-actin RNAs as is the case for the synthesis of the corresponding proteins, the adult phenotype being turned on later.

The radioactive bands corresponding to the specific messengers have been quantitated. In the case of α-actin mRNA, the accumulation of the messenger represents at least a 130 fold increase. The figure is less for myosin heavy chain and myosin light chain since the relative homology of the probes lowers the sensitivity of detection. The accumulation of the proteins in the same culture was followed by one and two dimensional gel analysis after labelling the cells 1 hour with ^{35}S met. The accumulation of proteins in every case follows closely the accumulation of the RNAs.

Our main conclusion is that there is no accumulation of non-translatable RNA prior to myoblast differentiation. Our results do not exclude the possibility that the genes are already transcribed in the myoblast but the RNA is rapidly degraded. We are currently looking at this by investigating the DNAse I sensitivity of the genes in the chromatin of the cell during differentiation.

ACKNOWLEDGMENTS

The authors are very grateful to Dr. Didier Montarras for helpful advice. The laboratory is supported by grants from the D.G.R.S.T., C.N.R.S., C.E.A., I.N.S.E.R.M. and the M.D.A. of America. M.C. was the recipient of a fellowship from the Swiss National Science Foundation and from INSERM A.M. from the M.D.A., P.D. from the Ligue Française contre le Cancer and S.A. from the D.G.R.S.T.

REFERENCES

Alwine J C, Kemp D J, Parker B A, Reiser J, Renart J, Stark G R & Wahl G M (1979). Detection of specific RNAs or specific fragments of DNA by fractionation in gels and transfer to diazobenzyloxymethyl paper. Methods in Enzymology 68 : 220

Britten R & Davidson E H (1971) Repetitive DNA sequences and a speculation on the origins of evolutionary novelty. The quarterly Review of Biology 46 : 111

Buckingham M E (1977). Muscle protein synthesis and its control during the differentiation of skeletal muscles in vitro. In Paul J (Ed) Biochemistry of cell differentiation II, Baltimore : University Park Press p 269.

Buckingham M E, Cohen A & Gros F (1976). Cytoplasmic distribution of pulse-labelled poly(A) containing RNA, particularly 26 S RNA, during myoblast growth and differentiation J Mol Biol 103 : 611.

Carmichael G C & McMaster G R (1980). The analysis of nucleic acids in gels using glyoxal and acridine orange. Methods in Enzymology 65 : 380.

Daubas P, Caput D, Buckingham M & Gros F (1981). A comparison between the synthesis of contractile proteins and the accumulation of their translatable mRNAs during calf myoblast differentiation. Develop Biol 84 : 133.

Devlin R B & Emerson C P (1978). Coordinate regulation of contractile protein synthesis during myoblast differentiation. Cell 13 : 599.

Devlin R B & Emerson C P (1979) Coordinate accumulation of contractile protein mRNA during myoblast differentiation. Develop Biol 69 : 202.

Frank G & Weeds A G (1974). The amino acid sequence of the alkali light chains of rabbit skeletal-muscle myosin. Eur J Biochem 44 : 317.

Garrels J I & Gibson W (1976). Identification and characterization of multiple forms of actin. Cell 9 : 793.

Jakob H, Buckingham M E, Cohen A, Dupont L, Fiszman M & Jacob F (1978). A skeletal muscle cell line isolated from a mouse teratocarcinoma undergoes apparently normal terminal differentiation in vitro. Exp Cell Res 114 : 403.

Minty, A J, Caravatti M, Robert B, Cohen A, Daubas P, Weydert A, Gros F & Buckingham M E (1981). Mouse actin messenger RNAs : construction and characterization of a recombinant plasmid molecule containing a complementary DNA transcript of mouse α-actin mRNA. J Biol Chem 256 : 1008.

Montarras D, Fiszman M Y & Gros F (1981). Characterization of the tropomyosin present in various chick embryo muscle types and in muscle cells differentiated in vitro. J Biol Chem 256 : 4081.

Nudel U, Katcoff D, Carmon Y, Zevin-Sonkin D, Levi Z, Shaul Y, Shani M & Yaffé D (1980). Identification of recombinant phages containing sequences from different rat myosin heavy chain genes. Nucl Acids Res 8 : 2133.

O'Farrell P Z, Goodman H M & O'Farrell P H (1977). High resolution two-dimensional electrophoresis of basic as well as acidic proteins. Cell 12 : 1133.

Pelham H R B & Jackson R J (1976). An efficient mRNA-dependent translation system from reticulocyte lysates. Eur J Biochem 67 : 247.

Robert B, Weydert A, Caravatti M, Minty A, Cohen A, Daubas P, Gros F & Buckingham M (1981). A cDNA recombinant plasmid complementary to mRNAs for both light chains 1 and 3 of mouse skeletal muscle myosin. Submitted for publication

Smith D F, Searle P F & Williams J G (1979). Characterization of bacterial clones containing DNA sequences derived from Xenopus laevis vitellogenin mRNA. Nucl Acids Res 6 : 487.

Vandekerckhove J & Weber K (1979). The complete amino acid sequence of actins from bovine aorta, bovine heart, bovine fast skeletal muscle, and rabbit Slow skeletal muscle. Differentiation 14 : 123.

Whalen R G, Butler Browne G S & Gros F (1978). Identification of a novel form of myosin light chain present in embryonic muscle tissue and cultured muscle cells. J Mol Biol 126 : 415.

Whalen R G, Schwartz K, Bouveret P, Sell S M & Gros F (1979) Contractile protein isozymes in muscle development : Identification of an embryonic form of myosin heavy chain. Proc Natl Acad Sci USA 76 : 5197.

Whalen R G & Sell S M (1980a). Myosin from fetal hearts contains the skeletal muscle embryonic light chain. Nature 286 : 731.

Whalen R G, Thornell L-E & Eriksson A (1980b). Heart purkinje fibers contain both atrial-embryonic and ventricular-type myosin light chains. Second International Congress on Cell Biology, Berlin (August 31-September 5 1980) Abstract C 948.

Whalen R G, Sell S M, Butler Browne G S, Schwartz K, Bouveret P & Pinset-Härström I (1981). Three myosin heavy-chain iso zymes appear sequentially in rat muscle development.Nature 292 : 805.

Yablonka Z & Yaffé D (1977). Synthesis of myosin light chains and accumulation of translatable mRNA coding for light chain like polypeptides in differentiating muscle cultures Differentiation 8 :133.

Yaffé D & Dym H (1972). Gene expression during differentiation of contractile muscle fibers. Cold Spring Harbor Symp.Quant Biol 37 : 543.

EARLY ALTERATION INDUCED BY TUMOR PROMOTERS ON CHICK EMBRYO MUSCLE CELLS IN CULTURE.

B.M.Zani and M.Molinaro

Inst.Histol.Gen.Embriol.Univ.of Rome

Via Scarpa 14, 00161 Roma,Italy.

SUMMARY

When differentiated, multinucleated cultured myotubes are treated with PMA (Phorbol-12-myristate-13-acetate), they display drastic morphological alterations and undergo inhibition of the expression of differentiative traits, without being induced to reenter the cell cycle. Differentiated myotubes, obtained after cytochalasin B treatment of primary chick embryo myoblast cultures, were treated with 1.6×10^{-7} M PMA for different times, labelled with ^{35}S-methionine and different fractions of cell extracts were analysed by SDS-PAGE followed by fluorography. The data presented here indicate that PMA treatment induces in myotubes increased synthesis of a 31.000 Mr polypeptide (31 K) within 4 hr of treatment, while the inhibition of the synthesis of contractile proteins, such as myosin and actin, occurs only after 8 hr of treatment. Morphological alterations of myotubes require longer incubation with PMA (15-20 hr). The reported effects of PMA are not induced by non tumor promoter analogs of the drug, and pulse chase experiments indicate that 31K stimulation is not the result of increased protein degradation induced by PMA. In addition the stimulation of 31K does not occur in cultured fibroblast indicating that this is a specific early response of differentiated myogenic cells preceeding the dedifferentiative effect of this tumor promoter.

INTRODUCTION

The tumor promoter PMA induces a variety of biological effects. Besides being involved in the phenomenon of tumor promotion, mostly studied in vivo skin carcinogenesis, it enhances the expression of transformation associated properties when administered to transformed cells, and induces reversible transformation-like changes in normal cells (Weinstein and Wigler, 1977). Furthermore PMA reversibly interferes with the differentiative program in a variety of differentiating and fully differentiated cell types (Rovera et al., 1977;Pacifici and Holtzer, 1977; Colburn et al., 1975; Balmain,1976). Differentiated myotubes undergo dedifferentiative changes when treated with PMA without being induced to reenter the cell cycle (Cohen et al., 1977). In this system PMA elicits conspicuous alterations of the myofibrillar structure and also reduced levels of acetylcholine receptor, increased levels of plasminogen activator and alterations of Ca^{++} fluxes and cyclic nucleotide levels (Toyama et al., 1979; Miskin et al., 1978; Grotendorst and Schimmel, 1980; Schimmel and Hallam, 1980). It has been found in our laboratory that the inhibition of acetylcholine receptor and creatine kinase activity, which occurs in PMA treated myotubes, is dependent on continuous protein synthesis (Cossu et al., 1981, submitted). We therefore studied wether specific proteins are synthesized in chick embryo cultured myotubes as an effect of PMA treatment.

MATERIALS AND METHODS

Materials. Cell culture media and sera were obtained from GIBCO. PMA was from Chemical Carcinogenesis. ^{35}S-methionine (1050 Ci/mMole) was from New England Nuclear. Cytochalasin B was from Sigma.

Culture conditions. Primary cultures of chick embryo myoblasts were prepared and grown as previously described (Adamo et al., 1976). 48 hr after plating, 5 ug/ml of cytochalasin B was added for 24 hr in order to obtain rounded postmitotic myoblasts, less adherent to the substrate (Holtzer et al.,1976).These cells were harvested,recultured in normal medium and pure myotubes were obtained after as little as 10 hr of culture.Fibroblast cultures were prepared from the **preplated muscle cell suspension** (Adamo et al.,1976),and subcultured twice.

PMA treatment and labelling conditions.PMA,1.6×10^{-7} M (in 0.1% DMSO final concentration),was added to the culture medium for the times indicated and,when necessary,was changed every 12 hr.Cultures were incubated with 50 uCi/ml ^{35}S-methionine during the last 4hr of PMA treatment,in Hank's containing 1% horse serum.

Cell fractionation.At the end of the incubation the cells were washed in PBS and scraped off in 10 mM Tris pH 7.4,2 mM PMSF(phenyl-metyl sulphonyl fluoride),(homogenization buffer). Cells were routinely disrupted by 10 sec.sonication at 0°C. Cells extracts were sedimented at 105,000xg for 1 hr and the soluble and the particulate fractions were analyzed by SDS-PAGE.

SDS analysis and fluorography.The particulate fractions were resuspended in homogenization buffer,sonicated and the same amounts of radioactivity were loaded on 7.5-15% logaritmic gradient polyacrilamide slab gels (Laemmli,1970;O'Farrell, 1975).The supernatant fraction was lyophilized,resuspended in Laemmli sample buffer and same amounts of radioactivity was analyzed by SDS-PAGE.

RESULTS

The effect of PMA on cell morphology was studied treating multinucleated myotubes in culture with 1.6×10^{-7} M PMA.In these conditions myotubes loose their spindle shape and acquire a typical myosac shape within 15-20 hr of treatment(Figure 1). Upon removal of PMA cells resume the normal morphology within the same period of time.The reported effect of PMA on myotube **morphology** appears to be similar to that observed after colcemid treatment (Bishoff and Holtzer,1967).
In order to study the early alterations induced by PMA,we analyzed the overall protein synthesis of myotubes treated for

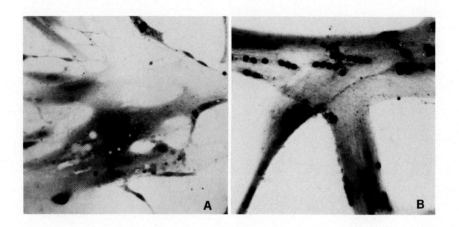

Fig.1.Ligth micrographs of stained cultures of myogenic cells 72 hr after reculturing (see"Materials and Methods").
A,cells treated with PMA;B,control cells.

different times (4,8,24 hr)with PMA and pulse labelled with ^{35}S-methionine during the last 4 hr of PMA treatment.
The pattern of total proteins ,synthetized by PMA-treated and control cells and separated by one dimensional gel electrophoresis,is presented in Figure 2.
The major bands in the fluorogram have Mr of 200 K (myosin haevy chain), 150 K,94 K,90 K,two haevily labelled bands in the region of Mr 55 K,likely corresponding to intermediate filament proteins,43 K (actin)and a 36 K protein likely corresponding to tropomyosin.
Distinct differences between treated and control cell extracts can be seen after 8 hr of PMA exposure.At this time of PMA treatment the incorporation of the label into several polypeptide is inhibited,among which myosin haevy chain and two polypeptide of Mr 105 K and 90 K;also the actin band is greatly reduced.
A sligth increase of the labelling of a polypeptide of Mr 31 K is evident after as little as 4 hr of PMA treatment (Figure 2).

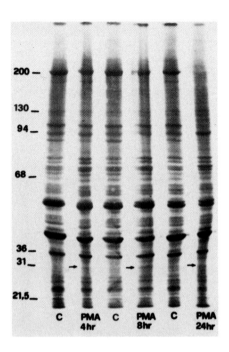

Fig.2. fluorograph of one dimensional SDS-PAGE of control
and PMA-treated myotubes.PMA treatment was performed for 4,
8,24 hr as indicated .Cells were labelled with ^{35}S-methioni-
ne during the last 4 hr of PMA treatment in parallel with
control cultures .Total homogenate was processed for SDS-PAGE
as described in "Materials and Methods".Molecular weight were determined by electrophoresis of the following standards:
myosin,200,000; -galactosidase,130,000;phosphorylase A,94,000;
BSA,68,000;actin,43,000;glyceraldeyde 3-phosphate deydrogenase,
36,000;DNAse,31,000;trypsin inhibitor,21,500.

In order to investigate in more detail the effect of PMA
treatment, the ^{35}S-methionine labelled cell homogenates were
separated into a particulate and soluble fraction (see Methods).
The analysis of the particulate fraction evidentiates similar PMA-dependent changes as observed in total cell extracts.

The PMA-dependent stimulation of the incorporation of 31 K
barely detectable in total cell extracts is more clearly e-
videntiable in the particulate fraction,after as little as
4 hr of PMA treatment (Figure 3).

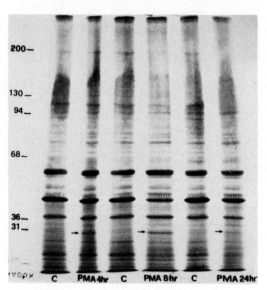

Fig.3. Fluorograph of the particulate fraction from myotubes
treated with PMA for different time periods and relative con-
trols.Same amounts of TCA-precipitable counts were used for
different samples.

The analysis of soluble fraction of control and PMA-treated
cells indicates that PMA inhibits the incorporation of the
label into a number of polypeptides,among which three polypeptides of Mr 105 K,90 K and 38 K respectively are evidently
inhibited after 24 hr of PMA treatment. It is important to
notice that no incorporation of the label in a definite band
at the level of 31 K,is evident in the soluble fraction after
SDS-PAGE,suggesting that this polypeptide is associated with
the particulate fraction (Figure 4).

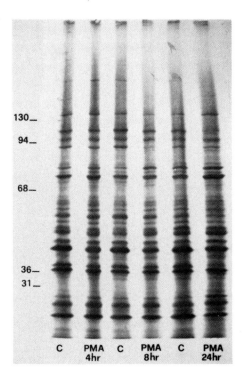

Fig.4. Fluorograph of the soluble fraction from labelled control and PMA-treated myotubes.The lenght of PMA treatment is indicated in the figure.

The non tumor promoter analog of PMA,4 -phorbol-12-13-didecanoate (4α-PDD),used at the same concentration of PMA, failed to induce the alterations observed in the particulate fraction of PMA-treated cells.Figure 5 shows that PMA-treated cells display the above described inhibitory pattern comparable to that of control cultures. Furthermore treatment of fibroblasts with PMA does not result in either the expression of 31 K nor inhibition of myosin (Figure 5).
Although PMA-treated fibroblasts do present several morphological alterations.

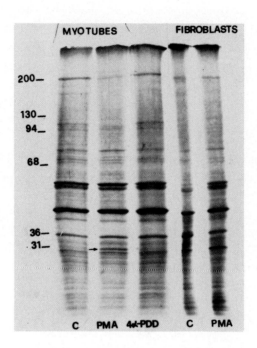

Fig.5. Fluorograph of particulate fraction from control myotubes,myotubes treated with PMA for 24 hr,and myotubes treated with 4α-PDD for 24 hr,and from control and PMA treated fibroblast cultures(24 hr of PMA treatment).

The possibility that the 31 K could be a result of increased protein degradation after PMA treatment of myotubes has been tested performing a chase experiment. ^{35}S-methionine prelabelled control cultures were chased with cold methionine for different time period (4,8,24 hr),in the presence or in the absence of PMA.No increase of 31 K occurs during the period of the chase in either case ,ruling out the possibility that 31 K is a product of PMA-induced protein breakdown. (Figure 6).
Similar results are obtained with shorter period of chase (30')(data not shown).

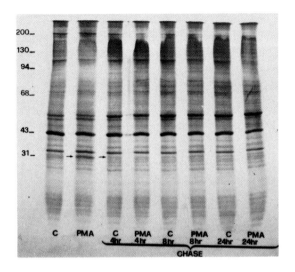

Fig.6. Fluorograph of the particulate fraction of ^{35}S-methionine labelled myotubes. All samples were labelled for 4 hr. 1st lane: control cells processed at the end of the labelling period. 2nd lane: cells treated with PMA during labelling and processed immediately thereafter. All subsequent samples were labelled in the absence of PMA and chased in the presence or in the absence of PMA for the indicated times.

DISCUSSION

The data presented demonstrate that PMA induces an early characteristic increase of the synthesis of a specific protein in multinucleated myotubes. Our previous observations have shown that the inhibitory effect of PMA on CPK activity and acetyl-choline receptor level requires protein synthesis(Cossu et al.,1981,submitted).This prompted us to test the hypothesis that the dedifferentiative effect of PMA in myotubes occurs through a sequence of events following the synthesis of a specific protein,likely triggered by the interaction of the promoter with plasma membrane receptors (Driedger and Blumberg,1980;Delclos et al.,1980;Mohammed and Todaro,1980)

We observed a series of late inhibitory effects, and particularly that myosin and actin synthesis is affected 8 hr after PMA addition and the same time of exposure to the tumor promoter is required to observe inhibition of α-bungarotoxin binding and CPK activity in PMA-treated myotubes. A more prolonged (15 hr) incubation with PMA is required to observe drastic morphological alterations of myotubes. On the other hand the increased synthesis of a 31 K polypeptide is detectable as an early effect after 4 hr of PMA treatment, at a time when the alterations on the expression of the differentiated phenotype are not evidentiable. The relatively early stimulation of 31 K, compared with the delayed inhibitory effect of PMA on specific products, suggest a peculiar role of this polypeptide. On the other hand the pulse chase experiments indicate that 31 K is not a product of a PMA-induced protein degradation(Wigler and Weinstein,1976;Vassalli et al.,1977; Miskin et al.,1978;Miskin et al.,1978).

The synthesis of 31 K appears to be a specific response of muscle cells since it is absent in PMA-treated fibroblasts. PMA-induced early synthesis of specific proteins occurs also in other systems(Cabral et al.,1981;Balmain,1978;Scribner et al.,1972).This suggests that early stimulation of proteins could be a general event in the mechanism of action of PMA. On the other hand the finding that the synthesis of specific products is stimulated by PMA in different cell types suggests that the response to this tumor promoter is cell type specific. Moreover during myogenesis such early response appears to be related to the differentiated state,since our preliminary data indicate that no stimulation of 31 K synthesis occurs in PMA treated replicating myoblasts(data not shown).

In line with our previous observations showing that PMA-induced inhibition of differentiative products is prevented in the absence of protein synthesis(Cossu et al.,1981,submitted),it can be hypothesized that the synthesis of 31 K triggers the PMA-induced dedifferentiative process.

REFERENCES

1)Adamo S,Zani B,Siracusa G,Molinaro M(1976).Expression of

differentiative traits in the absence of cell fusion during myogenesis in culture. Cell Diff 5:53.
2) Balmain A,(1976).The synthesis of specific proteins in adult mouse epidermis during phases of proliferation and differentiating layers of neonatal mouse epidermis.J Invest Dermatol 67:246.
3) Balmain A(1978).Synthesis of specific proteins in mouse epidermis after treatment with tumor promoter TPA.In:Slaga T J, Sivak A and Boutwell R K(eds)"Carcinogenesis"Vol 2 p.133,New York:Raven Press.
4) Bishoff R,Holtzer H,(1967).The effect of mitotic inhibitors on myogenesis in vitro.JCell Biol 36:111.
5) Cabral F,Gottesman M M,Yuspa S H (1981).Induction of specific protein synthesis by phorbol esters in mouse epidermal cell culture.Cancer Res 41:2025.
6) Cohen R,Pacifici M,Rubinstein N,Biehl J,Holtzer H (1977). Effect of tumor promoter on myogenesis.Nature 266:538.
7) Cossu G,Pacifici M,Adamo S,Bouché M,Molinaro M(1981).TPA-induced inhibition of the expression of differentiative traits in cultured myotubes:dependence on protein synthesis(submitted).
8) Driedger P E,Blumberg P M(1980).Specific binding of phorbol ester tumor promoters.Proc Nat Acad Sci USA 77:567.
9) Delclos B K,Nagle D S,Blumberg P (1980).Specific binding of phorbol ester tumor promoter to mouse skin. Cell 19:1025.
10) Grotendorst R G,Schimmell S D(1980).Alteration of cyclic nucleotide levels in phorbol 12-myristate-13-acetate treated myoblast.Biochem Biophys Res Comm 93:301.
11) Holtzer H. Croop J,Dienstman S,Ishikawa H,Somylo A(1976). Effect of cytochalasin B and colcemide on myogenic cultures. Proc Nat Acad Sci USA 72:513.
12) Laemmli U,(1970).Cleavage of structural proteins during the assembly of the head of bacteriophage T4.Nature(Lond.) 227:680.
13) Miskin R, Easton T G,Reich E,(1978).Plasminogen activator in chick embryo muscle cells:induction of enzyme by RSV,PMA and retinoic acid.Cell 15:1301.
14) Miskin R,Easton T G,Maelicke A,Reich E(1978).Metabolism of acetylcholine receptor in chick embryo muscle cells:effect of RSV and PMA.Cell 15:1287.

15) Moammed S,Todaro G J,(1980).Specific high affinity cell membrane receptors for biologically active phorbol and ingenol esters.Nature 288:451
16) O'Farrell H P,(1975).High resolution two-dimensional electrophoresis of proteins.J Biol Chem 250:4007.
17) Pacifici M,Holtzer H(1977).Effects of a tumor-promoting agent on chondrogenesis.Am J Anat 150:207.
18) Rovera G,O'Brien T G,Diamond L(1977).Tumor promoters inhibit spontaneous differentiation of Friend erythroleukemia cells in culture.Proc Nat Acad Sci USA 74:2894.
19) Schimmel S D,Hallam T(1980).Rapid alteration in Ca^{++} content and fluxes in phorbol 12-myristate-13-acetate treated myoblasts.Biochem Biophys Res Comm 92:624.
20) Scribner J D,Boutwell R K,(1972).Inflammation and tumor promotion:selective protein induction in mouse skin by tumor promoters.Eur J Cancer 8:617.
21) Toyama Y,West C M;Holtzer H(1979).Differential response of myofibrils and 10nm filaments to a carcinogen.Am J Anat 156:131.
22) Vassalli J D,Hamilton J,Reich E,(1977).Macrophage plasminogen activator :induction by concanavalin A and phorbol myristate acetate.Cell 11:695.
23) Weinstein I B,Wigler M,(1977).Cell culture studies provide new information on tumor promoters.Nature 270:659.
24) Wigler M,Weinstein I B,(1976).Tumor promoter induced plasminogen activator.Nature 259:232.

ACKNOWLEDGEMENTS

The authors are indebted to Miss A.Rando,L.Diana,for their skillfull collaboration,and to Dr.S.Adamo for helpfull discussion and critical reading of the manuscript.
This work has been supported by grant N°80.01594.96 of the CNR Finalized Project"Control of Tumor Growth".

A NEW PROCEDURE TO ANALYSE THE ROLE OF ADHESION TO SOLID
SUBSTRATA AND OF MOTILITY IN CHEMOKINESIS OF NEUTROPHIL
GRANULOCYTES

H.U. Keller and H. Cottier

Institute of Pathology, University of Berne,
Freiburgstrasse 30, 3010 Berne, Switzerland

SUMMARY

The role of crawling movements and of adhesion to the substratum in chemokinetic responses of neutrophil granulocytes is determined by a new procedure. The results show that this type of analysis is required to understand complex chemokinetic responses. Human serum albumin (HSA) was found to exert its chemokinetic effect exclusively by modulating adhesion to the substratum. In contrast, the positive chemokinetic effect of f-Met-Leu-Phe (10^{-8}M) is essentially due to its capacity to induce crawling-like movements, but depending on the test conditions it may or may not affect adhesion as well. The findings indicate that different mechanisms controlling chemokinesis may produce antagonistic effects on the rate of dislocation.

INTRODUCTION

The rate of locomotion is presumably controlled by a variety of mechanisms and mediators (Keller, Wilkinson, Abercrombie, Becker, Hirsch, Miller, Ramsey, Zigmond, 1977). It seems reasonable to assume that the activity of the locomotor apparatus and adhesion to the substratum are two major determinants in chemokinesis. Previous studies have shown that adhesion to the substratum has a considerable influence on the average speed of locomotion of neutrophils (Keller, Hess, Cottier, 1977; Keller, Barandun, Kistler, Ploem, 1979; Lackie, Smith, 1980) and other cells

(Wolpert, Gingell, 1968). Adhesion to the substratum is a prerequisite for locomotion of most metazoan cells. Therefore, changes in the rate of locomotion can only be determined in adherent cells.

It is conceivable that adhesion and the activity of the locomotor apparatus are regulated independently. But if cells are moving on solid substrata the relative role of adhesion and the activity of the locomotor apparatus can not be properly dissociated. We recently found that neutrophils in suspension can perform crawling-like movements without contact with any substratum (Keller, Cottier, 1981). This suggested that the activity of the locomotor apparatus can be judged in non-adherent cells and that the influence of adhesion to the substratum can be assessed after the cells had settled on the substratum. The present paper shows that this type of analysis is in fact required to understand chemokinetic responses. The results suggest that changes in motility and adhesion are independent phenomena. They may or may not have antagonistic effects on the net chemokinetic effects measured.

MATERIALS AND METHODS

Gey's solution was prepared as previously described (Keller, Wissler, Damerau, Hess, Cottier, 1980). Human serum albumin (HSA) was obtained from Behring-Werke Marburg (FRG), human standard gamma globulin (SGG) prepared by the method of Kistler and Nitschmann (1962) from the Swiss Red Blood Transfusion Service, Berne, Switzerland and the chemotactic peptide f-Met-Leu-Phe (FMLP) from Sigma, St. Louis, USA.

Neutrophils were prepared from heparinized human blood (10 units/ml) by a two-step procedure involving first separation from red cells using Isopaque-Methocel and second separation from monoculear cells using a Ficoll-Hypaque gradient (Böyum, 1968). Only cell preparations showing a low proportion of motile control cells in suspension were chosen for the present experiments.

Assessment of motility and adhesion to the substratum

Cell were suspended in the desired media and shaken at 37°C for 30 minutes. The suspension was filled into

Sykes-Moore chambers and placed immediately on the stage of a heated (37°C) inverted microscope (Leitz Diavert). Floating cells were recorded by means of a Newvicon television camera and a JVC tape recorder using the 100x objective for phase-contrast and reflection-contrast microscopy (Keller, Barandun, Kistler, Ploem, 1979). Cells which had settled on the substratum were again recorded using phase-contrast and reflection-contrast microscopy 30 minutes after the preparation had been set up. The path of the cells was traced for 15 minutes.

The velocity (net displacement against time) of individual cells was determined by analysing the path. The average values for all cells were calculated. Cells performing the type of shape changes associated with crawling-movements (formation of lamellipodia, polarisation and streaming of cytoplasm from tail to front, formation of constriction rings and a tail-knob) that were detectable without time lapse procedures were classified as "motile", those which showed no such shape changes as "non-motile". Non-motile cells were either spherical or spread on the substratum. Adhesion was determined by morphometric measurement of the grey and black areas of contact detectable by reflection-contrast microscopy using a MOP/AM01 morphometry system (Kontron AG, Zürich).

RESULTS

Effect of human serum albumin (HSA) and f-Met-Leu-Phe (FMLP) on motility, adhesion to glass and chemokinesis

The results presented in Table 1(A) show that FMLP (10^{-8}M) produced a striking increase in the proportion of motile cells in suspension provided the control cells were mostly spherical. FMLP-stimulated cells in Gey's solution alone perform crawling-like movements while floating but they loose motility following contact with the substratum. Polarity was often but not always lost concomitantly. This loss of crawling-movements was largely prevented by the addition of 2 % HSA to the medium (Fig. 1). Adhesion to the substratum was also reduced but not abrogated in the presence of HSA. Under these conditions the cells continued to perform crawling movements on the substratum and they were locomoting. But HSA alone did not significantly

Table 1

Effect of HSA, immunoglobulin (SGG) and f-Met-Leu-Phe (FMLP) on motility, adhesion and chemokinesis of neutrophil granulocytes

MEDIUM (Agent in Gey's solution)	% MOTILE NEUTROPHILS		LOCOMOTION	ADHESION
	floating	on glass	velocity (μm/min) ± SDM	average contact area*(μm²) per neutrophil (mean ± SDM)
Experiment A				
none	0	3	0	92 ± 8
FMLP 10^{-8}M	94	3	0	544 ± 32
2 % HSA	0	9	0.05 ± 0.03	28 ± 5
2 % HSA + FMLP 10^{-8}M	90	92	1.5 ± 0.1	26 ± 6
Experiment B				
none	3	6	0.15 ± 0.1	276 ± 35
2 % SGG	97	13	0.46 ± 0.2	650 ± 29
2 % HSA + FMLP 10^{-8}M	96	94	2.3 ± 0.3	76 ± 20

* determined with reflection-contrast microscopy

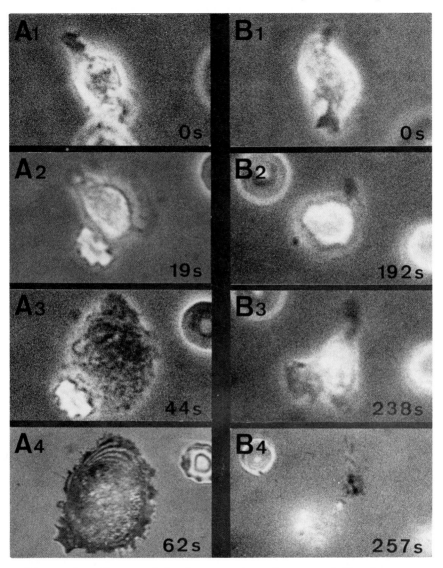

Figure 1: Sequential photographs of neutrophils in suspension (1) and after contact with the substratum (2-4) A: A cell in Gey's solution containing FMLP (10^{-8}M). B: A cell in Gey's solution containing 2 % HSA and FMLP (10^{-8}M). The time is given in seconds (s). The first three pictures of each sequence were taken by phase contrast, the last one by reflection-contrast-microscopy.

increase the proportion of cells performing crawling-like movements in suspension. This indicates that HSA had no direct stimulating effect on the locomotor apparatus.

While FMLP consistently stimulated the proportion of cells performing crawling-like movements in suspension, its effect on adhesion varied with the test conditions. In the experiment shown in Table 1(A), FMLP increased adhesion of neutrophils in Gey's solution alone but not of cells in Gey's solution containing 2 % HSA.

Effects of standard immunoglobulin (SGG) on neutrophil motility, adhesion to the substratum and chemokinesis

Experiments with ethanol-fractionated immunoglobulin (SGG) showed that this preparation produced a marked increase in the proportion of motile cells in suspension. In the experiment shown in Table 1(B), motility and polarisation was reduced following contact with the substratum. The average velocity remained very low. The cells were spread and exhibited large areas of contact. In contrast, FMLP-stimulated cells in Gey's solution were less adhesive but capable to express motility in the form of stimulated locomotion.

DISCUSSION

Chemokinetic responses of neutrophils to human serum albumin, immunoglobulin and f-Met-Leu-Phe were analysed by relating the velocity to 1) shape changes reflecting the activity of the locomotor apparatus and 2) adhesion to the substratum. Other investigators (Smith, Hollers, Patrick, Hasset, 1979) used cytotaxin-induced polarisation determined in fixed cell preparations as a correlate for motility. Direct observation of live cells in the present study confirmed that polarized cells in suspension are highly motile. We found, however, that this does not necessarily apply to cells on the substratum. Spread cells can exhibit a clear-cut front to tail-polarity without significant motility (Keller, unpublished observations). Therefore, polarity was not used as a single conclusive criterium for motility in the present study.

The results show that chemokinesis can be independently influenced by factors stimulating locomotion on the one hand and by chemicals modulating adhesion to the substratum on the other hand. Native HSA alone had no significant stimulating effect on motility. Its chemokinetic effect was exclusively due to its capacity to reduce excessive cell adhesion to the substratum. These results confirm and extend earlier studies (Keller, Hess, Cottier, 1977; Smith, Hollers, Patrick, Hassett, 1979). In contrast to HSA, FMLP produced a striking increase in the proportion of motile cells in suspension, but not necessarily a significant change in cell adhesiveness. Adhesion was increased by FMLP in plain Gey's solution but not in Gey's solution containing 2 % HSA. SGG stimulated both the proportion of cells performing crawling-like movements and their adhesiveness to the substratum. These results show that different chemicals produced different patterns of responses. HSA represents a class of chemicals which act exclusively by modulating adhesion. In contrast, the chemokinetic response to FMLP at a concentration of 10^{-8}M is essentially due to a stimulation of the locomotor apparatus.

The net chemokinetic effect depends on changes in adhesion and on the stimulation of crawling-like movements. Even weakly adherent cells do not migrate unless stimulated to perform crawling-like movements. But stimulation of crawling-like movements does not guarantee that the cells will actually locomote to any significant extent. These movements performed in suspension may or may not disappear when the cells make contact with the substratum. The degree of adhesion and sequent immobilisation was in the present experiments largely dependent on the presence or absence of HSA. In absence of HSA, FMLP-stimulated cells lost the polarized shape partially or totally following contact with the substratum. This is a sort of contact inhibition induced by an anorganic substratum. The stimulated crawling movements and the potential positive chemokinetic effect associated with it became totally obscured. In presence of HSA the cells continue to perform crawling movements after having made contact with the substratum and locomote. Thus, HSA and FMLP produce different cellular responses with an synergistic effect on net chemokinesis under the present est conditions. In absence of HSA, SGG or FMLP can stimulate both adhesion to the substratum and crawling movements. As a result the two different responses induced by the same chemical can exert antagonistic effects on chemokinesis. Stimulation of crawling movements does not necessarily result in dislocation under these conditions because locomotion is inhibited by excessive adhesion. Therefore, the stimulating effect on crawling movements is largely underrated because it becomes obscured and is no longer detectable by measuring chemokinesis. These examples show that the new procedure can produce new information on different aspects of the chemokinetic response. It becomes possible to characterise chemokinetic factors more precisely on the basis of the pattern of responses which they induce. The results show that responses such as adhesion and the control of crawling movements can be determined independently. This is necessary in order to understand their precise role in chemokinesis.

We do not as yet know whether this procedure which has been worked out with leucocytes can be applied to other locomoting cells such as fibroblasts or tumor cells. This should be feasable, provided the mechanism of locomotion of these cells is similar to that of leucocytes.

ACKNOWLEDGEMENTS

We thank Miss E. Ochsner and Mr. B. Haenni for excellent technical assistance. The work was supported by the Swiss National Science Foundation.

REFERENCES

Böyum A (1968). Separation of leucocytes from blood and bone marrow. Scand J clin Lab Invest 21:1.
Keller HU, Barandun S, Kistler P, Ploem JS (1979). Locomotion and adhesion of neutrophil granulocytes: Effects of albumin, fibrinogen and gamma globulins studied by reflection contrast microscopy. Exp Cell Res 122:351.
Keller HU, Cottier H (1981). Crawling-like movements and polarisation in non-adherent leucocytes. Cell Biol Int Reports 5:3.
Keller HU, Hess MW, Cottier H (1977). The chemokinetic effect of serum albumin. Experientia 33:1386.
Keller HU, Wilkinson PC, Abercrombie M, Becker EL, Hirsch JG, Miller ME, Ramsey WS, Zigmond SH (1977). A proposal for the definition of terms related to locomotion of leucocytes and other cells. Clin exp Immunol 27:377.

Keller HU, Wissler JH, Damerau B, Hess MW, Cottier H (1980). The filter technique for measuring leucocyte locomotion in vitro. Comparison of three modifications. J Immun Meth 36:41.

Kistler P, Nitschmann HS (1962). Large scale production of human plasma fractions. Vox Sang 7:414.

Lackie JM, Smith RPC (1980). Interactions of leucocytes and endothelium. In Curtis ASG, Pitts JD (eds.): "Cell Adhesion and Motility (Third Symposium of the British Society for Cell Biology)," Cambridge: University Press, p. 235.

Smith CW, Hollers JC, Patrick RA, Hassett C (1979). Motility and adhesiveness in human neutrophils. Effects of chemotactic factors. J Clin Invest 63:221.

Wolpert L, Gingell D (1968). Cell surface membrane and amoeboid movement. "Symposium of the Society for Experimental Biology. XXII. Aspects of Cell Motility," Cambridge: University Press, p. 169.

RESTRICTED DEVELOPMENTAL OPTIONS
OF THE METANEPHRIC MESENCHYME

Hannu Sariola, Peter Ekblom and Lauri
Saxén
Department of Pathology, University of
Helsinki
Haartmaninkatu 3, SF-00290 HELSINKI 29
Finland

For more than a century the developing metanephric kidney has been the subject of intensive research and speculation. In 1865, Kupffer showed that the Wolffian duct forms an evagination which he designated as the "Nierenkanal", associating this with the development of the permanent kidney, the metanephros (Kupffer 1865). For a long time there were arguments whether the branching bud forms all epithelial parts of the nephron or whether it connects with tubules derived from the mesenchyme. Carl Huber described the morphological steps of renal development, and his classic conclusion in 1905 was that the ureteric bud forms the collecting ducts whereas the nephrogenic mesenchyme forms the secretory nephrons (Huber 1905). The origin of the kidney vasculature and the glomerular endothelium has, however, remained unknown.

Grobstein showed half a century later with his separation and recombination experiments that the ureteric bud induces the mesenchymal cells which then become epithelial. The ureteric bud can be replaced by various heterologous tissues (Grobstein 1955). The nature of the induction is permissive, and the ureteric bud cannot induce other than the metanephric mesenchyme to form nephric tubules (Saxén 1970). This suggests that the metanephric mesenchyme is already predetermined to form kidney tubules before the ureteric bud grows into the blastema. Morphologically the metanephric mesenchyme is homogenious, and it has also been shown recently in immunology that the blastema uniformly expresses interstitial proteins (Ekblom et al. 1981a).

The nature of the induction is still unknown, but it has been shown that the mesenchyme needs a close contact with the inductor. This was demonstrated in transfilter experiments, where the mesenchyme and an inductor were separated by a filter. The induction takes about 24 hours (Saxén and Lehtonen 1978), but the mesenchyme does not form tubules until another 24 hours. Filters preventing cell contacts block this inductive interaction (Wartiovaara et al. 1974, Lehtonen 1976).

Chick embryonic mesonephros, the transient kidney, has a granulopoetic and a chondrogenic capacity (Romanoff 1960, Lash 1963). Similarly, human mesonephric mesenchyme can be converted into cartilage, but it is not known whether the mesenchyme of the permanent kidney, the metanephros, has options other than to become nephric epithelium (Lash and Saxén 1972).

Emura and Tanaka have combined the metanephric mesenchyme with various haematopoietic tissues, and they report haematopoietic islands in the mesenchyme, surrounded by endothelial-like cells (Emura and Tanaka 1972). This study supports the suggestions that the mesenchyme also forms endothelial cells.

Our own observations do not, however, support these suggestions. We have failed to reproduce the results of Emura and Tanaka by using the same experimental conditions (Fig. 1). The experimentally induced mesenchymes form in vitro all epithelial structures of the secretory nephron, tubules and glomerular epithelial cells, but no vascular elements including glomerular endothelial and mesangial cells or the juxtaglomerular apparatus (Bernstein et al.1981, Ekblom 1981, Ekblom et al. 1981b). In direct in vivo observations of the 11-day mouse kidneys, we failed to detect any vascular or haematopoetic elements within the mesenchymal blastema (Fig. 2). During the following day, when the ureter invades the mesenchyme, vascular elements are seen within it. Hence we concluded that the metanephric mesenchyme might not have haematopoietic or endothelial forming bias, and these elements could be derived from a different cell lineage outside the blastema.

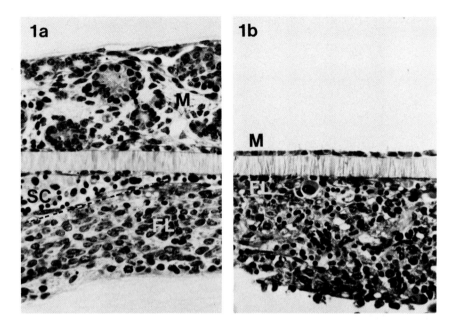

Figure 1. Micrographs of explants where the metanephric mesenchyme has been combined transfilter with a combination of liver+spinal cord (a) and with foetal liver alone (b).
1 a. The foetal liver (FL) + spinal cord (SC) combination induces differentiation of the mesenchyme (M) which remains avascular.
1 b. The foetal liver cannot induce differentiation and the mesenchyme becomes flattened.

To test this hypothesis we made interspecies grafts between both mouse and quail and chick and quail by transplanting undifferentiated kidneys onto chorioallantoic membranes. Since quail has a biological marker, an intensively staining nucleolus, the cells can be traced in these hybrid grafts. It turned out that the glomerular endothelium regularly expressed the character of the host. The entire kidney vasculature came from the chorioallantoic membrane (Fig. 3).

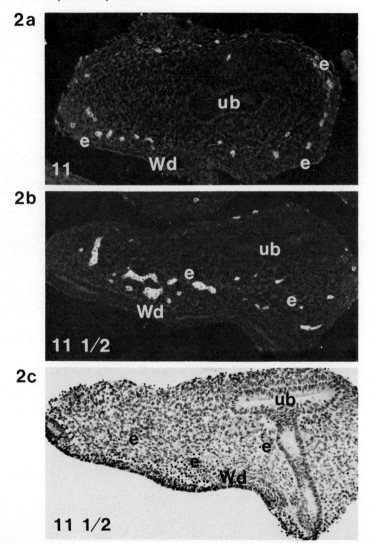

Figure 2. Localization of erythrocytes at the early stages of the metanephric kidney.
2 a. The undifferentiated 11-day mouse metanephros shows erythrocytes (e) only around the mesenchymal blastema. (Wd=Wolffian duct, ub=ureter bud). Anti-erythrocyte staining.
2 b. Half a day later erythrocytes invade the blastema.
2 c. Hematoxylin-eosin staining shows the ingrowing ureter bud and erythrocytes, but no haematopoesis.

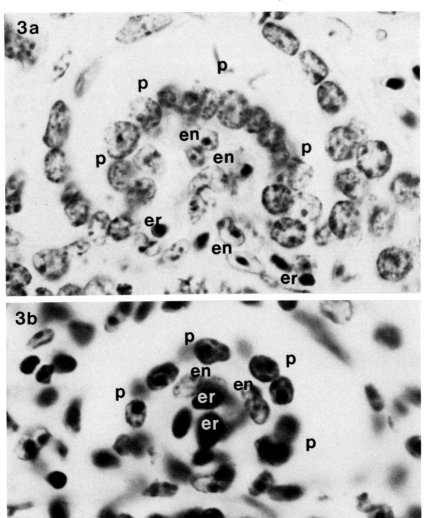

Figure 3. Micrographs of glomeruli in grafting experiments, where undifferentiated kidneys were transplanted on chorioallantoic membranes (CAM).
3 a. Mouse kidney transplants on the quail CAM have endothelial cells of host type in glomeruli.
3 b. Quail kidney transplants on the chick CAM show identically host endothelial cells in glomeruli. The quail cells can be identified by the big dark staining nucleolus, "the quail marker". p=podocytes, er=erythrocytes, en=endothelial cells. Feulgen staining.

We conclude that endothelial cells from outside vasculature are capable of invading the kidney anlage, where they homed to the developing glomeruli. Most likely this is the normal mechanism by which the metanephric kidney becomes vascularized in vivo. Hence, the mesenchymal cells, predetermined to become epithelial, might not have other developmental options.

REFERENCES

Bernstein J, Cheng F, Roszka J (1981). Glomerular differentiation in metanephric culture. Lab Invest 45:183.

Ekblom P (1981). Formation of basement membranes in the embryonic kidney. An immunohistological study. J Cell Biol (in press).

Ekblom P, Lehtonen E, Saxén L, Timpl R (1981a). Shift in collagen type as an early response to induction of the metanephric mesenchyme. J Cell Biol 89:276.

Ekblom P, Miettinen A, Virtanen I, Wahlström T, Dawnay A, Saxén L (1981b). In vitro segregation of the metanephric nephron. Dev Biol 84:88.

Emura M, Tanaka T (1972). Development of endothelia and erythroid cells in mouse metanephric mesenchyme cultured with fetal liver. Dev Growth Differ 14:237.

Grobstein C (1955). Inductive interaction in the development of the mouse metanephros. J Exp Zool 130:319.

Huber GG (1905). On the development and shape of uriniferous tubules of certain of the higher mammals. Am J Anat, Suppl 4:1.

Kupffer C (1865). Untersuchungen über die Entwicklung der Harn- und Geschlechtssystems. Arch mikr Anat Bd I.

Lash JW (1963). Studies on the ability of embryonic mesonephros explants to form cartilage. Dev Biol 6:219.

Lash JW, Saxén L (1972). Human teratogenesis: In vitro studies on thalidomide-inhibited chondrogenesis. Dev Biol 45:183.

Lehtonen E (1976). Transmission of signals in embryonic induction. Med Biol 54:108.

Romanoff AL (1960). "The Avian Embryo". New York: The Macmillan Company, p. 580.
Saxén L (1970). Failure to demonstrate tubule induction in a heterologous mesenchyme. Dev Biol 23:511.
Saxén L, Lehtonen E (1978). Transfilter induction of kidney tubules as a function of the extent and duration of intercellular contacts. J Embryol Exp Morphol 47:97.
Wartiovaara J, Nordling S, Lehtonen E, Saxén L (1974). Transfilter induction of kidney tubules: correlation with cytoplasmic penetration into Nuclepore filters. J Embryol Exp Morphol 31:667.

IN SITU RECORDING OF THE MECHANICAL BEHAVIOUR OF CELLS IN
THE CHICK EMBRYO

P. Kucera and Y. de Ribaupierre

Institute of Physiology, University of Lausanne

CH - 1011 Lausanne, Switzerland

INTRODUCTION

The early morphogenetical phenomena which can be studied directly under the microscope are mainly those based on the mechanical activity of the embryonic cells. Some events, for example the heart contractions, are rapid and therefore observable quite easily. A precise evaluation of slow events such as the cell proliferation and migration necessitates however the use of some tracing technique. A given cell population is at first labeled (dyes, adhering opaque particles, radioisotopes) at the initial position and the displacements of the labeled cells traced by successive drawings, microcinematography or autoradiography (see e.g. Spratt and Haas 1960; Nicolet 1971).

Such studies, although very valuable for the description of the morphogenetical movements, give however only a remote information about the physiological properties of the embryonic cells, namely the contractility and adhesiveness, both closely involved in the cell shaping and motility. Physiological studies of the active and passive mechanical behaviour of cells within an intact, normally developing embryo have not been possible because of lack of an appropriate sensitive and non-invasive technique.

We have developed a new method for continuous recording of tissue viscosity and elasticity changes which are known to be directly linked to the changes of the cell contraction state and adhesiveness. The viscosity is of special interest as it determines the tissue mechanical impedance - a parameter which can be measured *in situ* (Kucera and de Ribaupierre 1981). Some results obtained by this technique in the entire young chick blastodisc incubated *in vitro* are reported.

TECHNIQUES

The recordings of the tissue viscosity.

They are achieved by measuring the damping of submicroscopical mechanical waves created in the tissue by a very fine tungsten needle driven by a small piezoelectric oscillator (fig. 1). The amplitude and frequency of the oscillator-needle-tissue system depend on the tissue mechanical impedance, i.e. on its viscosity and elasticity. If the driving frequency is stabilized at the resultant resonance frequency of the oscillating system the amplitude of the recorded signal (an electrical potential) reflects mainly the viscosity changes. The sequential excitation-detection method results in a good signal-to-noise ratio in spite of a very small excitation amplitude used to preserve the tissue integrity. The parameters used are : excitation frequency and amplitude respectively 2kHz and 0.2µm (outside the tissue); sampling frequency 200Hz; frequency bandwidth 50Hz; spatial resolution 50-100µm.

The chick embryo preparation.

The eggs are preincubated to the desired stage of development (usually stage 3-4 HH) and the blastodisc with the surrounding vitelline membrane excised and clamped between two metallic concentric rings according to New (1955). Such a preparation is then mounted in a special metallic incubation chamber and placed onto the stage of an inverted microscope situated in a thermostabilized box. The incubation chamber, perfused with defined media allows for rapid expansion of the disc and normal development of the embryo for more than two days (Kucera and Raddatz, 1980). The development is followed through the central window of the chamber

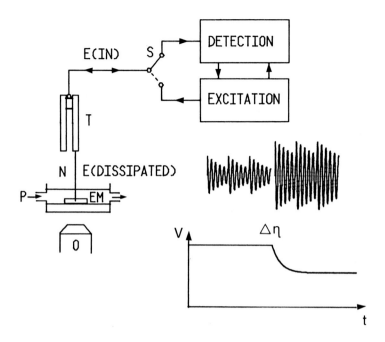

Fig. 1. A diagram of the experimental set-up. EM: embryo mounted in the chamber, P: controlled push-pull perfusion, O: objective of the microscope. Pulses of submicroscopical oscillations are periodically applied to the embryo by a microneedle (N) coupled to a small piezoelectric tuning fork (T), excited by a voltage controlled oscillator. The dissipation of the mechanical energy in the embryo is detected by a lock-in amplifier. The excitation/detection time ratio is determined by an electronic switch (S). The detected damped oscillations are integrated and an electrical signal (V) proportional to the viscosity changes thus obtained. A high-to-low viscosity transition ($\Delta\eta$) is schematically shown.

by time-lapse photography. The recording needle(s) and the stimulating electrode and/or pipette are positioned under the microscope using micromanipulators. The reference electrode is connected to the metallic chamber. The changes of viscosity appearing either spontaneously under constant physico-chemical conditions or evoked by stimulations (electrical, mechanical, chemical) are recorded continuously. The fig. 2 shows such a preparation ready for experiments.

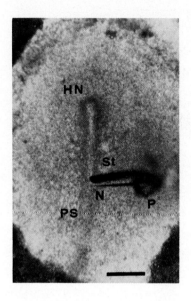

Fig. 2. The area pellucida of the chick embryo (stage 4 HH) mounted in the chamber. The needle (N) is inserted into the primitive streak (PS). The stimulating electrode (St) and the tip of a small pipette (P) are seen through the embryo and are therefore out of focus. HN- the Hensen's node. Calibration bar 0.5mm.

RESULTS AND COMMENTS

Spontaneous activity in the heart-forming regions.

The first rapid and synchronized cell contractions in the embryo are visible in the cardiac primordia at about 35 hours of incubation i.e. at the stage of 9 to 10 somites (Sabin 1920, Patten 1949, DeHaan 1965). We have repetitively recorded the very first (and invisible) beats already at the stage of 8 and sometimes even 7 somites. The fig. 3 A shows that such contractions appear rather suddenly, being often grouped by two or three. The initial frequency, about 1-3/min, increases very rapidly (up to 8/min within 20 min after the onset of activity) and becomes fairly regular. Occasionally the rhythm may stop for some minutes but recovers at the previous frequency and amplitude. It is interesting to compare these results to the optical signals obtained from the 8 somite chick heart isolated in vitro and stained with potential sensitive dyes (fig. 4, 7 in Fujii et

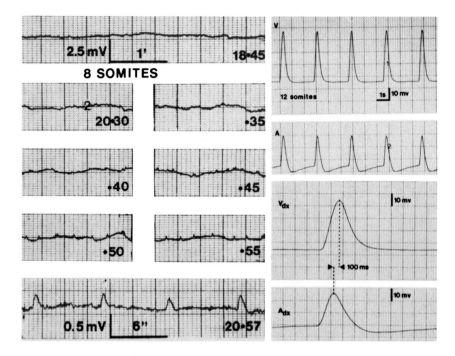

Fig. 3. Spontaneous activity in the intact embryonic heart. Left: Continuous recordings before (top) and after the formation of the 8th somite with respect to the real time. The very first beats are evident between 20h40 and 20h45. The rhythm rapidly accelerates and becomes regular -8/min (bottom). Right: Simultaneous records from two needles inserted in the right atrial (A) and ventricular (V) walls 10 hours later. Regular rhythm at 22/min, about 160 fold increase in amplitude, slow drift of viscosity preceding the atrial beat and clear atrio-ventricular delay are evident.

al. 1980). The correspondence of these records with ours is almost perfect. Thus, on the contrary to what state these authors, the mechanical activity of the embryonic heart does not seem to be delayed with respect to the onset of the electrical activity but starts practically at the same time.

As the heart differentiates the amplitude steeply rises (cf. the calibrations in fig. 3 A and B) and becomes visible in the microscope. At the stage of 12 somites, the simultaneous recordings obtained by two needles placed in the proximal and distal parts of the heart tube show two interesting phenomena (fig. 3 B). First, the viscosity changes of the atrial region resemble closely to records of the pacemaker electrical activity. The ventricular region does not show this feature. Second, the former region contracts about 100ms before the latter. This delay (corresponding to a propagation velocity of 5 mm/s) indicates that the atrio-ventricular hierarchy is present already at the very early stage of the heart differentiation. Moreover, the left heart contracts some 40 to 120ms later than the right

Spontaneous activity in the Hensen's node.

The Hensen's node (HN) is an important proliferation center and, together with the primitive streak, a zone of ingression of the epiblastic cells into the deeper layers of the embryo. From the stage 4 onward, as it gives off the cells forming the axial mesoderm and inducing the differentiation of the nervous system, the HN regresses and is thus found more and more posteriorly in the area pellucida. The viscosity when recorded continuously in the HN displays a clear oscillatory behaviour (fig. 4).

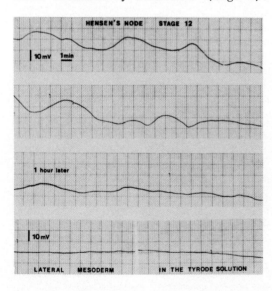

Fig. 4. Spontaneous activity of the Hensen's node. The periodical oscillations found in the node are compared to a stationary signal obtained elsewhere in the same embryo and to the background noise (bottom).

This spontaneous activity (as if the nodal cells were contracting periodically) appears at intervals from 2 to 7 min with a mean period of 3.5 min. The needle apparently does not impede the regression of the HN. Under the microscope, the nodal cells actually seem to flow around the needle leaving it finally behind. In parallel, the amplitude of the oscillations progressively decreases. In the young embryo (stage 4 HH), such a periodical activity can be recorded in various regions of the area pellucida whilst in the older embryos (fig. 4: lateral mesoderm area) it does not seem to be the case (preliminary observations).

The significance and the origin of this slow periodical behaviour of the embryonic cells are open to speculations and new experiments. The importance of oscillatory phenomena in the morphogenesis has been beautifully documented in studies of Dictyostelium (see e.g. Gerisch 1978). Goodwin and Cohen (1969) proposed a general hypothesis of the embryonic pattern formation based on propagated periodic signals which could occur in the embryos. In the young chick embryo, this idea seems to be supported by the work of Stern and Goodwin (1977) who have found that the migratory movements are discontinuous appearing in saccades at a mean period of 2.6 min. Thus the viscosity oscillations could reflect such periodical cell movements. They could however equally well represent variations in the active mechanical tension present in the expanding blastodisc (Downie 1976).

Responses to electrical stimulation.

A single brief pulse of electrical current (at least 1.5 to 2 mA for 0.5ms) is able to evoke a mechanical response in any region of the area pellucida so far investigated. Such a response invariably consists of a temporary increase of tissue viscosity. The amplitude, complexity and time characteristic of the response depend to some extent on the intensity of stimulus. Indeed, as the density of the stimulating current exponentially fades off with distance from the electrode tip, and, as the needle records from cells within a limited surrounding area, greater than liminar currents are necessary to recruit all these cells and thus obtain a fully developed and maximal response. Such a response is generally biphasic and lasts for several minutes (fig. 5). The

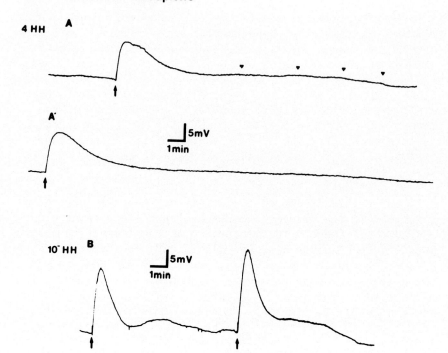

Fig. 5. Viscosity variations in response to a field stimulation. A,A': transient responses followed by sustained, sometimes oscillating activities (triangles). Hensen's node, anodic stimulation, 3.5 mA, 0.5ms (arrows). B: Similar responses from the posterolateral margin of the area pellucida in an older embryo. Cathodic stimulation, 5 mA, 50ms.

first transient phase is an immediate and rapid viscosity increase culminating in 30 s and then falling gradually with a time constant of about 2-4 min. This phase is followed after 2 to 6 min by a longer tonic plateau the amplitude of which amounts to some 15-25 % of the initial peak and presents sometimes several oscillations. The overall duration of the response is usually 10 to 20 min. A second stimulation at various intervals from the first peak produces a second and similar response, sometimes with a slight summation effect (fig. 5 B). A well-developing embryo shows no or little signs of fatigue to repetitive stimulations at 0.5/min.

Anoxic or dead embryos do not respond to electrical stimulation.

Do the viscosity variations induced by the stimulation result in cell movements? In the microscope, nothing seem to happen in the stimulated area. In order to answer this question properly the cells will be labeled and their behaviour evaluated by microcinematography done in parallel to stimulations.

Does the response spread to cells far from the place of stimulation? When the recording needle is progressively displaced away from the stimulating electrode or vice versa the response rapidly disappears. Thus the mechanical response itself does not seem to propagate through the undifferentiated embryonic tissues. This however does not exclude the use of this technique in investigations of other possible intercellular communications (e.g. electrical or chemical) provided they are accompanied by changes in cell mechanical activity. In this respect, the second slow response, for example, could be interpreted as a one to a message travelling from distant parts of the embryo, themselves modulated by signals emitted from the initially stimulated cells. Alternatively, it could reflect some metabolic changes (osmotic work?) resulting from the first rapid mechanical response.

Responses to mechanical stimulation.

The mechanical stimulus consists of a very short transient increase of the amplitude at the tip of the oscillating needle. The effect can be visible in the microscope as a tiny brief jig of the surrounding cells. Such stimuli, no matter where applied, produce an immediate and important decrease of viscosity. The recovery to the previous level is slow (5-10 min) and roughly exponential (fig. 6 on the right). In the Hensen's node the response is quite different (fig. 6 on the left). During the first minute after the stimulation the viscosity increases much more rapidly, reaches a short plateau and only then slowly recovers as in the other regions.

It is tempting to speculate that, although no visible

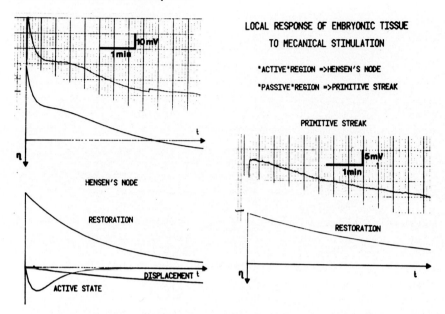

Fig. 6. Viscosity variations in response to a brief increase of the amplitude of vibration at the needle tip (100ms, + 40dB). Stage 6 HH embryo. Viscosity increase recorded downwards. The actual records, obtained in the two indicated regions, have been simulated numerically and, in the case of the HN, decomposed in three hypothetical phases.

damage of the cells has been observed so far, the slow return of the signal, common to all responses, represents a restoration of intercellular links disturbed by the stimulation. If this is true, the cell to cell adhesiveness *in situ* could be evaluated by these experiments. On the other hand, the transient initial viscosity variation found in the HN could reflect a change in the cell contraction state, i.e. a true active tension or possibly movement.

The recordings in isolated tissue fragments.

In addition to the *in situ* measurements presented above the viscosity changes have been successfully recorded also in small fragments dissected out of the embryonic heart. Such fragments are simply impaled by the recording needle and incubated in a perfusion chamber. Fig. 7 shows such records.

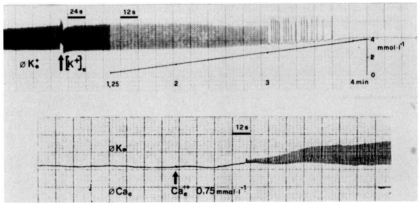

Fig. 7. Spontaneous activity of a small fragment of the right heart atrium impaled by the recording needle. Top: the arrest of activity due to a linear increase of the extracellular potassium concentration applied at the arrow and monitored against the time. Bottom: Complete and reversible suppression of activity due to absence of the extracellular calcium.

SUMMARY

The changes of viscosity in different regions of the intact living early chick embryo were recorded by a new sensitive and non-invasive technique

In the heart, the very first yet invisible mechanical activity has been clearly detected and its further differentiation followed. In other regions, especially in the Hensen's node, spontaneous slow and oscillating mechanical activity has been observed. It has been shown that the undifferentiated embryonic tissue can as well respond to electrical, mechanical or chemical stimulations. The recorded changes have been interpreted as manifestations of the cell mechanical work i.e. active tension, cell shaping or movement.

The technique represents a new tool for physiological and pharmacological studies of the undifferentiated embryonic cells and the early morphogenetic phenomena.

DeHaan RL (1965). Morphogenesis of the vertebrate heart. In DeHaan RL, Ursprung H (eds): "Organogenesis", New York: Holt, Rinehart and Winston, p 377.

Downie JR (1976). The mechanism of chick blastoderm expansion. J Embryol exp Morph 35: 559.

Fujii S, Hirota A, Kamino K (1980). Optical signals from early embryonic chick heart stained with potential sensitive dyes: evidence for electrical activity. J Physiol 304: 503.

Gerisch G (1978). Cell interactions by cyclic AMP in Dictyostelium. Biol. Cellulaire 32: 61.

Goodwin BC, Cohen MH (1969). A phase shift model for the spatial and temporal organization of developing systems. J Theoret Biol 25: 49.

Kucera P, Raddatz E (1980). Spatio-temporal micromeasurements of the oxygen uptake in the developing chick embryo. Respiration Physiology 39: 199.

Kucera P, de Ribaupierre Y (1981). In situ recording of cell mechanical properties. J Physiol 318: 5P

New DAT (1955). A new technique for the cultivation of the chick embryo in vitro. J Embryol exp Morph 3: 326.

Nicolet G (1971). Avian gastrulation. Adv Morphogen 9: 231

Patten BM (1949). Initiation and early changes in the character of the heart beat in vertebrate embryos. Physiol Rev 29: 31.

Sabin RF (1920). Studies on the origin of blood-vessels and red blood-corpuscules as seen in the living blastoderm of chicks during the second day of incubation. Contrib Embryol 9: 213.

Spratt NT Jr, Haas H (1960). Morphogenetic movements in the lower surface of the unincubated and early chick blastoderm. J exp Zool 144: 139.

Stern CD, Goodwin BC (1977). Waves and periodic events during primitive streak formation in the chick. J Embryol exp Morph 41: 15.

CELL CONTACT-DEPENDENT REGULATION OF HORMONAL INDUCTION OF GLUTAMINE SYNTHETASE IN EMBRYONIC NEURAL RETINA

P. Linser, A.D. Saad, B.M. Soh and A.A. Moscona

Developmental Biology Laboratory, Cummings Life Science Center, University of Chicago, Chicago, Illinois 60637, U.S.A.

ABSTRACT

Glutamine synthetase (GS) is a differentiation marker in the neural retina of the chick embryo. GS is localized specifically in Müller glia cells, and it can be precociously induced by adrenal corticosteroids (such as cortisol). The induction depends on cortisol-elicited gene expression and results in de novo synthesis of GS and in a multifold increase in its level. GS is inducible only when Müller cells are closely associated with retina neurons. When retina tissue from 10-day embryos is dissociated into single cells and these are maintained either in suspension or in monolayer culture, GS cannot be induced. However, if identically prepared cells are reaggregated and allowed to reconstruct retinotypic associations, they are inducible for GS. Measurements of cytoplasmic cortisol-receptors showed that cell dissociation results in a rapid and marked reduction in the level (or activity) of these receptors. Their low level persists if the cells are maintained in a dispersed state. However, if the cells are reaggregated and reestablish tissue-like contacts, the level of cortisol receptors increases, as does GS inducibility. The results indicate that, in the embryonic neural retina, histotypic cell contacts are involved in regulating the level of cytoplasmic cortisol receptors and of the responsiveness of Müller glia cells to the induction of GS. Whether the two aspects are causally related is a matter for future study.

INTRODUCTION

Adrenal corticosteroid hormones function as intercellular signals and inducers in the developing embryo and play a major role in the regulation of embryonic differentiation. Studies on biochemical differentiation of the neural retina (NR) of chick embryo have shown that the development of the enzyme glutamine synthetase (GS), whose level is very high in the adult NR, is regulated by hormonal induction (Moscona, 1972; Moscona et al., 1980). GS is involved in several major metabolic pathways; in the nervous system it plays a special role in recycling of neurotransmitters and derivatives (Hamberger et al., 1979). During the early embryonic development of the chick NR, the level of GS is very low; however, on the 16th day of embryonic development GS in the NR begins to rise sharply, plateauing in a few days after a 100-fold increase (Moscona, 1972). This rise of GS is preceded by maturation of the adrenal cortex and consequent elevation of systemic corticosteroids that induce GS in the NR (Piddington, 1970).

What makes this system especially interesting is that, GS can be induced in the NR precociously long before it normally begins to rise sharply in the embryo. This induction can be accomplished _in vivo_ by supplying the embryo prematurely with certain adrenal corticosteroids such as cortisol, and _in vitro_ by culturing isolated early embryonic retina tissue in medium with cortisol (Moscona, 1972; Moscona, M and Moscona, 1979). Studies on cultures of NR tissue demonstrated that the hormone promptly elicits accumulation of mRNA for GS (Sarkar and Moscona, 1973; Moscona and Wiens, 1975), resulting in a rapid increase in the rate of enzyme synthesis and in the enzyme level (Soh and Sarkar, 1978).

In this paper we briefly discuss two aspects of this system: the cellular location of GS in the NR, and the role of cell-cell interactions in the regulation of responsiveness to the hormonal stimulus (for a detailed review of this system see Moscona et al., 1980; Linser and Moscona, in press).

Induction of GS

Fig. 1 shows the age-dependent increase in responsive-

ness of the NR of the chick embryo to precocious induction
of GS by cortisol. The NR becomes programmed with competence for this induction between the 5th to 7th day of
embryonic development; during this time, competence acquisition can be prevented by BrdU (Moscona, M and Moscona,
1979). From the 7-8th day on, the NR is competent and its
responsiveness to GS induction increases with age. Induction of GS in competent NR is not blocked by BrdU (Moscona,
M and Moscona, 1979). As is also evident from Fig. 1, the
progressive increase in responsiveness to GS induction
represents cellular maturation, rather than increase in
cell number since the rate of cell growth in the NR declines
after the 8th day and virtually ceases after the 12th day
of embryonic development (Kahn, 1974; Moscona, M. and
Moscona, 1979). Thus, three phases in the development of
GS inducibility in the NR can be distinguished: 1) pre-
competence phase; 2) competence acquisition phase; 3) maturation of inducibility phase (Moscona, M and Moscona, 1979).
It is assumed that these phases reflect molecular-develop-
mental changes in those cells in which GS is inducible.

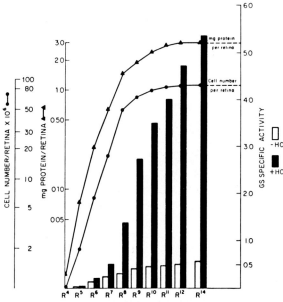

Fig. 1. Development of inducibility for GS in the neural retina (NR) of the chick embryo between 5 and 14 days of incubation: relation to embryonic age and to changes in cell number and in total protein per retina. Retinas dissected from embryos were cultured for 24 hrs in medium with cortisol (+HC) or without it (-HC). The black bars show the levels of GS activity induced in retinas of different embryonic ages. The white bars show the levels of GS in the absence of the steroid inducer. (From Moscona, M and Moscona, 1979).

Cellular Localization of GS

Analysis of this system was greatly advanced by the finding that in the mature NR, GS is localized in a specific type of cells, Müller glia cells, and that its precocious induction in embryonic NR also is confined to this particular cell type. This was determined by immuno-histochemical studies in which GS-specific antiserum, generated to GS purified from chicken NR (Sarkar et al., 1972), was used to detect the cells that contain GS in adult and in precociously induced embryonic NR (Linser and Moscona, 1979).

Fig. 2 shows that in the adult NR, GS is confined exclusively to Müller glia cells. They are the only kind of gliocytes present in the avian NR. They are identifiable by being the only cells that span the entire width of the NR, from the outer limiting membrane to the inner limiting membrane; their perikarya are located in the bipolar cell layer; their numerous fine arborizations extend through the inner plexiform layer and the ganglion cell layer, and terminate in characteristic end-feet anchored in the inner limiting membrane (Linser and Moscona, 1979).

Fig. 2 also shows that accumulation of GS in embryonic NR as a result of precocious induction in vivo (2E), or in vitro (2F) takes place exclusively in Müller glia cells. Thus, the gene-controlled response to the hormonal inducer (i.e., synthesis and accumulation of GS) is a function of a single cell type in the NR, the Müller glia cells.

This finding raised the following question: is the

responsiveness of Müller cells to the hormonal induction of GS independent of other cells in the retina, or does it involve interactions with NR neurons?

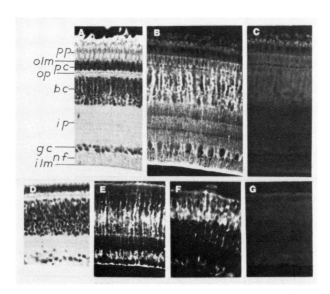

Fig. 2. Cellular localization by immunofluorescence of GS in mature chicken neural retina (A-C) and in cortisol-induced embryonic retina (D-G). (A) Section of adult retina stained with hematoxylin and eosin. pp, photoreceptor cell processes; olm, outer limiting membrane; pc, photoreceptor cell layer; op, outer plexiform layer; bc, bipolar cell layer; ip, inner plexiform layer; gc, ganglion cell layer; nf, nerve fiber layer; ilm, inner limiting membrane. (B) Section similar to that in A, immunostained with anti-GS serum and FITC-goat anti-rabbit IgG. Light areas represent immunofluorescence in Müller fibers. (C) Control section stained with preimmune serum and FITC-goat anti-rabbit IgG. (D) Section of 13-day embryo retina stained with hematoxylin and eosin. (E) 13-day retina induced for GS *in vivo* by injection of 1 mg cortisol into the egg at day 11; immunofluorescent staining for GS present only in Müller cells. (F) Section of 11-day retina after 2 days

in organ culture; immunofluorescence again localized to Müller cells. (G) Section of control noninduced 13-day retina treated for GS localization; no immunofluorescence. (From Linser and Moscona, 1979).

GS Induction and Cell Interactions

Earlier studies (Moscona, 1968; Morris and Moscona, 1970) have raised the possibility that responsiveness to GS induction by cortisol in the embryonic NR requires cellular interactions of the kind dependent on histotypic cell juxtapositions and associations (Moscona, 1972).

If embryonic NR tissue (of an age competent for GS induction) is dissociated by gentle trypsinization into a single cell suspension and the cells are plated at a relatively low density in monolayer cultures under conditions that minimize cell reaggregation (so that the cells persist in a randomly dispersed state), little or no GS induction can be elicited by cortisol (Moscona, 1972; Linser and Moscona, 1979). The cells are not inducible, even though they take up the hormone. Neither enzyme activity assays nor immuno-histochemical studies detected inducible GS accumulation in the dispersed gliocyte-derived cells at any time during cultivation in monolayer.

The situation is different if the dissociated cells are reaggregated by gently rotating the cell suspension in flasks on a gyratory shaker (Moscona, 1972); the cells reassociate into three dimensional spherical aggregates within which they reestablish histotypic associations and reconstruct a retinotypic tissue pattern (Moscona, 1972; Linser and Moscona, 1979), as shown in Fig. 3. In such histotypic cell aggregates GS is inducible; their inducibility increases in direct correlation with progressive restitution of histotypic cell organization. In cell aggregates, as in intact NR tissue, GS is immuno-histochemically detectable only in Müller cells. Furthermore, maturation of GS inducibility in aggregates derived from early embryonic NR cells follows a temporal pattern similar to that of intact NR tissue (Linser and Moscona, 1981).

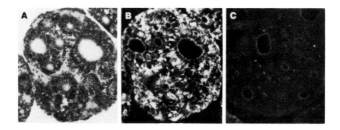

Fig. 3. Cellular localization of GS induced in retina cell aggregates formed by dissociated 6-day retina cells and cultured for 7 days; the final 48 hrs in culture were with cortisol (B) or without (C) cortisol in the medium. (A) Section stained with hematoxylin and eosin showing organization of cells in retinal rosettes and radial laminations. (B) Section similar to A, immunostained for GS; immunofluorescence shows localization of induced GS in Müller glia cells. (C) Section of a noninduced aggregate stained as in B. (From Linser and Moscona, 1979).

Fig. 4 compares GS induction in organ cultures of NR tissue, in cultures of cell aggregates, and in monolayer cultures of NR cells.

These and related other results have demonstrated convincingly that GS cannot be induced by cortisol in separated NR cells, and that retinotypic cell associations of the kind present in intact NR tissue, or reestablished in histotypic aggregates of induction-competent NR cells (of the appropriate embryonic age) are prerequisite for responsiveness of Müller cells to GS induction. This suggests that some kind of contact-dependent interactions between glia cells and retina neurons are essential for enabling the hormonal inducer to elicit GS induction.

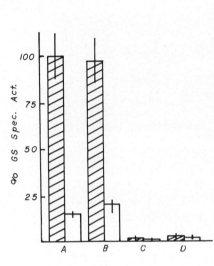

Fig. 4. Specific activity of GS induced in organ cultures (bars A); in cell aggregates (bars B); and in monolayer cell cultures (bars C and D). Hatched bars - plus cortisol; open bars - minus cortisol. A - GS activity in 11-day retinas cultured for 2 days. B - GS activity in aggregates of 6-day retina cells cultured for 5 days and then induced with cortisol. C and D - monolayer cultures of retina cells from 6-day (C) and 10-day (D) embryos, after 7 and 3 days in culture, respectively, following 48 hr treatment with cortisol. (From Linser and Moscona, 1979).

This suggestion is further supported by results obtained with monolayer cultures in which the cells were plated at higher densities to enhance, in a controllable manner, formation of cell-cell associations. At increasingly higher cell densities, varying degrees of cell aggregation occur, ranging from cell assembly into compact spherical aggregates, to formation of contiguous sheets of epithelioid glia-derived cells (Linser and Moscona, 1981) overlayed with single and/or multiple neurons. Fig. 5 shows that when such higher density cultures, maintained in medium with cortisol, are examined immunohistochemically for localization of GS, the epithelioid gliocytes show staining for GS only if they are in close contact with overlying neurons. Therefore, contact dependent interactions between retina gliocytes and retina neurons are in some way involved in maintaining in the gliocytes responsiveness to GS induction by cortisol.

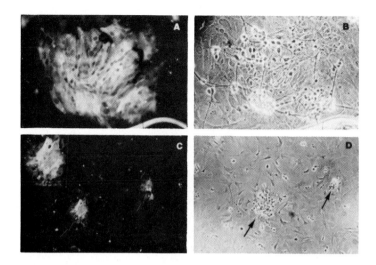

Fig. 5. Immunofluorescent analysis of GS localization in monolayer cultures of neural retina cells from 10-day (A, B) and 18-day (C, D) embryos. Trypsin dissociated suspensions of single cells were plated at 1.3×10^6 cells/cm^2 in Medium 199 with 10% fetal bovine serum and cortisol (0.33 µg/ml). Medium was changed daily; after 7 days, the cultures were examined by indirect immunofluorescence for localization of GS (A, C). Note staining for GS only in epithelioid cells (A, C and inset to C) and only when those cells are in contact with overlying neurons, as evidenced by phase contrast micrographs of the same field (B, D). (From Linser and Moscona, in press).

Cell Interactions and Cortisol Receptor Levels

Recent work in our laboratory has indicated that histotypic contacts among NR cells may be essential for maintaining in the cells normal levels of cytoplasmic receptors for cortisol (Saad et al., 1981). As in other inductions by steroid hormones (Baxter and Tomkins, 1971; Jensen and DeSombre, 1972; O'Malley and Means, 1974), also in the NR cortisol has been shown to bind to specific cytoplasmic receptors; the hormone-receptor complexes are rapidly translocated into the nucleus where they bind to chromatin

(Sarkar and Moscona, 1974, 1975, 1977; Koehler and Moscona, 1975) and elicit gene response that results in GS induction. Fig. 6 shows the levels of cortisol receptors detectable in intact NR tissue, in freshly dissociated NR cells, and in aggregates formed from the dissociated cells, as a function of time in culture. It is apparent that dissociation of NR tissue into single cells results in a rapid decline in the level of detectable receptors; after 8 hrs in monolayer culture such separated cells show only 20-30% of the initial level of receptor activity. However, if the cells are re-aggregated, receptor level in the aggregates rebounds with time, eventually approaching that of the control tissue. On the other hand, if the cells are continuously maintained in monolayer no such recovery of cortisol receptor activity is seen. In the monolayer cultures we were unable to detect any artifactual loss or "leakage" of receptors (Saad et al., 1981).

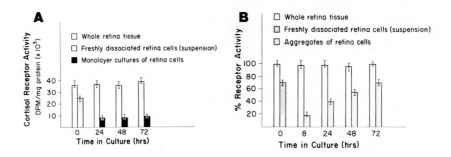

Fig. 6. Cortisol receptor activity in whole 10-day embryo retina tissue, in monolayer cultures of trypsin-dispersed cells, and in cell aggregates. A-Receptor decline in dissociated and monolayer-cultured cells. B-Receptor recovery in cell aggregates. Receptor activity was measured in cytosols by the charcoal assay. (From Saad et al., 1981).

   Our present working hypothesis is that in embryonic NR cells the level of cortisol receptors, or their specific cortisol-binding activity, is susceptible to regulation by cell-cell contacts. Whether the apparent correlation between cell-contact mediated effects on cortisol receptors and on GS inducibility are indicative of a direct cause-and-effect relationship, remains to be explored.

COMMENTS

Earlier work has shown that the corticosteroid-mediated induction of GS in the NR of the chick embryo involves several remarkably specific aspects. Only in the NR can GS be hormonally induced to such a high level and so rapidly; only 11-β-hydroxycorticosteroids (cortisol, corticosterone) are effective as inducers of GS in the NR.

The induction of GS involves specific and differential gene expression. The NR of the chick embryo becomes responsive to the hormonal triggering of this differential gene expression several days before the hormone is effectively available in the embryo. We have demonstrated that GS is induced and localized in the glial compartment of the NR, i.e., in the Müller glia cells. However, Müller cells are not inducible for GS if they are separated from retina neurons. Thus, in addition to the role of the hormonal inducer, induction of GS in Müller cells may depend on yet another "signal" communicated to them by contact with neurons. It is presently not certain if the hormone acts directly or exclusively on Müller cells, or if it elicits a "signal" from neurons which renders Müller cells inducible for GS.

Another heuristic possibility is raised by the finding that cell separation rapidly reduces the level of detectable cytoplasmic receptors for the corticosteroid hormone. This could indicate that specific cell-cell interactions are necessary for maintaining the relevant receptors in a state of activity or at a level required for GS induction. Further exploration of these possibilities could provide valuable information on the mechanism by which gene expression is differentially controlled by corticosteroids in neural systems, and on the role of cell interactions in the biochemical differentiation of glia cells.

ACKNOWLEDGMENT

Research supported by grant HD01253 from the National Institute of Child Health and Human Development and grant

1-733 from the March of Dimes-Birth Defects Foundation.

REFERENCES

Baxter J and Tomkins G (1971). Specific cytoplasmic glucocorticoid hormone receptors in hepatoma tissue culture cells. Proc Natl Acad Sci USA 68:932.

Hamberger AC, Chiang GH, Nylen ES, Scheff SW and Cotman CW (1979). Glutamate as a CNS transmitter I. Evaluation of glucose and glutamate as precursors for the synthesis of preferentially released glutamate. Brain Res 168:513.

Jensen EV and DeSombre ER (1972). Mechanism of action of the female sex hormones. Ann Rev Biochem 41:203.

Kahn AJ (1974). An autoradiographic analysis of the time of appearance of neurons in the developing chick neural retina. Develop Biol 38:30.

Koehler DE, and Moscona AA (1975). Corticosteroid receptors in the neural retina and other tissues of the chick embryo. Arch Biochem Biophys 170:102.

Linser P and Moscona AA (1979). Induction of glutamine synthetase in embryonic neural retina: localization in Müller fibers and dependence on cell interactions. Proc Natl Acad Sci USA 76:6476.

Linser PJ and Moscona AA (1981). Induction of glutamine synthetase in embryonic neural retina: its suppression by the gliatoxic agent α-aminoadipic acid. Develop Brain Res 1:103.

Linser P and Moscona AA. Cell interactions in embryonic neural retina: role in hormonal induction of glutamine synthetase. In Brown IR (ed): "Molecular Approaches to Neurobiology," New York: Academic Press, In Press.

Morris JE and Moscona AA (1970). Induction of glutamine synthetase in embryonic retina: its dependence on cell interactions. Science 167:1736.

Moscona AA (1968). Induction of retinal glutamine synthetase in the embryo and in culture. In "Molecular Basis of Differentiation", Vol 104, Milan: Accademia Nazionale dei Lincei, p 237.

Moscona AA (1972). Induction of glutamine synthetase in embryonic neural retina: a model for the regulation of specific gene expression in embryonic cells. In Monroy A and Tsanev R (eds): "Biochemistry of Cell Differentiation" FEBS Proc 24; London: Academic Press, p 1.

Moscona AA, Linser P, Mayerson P and Moscona M (1980). Regulatory aspects of the induction of glutamine synthetase in embryonic neural retina. In Mora J and Palacios R (eds) "Glutamine: Metabolism, Enzymology and Regulation" New York: Academic Press, p 299.

Moscona AA, Moscona M and Linser P (1980). Induction of glutamine synthetase in embryonic neural retina: role of cell interactions. In Richards RJ and Rajan KT (eds) "Tissue Culture in Medical Research (II)," Oxford and New York: Pergamon Press, p 59.

Moscona AA and Wiens AW (1975). Proflavine as a differential probe of gene expression: inhibition of glutamine synthetase induction in embryonic retina. Develop Biol 44:33.

Moscona M and Moscona AA (1979). The development of inducibility for glutamine synthetase in embryonic neural retina: inhibition by BrdU. Differentiation 13:165.

O'Malley B and Means A (1974). Female steroid hormones and target cell nuclei. Science 183:610.

Piddington R (1970). Steroid control of the normal development of glutamine synthetase in the embryonic chick retina. J Embryol exp Morph 23:729.

Saad AD, Soh BM and Moscona AA (1981). Modulation of cortisol receptors in embryonic retina cells by changes in cell-cell contacts: correlations with induction of glutamine synthetase. Biochem Biophys Res Commun 98:701.

Sarkar PK, Fischman DA, Goldwasser E and Moscona AA (1972). Isolation and characterization of glutamine synthetase from chicken neural retina. J Biol Chem 247:7743.

Sarkar PK and Moscona AA (1973). Glutamine synthetase induction in embryonic neural retina: immunochemical identification of polysomes involved in enzyme synthesis. Proc Natl Acad Sci USA 70:1667.

Sarkar PK and Moscona AA (1974). Binding of receptor-hydrocortisone complexes to isolated nuclei from embryonic neural retina cells. Biochem Biophys Res Commun 57:980.

Sarkar PK and Moscona AA (1975). Nuclear binding of hydrocortisone-receptors in the embryonic chick retina and its relationship to glutamine synthetase induction. Am Zool 15:241.

Sarkar PK and Moscona AA (1977). Glutamine synthetase induction in embryonic neural retina: interactions of receptor-hydrocortisone complexes with cell nuclei. Differentiation 7:75.

Soh BM and Sarkar PK (1978). Control of glutamine synthe-
messenger RNA by hydrocortisone in the embryonic chick
retina. Develop Biol 64:316.

CARBONIC ANHYDRASE-C IN NEURAL RETINA OF THE CHICK EMBRYO: DEVELOPMENTAL CHANGES IN CELLULAR LOCALIZATION

Paul J. Linser and A. A. Moscona

Developmental Biology Laboratory, Cummings Life Science Center, University of Chicago, Chicago, Illinois 60637 U.S.A.

ABSTRACT

Using quantitative immunoelectrophoresis we have established the developmental profile of carbonic anhydrase-C (CA-C) in the neural retina of the chick embryo: CA-C rises sharply early in development and peaks by the 5th day; its level then declines, but resumes by the 10th day a steady increase which continues through hatching and plateaus later.

Immunohistochemical studies showed that CA-C, when first detectable in the eye of 3-day embryos, is localized in the dorsal region of the retina adjacent to the developing lens. During the period of its early sharp rise, CA-C becomes present throughout the retina. Later, this generalized distribution is replaced by progressive confinement of CA-C to certain types of cells. Overall, it increases in Müller glia cells and declines in neurons. An exception are stratified amacrine neurons of the third level which transiently show intense immunostaining for CA-C. In the fetal and mature retina, CA-C is detectable only in Müller glia cells. Its specific localization in these cells - the only kind of glia in the avian retina - coincides with that of glutamine synthetase (GS), a previously established enzyme marker of Müller glia.

Although in the mature avian retina both CA-C and GS are confined to Müller cells, the developmental-temporal profiles and the expression of these two enzymes are independently regulated. GS begins to increase in the avian

retina at a much later embryonic age than CA-C and is
always localized only in Müller cells; it is subject to
inductive regulation by corticosteroids (cortisol). CA-C
first appears very early; it is transiently present also in
retina neurons, before it becomes restricted to Müller
cells; its expression is not influenced by cortisol. In
addition, while the inducibility of GS involves contact
interactions between glia cells and neurons, control of
CA-C levels appears to be independent of such interactions.

INTRODUCTION

Studies on biochemical and morphological differentiation of the nervous system have recently become increasingly concerned with the role of the "supportive" cells, i.e., the neuroglia cells. Our work on differentiation of neural retina (NR) of the chick embryo has drawn our attention to the characteristics of the retinal glia, the Müller cells. A prominent feature of these cells is their compartmentalization of glutamine synthetase (GS). This enzyme plays an important role in various metabolic pathways, including recycling of neurotransmitter metabolites (Hamberger et al., 1979), and is a specific marker for Müller glia cells. Our studies described the developmental profile of this enzyme in the avian neural retina (Moscona, M and Moscona, 1979; Moscona and Degenstein, 1981; see Linser et al., this volume); they also revealed that the synthesis of GS in Müller cells is subject to two kinds of control mechanisms: this enzyme is inducible by adrenal corticosteroids and its inducibility is dependent on contact relationships between Müller cells and neurons (Moscona, 1972; Linser and Moscona, 1979; see Linser et al., this volume).

In seeking further information about the developmental properties of Müller cells, we have recently extended our studies to another enzyme present in the NR, carbonic anhydrase-C (CA-C). In the mammalian brain CA-C is reportedly localized in oligodendroglia (Ghandour et al., 1979), unlike GS, which is confined mostly to astroglia (Norenberg, 1979). From earlier work it was known that in the NR of the chick embryo, carbonic anhydrase enzyme activity is detectable already very early in development (Clark, 1951); in adult retina of several non-avian species this enzymic activity was found to be associated with Müller glia cells (Musser and Rosen, 1973; Bhattacharjee, 1976). Using *immunohistochemical methods, we examined whether CA-C is, in fact, a specific marker for Müller cells and can be a useful probe for studying the differentiation of NR and especially of Müller cells; and whether the development of CA-C is regulated similarly to that of GS.*

DEVELOPMENTAL PROFILE AND CELLULAR LOCALIZATION OF CA-C IN THE NR OF CHICK EMBRYO

Using mono-specific antiserum produced in rabbits against CA-C purified by affinity chromatography from chicken erythrocyte (Osborne and Tashian, 1975), we measured by quantitative immunoelectrophoresis (Norgaard-Pedersen, 1973) the concentration of CA-C in the NR of chick embryos as a function of developmental age. Fig. 1 shows the developmental profile of CA-C in the embryonic NR. For comparison, this figure also shows the developmental profiles in the NR of GS specific activity and of total protein content. It is evident that the developmental program of CA-C is biphasic and very different from that of GS. There is an early sharp increase in CA-C concentration that peaks on the 5th day of embryonic development, followed by a decline; later, CA-C gradually increases again and plateaus after hatching. In the mature chicken NR the concentration of CA-C is approximately 30 µg/mg protein, i.e., 3% of the total protein of the NR.

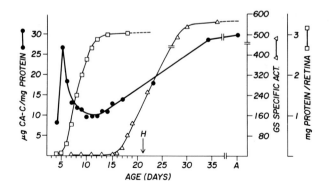

Fig. 1. The developmental profile of CA-C in the NR of chick embryos compared with that of GS and with growth of the retina (mg protein). Results are expressed as μg CA-C/mg total retina protein (●――●). mg protein/retina (□――□) and GS specific activity (△――△ ; Moscona, M and Moscona, 1979). Arrow (H) indicates time of hatching; A-adult stage. (From Linser and Moscona, 1981, in press).

Fig. 2 illustrates the cellular localization of CA-C (Fig. 2A), and of GS (Fig. 2C) in the NR of adult chicken, detected by double-label indirect immunofluorescence. The results demonstrate that, CA-C is localized apparently only in Müller glia cells. As reported earlier (Linser and Moscona, 1979; see Linser et al., this volume) also GS is confined only to Müller glia cells. Therefore, in the mature avian NR, CA-C is a glia cell marker and its cellular localization coincides with that of GS.

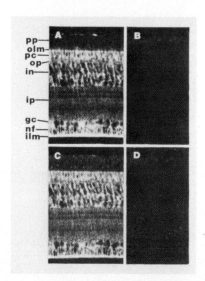

Fig. 2. Simultaneous localization of CA-C and GS in Müller glia cells of adult chicken NR by double-label indirect immunofluorescence. Reaction of tissue sections with rabbit anti-CA-C antiserum, and with mouse anti-GS antiserum was followed by rhodamine-conjugated goat anti-rabbit IgG Fab and FITC conjugated rabbit anti-mouse IgG (see Linser and Moscona, 1981, in press). A. Rhodamine fluorescence shows identical staining to that of FITC in C, demonstrating that both CA-C (A) and GS (C) are localized in Müller glia cells. B and D show fluorescence controls for rhodamine and FITC, respectively. pp-photoreceptor cell processes; olm-outer limiting membrane; pc-photoreceptor cell layer; op-outer plexiform layer; in-inner nuclear layer; ip-inner plexiform layer; gc-ganglion cell layer; nf-nerve fiber layer; ilm-inner limiting membrane. (From Linser and Moscona, 1981, in press).

Next, we examined immunohistochemically, changes in the cellular localization of CA-C in the embryonic NR as a function of development. Fig. 3 shows that in the NR of very early embryos CA-C is present in virtually all the cells; however, with progressing cytodifferentiation CA-C becomes increasingly restricted to Müller cells. CA-C is immunohistochemically detectable already early on the 3rd day. At this age, the NR still consists of an undifferentiated neuroepithelium and is continuous with the pigment epithelium (PE) as a double lamina (Fig. 3A). CA-C first appears at the fold of the lamina adjacent to the lens vesicle, in the upper temporal quadrant of the eye; subsequently, it extends in both the NR and the PE, around the margin of the lens and spreads gradient-like from the radial boundary towards the fundus. On the 4th day, CA-C is present throughout the NR and PE; however, its immunostaining in the NR is still most intense in the upper temporal quadrant (Fig. 3B). It is noteworthy that CA-C is also present in the developing lens. By the 5th day, CA-C is found throughout the NR in practically all the cells, and also in the now pigmented PE (Fig. 3C).

By the 6th day of embryonic development, the first definitive neurons, the ganglion cells, are discernible in the fundus (Meller and Glees, 1965; Kahn, 1974), and they do not immunostain for CA-C (Fig. 3D). As the other types of neurons become morphologically and topographically distinguishable, they also fail to immunostain for CA-C. Indeed, a characteristic feature of neuron differentiation in the chick NR appears to be loss of immuno-detectable CA-C previously present in their undifferentiated progenitors. An interesting exception are certain amacrine neurons that continue to immunostain intensely for a few days after becoming morphologically distinct and positioned. These neurons first appear as a double cell layer (Fig. 3D) bordering the arising inner plexiform layer (IP); as the IP expands, the perikarya of these amacrine neurons become vertically displaced, but their processes remain in contact with the IP and their lateral arborizations form a distinct lamina (Fig. 3E, F). They were identified as stratified amacrine neurons of the 3rd level and displaced amacrine neurons of the 4th level (Cajal, 1973). Eventually, immunostaining of these, as of all other neurons, declines to background level.

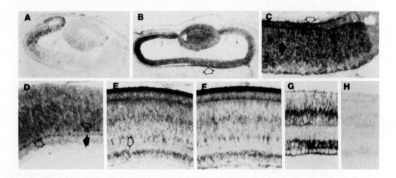

Fig. 3. Localization by immunoperoxidase of CA-C in embryonic retina during development; progressive changes in regional and cellular compartmentalization of the enzyme. A. Section through the eye of a 3-day chick embryo showing onset of detectable immunostaining for CA-C in the dorsal aspect of NR and PE, near the developing lens. B. Section of eye of 4-day embryo showing staining of NR, most intense in the dorsal aspect; also note generalized staining of PE (arrow), and of lens. C. Retina of 5-day embryo showing generalized staining for CA-C in most cells of the NR (closed arrow) and PE (open arrow). D. Section of 8-day NR showing reduction or absence of staining in ganglion cells (closed arrow). Note staining in double row of prospective amacrine neurons (open arrow) that border the inner plexiform layer. E. 13-day embryonic NR sectioned in a region close to the ciliary margin. Note immunostaining for CA-C in amacrine neurons; their lateral arborizations form a continuous line running horizontally through the IP (arrow). Also note intensified immunostaining of Müller cell perikarya in the central region of the IN. F. Same NR as in E, but sectioned closer to the fundus. Note still intense staining of amacrine neurons and their arborizations. Elsewhere staining is reduced, except in the perikarya of Müller cells. G. 16-day NR showing staining for CA-C confined to Müller cells. H. Control section for G, stained with pre-absorbed CA-C antiserum. (From Linser and Moscona, 1981, in press).

In contrast to neurons, immunostaining of Müller glia cells increases in intensity with development and spreads from their perikarya into the radial processes (Fig. 3E, F, G). In the fundus (developmentally the most advanced region of the NR) CA-C is confined predominantly to Müller cells already by the 11th day of embryonic development. However, even in 13 day NR there still are regional differences in CA-C localization; at the ciliary margin (developmentally the least advanced region of the NR) CA-C is still detectable in most of the cells, while in the fundus it is localized mostly in Müller cells. By the 16th day, CA-C is confined to Müller cells virtually throughout the NR. After hatching and later, it is detectable only in Müller cells.

REGULATION OF CA-C AND GS

Cortisol can precociously induce GS in Müller cells in the NR of the chick and quail embryo by differentially affecting gene expression (Moscona, 1972; Moscona and Degenstein, 1981). We examined if the level of CA-C or its cellular localization also could be influenced by this hormone. 11-day chick embryos were injected with cortisol; two days later the NR were isolated and the levels of CA-C and GS were determined. As expected, GS was induced and its specific activity level was 25-35-fold higher than in untreated 13-day controls. In contrast, the level of CA-C was essentially the same as in the controls. Immunohistochemical examination showed that the cellular localization of CA-C also was unaffected by cortisol and was typical for 13-day NR: in the fundus CA-C was present mostly in Müller cells, while in the peripheral regions it was present also in amacrine neurons. Fig. 4 illustrates simultaneous detection of CA-C and GS by double-label immunofluorescence in cortisol treated retina; it shows CA-C staining in both Müller cells and amacrine neurons, and GS staining confined only to Müller cells.

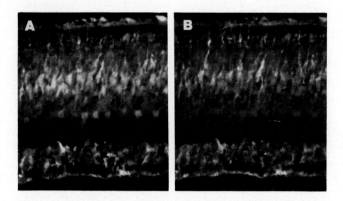

Fig. 4. CA-C and GS localization by double-label immunofluorescence (see legend to Fig. 2) in 13-day embryonic NR pretreated in ovo with cortisol (1 mg/egg) to induce GS. A. Rhodamine fluorescence for CA-C, showing presence of CA-C predominantly in Müller cells, but also in amacrine neurons (at the upper border of the inner plexiform layer). B. FITC fluorescence showing localization of induced GS only in Müller cells.

Cortisol-mediated inducibility of GS in Müller cells and the persistence of the induced high level of GS require contact-dependent interactions of Müller cells with neurons (Linser and Moscona, 1979; see Linser et al., this volume). Pre-induced GS declines rapidly if NR tissue is separated into single cells and these are maintained monodispersed in monolayer cultures; this decline takes place regardless of cortisol presence in the culture medium (Linser and Moscona, 1979). We examined if the control of CA-C also was dependent on normal cell contacts. 13-day NR were pretreated with cortisol for 2 days to induce GS; then the fundal region was isolated and was dissociated into single cells. The cells were plated as a monolayer in cortisol-containing medium and were immunohistochemically examined at intervals for CA-C and GS (Linser and Moscona, 1981, in press).

After 3 hrs in monolayer culture, gliocytes that immunostained for CA-C simultaneously immunostained also for GS. Thereafter, stainability for GS declined rapidly and was at basal level before 48 hrs. In contrast, CA-C was detectable in most of the gliocytes even after 6 days in culture (Fig. 5). Therefore, at the developmental stages and conditions examined, control of CA-C levels in the NR is apparently independent of the kind of cell interactions implicated in the regulation of GS levels in the same cells.

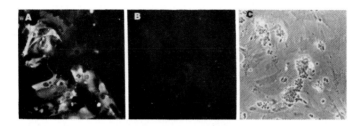

Fig. 5. Double-label immunohistochemical examination of CA-C and GS in monolayer cultures derived from embryonic NR cells. The fundal region of cortisol-pretreated (induced for GS) day-13 embryonic NR was dissociated into single cells. The cells were plated on glass cover-slips and maintained in medium with cortisol. After 6 days in culture, CA-C is present in most of the glia-derived epithelioid cells (A), but GS is not detectable (B). C. Phase contrast micrograph of the same field as shown in A and B. (From Linser and Moscona, 1981, in press).

DISCUSSION

We report here that, in late fetal and adult avian (chicken) NR, CA-C is a characteristic marker for Müller glia cells. This conclusion, based on immunological localization of this enzyme, agrees with earlier findings concerning cellular localization of carbonic anhydrase enzyme activity in mature retinas of other species (Musser and Rosen, 1973). In the early embryonic retina a different situation was found: CA-C is, at first, present in all the undifferentiated retinocytes and becomes gradually lost from emerging neurons and confined to differentiating glia cells.

It was reported that in the neural retina of fetal and neonatal mice carbonic anhydrase enzyme activity is always found only in Müller cells (Bhattacharjee, 1976). However, also in mice, the author noted a transient presence of enzyme activity in certain round cells proximal to the inner plexiform layer; these cells could correspond to the amacrine neurons described by us above, rather than represent migrating Müller cells, as suggested by the author. The apparent differences between the development of CA-C in avian and in mouse retina call for detailed re-investigation by immunological methods of mammaliam embryonic retinas.

Even though in the mature retina both CA-C and GS are characteristic of Müller cells, the developmental programs of these two proteins are evidently disparately regulated during retina differentiation. Unlike CA-C, GS begins to rise sharply only late in development and is **always** confined only to Müller glia cells; whereas GS is subject to induction by corticosteroids, control of CA-C levels and its cellular localization appear to be unaffected by this hormone. Inducibility and expression of GS in Müller cells requires glia-neuron contact-interactions, whereas CA-C can continue to be expressed in isolated retina gliocytes.

Since CA-C and GS represent approximately 3 and 0.5%, respectively, of the total protein content of adult NR, their restriction to Müller cells means that their concentration in these cells is very high. Their high level within a single cell type, and the differences in their developmental regulation make CA-C and GS an attractive "marker team" for comparatively investigating mechanisms controlling differential gene expression.

It is worthy of note that in the central nervous system CA-C and GS have been shown to be compartmentalized in different classes of glial cells: CA-C is found primarily in oligodendroglia (Ghandour et al., 1979), whereas GS is found primarily in astroglia (Norenberg, 1979). Since mature avian Müller cells contain both enzymes, functions that in the CNS are relegated to disparate types of glia, are evidently combined in the only type of gliocytes present in the avian NR.

The different functions of these two enzymes may be reflected in their temporally different programs during the development of the NR: GS is involved in the turnover of aminoacid neurotransmitter molecules (Hamberger et al.,1979) and hence is normally "turned on" only late in development when neurons become functional. CA-C controls the exchange of CO_2, ionic balance and fluid secretion (Giacobini, 1962; Maren, 1967; 1976) and, therefore, may be required already early in development for general homeostasis of the NR and the eye. In fact, we noticed that in the avian embryo the eye was apparently unique among other organs in its very early expression of a very high level of CA-C; this raises the possibility that CA-C may play some more general role in the overall development of the embryonic eye.

ACKNOWLEDGMENT

Research supported by grant HD01253 from the National Institute of Child Health and Human Development and grant 1-733 from the March of Dimes-Birth Defects Foundation.

REFERENCES

Bhattacharjee J (1976). Developmental changes of carbonic anhydrase in the retina of the mouse: a histochemical study. Histochem J 8:63.

Cajal SR (1973). In "The Vertebrate Retina", Rodieck RW (ed) San Francisco: Freeman, p 838.

Clark AM (1951). Carbonic anhydrase activity during embryonic development. J Exptl Biol 28:332.

Ghandour MS, Langley OK, Vincendon G and Gombos G (1979). Double labeling immunohistochemical technique provides evidence of the specificity of glial cell markers. J Histochem Cytochem 27:1634.

Giacobini E (1962). A cytochemical study of the localization of carbonic anhydrase in the nervous system. J Neurochem 9:169.

Hamberger AC, Chiang GH, Nylén ES, Scheff SW and Cotman CW (1979). Glutamate as a CNS transmitter. I. Evaluation of glucose and glutamine as precursor for the synthesis of preferentially released glutamate. Brain Res 168:513.

Kahn AJ (1974). An autoradiographic analysis of the time of appearance of neurons in the developing chick neural retina. Develop Biol 38:30.

Linser P and Moscona AA (1979). Induction of glutamine synthetase in embryonic neural retina: localization in Müller fibers and dependence on cell interactions. Proc Natl Acad Sci USA 76:6476.

Linser P and Moscona AA (1981). Carbonic anhydrase-C in the neural retina: transition from generalized to glia-specific cell localization during embryonic development. Proc Natl Acad Sci USA, in press.

Linser P, Saad AD, Soh BM and Moscona AA. Cell contact-dependent regulation of hormonal induction of glutamine synthetase in embryonic neural retina. This volume.

Maren TH (1967). Carbonic anhydrase: chemistry, physiology and inhibition. Physiol Rev 47:595.

Maren TH (1976). The rates of movement of Na^+, Cl^- and HCO_3^- from plasma to posterior chamber: effect of acetazolamide and relation to the treatment of glaucoma. Investig Ophthalm 15:356.

Meller K and Glees P (1965). The differentiation of neuroglia-Müller cells in the retina of chick. Zeitschrift für Zellforschung 66:321.

Moscona AA (1972). Induction of glutamine synthetase in embryonic neural retina: a model for the regulation of specific gene expression in embryonic cells. In Monroy A and Tsanev R (eds): "Biochemistry of Cell Differentiation" FEBS Proc 24; London: Academic Press, p 1.

Moscona AA and Degenstein Linda (1981). Normal development and precocious induction of glutamine synthetase in the neural retina of the quail embryo. Develop Neurosci 4:211.

Moscona M and Moscona AA (1979). The development of inducibility for glutamine synthetase in embryonic neural retina: inhibition by BrdU. Differentiation 13:165.

Musser GL and Rosen S (1973). Localization of carbonic anhydrase activity in the vertebrate retina. Exptl Eye Res 15:105.

Norenberg MD (1979). The distribution of glutamine synthetase in the rat central nervous system. J Histochem Cytochem 27:756.

Norgaard-Pedersen B (1973). In Axelsen NH, Kroll J and Weeke B (eds): "A Manual of Quantitative Immunoelectrophoresis"; Oxford: Blackwell Scientific Publications, p 125.

Osborne WRA and Tashian RE (1975). An improved method for the purification of carbonic anhydrase isozymes by affinity chromatography. Analys Biochem 64:297.

PROGRESS IN THE PURIFICATION OF A FACTOR INVOLVED IN THE NEUROTRANSMITTER CHOICE MADE BY CULTURED SYMPATHETIC NEURONS.

Michel J. Weber and Agathe Le Van Thai

LABORATOIRE DE PHARMACOLOGIE ET DE TOXICOLOGIE FONDAMENTALES
205 route de Narbonne 31400 TOULOUSE, France.

During development, neurons from the autonomic nervous system are plastic with respect to their neurotransmitter functions (Patterson, 1978). The developmental fate of neurons from the avian neural crest can be modified from adrenergic to cholinergic or *vice-versa* when their environment is experimentally altered by a suitable transplantation (Le Douarin, 1980). Evidence has also been presented showing that during post-natal development, sympathetic neurons which innervate rat sweat glands, first synthetize catecholamines (CA) and then switch to acetylcholine (Ach)(Landis and Keefe, 1980). In rat and mouse embryos, cells which exhibit transiently a catecholaminergic phenotype have been found in various structures (gut, kidneys, spinal cord) where no CA producing neurons exist in the adult (Cochard et al, 1978, Teitelman et al, 1978, 1981). The developmental fate of these transient catecholaminergic cells is not clear : the cells are probably not eliminated (Jonakait et al, 1979) and may possibly engage in the production of another neurotransmitter and/or neuropeptide. This hypothesis is made more attractive by the finding that precursors of glucagon-containing cells in mouse pancreas also express transiently a catecholaminergic phenotype (see Teitelman et al, 1981).

In order to study at the molecular level the mechanisms by which immature neurons are influenced towards cholinergic differentiation by their environment, primary cultures of dissociated neurons offer the advantage of experimental control of their cellular and fluid environment. When grown in the virtual absence of non-neuronal cells, sympathetic neurons from new-born rat superior cervical ganglia can acquire many traits characteristic of mature adrenergic neurons. (Mains and Patterson, 1973). However, in the presence of appropriate non-neuronal cells, or in the presence of a culture medium conditioned by these non-neuronal cells, these same neurons lose the ability to synthetize and store CA, and produce instead large amounts of Ach (Patterson and Chun, 1977 a,b). Such cholinergic sympathetic neurons can form functional synapses with appropriate targets (for a review, see Patterson, 1978). The electrophysiological examination of the neurotransmitter status of sympathetic neurons grown in isolation on cardiac myocytes has revealed that many neurons can secrete both Ach and CA on target cells. Transitions from purely adrenergic to dual-function and from dual-function to purely cholinergic have been observed during time in culture, suggesting that dual-function neurons may be in a transitory state during a switch from adrenergic to cholinergic metabolism (Potter et al, 1980). This switch is accompanied by changes in the cytochemistry of synaptic vesicles (Landis, 1976, 1980 ; Johnson et al, 1976 ; Potter et al, 1980 ; Johnson et al, 1980).

Conditioned medium (CM) from non-neuronal cells is not the only signal which can influence neurotransmitter choice in cultured sympathetic neurons. Human placental serum and/or embryonic extracts favor the differentiation of cholinergic or dual-function neurons (Johnson et al, 1976 ; Higgins et al, 1981), whereas elevated extracellular K^+ strongly fosters adrenergic differentiation as a consequence of an increased influx of Ca^{++} (Walicke et al, 1977 ; Walicke and Patterson, 1981). Similarly, chronic electrical stimulation of the cultures favors adrenergic phenotype and can counteract the cholinergic influence of CM, suggesting that during development the onset of electrical activity elicited by growing preganglionic fibers may play a key role in the neurotransmitter choice made by sympathetic neurons (Walicke et al, 1977). On the other hand, neurons grown on a monolayer of p.formaldehyde-,but not ethanol-fixed heart cells display a significant cholinergic induction, suggesting that surface associated macromolecules influence the neurotransmitter status of nerve cells (Hawrot, 1980). The relationships between membrane bound factor(s) and the diffusible factor from CM are not clear.

CM seems to exert a rather pleiotropic effect on cultured sympathetic neurons. As already mentionned, CM decreases CA synthesis and accumulation, and increases Ach production.

This last effect is at least in part attributable to a 100-1000 fold increase in the *in vitro* activity of choline acetyltransferase (Patterson and Chun, 1977 a, b). Neurons grown in CM contain more small agranular synaptic vesicles of the cholinergic type and less dense core vesicles containing CA than when they are grown in the absence of CM (Landis, 1980). In addition, CM modifies the pattern of the proteins spontaneously secreted by the neurons (Sweadner, 1981) and causes changes on the cell surface, revealed by the binding of lectins (Schwab and Landis, 1981) and monoclonal antibodies (Chun et al, 1980). Among these modifications, the stimulation of the Ach/CA ratio of the neurotransmitters synthetized and accumulated by 2-3 weeks old cultures can be used as a quantitative index of the biological activity of CM (Patterson and Chun, 1977) and has been used to monitor a partial purification of the active factor in CM from C_6 glioma cells or primary cultures of heart cells (Weber, 1981).

The biological assay for this cholinergic factor was performed in the following manner: neurons from new-born rat superior cervical ganglia were first cultured for a 10 day period in a culture medium which favors adrenergic differentiation (Hawrot and Patterson, 1979). The proliferation of ganglionic non-neuronal cells, which foster cholinergic phenotype, was prevented by treating the cultures between days 2-8 with cytosine-arabinoside. Between days 10-20, neurons were given, every second day, culture medium supplemented with material purified from CM. At day 20, the neurons were incubated for 4-6 hrs with [^{3}H]-tyrosine and [^{3}H]-choline. [^{3}H]-CA and [^{3}H]-Ach synthetized and stored by the neurons were then separated by high voltage paper electrophoresis, and counted by liquid scintillation. The Ach/CA ratio of cultures grown in the presence of purified material was always compared to the ratio of sister cultures grown in the absence of purified material. The starting material for the purification was CM from C_6 glioma cells or confluent rat heart cells grown in the presence of rat or fetal calf serum. Another source of cholinergic factor is the serum-free CM from heart cells described by Fukada (1980).

CM was first fractionated by back-washes of ammonium sulfate precipitates. Solid salt was added to reach saturation, and the precipitate was sequentially resolubilized in decreasing concentrations of ammonium sulfate. Activity was found only in the fraction which resolubilized when salt concentration was decreased from 70 to 60 % of saturation. This fraction contained 60 % of CM protein (fig. 1).

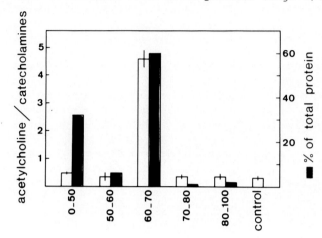

Fig. 1 : Ammonium sulfate precipitation of CM. CM from heart cells grown in L15-CO$_2$, 10 % FCS was supplemented with 2.5 mM Na$_2$ EGTA, 0.4 mM phenylmethylsulfonylfluoride and ammonium sulfate to saturation. CM was then centrifuged and the pellet resuspended in 80 % saturation in salt. After centrifugation, the supernatant (fraction 80-100) was reprecipitated with saturated ammonium sulfate, extensively dialysed and mixed with culture medium. The pellet was resuspended in 70 % saturation in salt and the sequence of operations repeated. The fraction 0-50 is the material still insoluble at 50 % saturation. Control cultures were grown in the absence of purified material.

The biological activity of CM was quantitatively recovered in a more rapid manner in the material (P100) which precipitated when ammonium sulfate concentration was increased from 60 to 100 % of saturation (Weber, 1981). As shown on Fig. 2, the P100 fraction increased, in a dose-dependent manner, the synthesis and accumulation of [^{3}H] Ach by the neurons, with half maximal effect around 0.8 mg protein/ml. A simultaneous decrease in [^{3}H] CA production was observed to 50 % of the control value, with maximal effect at 2 mg/ml. The amplitude of this decrease varied between experiments and was more marked when neurons were given purified material from day 2 on, instead of day 10 (see Table 1 B and Patterson and Chun, 1977 a). When the P100 fraction was submitted to gel filtration on Sephadex G-150, the biological activity was eluted as a single, symmetrical peak of apparent Mr 40-45,000 (Weber, 1981).

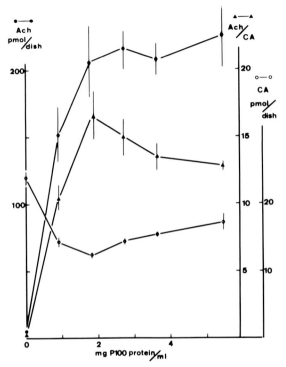

Fig. 2 : Dose-response curves for the P100 fraction. The P100 fraction was diluted in culture medium to the indicated protein concentration and given the neurons between days 10 and 20. Neurotransmitter assay was performed on day 21. Ach : acetylcholine - CA : catecholamines.

The biological activity of the P100 fraction was not retained by a DEAE-cellulose column in 5 mM phosphate buffer pH 7.0, which bound about 99.5 % of the input protein. This step thus resulted in a 200-fold purification with 75-100 % recovery of activity. This material was then applied on a CM-cellulose column in the same buffer, and the biological activity was eluted with 0.2 M NaCl, 5 mM Na phosphate pH 7.0. This fraction was further purified on a Sephadex G-100 column. The peak fractions caused a 20-fold increase in the Ach/CA ratio at 1 µg protein/ml. This succession of fractionation steps resulted in a 1500-fold purification over the P100 fraction with an overall recovery of 10-20 %. The pufied material still appeared largely heterogeneous (Weber, 1981).

We recently characterized the interaction of the cholinergic factor with hydroxyapatite. Material unretained by a DEAE-cellulose column in 5 mM Na phosphate pH 7.0 was applied on a hydroxyapatite column in the same buffer. The unadsorbed fraction (6 % of input protein) was inactive. Biological activity was eluted with 0.25 M NaCl, 5 mM phosphate (33 % of input protein)(Table 1 A).

	Ach pmol/dish	CA pmol/dish	Ach/CA
A			
Control	4.6 ± 1.2	5.2 ± 0.4	0.9 ± 0.2
Unretained fraction (2.5 µg/ml)	5.0 ± 1.8	7.6 ± 0.2	0.6 ± 0.2
Fraction eluted with 0.25 M NaCl (10.7 µg/ml)	20 ± 3	2.8 ± 0.6	7.2 ± 0.6
B			
Control	31.3 ± 1.7	59.4 ± 3.1	0.53 ± 0.02
Unretained fraction in 0.25 M NaCl (4 µg/ml)	251 ± 3	7.4 ± 0.2	33.9 ± 0.4

Table 1 : Interaction of cholinergic factor with hydroxyapatite.
A - The fraction unretained by a DEAE-cellulose column in 5 mM Na phosphate buffer pH 7.0 was applied on a hydroxyapatite column in the same buffer, which was then eluted with 0.25 M NaCl, 5 mM Na phosphate buffer pH 7.0. The unretained and eluted fractions were diluted in culture medium to the indicated protein concentration and given to neurons between days 6-16.

B - Conditioned medium was purified by ammonium sulfate precipitation and chromatography on DEAE- and CM-cellulose columns as described (Weber, 1981). Material eluted from the latter column with 0.25 M NaCl, 5 mM phosphate buffer pH 7.0 was applied on a hydroxyapatite column in the same buffer. The unretained fraction was diluted in culture medium and given to neurons between days 2-15.

In one other experiment, material purified through the CM-cellulose step (Weber, 1981) was applied on a hydroxyapatite column in 0.25 M NaCl, 5 mM phosphate pH 7. As expected, the activity was not retained by the gel in this buffer (Table 1 B). These four purification steps resulted in a 2500-fold purification over CM, and proved to be so reproducible that the entire procedure could be run without an activity test, a condition made necessary by the length of the biological assay. Interestingly, cholinergic factor purified through the hydroxyapatite step both increased [^{3}H] Ach production and decreases [^{3}H] CA production, suggesting that both activities are carried by the same macromolecule.

Experiments have been carried out to characterize the biochemical nature of the factor purified through the DEAE- or CM-cellulose step (Weber 1981). The activity in the DEAE-fraction was resistant to treatments with reducing agents (0.01 M dithioerythritol, 0.2 M 2-mercaptoethanol) as well as with denaturing agents (6 M guanidine-HCl, 8 M urea). The activity was also left intact by a treatment with 8 M urea in the presence of dithioerythritol followed by extensis dialysis. Thermal denaturation has so far given erratic results. The experiment of Table 2 demonstrates that the activity was stable at pH 3.5 and 10.5. Preliminary experiments suggested that the activity was even stable after dialysis against formiate-acetate buffer at pH 1.9. Material purified through the CM-cellulose step contained large amounts of trypsin- and chymotrypsin- inhibiting activities which prevented the hydrolysis by these enzymes.

	Ach pmol/dish	CA pmol/dish	Ach/CA
Control	2.8 ± 0.4	46.5 ± 8.8	0.06 ± 0.00
DEAE-fraction treated at pH 7.0	208 ± 28	8.5 ± 2.5	26 ± 11
pH 3.5	254 ± 28	8.8 ± 2.5	30 ± 5
pH 10.5	272 ± 10	10.3 ± 2.7	27 ± 6

Table 2 : pH-stability of the cholinergic factor.
Material purified through the DEAE-cellulose step (95 µg protein/ml) was supplemented with 0.1 M acetic acid (final pH 3.5) or with 0.15 M NH$_4$OH (final pH 10.5). An aliquot was kept in 5 mM Na phosphate pH 7.0. The fractions were incubated for 2 hrs at 4°C, extensively dialyzed against 5 mM phosphate buffer pH 7.0 and mixed to culture medium (20 µg protein/ml). These fractions were given to neurons between days 10 and 20.

The biological activity was partially decreased by an extensive treatment with immobilized Pronase (Fig. 3A). On the other hand, the activity was completely destroyed by a brief treatment with 1 mM Na periodate (Weber, 1981). This suggested that sialic acid residues may be important for the biological activity of cholinergic factor. However, the activity of the CM-cellulose fraction was unaffected by a treatment with immobilized Neuraminidase (Fig. 3B).

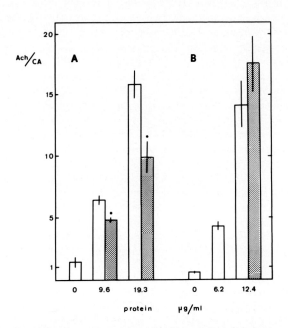

Fig. 3 : Treatment of purified cholinergic factor with Pronase and Neuraminidase.

A - Protein (340 µg/ml) eluted from a CM-cellulose column was incubated for 14 hrs at 25° with 400 BAEE units/ml of immobilized Pronase in 150 mM NaCl, 5 mM Na phosphate pH 7. The reaction was stopped by ultrafiltration. The fraction was diluted in culture medium to the indicated protein concentration and given to neurons between days 10 and 20. Control fraction was treated the same way except for the addition of enzyme. Clear bars = control fraction. Shaded bars = enzyme-treated fractions. (●, $p < 0.05$, t test)

B- Protein (340 µg/ml) in 150 mM NaCl, 60 mM phosphate buffer pH 5.7 was supplemented with 160 mU/ml of immobilized Neuraminidase from *C. perfringens* and treated as in A.

Although a significant purification of the cholinergic factor has been achieved, its biochemical nature thus remains rather elusive. The factor resisted treatments known to modify the structure of proteins, but one can not exclude the possibility that renaturation occured during dialysis. The sensitivity to low concentrations of periodate can not be taken as evidence that carbohydrate are involved in the activity of the cholinergic factor. For exemple, periodate can bind to the phosphate-binding sites of many enzymes, leading to their inactivation by oxidation of vicinal thiols (Rippa et al, 1981). An extensive characterization, including the study of the sensitivity to heat and proteases, thus remains to be done with a more purified material.

Acknowledgments :

I thank Dr. P.H. Patterson, under whose sponsorship this work was initiated, for advices and support, and Dr. R.C. Johnson for help with the manuscript. This work was supported by funds from the Centre National de la Recherche Scientifique, the Délégation Générale de la Recherche Scientifique et Technique and the Institut National de la Santé et de la Recherche Médicale.

REFERENCES :

Cochard P, Goldstein M, Black IB (1978) Ontogenic appearance and disappearance of tyrosine hydroxylase and catecholamines in the rat embryo. Proc. Natl. Acad. Sci. USA 75 : 2986

Chun LLY, Patterson PH, Cantor H (1980) Preliminary studies on the use of monoclonal antibodies as probes for sympathetic development J. exp. Biol. 89 : 73

Fukada K (1980) Hormonal control of neurotransmitter choice in sympathetic neurons cultures.Nature 287 : 553

Hawrot E (1980) Cultured sympathetic neurons : effects of cell-derived and synthetic substrata on survival and development.Dev. Biol. 74 : 136

Hawrot E, Patterson PH (1979) Long-term culture of dissociated sympathetic neurons. Methods Enzymol. 58 : 574

Higgins D, Iacovitti L, Joh TH, Burton H (1981) The immunocytochemical localization of tyrosine hydroxylase within rat sympathetic neurons that release acetylcholine in culture.J. Neurosci. 1 : 126

Jonakait GM, Wolf J, Cochard P, Goldstein M, Black IB (1979) Selective loss of noradrenergic phenotypic characters in neuroblasts of the rat embryo. Proc. Natl. Acad. Sci. USA 76 :4683

Johnson M, Ross D, Meyers M, Rees R, Bunge R, Wakshull E, Burton H (1976) Synaptic vesicle cytochemistry changes when cultured sympathetic neurons develop cholinergic interactions Nature 262 : 308

Johnson MI, Ross CD, Meyers M, Spitznagel EL, Bunge RP (1980) Morphological and biochemical studies on the developmennt of cholinergic properties in cultured sympathetic neurons I. Correlative changes in choline acetyltransferase and synaptic vesicle cytochemistry. J Cell Biol 84 : 680

Landis SC (1976) Rat sympathetic neurons and cardiac myocytes developing in microcultures : correlation of the fine structure of endings with neurotransmitter function in single neurons. Proc. Natl. Acad. Sci. USA 73 : 4220

Landis SC (1980) Developmental changes in the neurotransmitter properties of dissociated sympathetic neurons : a cytochemical study of the effects of medium. Dev. Biol. 77 : 349

Landis SC, Keefe D (1980) Development of cholinergic sympathetic innervation of eccrine sweat glands, in rat footpad. Soc. Neurosci. Symp. 10 Abst. 131.20

Le Douarin NM (1980) The ontogeny of the neural crest in avian embryo chimaeras. Nature 286 : 663

Mains RE, Patterson PH (1973) Primary cultures of dissociated sympathetic neurons. J. Cell Biol. 59 : 329

Patterson PH (1978) Environmental determination of autonomic neurotransmitter functions. Ann. Rev. Neurosci. 1 : 1

Patterson PH, Chun LLY (1977a) The induction of acetylcholine synthesis in primary cultures of dissociated rat sympathetic neurons I. Effects of conditioned medium. Dev. Biol. 56 : 263

Patterson PH, Chun LLY (1977b) The induction of acetylcholine synthesis in primary cultures of dissociated rat sympathetic neurons II Developmental aspects Dev. Biol. 60 : 473

Potter DD, Landis SC, Furshpan EJ (1980) Dual function during development of rat sympathetic neurons in culture J. exp. Biol. 89 : 57

Rippa M, Bellini T, Signorini M, Dallocchio F (1981) Evidence for multiple pairs of vicinal thiols in some proteins J. Biol. Chem. 256 :451

Schwab M, Landis S (1981) Membrane properties of cultured rat sympathetic neurons : morphological studies of adrenergic and cholinergic differentiation Dev. Biol. 84 :67

Sweadner KJ (1981) Environmentally regulated expression of soluble extracellular proteins of sympathetic neurons. J. Biol. Chem. 256 : 4063

Teitelman G, Joh TH, Reis DJ (1978) Transient expression of a noradrenergic phenotype in cells of the rat embryonic gut. Brain Res. 158 : 229

Teitelman G, Gershon MD, Rothman TP, Joh TH, Reis DJ (1981) Proliferation and distribution of cells that transiently express a catecholaminergic phenotype during development in mice and rats. Dev. Biol. 86 : 348

Walicke PA, Campenot RB, Patterson PH (1977) Determination of transmitter function by neuronal activity. Proc. Natl. Acad. Sci. USA 74 : 5767

Walicke PA, Patterson PH (1981) On the role of Ca^{++} in the neurotransmitter choice made by cultured sympathetic neurons. J. Neurosci. 1 : 343

Weber MJ (1981) A diffusible factor responsible for the determination of cholinergic functions in cultured sympathetic neurons. Partial purification and characterization J. Biol. Chem. 256 : 3447

CLONAL ANALYSIS OF THE QUAIL NEURAL CREST: PLURIPOTENCY, AUTONOMOUS DIFFERENTIATION, AND MODULATION OF DIFFERENTIATION BY EXOGENOUS FACTORS

Maya Sieber-Blum

Department of Cell Biology and Anatomy, The Johns Hopkins University School of Medicine, Baltimore, Maryland 21205 U.S.A.

INTRODUCTION

During embryonic development, cells proliferate, undergo morphogenetic movements, interact with other cells, with extracellular matrices and soluble factors, and become progressively restricted in their differentiative potentialities. The complex mechanisms which regulate cell differentiation are not yet understood. The neural crest of the vertebrate embryo is an attractive experimental system for studying differentiation at a cellular and molecular level, since crest cells differentiate into several different cell types, some of which are also phenotypically expressed in vitro. The neural crest originates as a thickening of the neural folds. The crest cells leave the developing neural tube and migrate into the embryo. There they localize in different areas and give rise to a wide range of cell types and tissues. The cranial neural crest forms visceral cartilage, head mesenchyme, odontoblasts, neurons, nerve supportive cells and melanocytes. At lower axial levels, the crest cells contribute to the meninges and vagal root ganglia; they give rise to sympathetic and parasympathetic ganglia, adrenal medulla, melanocytes, oligodendroglia, Schwann sheath cells and melanocytes (e.g., see review by Weston, 1970).

In tissue culture, quail neural crest cells of the trunk region differentiate into melanocytes and neuronal cells (Cohen, 1977). In order to gain insight into the underlying regulatory mechanisms, we developed a method for the cloning in vitro of neural crest cells at the onset of their migration and analyzed their properties and their response to exogenous regulators. The mixed clones containing neurons and melanocytes that we regularly observed under these conditions demonstrate a) that there are pluripotent neural crest cells in the premigratory quail crest, and b) that they can differentiate into these two cell types in the absence of noncrest cells (Sieber-Blum, Cohen, 1980). Although this suggests that commitment to the neuronal and melanogenic lineages are to a large extent preprogrammed, the expression of the differentiated phenotypes may still be influenced by exogenous regulators such as fibronectin or heart cell conditioned medium (Sieber-Blum, Kahn, Cohen, 1979; Sieber-Blum, Sieber, Yamada, 1981).

Additional insight into the mechanisms controlling cell differentiation may be obtained by perturbing cell development by well defined nonphysiological substances. We have recently shown that tumor-promoting phorbol esters induce quail neural crest cells to differentiate preferentially along the melanocytic pathway and inhibit expression of the adrenergic phenotype (Sieber-Blum, Sieber, 1981).

EXPERIMENTAL SYSTEM

For primary cultures (Cohen, Konigsberg, 1975), the last six segments of stage 14-15 embryos are excised and trypsinized. The released neural tubes are then placed in collagen coated culture dishes. Three to four hours after explantation, the neural tubes are removed. The crest cells remain on the substratum, proliferate rapidly, and differentiate into neuronal cells and melanocytes within four days.

For secondary cultures (Sieber-Blum, Cohen, 1980), crest cells in primary culture are detached by trypsinization approximately 18 hours after explantation of the neural tubes, resuspended into single cells, and replated at a concentration of 50-100 cells per 35 mm plate. The effect of adding different compounds to the culture medium on cell differentiation is evaluated after 12-14 days in secondary culture by enumerating the different types of colonies. For cloning, single cells are pipetted into small wells. Repeated screening of the wells during the first two hours after plating ensures that only true clones are evaluated.

Clonal Analysis of Neural Crest Cells / 487

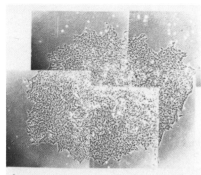

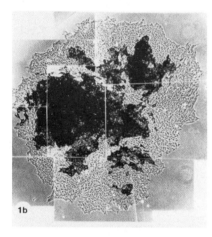

We have shown that single neural crest cells grown in isolation give rise to three different types of clones: unpigmented, pigmented, and mixed clones (Sieber-Blum, Cohen, 1980; Fig. 1). Whereas the pigmented clones consist of melanocytes only, up to 20% of the unpigmented and mixed clones contain adrenergic cells (Fig. 2). Since some mixed clones contain adrenergic neurons and melanocytes, we can conclude 1) that pluripotent cells are present in the early neural crest and 2) that differentiation into neurons and pigment cells can occur in the absence of non-crest cells.

These findings are at variance with earlier work by Cohen (1972), Newgreen, Jones (1975) and Teillet, Cochard, LeDouarin (1978) who reported that, in organ culture, neural crest cells require the presence of somites to differentiate into adrenergic neurons. Later, Cohen (1977) described the differentiation of quail crest cells into adrenergic neurons in tissue culture in

Fig. 1. Clones derived from single neural crest cells. a, unpigmented; b, mixed; and c, fully pigmented clones. The clones were grown for 9 days on collagen. Unpigmented cells tend to aggregate somewhat in the middle of the colony (a). Pigmented cells form a monolayer and stay flattened (c). In mixed clones, pigmented cells are always located on top of the unpigmented cells (b). Bar, 1 mm.

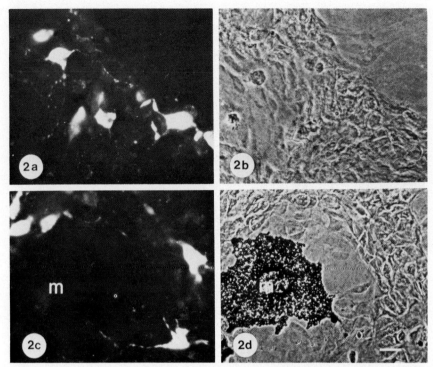

Fig. 2. Presence of catecholamine-producing cells in unpigmented and mixed clones. These clones were grown on collagen with fibronectin for 12 days and characterized by formaldehyde-induced histofluorescence. (a) Adrenergic cells of part of an unpigmented clone. (b) Same section with phase-contrast optics. (c) Part of a mixed clone grown on collagen with fibronectin. Fluorescent cells with axonal processes and one melanocyte (m) are shown. (d) Same section under phase-contrast optics. Note the large difference in size of melanocytes and adrenergic cells. The phase-contrast micrographs were taken after processing for histofluorescence. Bar, 10 μm.

the absence of somites. In these dense multilayered cultures, however, direct interactions with other crest derivatives and with contaminating noncrest cells may still occur. Our observation that fibronectin, which is synthesized by somitic cells, promotes expression of the adrenergic phenotype may explain the conflicting results.

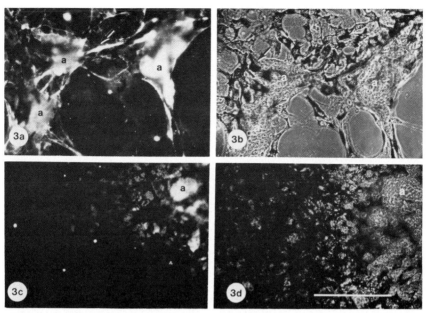

Fig. 3. Presence of fibronectin in the neural crest outgrowth from neural tube explants (primary cultures) on the 6th day of culture as detected by indirect immunofluorescence. (a) Intense fluorescein-fluorescence on aggregates of unpigmented cells (a) and interconnecting axons indicated by the arrow. (b) The same field under phase contrast optics showing unpigmented and pigmented (arrow) cells. (c) Field in the culture showing largely pigment cells: (left) note no fluorescein-fluorescence is observed on melanocytes; (right) an aggregate of unpigmented cells (a) which stained positively for CFN. (d) The same field under phase contrast optics showing an aggregate of unpigmented cells (a) and numerous melanocytes. Bar, 100 μm.

MODULATION OF DIFFERENTIATION BY EXOGENOUS FACTORS

Cellular Fibronectin (CFN)

Fibronectin is a high molecular weight glycoprotein that has been implicated in cell-cell and cell-substratum adhesion (e.g., see review by Yamada, Olden, 1978). In primary cultures, CFN is always associated with unpigmented cells and cell aggregates but not with melanocytes (Fig. 3; Sieber-Blum, Sieber, Yamada, 1981). The strands of fibronectin always appear to originate from flattened noncrest cells which contaminate the primary cultures in varying amounts. It seems likely that these are somitic cells because of their proximity to the neural crest in vivo. When grown in culture, somitic cells from the same embryonic age and axial level as the crest cells do indeed synthesize large

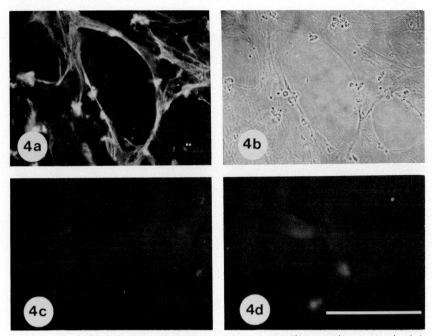

Fig. 4. Indirect immunofluorescence to detect fibronectin synthesized by somitic cells and neural crest clones of the same embryonic age and axial level.

Upper, somitic cells derived from the trunk region of stage 14-15 quail embryos synthesizing fibronectin in vitro. (a) Large amounts of fibronectin (some of it organized into bundles) detected by indirect immunofluorescence on the 4th day of culture; (b) same area under phase contrast optics. Somitic cells are extremely flattened and transparent when grown on a collagen substratum.

Lower, clones of neural crest cells grown in isolation. Fibronectin is detected neither on unpigmented clones (c) nor on mixed clones (d) by indirect immunofluorescence. The photomicrographs were overexposed in order to reveal the outline of the cells. Bar, 100 μm.

amounts of CFN (Fig. 4a, b). In contrast, crest cell clones do not produce any fibronectin (Fig. 4c, d). When purified CFN is added to crest cells in sparse secondary culture, it not only increases the plating efficiency of the progenitor cells but also promotes expression of the adrenergic phenotype as indicated by a significant increase in catecholamine-positive colonies (Table 1). The observation that CFN promotes adrenergic differentiation, although crest cells themselves do not synthesize this glycoprotein, has several implications for our understanding of neural crest cell development in vivo: It could explain the "induction" of adrenergic differentiation by somites reported by other investigators. Pouysségur, Willingham, Pastan (1977), Ali, Hynes (1978), and Yamada, Olden, Pastan (1978) have shown

Table 1. Plating Efficiency and Differentiation in the Presence and Absence of Added Cellular Fibronectin.

Colonies	Number of colonies per plate (mean ± S.E.)		
	Without CFN	With CFN	p
Unpigmented	4.3 ± 0.3	4.8 ± 0.5	0.5 (NS)
Mixed	12.0 ± 0.7	16.5 ± 1.1	0.004
Pigmented	38.7 ± 0.9	39.7 ± 1.6	0.6 (NS)
Catecholamine-positive	3.4 ± 0.2	7.1 ± 0.5	<< 0.001
Catecholamine-positive per unpigmented plus mixed	0.21 ± 0.02	0.34 ± 0.03	0.002

One milliliter of the same cell suspension was pipetted into each of 10 plates coated with collagen and 10 plates with collagen plus cellular fibronectin. After 12 days of culture, the dishes were processed for histofluorescence, the different types of colonies were counted, and the results were subjected to statistical analysis (two-sided Student's t-test). In an independent experiment, fibronectin treatment increased the number of adrenergic colonies per plate from 0.6 ± 0.2 to 1.8 ± 0.4 (p = 0.02). NS: not significant.

that fibronectin stimulates the migration of several cell types in vitro. Weston (1963) noted that neural crest cells giving rise to sensory ganglia migrate in vivo through the somitic mesenchyme where segmentation of the crest cell population occurs, probably due to accelerated migration within somites. Thus, fibronectin may have a regulatory function in adrenergic differentiation and it may also be involved in metamerization and ganglion formation of neural crest derivatives.

Heart Cell-conditioned Medium (HCM)

Another activity that modifies crest cell differentiation is present in medium conditioned on chicken embryo heart cell cultures (Sieber-Blum, Kahn, Cohen, 1978). Supplementation of crest cell primary cultures with HCM stimulates cholinergic differentiation as indicated by a significant increase in the specific activity of the enzyme acetylcholine transferase. It also blocks melanotic and adrenergic differentiation as indicated by a lack of melanin granules and catecholamine-specific histofluorescence. Inhibition of pigmentation is fully reversible for at least eight days whereas inhibition of adrenergic differentiation becomes irreversible when the cultures are exposed to HCM for more than two days (Fig. 5). HCM also promotes cell attachment and

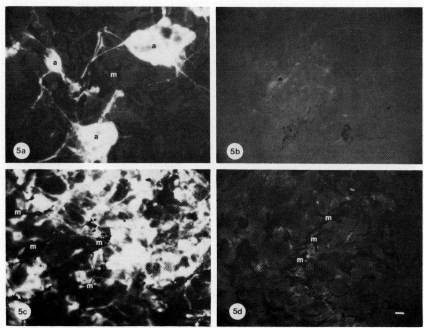

Fig. 5. Effect of HCM on differentiation of neural crest cells. (a) Neural crest cell culture grown for 7 days in regular culture medium. Aggregates of adrenergic cells (a) with axons and numerous melanocytes (m). (b) Neural crest cell cultures at day 7 that were grown in the presence of heart conditioned medium. No adrenergic cells and few melanocytes are present. (c) Culture of crest cells grown in the presence of heart conditioned medium for 2 days (from day 1 to day 3). Note numerous flattened adrenergic cells and melanocytes (m). (d) Crest cell culture grown in the presence of heart conditioned medium for 3 days (from day 1 to day 4). No adrenergic cells but normal amounts of melanocytes are observed. Bar, 10 µ.

spreading on the collagen substratum. It is not yet known whether HCM stimulates an existing population of cholinergic progenitors or whether it diverts adrenergic cells to the cholinergic pathway in a manner similar to the stimulation of cholinergic differentiation in postmitotic rat sympathetic neurons (e.g., see review by Patterson, 1978).

Tumor-promoting Phorbol Esters

Tumor-promoting phorbol esters have been shown to affect proliferation and differentiation of a number of cultured cell types (reviewed by Diamond, O'Brien, Baird, 1980; Slaga, Sivak, Boutwell (eds.), 1978). It has been proposed that

Table 2. Influence of Tumor-promoting and Nonpromoting Phorbol Esters on Neural Crest Cell Differentiation.

Additions	Number of colonies (mean ± S.E.)				
	Unpigmented	Mixed	Pigmented	Adrenergic	Total
TPA (10^{-7}M)	2.5*+0.27	3.5*+0.43	36.1*+1.10	0*	42.1+1.12
None	7.3 +0.56	10.1 +0.67	24.0 +1.38	4.5+0.45	41.4+0.88
PDD (10^{-7}M)	1.8*+0.49	1.8*+0.33	52.8*+2.59	0*	56.4+2.65
None	8.0 +0.95	8.5 +0.95	39.0 +2.26	3.2+0.42	55.5+2.75
PDA (10^{-7}M)	5.6 +0.58	9.1 +0.85	27.4 +1.89	2.8+0.42	42.1+1.73
None	7.3 +0.56	10.1 +0.67	24.0 +1.38	4.5+0.45	41.4+0.88
4-0-Me-TPA (10^{-7}M)	3.5 +0.65	3.9 +0.62	18.2 +1.59	3.7+0.79	25.6+1.78
None	2.8 +0.74	3.4 +0.72	19.6 +1.59	2.6+0.52	25.8+1.96
4α-PDD (10^{-7} M)	7.0 +0.61	9.1 +0.60	25.1 +2.00	5.3+0.67	41.2+1.70
None	7.3 +0.56	10.1 +0.67	24.0 +1.38	4.5+0.45	41.4+0.88
PHR (10^{-7}M)	8.3 +0.82	8.1 +0.64	37.7 +1.89	3.6+0.54	54.1+2.41
None	8.0 +0.95	8.5 +0.95	39.0 +2.26	3.2+0.42	55.5+2.75
DMSO (1.4×10^{-4}M)	7.6 +0.76	8.2 +0.63	40.0 +2.74	3.2+0.42	55.8+2.15
None	8.0 +0.95	8.5 +0.95	39.0 +2.26	3.2+0.42	55.5+2.75

Between 30 and 60 cells were seeded into each dish and the medium was supplemented as indicated. The different types of colonies were enumerated after 12 days in secondary culture. It was technically not possible to test all phorbol esters in the same experiment. Furthermore, a normalization of the data was not feasible because the relative frequency of the different colony-forming cells is known to vary somewhat between different cell preparations. Each experimental series, therefore, is listed separately together with the corresponding control (None = no additions) series using the same preparation of cells. Data are expressed as means of 10 replicate plates. Experimental values which were significantly different from the control series as judged by a 2-sided Student's t-test are marked with an asterisk ($p < 0.001$).

they achieve this by interfering with physiological regulatory mechanisms (Weinstein, Wigler, Pietropaolo, 1977). Phorbol esters, therefore, are considered useful tools for the analysis of growth control in normal cells at the cellular and molecular level.

We have shown that a clear correlation exists between the ability of phorbol esters to promote skin tumors in mice and their ability to interfere with the development in vitro of quail neural crest cells. The potent promoters, 12-0-tetradecanoyl-phorbol-13-acetate (TPA) and phorbol-12,13-didecanoate (PDD) are the most effective. Phorbol 12,13-diacetate (PDA) is considerably less effective. The nonpromoting analogs, 4-0-methyl-12-0-tetradecanoyl-phorbol-13-acetate (4-0-Me-TPA) and 4α-phorbol-12,13-didecanoate (4α-PDD) and

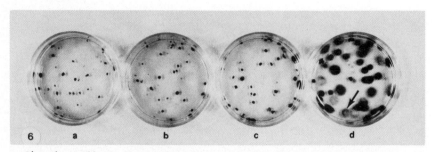

Fig. 6. Effect of tumor-promoting phorbol esters on neural crest cell proliferation. Macroscopic appearance of representative cultures after 10 days in secondary culture in 35 mm dishes. Only melanocyte-containing colonies are discernible since the cultures were not stained prior to being photographed. (a) Control culture (no additions to the culture medium). (b) With 10^{-7}M 4α-PDD. (c) With 10^{-7}M PDA. (d) With 10^{-7}M TPA. The arrow points at a colony which had initially assumed the morphology of an unpigmented colony. The grayish appearance indicates that the colony is still in the early phase of pigmentation.

the parent alcohol phorbol (PHR), have no effect except that both analogs cause a modest stimulation of adrenergic differentiation (Table 2; Fig. 6). The number of adrenergic cells, the number of axons per positive colony, and the number of adrenergic colonies are all elevated. Potent promoters also shorten the doubling time of the crest cells by 25%, which results in a considerable increase in the size of the three types of colonies (Fig. 6).

When the crest cells are grown in sparse secondary culture in the presence of potent promoters, the onset of pigmentation is delayed by about 36 hours and the expression of the adrenergic phenotype is suppressed. The number of unpigmented and mixed colonies is decreased while the total number of colonies stays the same (Table 2). The most plausible explanation for this latter observation is that potent promoters direct neural crest cells to differentiate preferentially along the melanogenic pathway (Sieber-Blum, Sieber, 1981).

CONCLUSIONS

Clonal analysis of the quail neural crest provides the first direct evidence a) that there are pluripotent neural crest cells in the early embryo and b) that they can differentiate in the absence of noncrest cells. These findings indicate that neural and melanocytic differentiation are preprogrammed in the neural crest cells at the onset of migration. The observation that ubiquitous and soluble substances such as fibronectin and the material(s) found in heart cell-conditioned medium influence expression of the differentiated phenotype does not contradict this view, since it is unlikely that they instruct crest cells to differentiate along the neural or melanogenic pathway. Rather, they provide a favorable environment for the expression of a particular program and the specificity of response lies within the responding cell and not in the signal. Defined substances such as tumor-promoting phorbol esters that interfere with cell differentiation hold promise as valuable tools for further evaluation of the mechanisms regulating neural crest cell differentiation.

ACKNOWLEDGMENTS

I thank Dr. Pamela Talalay and Mrs. Alverta E. Fields for their assistance in the preparation of the manuscript. This work was supported by U. S. Public Health Service grant HD-07389 to Dr. Alan M. Cohen, National Research Service Award HD-05646, and grant HD-15311 to M.S-B. Figures 1 to 4 and Table 1 were reproduced with permission from Academic Press, Inc., and Figure 6 and Table 2 from Springer Verlag.

REFERENCES

Ali IU, Hynes RO (1978). Effects of LETS glycoprotein on cell motility. Cell 14:439.
Cohen AM (1972). Factors directing the expression of sympathetic nerve traits in cells of neural crest origin. J Exp Zool 129:167.
Cohen AM (1977). Independent expression of the adrenergic phenotype by neural crest cells in vitro. Proc Natl Acad Sci USA 74:2899.
Cohen AM, Konigsberg IR (1975). A clonal approach to the problem of neural crest determination. Develop Biol 46:262.
Diamond L, O'Brien TG, Baird WM (1980). Tumor promoters and the mechanism of tumor promotion. Adv Cancer Res 32:1.
Newgreen DF, Jones RO (1975). Differentiation in vitro of sympathetic cells from chick embryo sensory ganglia. J Embryol Exp Morphol 33:43.

Norr SC (1973). In vitro analysis of sympathetic neuron differentiation from chick neural crest cells. Develop Biol 34:16.

Patterson PH (1978). Environmental determination of autonomic neurotransmitter functions. Ann Rev Neurosci 1:1.

Pouysségur J, Willingham M, Pastan I (1977). Role of cell surface carbohydrates and proteins in cell behavior: studies on the biochemical reversion of an N-acetylglucosamine-deficient fibroblast mutant. Proc Natl Acad Sci USA 74:243.

Sieber-Blum M, Cohen AM (1980). Clonal analysis of quail neural crest cells: they are pluripotent and differentiate in vitro in the absence of noncrest cells. Develop Biol 80:96.

Sieber-Blum M, Kahn CR, Cohen AM (1979). Influence of heart conditioned medium on quail neural crest cell differentiation. J Cell Biol 83:30a.

Sieber-Blum M, Sieber F (1981). Tumor-promoting phorbol esters promote melanogenesis and prevent expression of the adrenergic phenotype in quail neural crest cells. Differentiation, in press.

Sieber-Blum M, Sieber F, Yamada KM (1981). Cellular fibronectin promotes adrenergic differentiation of quail neural crest cells in vitro. Exp Cell Res 133:285.

Slaga TJ, Sivak A, Boutwell RD (eds) (1978). Carcinogenesis - a comprehensive survey. Mechanisms of tumor promotion and co-carcinogenesis, Vol 2, New York: Raven Press.

Teillet MA, Cochard P, LeDouarin NM (1978). Relative roles of the mesenchymal tissues and of the complex neural tube-notochord on the expression of adrenergic metabolism in neural crest cells. Zoon 6:115.

Weinstein IB, Wigler M, Pietropaolo C (1977). The action of tumor-promoting agents in cell culture. In Hiatt H, Watson TD, Winsten JA (eds): "The Origin of Human Cancer," Vol 4, New York: Cold Spring Harbor, p 751.

Weston JA (1963). A radioautographic analysis of the migration and localization of trunk neural crest cells in the chick. Develop Biol 6:279.

Weston JA (1970). The migration and differentiation of neural crest cells. Advan Morphogen 8:41.

Yamada KM, Olden K (1978). Fibronectin: Adhesive glycoproteins of cell surface and blood. Nature 275:179.

Yamada KM, Olden K, Pastan I (1978). Transformation-sensitive cell surface protein: isolation, characterization, and role in cellular morphology and adhesion. Ann N Y Acad Sci 312:256.

MECHANISM OF AVIAN CREST CELL MIGRATION AND HOMING

DUBAND, J.L., DELOUVEE, A., ROVASIO, R.A. and
THIERY, J.P.
Institut d'embryologie du C.N.R.S. et du collège
de France
49bis avenue de la Belle Gabrielle 94130 NOGENT
sur MARNE FRANCE

Morphogenesis involves a limited repertoire of cellular behavior: cell proliferation, cell movement, cell differentiation, cell adhesion and cell death. Studies on such coordinated mechanisms can be well approached on a limited number of embryonic model systems; in this respect, the neural crest offers a unique opportunity to study the mechanisms governing interactions between cells and their environment.

The neural crest arises at the dorsal dorder of the closing neural tube; its cells soon emigrate away from their source, localize in various sites of the embryo and give rise to the ganglia of the peripheral nervous system, endocrine cells, pigment cells and the mesectodermal derivatives in the head and neck (see Weston 1970, Noden 1978, Le Douarin 1980).

It is astonishing how precisely the crest cells reach their target sites. Many hypotheses have been proposed to account for this phenomenon: the first class of hypotheses emphasizes the role of intrinsec properties of crest cells (chemical repulsion between crest cells, Twitty and Niu 1948; contact inhibition of movement, Tosney 1978) whereas the others implicate the environment and more precisely the substrate of migration (contact guidance, Weiss 1958, Löfberg et al 1980; haptotaxy, Carter 1965; chemotaxy, Weston 1970). It should be stressed that crest cells are not predetermined to migrate in defined territories, but are provided with specific pathways along the neural tube axis (Le Douarin and Teillet 1974). This result suggest a key role for the morphogenesis of the migration pathways.

Crest cells migrate within an extracellular matrix (ECM) filled with a fibrillar meshwork (Bancroft and Bellairs 1976,

Tosney 1978, Hay 1978). Hyaluronic acid (HA) was shown to be an important component of crest migration pathways (Pratt et al 1975). In the trunk, spatial and temporal changes of the glycosaminoglycans (GAGs) composition are found in the ECM (Weston et al 1978).Toole et al(1972) postulated that a hydrated matrix of HA might serve as a suitable substrate for cell-migration and that an accumulation of HA produces space which allows cells to migrate into inaccessible areas. But recent studies of Newgreen (personal communication) show that HA alone cannot promote crest cell migration. Collagen has already been proposed as one of the major ECM components in avian and amphibian crest migration pathways (Hay 1978, Löfberg et al 1980). However, ultrastructurally it has been impossible to demonstrate the presence of numerous striated fibrils (Hay 1978).

Recent studies on tissue culture of fibroblasts have demonstrated that the glycoprotein Fibronectin (FN) is involved in cell to substrate adhesion (Yamada and Olden 1978). In addition, FN stimulates cell migration in vitro (Ali and Hynes 1978). In vivo, FN is expressed very early during embryogenesis and its appearance is correlated with morphogenetic movements (Crichley et al 1979, Mayer et al 1981).

In the present study, we have tested the in vitro ability of various tissues to synthesize FN and the in vitro behavior of crest cells on different substrata. Finally, we describe the distribution of FN at the time of crest cell migration. FN appeared to be a good marker of crest migration pathways. Our results allow us to suggest possible mechanisms that ensure proper directionality and homing of crest cells.

MATERIEL AND METHODS

In Vivo Localization of FN

Embryos: White Leghorn chick embryos were fixed in 3.7% formaldehyde in phosphate buffered saline (PBS) or in Zenker's fluid, washed in PBS and infiltrated in a gradual series of polyethylene-glycol 400 (PEG 400), transferred in PEG 1000 and finally embedded in PEG 1500. Serial sections were fixed to glass slides previously coated with rubber cement partially redissolved with ethyl-acetate.

Labelling of sections: Sections were incubated with 5-20 µg/ml of rabbit IgG anti-chick FN in PBS-0.5% bovine serum albumin (BSA) for 1h at room temperature. After washing in

PBS, sections were incubated with FITC-conjugated sheep anti-rabbit IgG in PBS-0.5% BSA for 30min, washed in PBS and then mounted in glycerol.

Chimaeric embryos: Isotopic and isochronic grafts of quail neural tube segments into chick host embryos were performed as previously described (Le Douarin and Teillet 1973). In the trunk region, they were performed at the 15th-17th somite level in 17-somite embryos. After 10-24h of incubation, chimaeric embryos were fixed in Zenker's fluid, labelled with anti-FN antibodies and stained with the Feulgen-Rossenbeck tecnique.

In Vitro Studies

Tissue culture: Rose chamber cultures of quail embryo tissues were established as described by Newgreen et al(1979) Tissues were cultured in DMEM with 10% Fetal Calf Serum (FCS) in some cases, FN was removed from FCS by affinity chromatography on a gelatin sepharose column (Engwall and Ruoslahti 1977).

In vitro localization of FN: The capacity of various embryonic tissues to synthesize FN was studied by immunofluorescence labelling for FN (see above).

In vitro migration of crest cells: Crest cells were obtained by explanting segments of dispase isolated neural primordia (Cohen and Konigsberg 1975) deposited on glass coverslips pretreated with a FN-solution (150 μg/ml).To compare the efficiency of FN and glass on crest cell migration and adhesion, we removed locally the FN substrate with a glass needle. Under these conditions, we obtained a matrix with alternating stripes of FN and glass substrate; the neural tube segments were deposited astride these streaks.

RESULTS

In Vitro Localization of FN

Different tissues were tested for their capacity to synthesize FN at the time of crest cell migration. Immunolabelling of ectoderm and endoderm showed an extraordinarily dense layer of fluorescent fibers under the cell sheets. The notochord was invasted in a mesh of fibers. Somitic cells were also surrounded by a meshwork of a great density. On the contrary, trunk neural tube segments were covered with a meshwork of fine, oriented fibers which disappeared after 48h in culture.

It was possible to distinguish two crest cell population with regard to their capacity to secrete FN: The peripheral cephalic crest cells of the outgrowth were large, flat and polygonal and when labelled with anti-FN antibodies showed cytoplasmic fluorescent granules and a complex extracellular meshwork of fibers. In contrast, the median cephalic crest cells of the outgrowth and all the trunk crest cells were small and multipolar (stellate morphology) and formed a dense population showing neither cytoplasmic nor extracellular fluorescence.

Comparison of Crest Cells Migration on Different Subtrata

On a glass-substrate, trunk crest cells were able to emigrate from a explanted neural tube segment. They had the typical stellate morphology, but rapidly rounded up, stopped moving and collected into small aggregates. After several hours in culture, many crest cells detached from the glass and died.

When cultured on a FN-substrate, trunk crest cells initiated their migration more rapidly, adhered very efficiently to the substrate and migrated very quickly (70 µm/h). We never observed the formation of aggregates on FN-matrices.

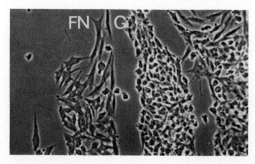

Figure 1. In vitro migration of crest cells on FN-stripes. Trunk crest cells only migrate on FN as a confluent layer of cells; those which are located onto glass have a round morphology and bleb actively. G: glass

When confronted to alternating rows of FN and glass, crest cells almost exclusively migrated on the FN-stripes as a dense cell population. The rare crest cells which migrated onto the glass rounded up and died (Fig.1).

In Vivo Migration of Crest Cells.

Cephalic levels. At the mesencephalic level, prior to crest cell migration, FN was present in the basement membranes of the epithelia, i.e., neural tube, ectoderm, notochord, endoderm and somatopleural-splanchnopleural mesoderm. Loose mesenchymal cells were surrounded by a poor FN-matrix. It should be noted that the upper side of the neural tube, i.e. the presumtive crest cells, was not limited by a complete basement membrane.

Slightly after the neural tube closure, crest cells initiated their migration between the basement membranes of the neural tube and the ectoderm. This phase of separation from the neural tube was characterized by an active crest cell proliferation, a disappearance of the epithelial structure within the crest population, and an appearance of FN-strands between the first migrating cells (Fig.2a). Within two hours crest cells reached a cell-free space between the ectoderm and the mesenchyme. The first crest cells which invade this space soon became surrounded by a FN-rich matrix.

As the crest cells continued their migration as a confluent multilayered cell population under the ectoderm, they could be distinguished from the mesenchymal cells by their higher cell density and a higher FN content (Fig.2b).

The sequence of events of early crest cell migration was similar in the anterior rhombencephalon. At the proencephalon and the median rhombencephalon levels, the neural tube basement membrane was tightly linked with that of the ectoderm of the optic vesicules and the otic placodes, respectively. When crest cells reached these structures, they could not progress any further ventrally and accumulated near their site of origin, between the ectoderm and the neural tube. Later, they bypassed these obstacles by migrating both rostrally and caudally.

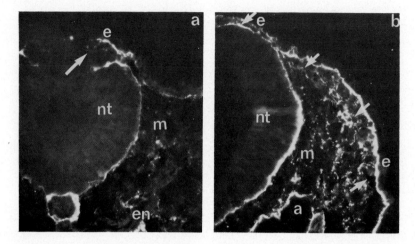

Figure 2. Transverse sections at the mesencephalic level. Immunofluorescence labelling for FN. a) 7-somite stage. Crest cells individualize at the dorsal aspect of the neural tube (nt). b) 12-somite stage. Crest cells migrate in a cell free space under the ectoderm (e). a: aorta, en: endoderm, m: mesenchyme, arrow: crest cells.

Trunk level. Prior to crest cell migration, the trunk level was exclusively composed of tightly linked epithelial structures. As opposed to the cephalic level, no space was available and the neural tube was completely closed and covered by the ectoderm.

Crest cells became separated from the neural tube by FN-strands and rapidly initiated their migration between the basement membranes of the neural tube and the ectoderm. They reached the apex of the somite as a monolayer, where they accumulated during the next 10h (Fig.3a). Crest cells resumed their migration only when the somite dissociated into sclerotome and dermomyotome. At that stage, they moved ventrally in a FN-rich ECM between the dermomyotome and the neural tube (Fig.3b). The sclerotome soon expanded ventrally and medially around the notochord, preventing a further progression of crest cells. The latter accumulated above the sclerotome and occupied a large space laterally. Progressively, FN disappeared among the crest cells, which then aggregated into the sensory ganglion (Fig.3c).

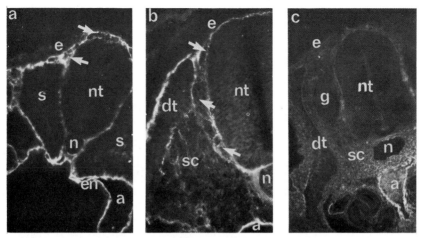

Figure 3. Transverse sections at the 15th-somite level. Immunofluorescence labelling for FN. a) 20-somite stage. Crest cells reached the apex of the somite (s). b) 32-somite stage Crest cells accumulate above the sclerotome (sc) between the dermomyotome (dt) and the neural tube (nt). c) 45-somite stage. Crest cell aggregate to form the sensory ganglion (g). a: aorta, en: endoderm, e: ectoderm, n: notochord, arrow: crest cells.

Due to the metameric structure of the trunk, we analyzed the migration pathways along two consecutive somites. It was found that crest cells could also use a very narrow pathway between two adjacent somites (Fig.4). Using the chick-quail chimaera technique, we observed that crest cells which moved between two somites, rapidly reached the aorta, became distributed rostrally and caudally along the surface of the aorta under the sclerotome, whereas at the somitic level, crest cells were still found at the dorsal border of the somite. The former differentiated into sympathetic ganglia at the intersomitic cleft, and into aortic plexuses along the aorta. At the end of crest cell migration, the narrow pathway between the two somites had become obliterated by the proliferation and invasion of the adjacent sclerotomes.

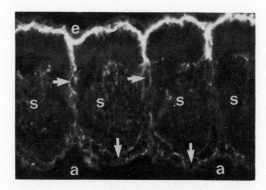

Figure 4. Saggital section through the 15th-17th somites of a 26-somite stage embryo. Crest cells migrate in a narrow pathway between two adjacent somites (s), and reach the aorta (a). e: ectoderm, arrow: crest cells.

DISCUSSION

Synthesis of FN

The in vitro and in vivo experiments showed that all the embryonic tissues deriving from the 3 primary germ layers can synthesize FN and organize it into matrices. The epithelia deposited FN in a basement membrane, while the mesenchymal cells were surrounded by a FN-rich ECM. However, in some embryonic tissues, the capability to organize FN in a ECM was transient. In the case of the neural tube, FN disappeared both in vivo and in vitro. Since FN has a long half-life in vitro (Yamada and Weston, 1975), this disappearance suggests an active removal of FN by the neural tube.

On the contrary, most of the crest cells lacked the ability to synthesize FN in vitro. Because this offers an interesting parallel with a variety of transformed cells (Hynes 1976), it is then tempting to relate the loss of FN to the invasive phase malignancy by comparison with the behavior of crest cells.

Some cranial crest cells can synthesize FN in vitro. In vivo, we observed that pioneer crest cells which entered the cell-free space between the ectoderm and the mesenchyme were surrounded by FN. Since this space is devoid of FN, this suggests that pioneer crest cells synthesize their own substrate.

Furthermore, the heterogeneity of cephalic crest cell population in their ability to synthesize FN may reflect the heterogeneity of cell lineages described by Le Douarin and Teillet (1974). Indeed, the cranial crest population includes the committed mesectodermal cells; the ability to secrete FN may be an early expression of a mesodermal phenotype.

Pathways of Migration

Crest cells exclusively migrate into morphologically defined pathways delimitated by the basement membranes of the epithelia. It is notable that at the onset of crest cell migration, epithelia are practically the exclusive morphological structures, whereas after the migratory phase, the mesenchymal structures predominate.

Space and substrate: Prior to trunk crest cell migration no space is available between the basement membranes of the epithelia. When crest cells migrate, the basement membranes are separated as a result of pressures exerted by an increase in cell number and in HA content secreted by the surrounding tissues and the crest cells themselves. In contrast, a large cell free space is available under the cephalic ectoderm, in which crest cells migrate and produce their own substrate for migration.

Invasive properties of crest cells: We never observed numerous crest cells invading mesenchymal tissues. In this respect, undifferentiated crest cells are much less invasive than melanoblasts or sarcoma cells.

Segregation of cell lineages: Since crest cells do not invade epithelial structures, they use most of the acellular spaces delimitated by the basement membranes, i.e. in the trunk, the ventral pathways between the neural tube and the somite, the pathway between two adjacent somites and a lateral pathway between the somite and the ectoderm. It appears from our study that the relative position of crest cells towards these pathways may, in some cases, be responsible of the segregation of crest cells, e.g., cells facing the somite differentiate into sensory ganglia, whereas those facing the intersomitic space give rise to adrenergic system in the trunk.

Evolution of pathways: Crest cells become progressively

separated from their site of origin by a basement membrane which completely covers the neural tube and fuses with that of the ectoderm. The pathways of migration completely disappear subsequent to crest cell migration. Indeed the sclerotomal cells invade the spaces between the somites and around the neural tube, preventing a backward migration of crest cells.

Homing

Prior to crest cell migration, the sites of arrest do not preexist, thus weakening the chemotactic hypothesis. For instance, the region where crest cells will differentiate into sensory ganglion is located in the middle of a migration pathway, i.e. the ventral one. As far as the peripheral nervous system is concerned, the sites of arrest of crest cells appear during morphogenesis and are delineated by physical barriers, such as a mesenchyme or an epithelium, that crest cells cannot traverse. Since crest cells continue to proliferate (Carr and Simpson 1978), and since the sclerotome progressively surrounds crest cells, the cell density may increase more rapidly than the space. Both these conditions may induce the aggregation of crest cells into cohesive ganglia. Furthermore, the disappearance of some ECM components (FN and HA) may also be responsible for the arrest of crest cells.

Conclusion

Based on our studies, we proposed that crest cells migrate rapidly as a confluent layer in defined transitory pathways by producing either their own space (in the trunk), or their own substrate (in the head). An active cell proliferation, combined with specific motility properties in an efficient ECM, are sufficient to explain a proper directionality. The final localization of crest cells is greatly influenced by the surrounding tissues, the development of physical barriers in front of migration, and the change in ECM chemical composition. (see fig.5).

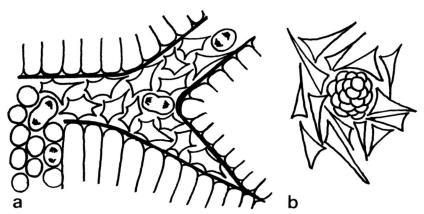

Figure 5. Model of the migration and homing of crest cells. a: individualization and migration; b: aggregation. ✧ crest cell; ◁ mesenchymal cell; ⊓ epithelial cell

REFERENCES

Ali IU, Hynes RO (1978). Effects of LETS glycoprotein on cell motility. Cell 14:439.

Bancroft M, Bellairs R (1976). The neural crest of the trunk region of the chick embryo studied by SEM and TEM. Zoon 4:73

Carr V, Simpson SB (1978). Proliferative and degenerative events in the early development of chick dorsal root ganglia. I:normal development. J Comp Neurol 182:727.

Carter SB (1965). Principles of cell motility: the direction of cell movement and cancer invasion. Nature 208: 1183.

Cohen AM, Konigsberg IR (1975). A clonal approach to the problem of neural crest determination. Dev Biol 46:262.

Critchley DR, England MA, Wakely J, Hynes RO (1979). Distribution of fibronectin in the ectoderm of gastrulating chick embryos. Nature 280:498.

Engvall E, Ruoslahti E (1977). Binding of soluble form of fibroblast surface protein to collagen. Int J Cancer 20:1.

Hay ED (1978).Fine structure of embryonic matrices and their relation to the cell surface in ruthenium-red fixed tissues Growth 42:399.

Hynes RO (1976). Cell surface protein and malignant transformation. Biochim Biophys Acta 458:73.

Le Douarin N (1980). Migration and differentiation of neural crest cells. Curr Topics Dev Biol 16:31.

Le Douarin N, Teillet MA (1973). The migration of neural crest cells to the wall of the digestive tract in the embryo. J Embryol exptl Morphol 30: 31.

Le Douarin N, Teillet MA (1974). Experimental analysis of the migration and differentiation of neuroblasts of the autonomic nervous system and of neurectodermal mesenchymal derivatives, using a biological cell marking technique. Dev Biol 41:162.

Löfberg J, Ahlfors K, Fällstom C (1980). Neural crest cell migration in relation to the extracellular matrix organization in the embryonic axolotl trunk. Dev Biol 75:148.

Mayer BW, Hay ED, Hynes RO (1981). Immunocytochemical localization of fibronectin in embryonic chick trunk and area vasculosa. Dev Biol 82:267.

Newgreen DF, Ritterman M, Peters EA (1979). Morphology and behaviour of neural crest cells of chick embryo in vitro. Cell Tiss Res 203:115.

Noden DM (1978). Interactions directing the migration and cytodifferentiation of avian crest cells. In Garrod DR (ed) Specificity of embryological interactions. Receptors and recognition. Series B Vol 4. London: Chapman and Hall, p 5.

Pratt RM, Larsen MA, Johnston MC (1975). Migration of cranial neural crest cells in a cell-free hyaluronate-rich matrix. Dev Biol 44:298.

Toole BP, Jackson G, Gross J (1972). Hyaluranate in chondrogenesis: Inibition of chondrogenesis in vitro. Proc natl Acad Sci USA 69:1384.

Tosney KW (1978). The early migration of neural crest cells in the trunk region of the avian embryo. Dev Biol 62:317.

Twitty VC, Niu NC (1948). Causal analysis of chromatophore migration. J Exptl Zool 108:405.

Weiss P (1958). Cell contact. Int Rev Cytol 7:391.

Weston JA (1970). The migration and differentiation of neural crest cells. Adv Morph 8:41.

Weston J, Derby M, Pintar J (1978). Changes in the extracellular environment of neural crest cells during their early migration. Zoon 6:103.

Yamada KM, Weston JA (1975). The synthesis, turnover and artificial restoration of a major cell surface glycoprotein. Cell 5:75.

Yamada KM, Olden K (1978). Fibronectins-adhesive glycoproteins of cell surface and blood. Nature 275:179.

BIOCHEMISTRY OF GRANULE CELL MIGRATION IN DEVELOPING MOUSE
CEREBELLUM

M.E. Hatten, D.B. Rifkin*, M.B. Furie*,
C.A. Mason and R.K.H. Liem
Departments of Pharmacology and Cell Biology*
New York University School of Medicine,
New York, USA

In the the developing mammalian brain, young neurons migrate from the germinal centers where they are generated to the sites where they form synaptic connections (Sidman and Rakic, 1973). Recently, Rakic (1972) has shown that many young, migrating neurons are closely associated with astroglial processes, processes that appear to guide the immature neurons through the tissue. Although the routes and timetables of neuronal migrations have been mapped for many brain regions, the biochemical mechanisms underlying these events have not been described. The present report presents a model system for the analysis of the biochemistry of the migration of the granule neuron in the developing mouse cerebellum.

The migration of the granule neuron occurs in two steps. The first step is a morphogenetic movement of precursor granule cells from the lateral caudal edge of the cerebellar plate up onto the pial surface. In the mouse, the first granule cells arrive on the pial surface at about embryonic day 13. Between that time and birth, a wave of cells spreads across the surface of the developing cerebellar anlage, forming the external granule cell layer (EGL) (Altman and Bayer, 1978; Miale and Sidman, 1961; Rakic, 1971).

Indirect cytochemical staining of tissue sections with antisera raised against human plasma fibronectin revealed a developmentally regulated pattern of endogenous fibronectin. At E13, dense, diffuse, specific staining of the ventral portion of the neuroepithelium, the site of emergence of the EGL, was observed (FIG.2). The choroid plexus was also stained and intermittent patches of staining were visible at the leading edge of the EGL.

Between E16 and birth, the region of intense staining was observed over the ventral portion of the cerebellum between the choroid plexus and the external surface. Intermittent patches of intense staining were also evident in the EGL located along the pial surface, but the dense region of staining was confined to the portion of the EGL which interfaced the brainstem.

These studies suggest that fibronectin might provide a transient guide that directs precursor granule neurons from the ventricular surface of the neuroepithelium where they are generated to the external surface of the developing cerebellum where they form the EGL.

This model is consistent with other recent studies of the role of fibronectin in morphogenetic movements during the early histogenesis of the nervous system. Thiery has shown that fibronectin plays a transient role in neural crest migration (Newgreen and Thiery, 1980) and Mayer et. al. (1981) have localized fibronectin in the embryonic chick trunk and area vasculosa.

In a separate study carried out to assess whether the specific associations observed between cerebellar astroglia and migrating neurons in vivo occurs in microwell cultures, cells dissociated from embryonic and early postnatal mouse cerebellum and plated on microwell surfaces coated with polylysine, were stained with antiserum raised against purified glial filament protein (AbGF) (Liem, 1981). Approximately 15 percent of the cells in microwell cultures of P7 cerebellar cells were intensely stained with AbGF (FIG.3). Stained cells were larger than the phase-bright, tetanus toxin positive neuronal cells (Mirsky et. al., 1978) in the cultures and had the morphological characteristics of astrocytes (Palay and Chan-Palay, 1974). Both the perikaryon and the processes of astrocytes were stained with AbGF.

Granule Cell Migration in Cerebellum / 511

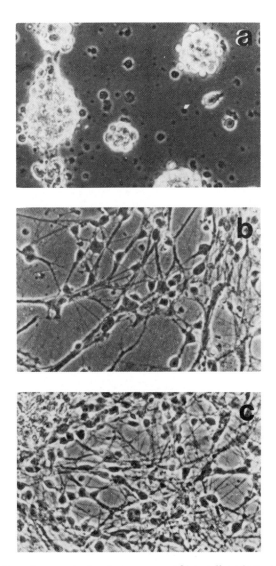

FIGURE 1: Developmental stage specific adhesion of cerebellar cells to a fibronectin-coated microwell surface. Cells dissociated from a)E13, b)P0 and c) P7 cerebellar tissue were plated as described (Hatten and Sidman, 1978; Hatten and Francois, 1981). After 48h in vitro x 160.

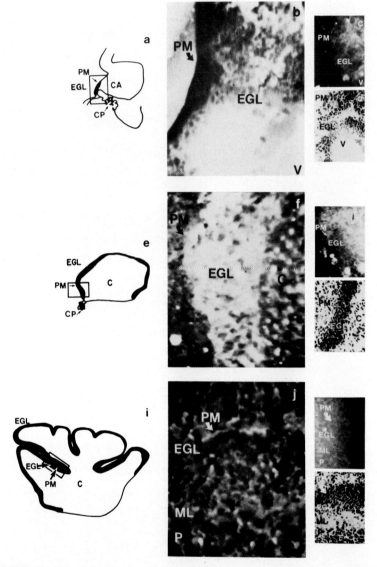

FIGURE 2: The binding of antiserum raised against purified human plasma fibronectin (AbCIG) to cryostat sections of cerebellar tissue from E13, P0 and P7 mouse cerebellum.

a) Drawing of the cerebellar anlage at E13. The first precursor granule cells are emerging (shaded area). b) staining of area in box with AbCIG. The ventral edge of the ECL is intensely stained and some areas of staining are visible at the leading edge of the EGL. The pial membrane (PM) and fourth ventricle (V) are marked for orientation. Control sections stained with c) AbCIG preabsorbed with CIG and d)toluidine blue to reveal cell bodies.

e) Drawing of the cerebellum at P0. The EGL (shaded area) now covers the surface of the cerebellar anlage. f) binding of AbCIG to the ventral portion of the EGL between the choroid plexus (not shown) and the external surface (PM). Control sections stained with g)AbCIG preabsorbed with CIG and h)toluidine blue to reveal cell bodies.

i) Drawing of the cerebellum at P7. The EGL (shaded) is visible along the external edge of the newly formed folia and migration down into the cortex has begun. j)staining of area in box with AbCIG. control sections stained with k)AbCIG preabsorbed with CIG and with toliudine blue to reveal cell bodies (Hatten and Liem, 1981).

More than 99 percent of the processes present in the P7 cultures were stained with AbGF, suggesting that they were glial in origin. After 48h in vitro, the stained astroglial processes had produced a thick network which appeared to provide a template for the attachment of phase-bright, tetanus toxin-positive neuronal cells in the culture (FIG.3a,b,c) (Mirsky et. al., 1978). More than 95 percent of the neuronal cells in the culture were positioned within 20μm of a stained astrocytic process (FIG 4). Measurements of the distance between the center of neuronal cells and the center of stained processes revealed that no cells were located more than 60μm from a glial process (FIG 4).

After five days in vitro (FIG 3d,e,f), the network of processes had thickened, but nearly all of the processes were glial in origin. After eight days in vitro (FIG.3g,h,i), a few unstained processes were observed which were stained with tetanus toxin. Although the vast majority of the neuronal cells were still associated with stained glial processes, a few cells were observed at distances greater than 40-60μm from a stained glial process, suggesting that some of the neuronal cells were dissociating from the glial processes.

The second step of granule cell migration occurs postnatally. Postmitotic precursor granule cells leave the ECL and descend into the deep layers of the cerebellar cortex, apparently using Bergmann glial fibers, a specialized form of astrocyte, as a guide (Rakic, 1971; Rakic and Sidman, 1973). Having reached their proper cortical address deep to the Purkinje cell layer, granule cells extend neurites and form synaptic connections.

As reported earlier, embryonic and early postnatal cerebellar cells do not adhere to untreated microwell tissue culture surfaces. In contrast, cells dissociated from embryonic or early postnatal cerebellar tissue formed a monolayer on microwell surfaces coated with polylysine (Hatten and Sidman, 1978), with certain lectins (Hatten and Francois, 1981) or insoluble carbohydrates (Hatten, 1981). The question was raised as to whether dissociated cerebellar cells would adhere to natural matrix components such a fibronectin (Hynes, 1976; Yamada and Kennedy, 1979).

The adhesion of cerebellar cells to microwell surfaces coated with purified human plasma fibronectin (Furie and Rifkin, 1980) is developmental stage specific. Cells dissociated from cerebellar tissue at embryonic day 13 (E13) formed cellular aggregates with no fiber outgrowth and low cell viability (Fig. 1a). After 48h in vitro, extensive cell death was observed. The poor survival of E13 cells on a fibronectin-coated substratum was consistent with E13 cell behavior on other substrata which do not support their adhesion.

In contrast, cells dissociated from cerebellar tissue harvested between embryonic day 16 (E16) and postnatal day 7 (P7) formed a monolayer with 95 percent plating efficiency and extensive fiber outgrowth on a fibronectin-coated microwell surface (FIG.1b,c). Time-lapse video recordings of the cultures revealed extensive cell movement on the fibronectin substratum (Hatten et al., 1981)).

The adhesion of late embryonic and early postnatal cerebellar cells to a fibronectin-treated culture surface depended on the amount of fibronectin coupled to the surface and was inhibited by antiserum raised against purified human plasma fibronectin (Hatten et. al., 1981).

Granule Cell Migration in Cerebellum / 515

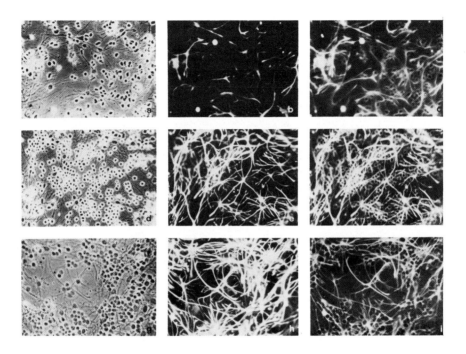

FIGURE 3: Immunofluorescence staining of P7 cerebellar cells in microcultures with antisera raised against purified glial filament protein (AbGF). The identical field was photographed with phase contrast (a,d,g), epifluorescence (b,e,h) or double image of phase contrast and epifluorescence (c,f,i) illumination. After 48h (a,b,c), 5 days (d,e,f) or one week (g,h,i) in vitro x 210.

Similar results were observed with cells dissociated from mouse cerebellum on the day of birth, P0. In contrast, when cells were dissociated from E14 cerebellum and plated as described, less than 40 percent of the phase-bright, tetanus toxin-positive cells were associated with stained astroglial processes (FIG. 4).

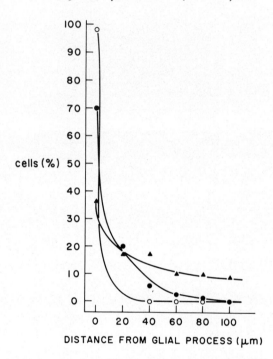

FIGURE 4: Distribution of neurons dissociated from E14, P1 or P7 cerebellum along astroglial processes after 48h in vitro. The distance between the center of any given glial cell and the center of the nearest stained glial process was measured as described (Hatten and Liem, 1981)
▲ E14, ● P1, ○ P7.

The characteristic morphology of stained astroglial cells from embryonic cerebellum differed from that of stained cells in cultures of postnatal cells. E14 cells had fewer, thicker, less branched processes than the highly branched, stellate cells observed in cultures of postnatal cells. The morphological characteristics of stained embryonic astroglia differentiated on approximately the same time schedule as occurred in vivo when the cells were kept in vitro for several weeks. (FIG.5).

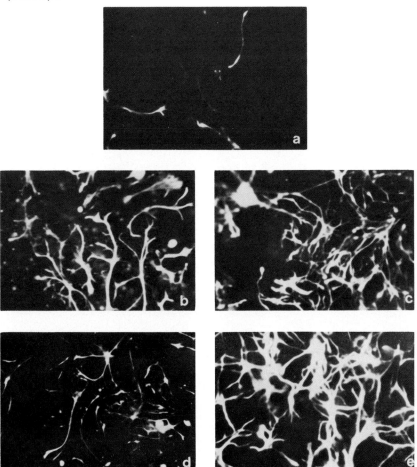

FIGURE 5: Morphological Development of cerebellar astroglia in vivo and in vitro. Immunofluorescence staining with AbGF of a) E14 cells after 48h in vitro b)P1 cells after 48h in vitro c)P7 cells after 48h in vitro d)E14 cells after 5 days in vitro and E14 cells after one week in vitro. x 210.

Localization of AbGF in tissue sections of cerebellum with the PAP technique (Sternberger, 1979) revealed specific staining of cerebellar astrocytes, the cell bodies and processes of Bergmann glial cells and fibrous astrocytes being intensely stained. No staining of cerebellar neurons was observed with AbGF (FIG. 6)

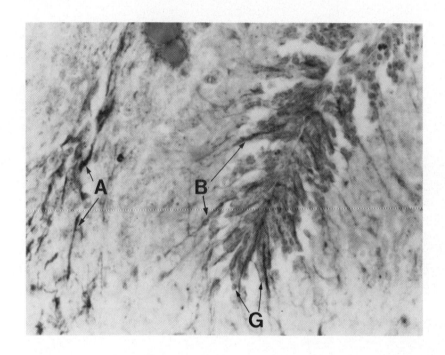

FIGURE 6: Localization of AbGF with the PAP technique in a 6 m paraffin section through the base of two cerebellar folia, postnatal day 7. Both astrocytes (A) in the white matter and Bergmann glia (B) are stained darkly. Granule cells (G), visible after counterstaining with cresyl violet, are closely apposed to Bergmann fibers. x 590.

These studies suggest that the two-step sequence of granule cell migration in the developing mouse cerebellum is governed by two separate mechanisms, both mediated by the cell surface. In the first step, the formation of the EGL, a transient role for fibronectin is proposed to provide a pathway for precursor granule cells to follow from the rhombic lip to the pial surface. The disappearance of fibronectin at the second step, namely migration into the cortex along astroglial processes, suggests that fibronectin does not play a role in this stage. Two possible mechanisms seem likely. First, other matrix components such as glycosaminoglycans could guide the granule cells down into the cortex. This is difficult to assess at present, since indirect probes for these macromolecules are not available. A second possibility, which seems more likely in view of Rakic's electron micrographs, showing granule cells in close apposition to Bergmann glial processes, is that granule cells bind to the glial processes via specific ligand receptor mechanisms.

The studies with AbCF suggest that the microwell tissue culture system provides a paradigm for specific associations between astroglial processes and immature granule neurons. This system should prove useful for studies of cell surface components that regulate cell-cell interactions between astroglia and granule neurons in the developing mouse cerebellum.

REFERENCES
Altman J, Bayer SA (1978). J Comp Neurol 145:353-398.
Furie MB, Pifkin DB (1980). J Biol Chem 225:3134-3140.
Hatten ME (1981) J Cell Biol 89: 54-61.
Hatten ME, Francois AM (1981). Develop Biol 87: 102-113.
Hatten ME, Furie MB, Pifkin DB (1981). submitted.
Hatten ME, Liem RKH (1981) J Cell Biol 90:622-630.
Hatten ME, Sidman PL (1978). Exp Cell Res 113:111-125.
Hynes RO (1976). Biochim Biophys Acta 458:73-107.
Liem RKH (1981). J Neurochem in press.
Mayer BW, Hay ED, Hynes RO (1981). Dev Biol 82:267-286.
Miale I, PL Sidman (1961). Exp Neurol 4:277-296.
Mirsky R, Wendon LMB, Black P, Stolkin C, Bray,D (1978). Brain Res 148:251-259.
Newgreen D, Thiery J-P (1980). Cell Tiss Res 211:269-291.
Palay SL, Chan-Palay V (1974)."Cerebellar Cortex" New York: Springer-Verlag.
Rakic P (1971). J Comp Neurol 141:283-312.
Rakic P (1972). J Comp Neurol 145: 61-83.
Rakic P, Sidman PL (1973). J Comp Neurol 152: 102-132.
Sidman PL, Rakic P (1973). Brain Res 62:1-35.
Sternberger LA (1979). "Immunocytochemistry" New York: Wiley
Yamada D, Kennedy DW (1979). J Cell Biol 80: 492-498.

STUDIES ON PLANT CELL LINES SHOWING TEMPERATURE-SENSITIVE EMBRYOGENESIS

M.Terzi*, G.Giuliano, F.Lo Schiavo, and
V.Nuti Ronchi
Istituto di Mutagenesi e Differenziamento
CNR, Via Svezia, 10, 56100-Pisa, Italy

INTRODUCTION

The early embryogenesis of higher plants differs, in many aspects, from many other well-known embryogenetic systems. The absence of complex morphogenetic movements of cells with respect to each other, the organization of apical meristems and the subsequent "open" development are characteristics peculiar to this system. This particular kind of development can probably not be explained exhaustively through the principles of animal development; specific experimentation in this field, using classical and original approaches alike, seems necessary for a further understanding.

The isolation of mutations that interfere with development, both constitutive and conditional, proved a powerful tool to elucidate genetic control of the development of other organisms such as the metazoans Drosophila (Gehring, 1976) and Caenorhabditis (Hirsch and Vanderslice, 1976), the colonial protozoan Volvox (Huskey et al., 1979), the slime mold Dictyostelium and the moss Physcomitrella (Ashton et al., 1979). In particular, conditionality of ts mutations allows for the resolution of the temporal activation pattern of the genes involved, what is called the "genetic dissection" of development.

* To whom correspondence should be addressed.

In vitro somatic embryogenesis of carrot has been known since the late fifties (Reinert, 1958). When a somatic cell culture is transferred to an auxin-free medium, single cells (Backs-Hüsemann and Reinert, 1970; McWilliam et al., 1974) give rise to globular embryoids, which follow the normal pathway giving heart-shaped, torpedo-shaped, and plantula stages.

Although some differences from zygotic embryogenesis have been described, viz. the pattern of the first two cellular divisions (McWilliam et al., 1974), and the shape and size of torpedoes (Halperin, 1966), the process resembles that in vivo enough to be chosen as a model system due to the many experimental advantages that it offers. This process is infact highly synchronizable (Komamine et al., 1978), it offers a high yield of free embryos without manipulation, and is reproducible enough to be used as a screening system for various substances supposed to interfere with embryogenesis (Fujimura and Komamine, 1975). Techniques are also available for obtaining highly pure globular, heart, and torpedo fractions for biochemical studies (Warren and Fowler, 1977).

The protocols used to isolate auxotrophs (Carlson, 1970) or ts growth mutants (Malmberg, 1979) from plant cell cultures comprise often, but not always (Savage et al., 1979), a selective step based on the 5-BUdR incorporation technique (Puck and Kao, 1967). This eliminates part of the hard screening work due to the rudimentary status of replics plating techniques (Schulte and Zenk, 1977). Unfortunately, 5-BUdR completely inhibits somatic embryogenesis, even at low concentrations (Dudits et al., 1979). The approach we took for performing this selection is based on the different dimensions reached by the embryoids with respect to undifferentiated cell clumps, if initial inoculum is standardized in small dimensions by double filtration. This is due to the fact that undifferentiated cells either do not divide, or do so very slowly, in basal medium. Therefore, embryoids can be separated from cell clumps in the suspension by simple filtration through nylon sieves with appropriate pore size.

RESULTS

Isolation of Mutants

Before starting the selection procedure it was ascertained that in the interval 24°-31°C embryogenesis of a wt culture was independent from temperature. Therefore the two extreme temperatures were chosen as permissive and non permissive respectively. Another experiment (Table 1), confirmed the complete independence of embryogenic efficiency from temperature, while it indicated a strong dependence from cell concentration, already reported (Hari, 1980).

Mutagenesis was performed with 0.8% (v/v) EMS for 1 hr as described (Lo Schiavo et al., 1981). This treatment reduced by 84% the number of colonies formed by a <80μm cell population. Unexpectedly the reduction of embryogenesis yield was only of 63%. This might be an indication that EMS induces the embryogenetic pathway, in some unknown way. A similar effect was reported after irradiation of Citrus calluses (Vardi et al., 1975).

The selection of the mutants (Giuliano et al., in preparation) was performed by differentiation of the culture at the non-permissive temperature for 14 days, followed by removal of the embryoids (negative selection) by filtration. The culture was then shifted to the permissive temperature and the newly generated embryoids were collected at various times (positive selection). The embryoids were then plated on solid B5+ medium. From these plates we established suspension cultures starting only from calluses originated from isolated embryoids. This procedure is operationally equivalent to cloning. These suspension cultures were then screened qualitatively for ts phenotype.

They were differentiated and put at the permissive and non-permissive temperatures. A first check after 15 days was followed by a definitive check after one month. All cell lines that proved positive on this preliminary analysis were then recloned

Table 1: Embryogenic efficiency (Embryogenic Units obtained after 20 days vs. no. of inoculated clumps).

Cell Density (Clumps/ml)	24°C	31°C
10^3	1.8×10^{-3}	1.7×10^{-3}
10^4	7.2×10^{-3}	7.1×10^{-3}

The 50-80μm fraction of an 8-days-old suspension culture was inoculated in Gamborg's B5 basal medium (3 ml in 35-mm tissue culture dishes under 1500 lux). The density of the initial inoculum is shown in the first column.

through differentiation until the plantula stages, followed by further dedifferentiation. This was done also to eliminate unstable phenotypes. Those cell lines were then checked again with a more accurate protocol. About half of them proved definitively ts for embryogenesis (Table 2). Instability mechanisms should not be necessarily invoked for this loss, due to the fact that we used rather generous criteria in the first screening to include also "leaky" phenotypes. The results indicate also that time between negative and positive selection is irrelevant to the selective efficiency of the protocol. On the other hand, a more stringent negative selection protocol, performed with a 150 μm screen, gave no positives on 45 lines analyzed.

Differentiation and Growth Phenotypes

The inhibition of embryogenesis and or growth was determined in highly reproducible conditions (Giuliano et al., in preparation). Fig. 1 shows the % inhibition of embryogenesis and of growth at the non-permissive temperature for some variants.

Table 2: No. of ts mutants recovered, after enrichment, in different fractions.

Fraction	Lines analyzed	1st screening (ts)	(const.)	2nd screening (ts)
A	45	0	0	-
B	80	11	2	5
C	115	17	4	8
Tot.	240	27	6	13

A first negative selection, performed by filtration with 250 μm screen was followed by regeneration at 24°C for 5 days (Fraction C) or 9 days (Fraction B) and collection of newly formed embryoids with a 300 μm screen. A stronger negative selection (150 μm screen) did not give positive results (Fraction A). The second screening was performed after one cycle of differentiation-dedifferentiation. In the first screening, const. is the number of lines inhibited in differentiation even in permissive conditions.

As can be seen the lesions affect the two phases (growth and differentiation) in widely different, mutant-specific ways.

For the lines tested (ts 2,6) the inhibition of embryogenesis proved to be perfectly reversible after shoft down of the culture at 24°C. The situation is somewhat different in the case of inhibition of growth: ts 5 regains immediately the ability to grow after shift-down, whereas ts 11 shows a large lag. Another experiment performed by growing the cultures at the permissive and non-permissive temperatures before differentiating them (again at the two temperatures, giving in total 4 different regenerating cultures) showed two kinds of effects after pregrowth at the non-permissive temperature: a) a clear inhibition of embryogenesis shown by growth-ts lines (such as ts 5, 11) even at the

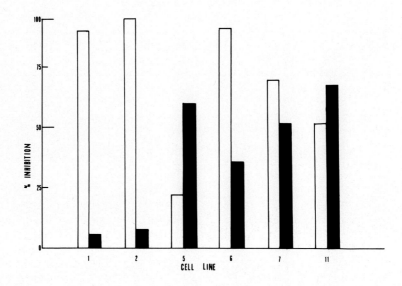

Fig. 1: Temperature sensitivity of the variants. % inhibition of embryogenesis (open bars) and of growth (solid bars) at the non-permissive temperature for some cell lines. Embryogenesis was induced as described in the legend to table 1 and the embryoids counted after 15 days. Growth was monitored by looking at the number of duplication of cell volume after 20 days in cultures obtained by initially inoculating 2cc of packed cells into 50 ml of medium. The cultures were kept under 500 lux on a 70 rpm gyratory shaker. The inhibition of growth normalized for wt, which shows itself a 30% inhibition at 31°C.

permissive temperature.
b) a mutant-specific alteration in the number of embryoids obtained at 24°C vs 31°C.

To understand something more about the time

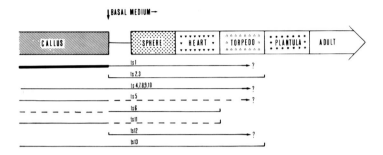

Fig. 2: Time patterning of expression of the ts phenotype in our variants. Continuous line: ts phenotype. Dashed line: reduced expression. Thick line: constitutive expression. Question mark: indetermination about the end of the temperature-sensitive stage (missing experiments).

patterning of the expression of the genes involved, single embryoids, belonging to various stages, were shifted up to 31°C. In fig. 2 we summarize these results, together with those on callus growth. All lines are blocked in the early embryogenesis, and a good fraction in callus growth as well. One interesting feature is that all variants tested seem able to overcome the block after a certain developmental stage (somewhat different for different variants) and are no longer temperature-sensitive in adult life.

Morphological analysis

A morphological analysis has been started on various structures resulting from embryoids shifted up at different stages: globular, heart- ,torpedo-shaped. Sofar 4 ts variants have been studied with the method described in the legend to fig. 4, which allows one to detect, for the same embryoid, the pattern of differentiation of different tissues. The preliminary results state that each apparently homogeneous stage (globular-heart-torpedo) may have physiologically distinct moments which respond differently to the shift-up. Such a situation is particularly evident at the heart stage, which, both in ts 2 and ts 6, when shifted up to the non-permissive temperature, may be either blocked immediately (13 out of 21 for ts 2 and 5 out of 26 for ts 6) or may proceed in differentiation to a torpedo-like stage. The procambium differentiation in late heart-shaped stage seems to have a key role and, once allowed to form, it proceeds to final maturation as far as vascular tissue is concerned, in parallel with torpedo formation. Nevertheless, the development is markedly altered, and irregular masses of tissues are produced.

For ts 2 the shift-up from the globular stage results in a complete block, whereas for ts 6 two out of ten embryoids could differentiate only a well developed root. In this case, and in the case of shift-up from heart stage as well, the cotyledons seemed arrested, the axis bearing a rosette of short cotyledon-like leaves (Fig. 4), whose bases are fused and which are apparently not developping from a plumula. (Esau, 1940).

The opposite was true for two other variants (ts 13 and ts 3), which, when shifted up from the heart stage, show a complete inhibition of root differentiation; the vascular system seems arrested at the transition region just below the branching of the cotyledonary traces and various out-growths of proliferating callus tissue develop instead of the root. (Fig. 5b,c,d). In most of the shift-ups so far analyzed, the apical meristems (either of the shoot, or the root or both) seemed absent, even if the cotyledonary traces or the root vascular tissue were well differentiated.

CONCLUSIONS

We think we have shown that mutants ts for

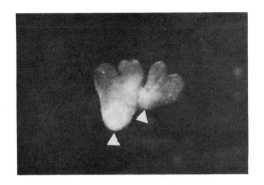
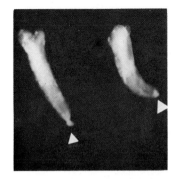

A   B

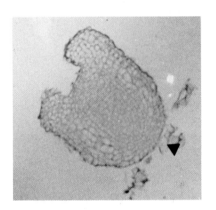
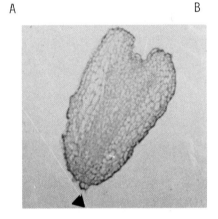

C   D

Fig. 3: Normal somatic embryogenesis. A: Heart and early torpedo (500x). B: Late torpedoes (200x). C and D: Safranin-fast green stained sections of heart and early torpedo, showing procambium differentiation (1000x). E: The same on late torpedo, (showing meristem and vascular differentiation (400x). Triangles indicate, in this and subsequent figures, the radical region.

E

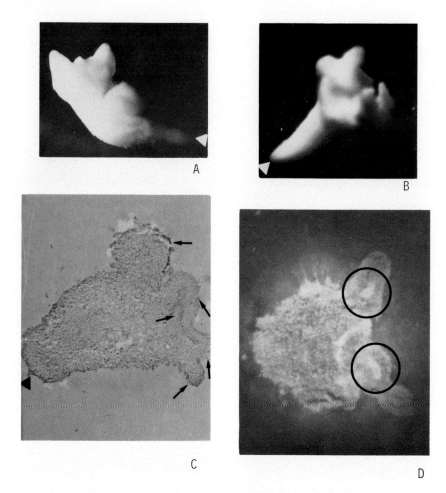

Fig. 4: A: Shift-up from sphere of ts 6 (200x). B: Shift-up from heart (200x). C: Safranin-fast-green stained section of B, showing multiple cotyledon-like leaves (arrows) (400x). D: Another section of B, observed in primary fluorescence, showing differentiation of vascular tissue in the leaves (circles) (400x). Embryoids were photographed and cryostat sectioned 20 days after shift-up. Sections were 20μm thick and were either stained with safranin-fast green or observed in primary or induced UV-fluorescence.

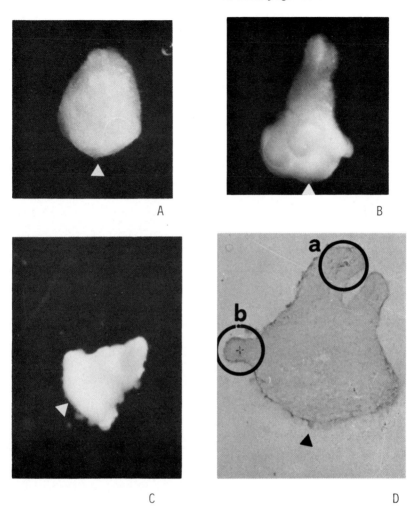

Fig. 5: A and B: Shift-ups from sphere and heart of ts 3 (200x). C: Shift-up from heart of ts 13 (200x). D: Safranin-fast-green section of C, showing callus proliferation instead of root formation, differentiation of vascular tissue in cotyledons (a), and adventive embryony (b) (400x).

development can be obtained with relative ease.
After enrichment 13 out of 195 embryoids analysed
(coming from the fractions B and C) proved ts. If
we compare this figure with the frequency with
which drug resistant mutants were obtained after
the same mutagenic treatment, we end up with a
rough estimate of the number of genes involved in
development which would be in the order of few hundreds.

However it should be stressed that the condition imposed for selection were rather strict: mutants should be reversible upon shift-down and they
should grow in size from <250μm to >300μm in 5 (or
9) days. When we throught that we might increase the
spectrum of our mutants by chooving earlier steps
(<150 μm) we found no mutants at all. By looking
at the other side of the analysed period of development, we notice that the ts function is always
expressed before plantula stage. Does this fact
depend on the way selection operated or is it because morphogenesis is virtually complete at torpedo stage?

Our temporal analysis performed with experiments of temperature shifts gave interesting results concerning different response patterns shown
by different mutants and pointed to a rôle, on differentiation, on the conditions of growth (pre-gro-
wth) unsuspected so far. The histological analysis
however showed an even greater resolving power: the
morphologically defined stages (heart, torpedo etc.)
can be further subdivided and these suddivisions
coincide in a better way with physiologically distinct phases. With this histological analysis,
still very incomplete, interesting observations were already possible: missing organs, the time if
differentiation of new tissues and so on.

We do not think this is the proper place where to discuss these results in detail. We feel however that we have shown that these ts mutants can
be of great help for understanding the various steps
of plant development and, possibly, its plan of
organization.

REFERENCES

Ashton NW, Cove DJ, Featherstone DR (1979). The isolation and physiological analysis of mutants of the moss, Physcomitrella patens, that over-produce gametophores. Planta 144:437.
Backs-Hüsemann D, Reinert J (1970). Embryobildung durch isolierte Einzelzellen aus Gewebenkulturen von Daucus carota. Protopl. 70:49.
Carlson PS (1970). Induction and isolation of auxotrophic mutants in somatic cell cultures of Nicotiana tabacum. Science 168:487.
Dudits D, Lazar G, Bajszar G (1979). Reversible inhibition of somatic embryo differentiation by bromodexyuridine in cultured cells of Daucus carota L. Cell Diff 8:135.
Esau K (1940). Developmental anatomy of the fleshy storage organ of Daucus carota. Hildgardia 13:175.
Fujimura T, Komamine A (1975). Effects of various growth regulators on the embryogenesis in a carrot cell suspension culture. Pl Sc Lett 5:359.
Gehring W (1976). Developmental genetics of Drosophila. Ann Rev Genet 10:319.
Hari V (1980). Effect of cell density changes and conditioned media on carrot cell embryogenesis. Z Pfl 96:227.
Hirsch D, Vanderslice R (1976). Temperature-sensitive developmental mutants of Caenorhabditis elegans. Develop Biol 49:220.
Huskey RJ, Griffin BE, Cecil PO, Callahan AM (1979). A preliminary genetic investigation of Volvox carteri. Genetics 91:229.
Komamine A, Morigaki T, Fujimura T (1978) Metabolism in synchronous growth and differentiation in plant tissue and cell cultures..In Thorpe T (ed):Frontiers of plant tissue culture, Calgary p. 159.
Lo Schiavo F, Biasini G, Giuliano G, Terzi M (1981). Development of a selective system for somatic cell hybrids In: IAEA International symposium on induced mutations as a tool for crop plant improvement, Vienna. In the press.
Malmberg RL (1979). Temperature-sensitive variants of Nicotiana tabacum isolated from somatic cell culture. Genetics 92:215.
McWilliam AA, Smith SS, Street HE (1974). The origin and development of embryoids in suspension

cultures of carrot (Daucus carota). Ann Bot 38:243.
Halperin W .(1966). Alternative morphogenetic events in cell suspensions. Am J Bot 53:443.
Puck TT, Kao FT (1967). Genetics of somatic mammalian cells V. Treatment with 5-Bromodeoxyuridine and visible light for isolation of nutritionally deficient mutants. PNAS 58:1227.
Reinert J (1958). Morphogenese und ihre Kontrolle an Gewebenkulturen aus Karotten. Naturwiss 45:244.
Savage A, King J, Gamborg O (1979). Recovery of a pantothenate auxotroph from a cell suspension culture of Datura innoxia mill. Pl Sc Lett 16:367.
Schulte U, Zenk MH (1977). A replica plating method for plant cells. Phys Plant 39: 139.
Vardi A, Spiegel-Roy P, Galun E (1975). Citrus cell culture: isolation of protoplasts, plating densities, effect of mutagens and regeneration of embryos. Pl Sc Lett 4:231.
Warren GS, Fowler MW (1977). A physical method for the separation of various stages in the embryogenesis of carrot cell cultures. Pl Sc Lett 9:71.

COMPARISON OF ENDOGENOUS LECTINS AND GLYCOSIDASES IN WHEAT GRAINS

Rosemary C. Miller and Dianna J. Bowles

Department of Biochemistry

University of Leeds, LS2 9JT, West Yorkshire, UK

INTRODUCTION

Lectins are proteins which bind specific carbohydrate residues reversibly, non-enzymically and with multiple binding sites. As such they have become extremely useful tools in biochemical research. However, it is only recently that the physiological role of the endogenous lectins has been studied with any great effect. Data from a variety of systems indicate that the synthesis of lectins can be developmentally regulated (Rosen et al., 1970; Barondes S.H., 1980; Nowak et al., 1976; Den et al., 1976). The transient presence of lectins at a particular stage of development suggests that they may be involved in a specific recognition phenomenon characteristic of that stage.

However in contrast, the function of lectins may not be an inherent property of the molecules but may depend rather upon the receptor to which the lectin is attached at any one time, i.e. one lectin could have several functions depending on the array of receptors in its immediate environment (Barondes, 1981).

Wheat germ agglutinin (WGA), originally isolated from commercial wheat germ, is a plant lectin, the physical and chemical properties of which have been well characterised (Aub et al., 1963; Burger and Nagata, 1967). It is an H_2O-soluble, dimeric, protein (subunit molecular weight 18,000) with four equivalent binding sites for residues of N-acetyl-D-glucosamine and its $\beta(1\to4)$-linked oligomers (Nagata and

Burger, 1974; Allen et al., 1973; Privat et al., 1974; Wright, 1981).

Most of the well-studied plant lectins have been isolated from the cotyledons of legumes and suggestions have arisen in the literature that these lectins may be necessary for correct symbiotic relationships between legumes and free-living or rhizobial nitrogen-fixing organisms. However, since both the distribution and developmental regulation of WGA is very different to that of the legume lectins, the biological role of WGA may also differ. Our own studies, described here and elsewhere, have shown that the lectin is localised in the embryo not in the storage tissue and yet there are no soluble glycoprotein receptors to the lectin in the embryo (Miller and Bowles, in preparation). WGA is found to be developmentally regulated since activity appears late in grain maturation and decreases after germination.

If the physiological function of the lectin is directly related to its carbohydrate-binding capacity, the presence or absence of complementary glycosidases in the endogenous environment could switch off or switch on lectin-receptor binding interactions. In this paper we describe experiments on the comparative localisation of the lectin and complementary glycosidases during germination of wheat.

MATERIALS

Wheat grains (*Triticum aestivum*) (Flanders variety) were obtained from University of Leeds, Headley Hall Farm, Tadcaster, Yorkshire, UK. All chemicals were of Analar grade and were obtained through Sigma Chemical Company Ltd., Fancy Road, Poole, Dorset BH1Y 7NH, UK).

METHODS

1. Dissection of Seeds and Preparation of Embryos

(a) For the dissection experiments 15 wheat grains were soaked for 1 hr, the testas removed and the embryo containing the scutellum, undeveloped cotyledons and axis, were dissected out by hand. The remaining endosperm and aleurone layer was sectioned into five pieces numbered 1 to 5 as shown in Fig.1. Respective sections were pooled in 1 ml of ice cold buffer containing 10 mM KPO_4 and 145 mM NaCl (PBS) with 5 mM PMSF prior to hand homogenisation in a pestle and mortar and sonication of the homogenates at 5 x 10 sec using a Dawe sonicator fitted with a microprobe. The suspension was then centrifuged at 10,000 g for 4 min at room temperature (Eppendorf microfuge) to give a supernatant which was assayed and an insoluble particulate fraction which was discarded.

(b) For 0 hr experiments dry seeds were used for the experimental material. The testa was not removed and a bulk sample of embryos was prepared using a modification of the sucrose flotation method as described by Johnston and Stern (1957).

For times up to 96 hr experiments the seeds were grown as described in the next section. By 96 hr the embryonic axis had developed into a root (1-2 cm) and shoot (0.5-1 cm) which was dissected by hand from the endosperm.

2. Growth Conditions of Wheat Seeds

Wheat seeds were surface sterilised by soaking in Milton's sterilising fluid for 0.5 hr before thoroughly rinsing with sterile water. The seeds were then soaked overnight in sterile water containing chloramphenicol (25 $\mu g.ml^{-1}$) before being planted in sterile Petri dishes lined with 2 layers of Whatman No.1 filter paper also soaked in sterile water containing chloramphenicol. After the required period of growth, in dark conditions, the seeds were dissected into embryo and endosperm.

3. Extraction Procedure

Tissues were extracted with PBS containing 5 mM benzamidine. In all experiments, the tissue/buffer ratio was 1 g fresh weight/10 ml extraction medium. After homogenisation in a pestle and mortar, and sonication (6 times for 10 seconds at maximum current on a Dawe sonicator fitted with a microprobe) the extract was centrifuged at 100,000 g for 1 hr (Beckman Model L Ultracentrifuge Ti50 rotor) at $4^{\circ}C$. The resultant supernatant ($\equiv$ soluble extract) was removed and analysed.

4. Protein Estimations

Unless stated otherwise protein estimations were performed by the method of Sedmark and Grossberg (1977) using Coomassie Dye (blue G90, BDH).

A modification of the Lowry method (1951) was also used for estimations of purified WGA and extracts containing Triton X-100. It was noted for WGA that the Lowry method gave equivalent protein as dry weight estimations whilst the Sedmark method gave only a quarter of that quantity (Miller and Bowles, unpublished).

5. Haemagglutination Assays

Haemagglutination assays were carried out using microtitre plates as in Bowles and Kauss (1976). Human red blood cells (of non-specified blood group) were treated for 5' at $37°C$ with pronase (Sigma), at a conc. of 1 mg pronase.ml^{-1} packed washed cells in a final vol. of 2 m PBS, and used in the assay at a final concentration of 0.33% washed RBCs.

6. Glycosidase Assays

p-Nitrophenol derivatives of monomeric sugars (Sigma) were dissolved in H_2O at a concn. of 1 mg.ml^{-1}. 300 μl of sugar solution was added to an Eppendorf tube containing 100 μl of H_2O and 100 ml of sodium acetate (500 mM, pH 5.5). This was preincubated at $37°C$ before addition of 500 μl of soluble seed extract in PBS at 1 mg.ml^{-1} equivalent protein. 200 μl of solution was removed at times 0, 3, 6 and 10 min. Each was added to 1.8 ml of sodium carbonate (50 mM) and the absorbance of the resultant solution read at 420 nm.

RESULTS

Localisation of WGA in Wheat Grains

(a) <u>Dry grains</u>. Wheat grains were sectioned as described in Methods and soluble extracts in PBS were made. Each extract was assayed for haemagglutination activity using pronase-treated human red blood cells. As shown in Table 1, all haemagglutination activity was found in the embryo or the section next to the embryo containing fragments of embryo from the dissection procedure. This activity was inhibited by 50 mM N-acetyl-D-glucosamine (GlcNAc) and was concluded to be WGA.

Fig.1   Sectioning of Grains

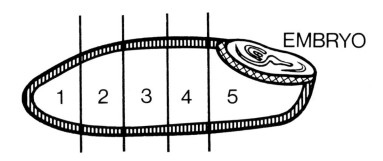

Table 1   Haemagglutination assays of dry wheat grain sections

Grain sectioning, preparation of soluble extracts and haemagglutination assays were carried out as described in Methods. All agglutination activity was inhibited by 50 mM N-acetyl-D-glucosamine.

Section of seed	Protein concentration mg/ml	Haemagglutination titre
Embryo	1.75	64
5	2.25	2
4	1.95	0
3	1.85	0
2	1.9	0
1	1.675	0
WGA	2.0	1600

(b) <u>Imbibed grains</u>. Wheat grains were imbibed under sterile conditions as described. At intervals of 24 hr grains were dissected into embryo and endosperm and soluble extracts were assayed for lectin activity. The results were as shown in Table 2. The amount of WGA in the embryo of wheat grains decreases in a uniform manner with respect to time over the first 96 hr of imbibition. No WGA was found in the endosperm at any time after imbibition.

Table 2   <u>Haemagglutination assays of imbibed wheat grain sections</u>

Grain sectioning, preparation of soluble extracts and haemagglutination assays were carried out as in Methods. All agglutination activity was inhibited by 50 mM N-acetyl-D-glucosamine. Extracts were assayed at 1 mg.ml^{-1} protein concentration.

Time (Hr)	Embryo H.U. for 1 mg.ml^{-1}	Endosperm H.U. for 1 mg.ml^{-1}
0	64	0
24	32	0
48	32	0
72	4	0
96	2	0

2. Glycosidase Activity in Wheat Grains

(a) <u>Dry grains</u>. Soluble extracts from embryo and endosperm were tested for glycosidase activity. The p-nitrophenyl-glycosides tested were related to the specificity of WGA and included p-nitrophenyl-β-N-acetyl-D-glucosaminide, p-nitrophenyl-β-N-acetyl-D-galactosaminide, p-nitrophenyl-α-D-glucoside, p-nitrophenyl-β-D-glucoside and p-nitrophenyl-α-D-mannoside. The results are shown in Fig.2. Under the conditions described, the highest glycosidase activity recorded was β-glucosidase. However, enzyme activities were higher in the endosperm tissue than in the embryo. N-acetyl-D-glucosaminidase activity was found in both the embryo and endosperm.

Fig.2

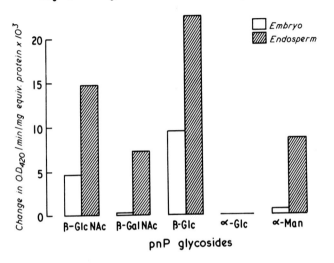

(b) <u>Imbibed grains</u>. As described before, wheat grains were imbibed under sterile conditions and dissected into embryo and endosperm at 24 hr intervals. Soluble extracts from the developing embryo and endosperm were assayed for certain glycosidase activities, i.e. β-N-acetyl-D-glucosaminidase, β-N-acetyl-D-galactosaminidase, β-D-glucosidase and α-D-mannosidase. It can be seen from the results shown in Figs.3A and 3B that the only glycosidase activity which is significantly increased is the β-D-glucosidase activity in the embryo tissue after 48 hr. Although other glycosidase activities exist in both embryo and endosperm, these are maintained at the low level found in the dry grain.

Fig.3.

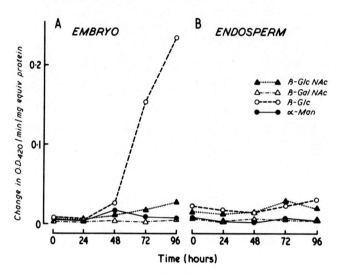

Activity of Soluble Glycosidases prepared from developing wheat grains measured at pH 5·0

DISCUSSION

Results from the haemagglutination assays indicate that WGA is present only in the embryo of a dry wheat grain. This is in direct contrast to the localisation of legume lectins such as those of Jack Bean and Soybean where the respective lectins (concanavalin A and soybean agglutinin) are found mainly in the storage tissue (the cotyledons) of the seed. Wheat, like all other grasses, is a monocot and its storage tissue, the endosperm, is composed mainly of dead cells, surrounded by a living aleurone layer which synthesises the necessary enzymes, in particular α-amylase, to catabolise the stored reserves. As shown in Table 2, no WGA was ever found in the storage part of the seed between 0 and 96 hr after imbition. Therefore, presumably, WGA is not involved in any of the processes of breakdown of stored reserves in the endosperm during germination.

The overall amount of WGA in the embryo decreases

uniformly with time after imbibition. This suggests that a prime function of the lectin is during grain maturation or during the very early stages of grain germination. Mishkind et al. (1980) have shown a similar overall decrease in the amount of the lectin during development; by 34 days, the remaining WGA (~50% of that in dry grain) was maintained at measurable quantities only in the rapidly dividing meristemic tissues of the roots and shoots.

Regulation of lectin function could be achieved by the regulation of the receptor(s) which are present. The types of glycoprotein receptors to WGA in the germinating wheat grain is currently under investigation (Miller and Bowles, in preparation). However, glycosidases capable of removing the specific sugar residues which the lectin recognises may be involved in the lectin-receptor interaction. As shown in Figs.1 and 2, a N-acetyl-D-glucosaminidase capable of hydrolysing p-nitrophenyl-β-N-acetyl-D-glucosamine does exist in the embryo and endosperm of the germinating wheat grain but is present only at comparatively low levels. Clearly, endogenous glycosidase may be present in the tissue but remain either undetected or measurable only at low levels due to the unsuitability of the synthetic substrate used in the assay. The α-amylase of wheat is a glaring example of such an α-glucosidic hydrolase.

At the present time, we can say that both a lectin and a glycosidase with parallel specificity exist in developing wheat grains. Their functional relationship and their potential involvement in grain development must await further investigation.

References

Allen AK, Neuberger A, Sharon N (1973). Biochem J 131:155.
Aub JC, Tieslau C, Lankester A (1963). Proc Natl Acad Sci USA 50:613.
Barondes SH (1980). Cell adhesion and motility. Curtis ASG, Pitts J (eds) Cambridge: Cambridge University Press p 309.
Barondes SH (1981). Ann Rev Biochem 50:207.
Bowles DJ, Kauss (1976). Biochim et Biophys Acta 443:360.
Burger MM, Goldberg AR (1967). Proc Natl Acad Sci USA 57:359.
Den H, Malinzak DA, Rosenberg A (1976). Biochem Biophys Res Commun 69:621.
Johnston FB, Stern H (1957). Nature 179:160.

Lowry OH, Rosebrough NJ, Farr AL, Randall RJ (1951). Journal of Biological Chemistry 193:265.
Mishkind M, Keegstra K, Palevitz BA (1980). Plant Physiology 66:950.
Nagata Y, Burger MM (1974). J Biol Chem 249:3166.
Nowak TP, Haywood PL, Barondes SH (1976). Biochem Biophys Res Commun 68:650.
Privat JP, Delmotte F, Monsigny M (1974). FEBS Letts 46:229.
Rosen SD, Rotherman RW, Barondes SH (1970). Exp Cell Res 95:159.
Sedmark JJ, Grossberg SE (1977). Analytical Biochemistry 79:544.
Wright CS (1981). J Mol Biol 145:453.

GLOBULINS OF DEVELOPING MAIZE SEEDS: PRELIMINARY
CHARACTERIZATION

Christa Dierks-Ventling

Friedrich Miescher-Institut
P.O.Box 273
CH-4002 Basel, Switzerland

Globulins of plants represent a class of proteins found in seeds of legumes and cereals from where they can be extracted by means of high salt. In legume seeds, globulins are the major reserve proteins; in cereals they are a minor class yet subject to substantial increases in certain mutations and therefore of interest to agricultural nutritionists as well as scientists. Cereal globulins furthermore contain <u>all</u> amino acids for human and animal nutrition while the major cereal storage protein class is deficient in one or more essential amino acids. In the case of maize, the world's third most important cereal crop, the predominant but deficient protein is zein which lacks lysine and is very low in tryptophan and methionine.

With the discovery of the high-lysine, low-zein opaque-2 mutant of maize in 1964 (Mertz et al 1964) a type of corn was found where the deficiencies were overcome. Not only was zein reduced in this mutant, but the water- and salt-soluble protein fraction was increased and thus nitrogen was preserved (Misra et al 1972). Our interest in storage protein regulation has led us to study the various protein classes in opaque-2 (Dierks-Ventling 1981) and other mutants (Dierks-Ventling 1982) for quantitative and qualitative differences with the hope of obtaining more specific information on certain proteins. In this paper I recall some of the findings and report preliminary results obtained on globulin proteins.

EXPERIMENTAL PROCEDURE

Globulins were extracted from endosperms of mature W 64A normal and opaque-2 mutant seeds. Both maize lines were also grown in the greenhouse, where they were selfed and ears were harvested every 10 days after pollination in order to obtain endosperms for a developmental study of storage proteins. The endosperms were ground to a fine flour (or homogenized if very young), defatted by means of acetone and then submitted to a series of extractions yielding quantitatively the water-soluble proteins (= albumins) first, then the 0.5 M NaCl soluble ones (= globulins), followed by the alcohol-soluble prolamines (= zeins) and last the alkali-detergent extractable proteins (= glutelins). Precise details of the extraction procedure as well as of the conditions employed for SDS-polyacrylamide gel electrophoresis are being published elsewhere (Dierks-Ventling 1981).

RESULTS AND DISCUSSION

A quantitative extraction of all endosperm protein classes, carried out with mature seeds of four different maize lines and their corresponding opaque-2 mutants showed that indeed the reduction in zein seen in all opaque-2 mutants, was accompanied by a 100% increase each of albumins and of globulins (Table 1). Since albumins and globulins represent families of proteins, the question arose whether among these families specific proteins would be increased in the opaque-2 mutation. For this purpose, albumins and globulins from mature normal and mutant endosperms were also analyzed on SDS-PAGE. The results showed (Fig. 1) that among globulins clearly certain polypeptides were increased, notably a 47 K, 52 K and 58 K, in one case also a 21 K and less clearly a 68 K polypeptide in all mutants. The latter polypeptide was also increased among albumins (Dierks-Ventling 1981) and may therefore be a protein carried over from the albumin into the globulin fraction.

In fact, albumins are not true storage proteins. This became evident when the accumulation of storage proteins was studied as a function of time of development of seeds (Fig. 2). Albumins consisting of water-soluble proteins and therefore a fair number of enzymes, increased to a peak coinciding with the cessation of cellular division. After this

Table 1. Quantitative distribution of endosperm proteins in in mature seeds. From the following normal genotypes W 64A, R 802A, B 37 and Oh 43 and from their corresponding opaque-2 mutants, at least 4 different endosperm preparations were made and subjected to a quantitative analysis of all major storage proteins. For details see Dierks-Ventling 1981. Average $\pm$ 1 SD is given.

Proteins	Normal endosperm		Opaque-2 endosperm	
	mg/g flour	% of total	mg/g flour	% of total
Albumins	9.3 $\pm$ 3.1	11.8	15.4 $\pm$ 2.7	20.8
Globulins	7.8 $\pm$ 2.7	9.9	12.9 $\pm$ 2.1	17.4
Zeins	30.0 $\pm$ 3.8	38.0	15.4 $\pm$ 5.4	20.8
Glutelins	31.9 $\pm$ 4.1	40.3	30.4 $\pm$ 4.3	41.0
Total	79.0	100	74.1	100

time albumins were removed faster than they were synthesized. Although albumins were found increased at maturity in opaque-2 mutant (cf. Table 1), this increase was only apparent resulting from a lesser degree of breakdown of albumins in the mutant toward maturation.

Globulins, on the other hand, are true storage proteins. This was borne out of the fact that they accumulated steadily in normal seeds just like zeins, but in the mutant they increased at a greater rate from the moment that zein synthesis stopped (around 30 days). This suggested that the regulation of globulin and zein synthesis may be linked.

In another maize mutant dependent solely on proline for growth, a proline-auxotroph, which segregates on normal ears as opaque-looking seeds, the zein content was also found to be reduced and this reduction was in part compensated for by an increase in globulins and albumins (Dierks-Ventling 1982). Furthermore, a 47 K globulin polypeptide was also enhanced in endosperms of pro mutant seeds.

These results, obtained with two unrelated maize mutants lead to the conclusion that the regulation of zein synthesis may be coupled to that of globulin synthesis in a as yet unknown fashion. Such a hypothesis requires bio-

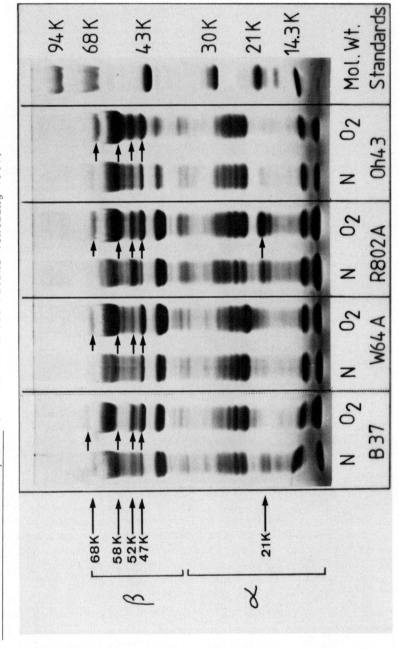

Fig. 1. SDS-PAGE of mature endosperm globulins from different opaque-2 mutants and their normal counterparts (For details see Dierks-Ventling 1981)

Fig. 2. Accumulation of storage proteins in W 64A normal and opaque-2 mutant endosperms during development (Fig. from Dierks-Ventling 1981). Albumins, globulins and zeins were extracted from developing endosperms at regular intervals after pollination. They were plotted as mg protein per endosperm as a function of age (days post pollination). From mature material, the average value from four different determinations carried out in duplicates $\pm$ 1 SD were plotted. Note that in contrast to globulins and zeins, a reduction of albumins occurred during the second half of development, which was less pronounced in normals than in mutants and which therefore led to greater values at maturity in the mutant line than in the normal. The apparent increase in albumins in mature mutant endosperm is therefore a pseudo-increase.

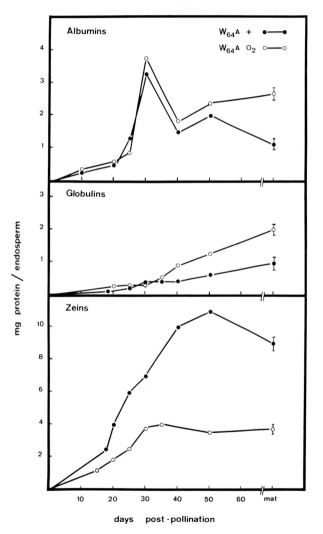

chemical studies of the properties of the respective proteins first and ultimately a structural study of the zein and globulin genes themselves.

As a beginning, we have started a purification of globulins concomitant with a study of the physical and chemical properties. We have analyzed the protein composition of globulins, isolated from endosperms of normal and opaque-2 mutant seeds. Results are seen on Table 2.

Table 2. Amino acid composition of endosperm globulins from W 64A normal and opaque-2 mutant seeds. Globulins (1 mg) were hydrolyzed in sealed tubes with 6 N HCl at 120°C for 24 hours. The HCl was removed by lyophilization and the amino acids were dissolved in 0.2 N citrate buffer pH 2.2 for analysis on a Technicon semi-automatic amino acid analyzer. No values are given for Tryp, Cyst and Met because these amino acids are destroyed during acid hydrolysis.

	mol/100 mol total amino acids	
	normal	opaque-2
Lys	8.3	8.1
His	4.0	5.2
Arg	10.2	12.0
Asp	7.2	7.2
Thre	3.9	3.6
Ser	5.2	5.1
Glu	12.0	11.4
Pro	6.2	6.1
Gly	9.8	9.1
Ala	9.7	8.8
Val	7.1	6.8
Ileu	4.2	4.0
Leu	7.1	6.7
Tyr	0.6	1.3
Phe	4.2	4.8

Globulins are rich in lysine compared to zeins (Gianezza et al 1977). Some differences exist in the amino acid composition of globulins: those from the mutant contain more His, Arg, Phe and Tyr and equal or less amounts of all the other amino acids than those from normal seeds. We do not know whether this results from the specifically increased globulin polypeptides in the mutant. Globulin analysis by analytical centrifugation produced three Svedberg constants whereby most of the material sedimented as a 1.76 S peak (Table 3).

Table 3. Svedberg constants for globulins. S_{20} values were obtained from an analysis in the analytical centrifuge (courtesy Dr. A. Lustig, Biozentrum, Basel) of globulins extracted from endosperms of W 64A normal seeds and dissolved in 0.5 M NaCl - 1 mM 2-mercapto-ethanol.

S_{20}	1.76	6.67	8.55
Distribution	60%	30%	10%

In 5-25% sucrose gradients, containing 0.5 M NaCl and 1 mM 2-mercapto-ethanol, run for 28 hours at 40,000 rpm and 4°C using a SW 40 rotor, globulins also produced three peaks of 0.7, 2.8 and 8.1 S. When the fractions under each peak were collected and analyzed on SDS-PAGE, all the major polypeptides were found just like in the unfractionated material. Differential aggregation is therefore not excluded. Danielsson (1949) found S values of 2.6 and 8.5 for maize globulins; it is not clear at the moment where the discrepancies originate from.

Maize globulins being very hydrophobic, can only be solubilized - once precipitated out - by means of reducing agents in conjunction with detergents or 6 M urea. Thus, a fractionation of the globulin class must be done in the appropriate solvent. We have fractionated globulins on DE-52 columns using 6 M urea - 10 mM 2-mercapto-ethanol - 10 mM Tris buffer pH 8.0 in a salt gradient from 0-0.5 M NaCl (Fig. 3) for elution of the polypeptides. The fractions underneath the peaks were pooled, dialyzed, lyophilized and analyzed on SDS-PAGE.

Fig. 4 shows that most fractions contain 1-2 major globulin polypeptides. Each fraction will be further fractionated using a different chromatography system until individual, purified polypeptides are obtained. Such experiments are under way.

Fig. 3. Fractionation of globulins from W 64A opaque-2 endosperms by DE-52 column chromatography.

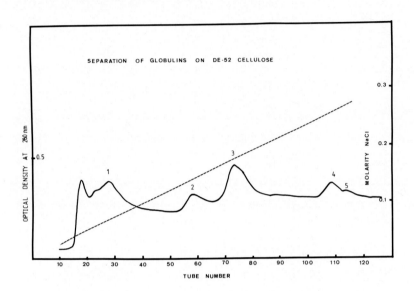

Fig. 4. SDS-PAGE of globulin fractions from Fig. 3.

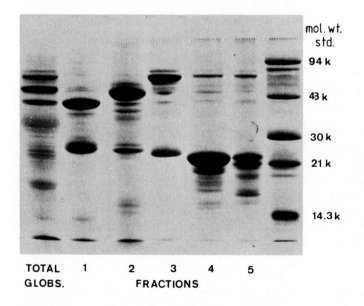

ACKNOWLEDGEMENT

I am most grateful to Dr. Ib Jonassen for his help with the column chromatography and to Ms. Sylvia Wirth for her excellent technical assistance.

REFERENCES

Danielsson CE (1949). Seed globulins of the Gramineae and Leguminosae. Biochem J 44:387

Dierks-Ventling C (1981). Storage proteins in Zea mays (L): Interrelationship of albumins, globulins and zeins in the opaque-2 mutation. Eur J Biochem in press.

Dierks-Ventling C (1982). Storage protein characteristics of proline-requiring mutants of Zea mays (L). Theoret Appl Genet in press

Gianezza E, Viglienghi V, Righetti PG, Salamini F, Soave C (1977). Amino acid composition of zein molecular components. Phytochemistry 16:315

Mertz ET, Bates LS, Nelson OE (1964). Mutant gene that changes protein composition and increases lysine content of maize endosperm. Science 145:279

Misra PS, Jambunatham R, Mertz ET, Glover DV, Barbosa HM, McWhirter KS (1972). Endosperm protein synthesis in maize with increased lysine content. Science 176:1425

STUDIES ON THE CHALCONE SYNTHASE GENE OF TWO HIGHER PLANTS:
PETROSELINUM HORTENSE AND MATTHIOLA INCANA.

V.Hemleben, M.Frey, S.Rall, M.Koch, M.Kittel
University of Tübingen, Institute of Biology II
74oo Tübingen, FRG
F.Kreuzaler, H.Ragg, E.Fautz, and K.Hahlbrock
University of Freiburg, Institute of Biology II
78oo Freiburg, FRG

INTRODUCTION

The regulation of coordinated gene expression in a specific developmental step or in a metabolic pathway is still an unsolved problem of cell biology. Higher plants are very sensitive to external factors as light, phytohormones, temperature or O_2-availability to summarize some of those stimuli which induce a defined coordinated reaction. Light causes the most dramatic effects in plants; in addition to chloroplast development and differentiation of the photosynthetic apparatus various other developmental processes follow light induction (photomorphogenesis; Smith, 1975). We are interested in the induction of flavonoid biosynthesis by light in seedlings, flowers or tissue culture cells of higher plants, a process which seems to be very appropriate to study gene regulation and coordinated gene expression (Mohr, 1970).

The pathway of flavonoid biosynthesis has been very well studied on the enzyme level (Hahlbrock and Grisebach, 1979; Fig.1).The first enzyme of this pathway, the phenylalanine ammonia-lyase (PAL) is extremely well investigated. It opens the first group of enzymes (I) involved in the general phenylpropanoid metabolism leading from phenylalanine to the activated cinnamic acid coumaroyl-CoA. This compound is combined with 3 malonyl-CoA to the C-15 chalcone molecule by the key enzyme of the flavonoid biosynthesis, chalcone synthase (the previous flavanone synthase),opening the second group of enzymes (II) producing the modified flavonoid pigments (Hahlbrock and Grisebach, 1979; Hahlbrock et al. 1980; Heller and Hahlbrock, 1980; Spribille and Forkmann,1981).

In the most thoroughly investigated system of parsley tissue culture cells (Petroselinum hortense) the concomitant expression of all group I and II enzymes following light induction could be shown (Hahlbrock et al. 1980) and evidence is given that the regulation takes place on the level of transcription (Schröder, 1978). The main question of our studies presented here, therefore, is how are genetically uncoupled genes activated and regulated on the transcriptional level. For this purpose firstly the gene structure and the regulatory regions of the genes coding for the basic enzymes of the flavonoid metabolism have to be analyzed. We are working with two systems: The group in Freiburg follows the light-induced processes in tissue culture cells of parsley (Petroselinum hortense, Apiaceae); the group in Tübingen is studying the genes involved in anthocyanin biosynthesis in Matthiola incana (Brassicaceae) from which different genetically well defined lines are available (Table 1; Seyffert, 1960, 1963, 1971, and in press) and various developmental stages can be investigated.

The first enzyme we are going to attack on the gene level is the chalcone synthase. A cDNA clone of the parsley gene has been constructed. This cDNA clone conferring the chalcone synthase gene sequences was used as hybridization probe to look for the corresponding gene in Matthiola incana. The structure of the wild type gene shall be compared with the parsley gene and with the gene of the white flowering mutant of Matthiola, line 18, exhibiting a genetic block in gene f, the gene locus responsible for the chalcone synthase. This mutant line shows no enzyme activity (Spribille and Forkmann, 1981). Therefore, it might be interesting to see where the mutation is localized and where the expression is blocked.

THE GENETIC SYSTEM OF MATTHIOLA INCANA

From Matthiola incana 16 different cyanic lines have been developed (Seyffert, 1960, 1963, 1971) homozygous for the three basic complementary acting genes involved in the synthesis of the anthocyanin compound (Table 1; genes f, e, and g). These lines differ only in their genetic constitution with respect to the modificationally acting genes b, l, u, and v (Fig.2). Lines 17 - 19 are white flowering mutants with a specific defect in one of the basic genes f, e, or g. Gene b is coding for a 3'-hydroxylase adding a hydroxy group to the 3' position of the flavonoid compound (Fig.2) therewith deciding on the cyanidin type of the pigment (Forkmann,

1980). Genes l, u, and v are responsible for further glycosylation and acylation reactions on the flavonoid molecule (Fig.2).

```
L-Phenylalanine
    ↓           ◄ Phenylalanine ammonia-lyase (PAL)  ⎤
 Cinnamate                                           |
    ↓           ◄ Cinnamate 4 - hydroxylase           ⎬  I
 4-Coumarate                                         |
    ↓           ◄ 4-Coumarate:CoA ligase              ⎦
 4-Coumaroyl - CoA + 3 Malonyl-CoA                    ⎤
    ↓           ◄ Chalcone synthase    gene f         |
 Naringenin - chalcone                                |
    ↓           ◄ Chalcone isomerase                  |
 Naringenin ────────────────► Eriodictyol             |
►   ↓  gene b    3'-Hydroxylase    ↓  ◄ 3-Hydroxylase  ⎬ II
 Dihydrokaempferol ──────────► Dihydroquercetin        |
                  genes e and g                        |
 Pelargonidin                       Cyanidin           |
    ↓             genes l, u, v        ↓               |
    ↓           modified Anthocyanins  ↓               ⎦
```

Figure 1. Scheme of the metabolic pathway of anthocyanin biosynthesis with the enzymes involved (◄) and the position of the genes known for Matthiola incana.

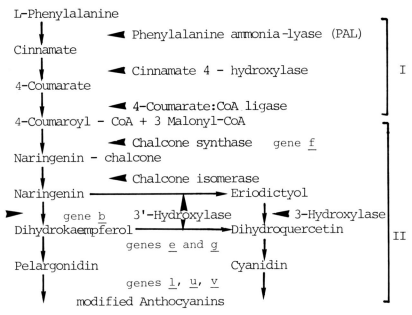

Figure 2. Modifying effects of the alleles l, u, and v, and b on the flavonoid molecule, according to Seyffert,1960).

Table 1. Genotypes of Matthiola incana, cyanic lines
1 - 16, and white mutant lines 17 - 19.

Line	Complementary factors for colour formation			Modificationally acting genes			
	f	e	g	b	l	u	v
01	+	+	+	+	+	+	+
02	+	+	+	+	l	+	+
03	+	+	+	+	+	u	+
04	+	+	+	+	+	+	v
05	+	+	+	+	l	u	+
06	+	+	+	+	l	+	v
07	+	+	+	+	+	u	v
08	+	+	+	+	l	u	v
09	+	+	+	b	+	+	+
10	+	+	+	b	l	+	+
11	+	+	+	b	+	u	+
12	+	+	+	b	l	u	+
13	+	+	+	b	l	+	v
14	+	+	+	b	l	+	v
15	+	+	+	b	+	u	v
16	+	+	+	b	l	u	v
17	+	e	+	b	+	+	+
18	f	+	+	+	+	+	+
19	+	+	g	+	l	u	+

Since gene f is known to code for the chalcone synthase (Fig.1; Spribille and Forkmann, 1981) the mutant line 18 (Table 1) is most interesting for our studies. Wild type lines (1 - 16) will be therefore compared with the mutant line 18.

The Matthiola system we use allows to look for gene structure and gene expression in various mutants and, in addition, on different levels of plant development such as dry seeds, seedlings, several stages of flower development, or tissue culture cells (Fig.3). In order to get a more defined light induction system and to become more indepent from green house or field material we developed callus and cell suspension cultures of the different Matthiola lines (1 - 10 and 17 - 19). Because anthocyanin synthesis was only observed in callus cultures grown on agar medium according to Seitz and Richter (1970) we used this medium for callus cultures derived from root tips of sterile Matthiola seed-

Figure 3. Different developmental stages of Matthiola incana. A. seedlings; B. flowering stem; C. callus culture (line O7).

lings. Under constant culture conditions (27°C, permanent irradiation with white light, 3000 lux) line specific anthocyanins are produced as already demonstrated by Leber (1977). In dark grown callus cells the enzymes of the flavonoid metabolism can be induced by light obviously with a very similar kinetic pattern as in parsley cells. PAL enzyme activity (Fig.4) reaches its maximum about 6 h (line 06 and 10) to 11 h (line 17) after onset of irradiation with white light. Therefore, tissue culture cells of Matthiola can be a suitable material to study processes and structures involved in anthocyanin bio**sy**nthesis activated by light in different genotypes.

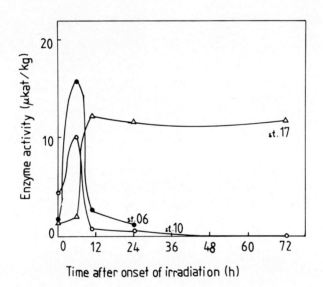

Figure 4. Time course of light-induced changes in PAL enzyme activity in different lines of Matthiola incana (line 06, 10, and 17). PAL activity was determined according to Rissland and Mohr (1967).

CHALCONE SYNTHASE GENE OF PETROSELINUM AND MATTHIOLA

Since the light induction process of flavonoid synthesis has been worked very efficiently for Petroselinum tissue culture cells on the level of enzyme activity and messenger RNA synthesis (Hahlbrock and Grisebach, 1979; Kreuzaler et al. 1979) the attempt of cloning a specific gene sequence coding for group I or group II enzymes (Fig.1), especially the chalcone synthase gene, could be made. Total polyribosomal RNA isolated from previously irradiated Petroselinum cells contained messenger RNA for the group I and the group II enzymes involved in the flavonoid metabolism. This RNA was copied into cDNA with reverse transcriptase and ligated into the restriction enzyme PstI site of the E.coli plasmid pBR322. After transformation of E.coli cells recombinant clones were selected and screened by colony hybridization with ^{32}P-labeled mRNA from light-induced parsley cells in comparison to mRNA from non-induced cells (Fig.5).

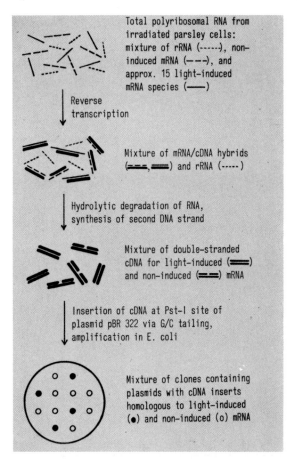

Figure 5. Scheme of the cloning procedure for the chalcone synthase cDNA from Petroselinum.

Positive clones were further analyzed in a hybrid-arrested
in vitro protein synthesis assay (modified after Paterson
et al.1977). The in vitro protein products were characterized
with antibodies raised specificly against chalcone synthase
(Kreuzaler et al. 1979). The clone pLF56 was able to remove
chalcone synthase specific mRNA sequences from the translation system by hybridization. These sequences are released
by heating and produce again chalcone synthase specific proteins demonstrated by immunoprecipitation with the antibody.
Further characterization of pLF56 DNA by restriction enzyme
analysis showed that the plasmid contains an insert 1500
base pairs (bp) in size with a PstI site. PstI digestion
therefore results in three fragments with 4360 (pBR322),
1000 and 500 bp (Fig.6)

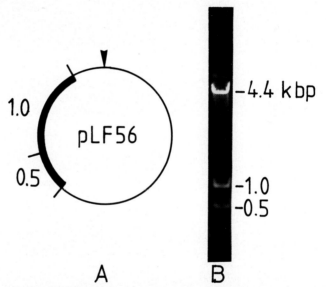

Figure 6. Characterization of the clone pLF56 DNA.
A. Scheme of the pLF56 sequences. | PstI sites; ▼ EcoRI
site. B. PstI digest of pLF56 DNA separated on 1% agarose.

This pLF56 DNA was used as hybridization probe to detect
chalcone synthase specific DNA sequences in Matthiola incana.
Successfull cross hybridization between these parsley sequences and Matthiola DNA could be shown by saturation hybridization with the ^{32}P-nick translated probe and Matthiola DNA
bound onto nitrocellulose filters (Grierson and Hemleben,
1977). To look for the specific genomic sequence high mole-

cular weight DNA was isolated from dry seeds, seedlings, and flower petals of different Matthiola lines. Previously isolated and proteinase K treated cell nuclei were fractionated on neutral caesium chloride gradients and subsequently by a second centrifugation step on actinomycin D-CsCl gradients to remove ribosomal DNA (Hemleben et al. 197). DNA from dry seeds was obtained by phenol extraction after homogenization in TNS medium (Grierson and Loening, 1974). The purified DNA was digested with the restriction endonuclease HindIII and separated on 1% agarose gels (Leber and Hemleben, 1979). The Southern blotted DNA was hybridized to ^{32}P-nick translated pLF56 DNA conferring the cDNA sequences of the parsley chalcone synthase gene. In the corresponding autoradiographs radioactivity was detectable in distinct DNA bands, 4×10^6 and 8×10^6 in size (Fig.7) Only in the lane with DNA isolated from Matthiola line O6 seedlings the high molecular weight band was not visible. In all other cases the hybridization pattern of parsley DNA, DNA from different Matthiola wild type lines (O4 and O6), and from the chalcone synthase mutant line 18 seems to be identical if DNA was digested with Hind III. Possibly, the 8×10^6 band is only an incomplete digestion product which does not show in the DNA preparation of line O6 seedlings.

This means that a DNA fragment with a molecular weight of 4×10^6 (6000 bp) containing the chalcone synthase gene flanked by HindIII cutting sites is present in Matthiola lines, even in line 18, and in the taxonomically far separated species Petroselinum hortense. However, since at least the smaller fragment belongs to the main size class of fragments generated randomly by HindIII this may be a simple case of coincidence: More detailed restriction mapping has to be carried out to decide whether there are differences in sequence organization and where they are located.

Our main effort, therefore, is to construct genomic clones of Matthiola wild type and mutant lines and select for chalcone synthase gene containing sequences. The gene structure of the parsley gene in comparison to the Matthiola gene will be investigated with respect to gene evolution. Gene expression in wild type and mutant cells will be followed on the level of transcription and translation. In addition, the Matthiola mutant line 18, deficient for chalcone synthase, may be a suitable system to study gene transformation in higher plants.

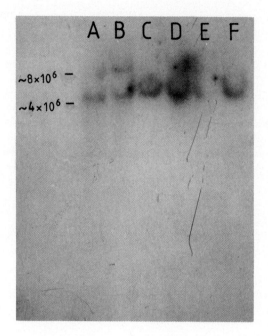

Figure 7. Hybridization of pLF56 DNA to HindIII digested Matthiola and Petroselinum DNA. High molecular weight DNA from seeds (lane A, line O6; lane D, line 18), seedlings (lane F, line O6), or flower petals (lane B, line O6; lane C, line O4), and from parsley tissue culture cells (lane E) were digested with Hind III, the fragments separated on 1% agarose gels, blotted onto nitrocellulose filters, and hybridized to ^{32}P-nick translated pLF56 DNA.

SUMMARY

Two higher plant systems are presented which allow to study coordinated gene expression of the light-induced metabolic pathway of flavonoid biosynthesis: tissue culture cells of Petroselinum hortense (Apiaceae) and different developmental stages of various genotypes of Matthiola incana (Brassicaceae). The gene structure of the chalcone synthase is mainly studied. A cDNA clone (pLF56) of parsley has been constructed and characterized conferring the chalcone synthase gene sequences. Strong cross hybridization between the parsley cDNA and Matthiola DNA allowed to identify

a HindIII fragment (6000 bp) identical in size for parsley
and different Matthiola wild type lines and a mutant line.

REFERENCES

Forkmann G (1980). The B-ring hydroxylation pattern of
 intermediates of anthocyanin synthesis in pelargonidin-
 and cyanidin producing lines of Matthiola incana.Planta
 148:157.
Grierson D, Loening UE (1974). Ribosomal RNA precursors and
 the synthesis of chloroplast and cytoplasmic ribosomal
 ribonucleic acid in leaves of Phaseolus aureus. Eur J
 Biochem 44:501.
Grierson D, Hemleben V (1977). Ribonucleic acid from higher
 plant Matthiola incana: molecular weight measurements and
 DNA-RNA hybridization studies. Biochim Biophys Acta 475:424
Hahlbrock K, Griesebach H (1979).Enzymic controls in the biosynthesis of lignin and flavonoids.Ann Rev Plant Physiol 30:105
Hahlbrock K, Schröder J, Vieregge J (1980). Enzyme regulation in parsley and soybean cell cultures. In Fiechter A
 (ed):"Advances in Biochemical Engineering, Plant Cell
 Cultures II," Vol 18
 New York,Berlin,Heidelberg: Springer, p 40.
Heller W, Hahlbrock K (1980). Highly purified "flavanone
 synthase" from parsley catalyses the formation of naringe-
 nin chalcone. Arch Biochem Biophys 200:617.
Hemleben V, Grierson D, Dertmann H (1977). The use of equi-
 librium centrifugation in actinomycin caesium chloride
 for the purification of ribosomal DNA. Plant Sci Let
 9:129.
Kreuzaler F, Ragg H, Heller W, Tesch R, Witt I, Hammer D,
 Hahlbrock K (1979). Flavanone synthase from petroselinum
 hortense. Eur J Biochem 99:89.
Leber B (1977). Entwicklung eines geeigneten pflanzlichen
 Donor-Rezipienten-Systems für den Nachweis der Aufnahme,
 Integration und Expression von exogener homologer DNS.
 Ph. D. Thesis, Tübingen.
Leber B, Hemleben V (1979). Structure of plant nuclear and
 ribosomal DNA containing chromatin.Nucl Acids Res 7:1263.
Mohr H (1970). Regulation der Enzymsynthese bei der höheren
 Pflanze.Naturwiss Rundschau 23:187.
Paterson BM, Roberts BE, Kuff EL (1977). Structural gene
 identification and mapping by DNA:mRNA hybrid-arrested
 cell free translation. Proc Natl Acad Sci U.S.A.74:4370

Rissland J, Mohr H (1967). Phytochrom-induzierte Enzymbildung (Phenylalanindesaminase), ein schnell ablaufender Prozeß. Planta (Berlin) 77:239.

Schröder J (1978). Lichtinduzierte Messenger RNA in Pflanzen. Biol in uns Zeit 8:147.

Seitz U, Richter G (1970). Isolierung und Charakterisierung schnell markierter, hochmolekularer RNS aus frei suspendierten Calluszellen der Petersilie (Petroselinum hortense). Planta (Berlin) 92:309.

Seyffert W (1960). Über die Wirkung von Blütenfarbgenen bei der Levkoje Matthiola incana R. Br. Z Pflanzenzüchtg 44:4

Seyffert W (1963). Spektrale Remission als objektive Methode zur Beschreibung und Klassifizierung von Phänotypen. Der Züchter 33:356.

Seyffert W (1971). Simulation of quantitative characters by genes with biochemically definable action. Theor Appl Genet 41:285.

Smith H (1975)."Phytochrome and Photomorphogenesis." London: McGraw Hill.

Spribille R, Forkmann G (1981). Genetic control of chalcone synthase activity in flowers of Mattiola incana R. Br. Z Naturforschung 36 C:619.

TUMOR REVERSAL IN PLANTS

Frederick Meins, Jr.

Friedrich Miescher-Institut, P.O.Box 273,
CH-4002 Basel, Switzerland

One of the basic problems in tumor biology is the relationship between the genetic constitution of the neoplastic cell and its capacity for autonomous growth. Plant tumor diseases, such as crown gall, provide experimental systems well suited for approaching this problem because the developmental potential and genetic constitution of totipotent cells exhibiting different phenotypes can be assessed by regenerating plants from individual cells. The thesis I wish to develop in this paper is that even when tumor transformation results from the introduction of foreign genes into the plant cell, the stable expression of these genes and the capacity of the cell for autonomous growth are ultimately controlled by developmental mechanisms of the host plant.

CROWN GALL TUMOR TRANSFORMATION

Tumors result when competent cells from a wide variety of dicotyledonous plants are exposed to a tumor-inducing principle (TIP) produced by the crown gall bacterium, Agrobacterium tumefaciens. Once tumor inception is complete, the plant cells are capable of non-self-limited growth in the absence of the inciting bacterium indicating that crown gall involves a true neoplastic transformation (reviewed by Braun, 1969).

There is growing evidence that TIP is a DNA molecule. First, DNA from tumor-inducing (Ti) plasmids residing in virulent strains of the bacterium is transferred to the host cell during tumor inception. This tumor DNA (T-DNA) is integrated into the genome, is transcribed, and induces the host cell to produce opine amino acids, such as nopaline or octopine (reviewed by Zambryski et al., 1980). Second, reversion of tumor cells to the normal state is accompanied by the loss of most, and in some cases all, of the T-DNA sequences from the plant cell (Yang et al., 1980; Yang and Simpson, in press). Finally, there are experiments suggesting that purified Ti-plasmids can induce crown gall transformation of cultured protoplasts (Davey et al.,1980). Apparently, crown gall transformation, like neoplastic transformation of animal cells induced by oncogenic viruses, depends upon the stable introduction of foreign genes into the host cell.

It is likely that T-DNA effects a change in growth regulation by inducing the host cell to produce substances needed for cell proliferation. For example, pith cells of tobacco show an absolute requirement for an auxin, e.g. indole-3-acetic acid, and for a cell-division factor (CDF), e.g. the cytokinin, kinetin, for continuous proliferation on an otherwise complete, basal medium. In striking contrast, transformed pith cells proliferate on a basal medium and contain sufficient amounts of auxin and CDF to support the proliferation of normal pith cells. Thus, it appears that tumor cells still require the growth factors, but have learned how to produce them (reviewed by Braun, 1969).

The specific factors required for growth and produced by tumor cells vary somewhat from species to species; however, in each case the tumor cells produce the same factors as the host plant from which they were derived. This observation and the fact that the capacity for autonomous growth and production of these factors are correlated regardless of the proximal cause of transformation provides strong evidence that tumor autonomy results from the activation of biosynthetic capacities necessary for cell proliferation, but repressed in cultured normal cells (Braun, 1958; Meins, 1974).

TUMOR SUPPRESSION AND TUMOR REVERSAL

Although crown-gall tumor cells are extremely stable -- some cell lines have been serially propagated since 1942 in culture -- they nevertheless retain a potential for normal development. The first experimental demonstration of tumor reversal was achieved by Braun (1959) using cloned, multipotential teratoma cells of tobacco. This study and more recent ones with teratoma cells bearing genetic markers show that loss of the tumor phenotype can occur in two ways (Braun and Wood, 1976; Turgeon et al., 1976). Cloned teratoma tissues grafted onto the cut-stem tip of tobacco plants from which the axillary buds have been removed form abnormal shoots which become progressively more normal in appearance, eventually flower, and set seed. The seed develop into plants composed of cells which have lost their T-DNA sequences (Yang et al., 1980) and capacity to produce opines and grow autonomously (Wood et al., 1978). At some point in the growth of the teratoma-derived shoot and formation of seed, progeny of highly autonomous teratoma cells revert to the normal state.

The teratoma state can also be suppressed phenotypically. For example, leaves on teratoma-derived shoots are often normal in appearance and histological detail but still contain T-DNA (Yang et al., 1980) and still produce the opine, nopaline (Wood et al., 1978). Moreover, tissues from these leaves can proliferate in culture on a basal medium and rapidly regain their teratomatous appearance. In striking contrast to leaf tissues from seed-grown plants of teratoma origin, teratoma-derived shoot tissues retain their capacity for autonomous growth.

Interpretation of what happens at the cellular level during tumour suppression and reversal requires knowledge of the cellular constitution of teratoma tissues and teratoma-derived shoots. For example, it may be argued that teratoma cells, although multipotential, are themselves incapable of normal differentiation or organogenesis, but give rise to revertant cells during prolonged culture. Mosaic tissues arising in this way would be expected to be quite stable and capable of abnormal organ formation since tumor cells produce auxins and CDF that promote both the proliferation and organogenesis of normal cells (Skoog and Miller, 1957).

To test this hypothesis, we isolated several clones of teratoma cells from teratomas induced on Havana 425 tobacco plants by inoculation with the T37 strain of crown gall bacterium. After serial propagation for about three years on a basal medium, the cloned tissues were subcultured for several transfers on medium containing auxin, and then were cloned in a medium containing both auxin and CDF. The supplemented media were used to select for normal cells, which require both growth factors, and to select against tumor cells, which are inhibited by these factors. We found that 1/50 of the subclones from clone HT37-5, 6/20 of the subclones from clone HT37-8, and 0/42 of the subclones from clone HT37-15 were revertants with an absolute requirement for auxin and CDF. These results show that complete reversion does occur in culture, confirming the recent findings of Yang and Simpson (in press), but that in spite of strong selection for normal cells the incidence of revertants is far too low to account for the proportion of differentiated cells found in teratoma tissues. The most direct interpretation of these findings is that teratoma cells are capable of differentiation and organogenesis while retaining their potential for autonomous growth.

It is likely that this interpretation applies to the suppressed teratoma shoots as well. Using the same cloned lines just described, Binns et al. (in press) examined a total of 560 clones isolated from highly differentiated leaf mesophyll cells present in suppressed teratoma shoots. All but three of these clones exhibited the teratomatous phenotype in culture. Thus, although reversion to the normal state does occur in culture or during shoot regeneration, this is a rare event; suppressed teratoma shoots consist primarily of differentiated teratoma cells with the potential for autonomous growth.

The experimental evidence reviewed here emphasizes that the regulation of tumour autonomy is a developmental problem. The organized structures formed by teratoma cells in culture and in the suppressed teratoma shoots consist of teratoma cells which retain their tumorous character in a covert form, i.e. the cells are still determined as tumor cells. The suppressed tumor state probably has as its basis an alteration in the expression of genes transferred to the cell during tumor inception. Thus, suppressed teratoma cells still express at least one T-DNA specified character, nopaline production, while those T-DNA specified functions concerned with autonomous growth are repressed.

Teratoma cells may also lose their tumorous character in a stable fashion. This occurs at some time during the meiosis of cells in flowering, suppressed teratoma shoots. It is not known whether meiotic reversion is a directed process, like tumor suppression, or the result of rare, spontaneous reversions of the type that occur when teratoma cells are serially propagated in culture.

EPIGENETIC CHANGES IN THE GROWTH FACTOR REQUIREMENT OF NORMAL CELLS

Tumor inception, suppression, and reversal involve stable changes in the nutritional requirement for, and presumably capacity to produce, CDF. This class of growth regulators also plays an important role in controlling normal growth and development. CDF habituation, a type of heritable, cellular change which occurs in culture provides a model system for studying how different states of CDF production arise and how this process is altered by tumor transformation.

On prolonged culture or following treatment with specific inducers, pith parenchyma tissues of tobacco may lose their requirement for exogenously supplied CDF in culture. Thereafter, such habituated tissues and most clones derived from these tissues can be serially propagated in the absence of CDF, which they now produce. There is compelling evidence that habituation has an epigenetic basis. It is a directed process that occurs at high rates; it is regularly reversible; it leaves the heritably altered cell totipotent; and, it results in the production of CDF of the type produced by certain tissues in the tobacco plant, but repressed in cultured pith cells (Meins and Binns, 1978; Meins, in press, a).

We also have strong evidence that stable, epigenetic changes in CDF requirement occur in the normal development of the plant (Meins and Lutz, 1979). Cells from different regions of the tobacco plant inherit different states of habituation and this persists in culture. Thus, cultured pith cells are initially CDF requiring but can be induced to habituate by CDF treatment, whereas leaf cells are CDF requiring, but not inducible. Stem-cortex derived cells exhibit a third phenotype: they are CDF habituated even when not induced. During the plant regeneration process, the different states are regularly erased. Plants derived from pith, cortex, or leaf cells are composed of tissues which exhibit the same CDF requirement and competence for habituation as comparable tissues from seed-grown plants. These findings show that the stable expression of the CDF autotrophic state is not unique to crown gall tumor cells. This state can arise both in normal development and in tissue culture without either the introduction of foreign genes into the plant cell or the permanent alteration of the plant-cell genome.

The important question that arises is: how are different states of habituation maintained when cells divide? Our working hypothesis, described in detail elsewhere (Meins and Binns, 1978), is that expression of the habituated state is stabilized by a positive-feedback loop in which CDF either induces its own synthesis or inhibits its own degradation. Competent pith cells are poised to make CDF but require CDF above a threshold concentration to trigger this process. When provided with CDF, cells shift to the producing state and express the habituated phenotype. Cells then remain in this state until the feedback loop is broken.

This hypothesis is supported by two observations. First, in the presence of high concentrations of auxin, the synthetic CDF, kinetin, induces habituation of pith cells. This is a cooperative process with a sharp threshold concentration of less than 10^{-9} M kinetin, about 100- to 1000-fold lower than the kinetin concentration optimal for the proliferation of non-competent, CDF-requiring pith cells (Meins and Lutz, 1980a). Second, when the production of CDF is blocked by cold treatment (Binns and Meins, 1979), or by incubation on a low-auxin medium (Binns, 1979), the putative feedback loop is broken and the tissues revert to the non-habituated state. These revertants, like the competent pith cells, rapidly habituate when treated with kinetin under permissive conditions. These experiments clearly show that some type of auto-catalytic relationship is involved in maintaining the habituated state. How this is accomplished at the molecular level is not known.

GENETIC CHANGES IN HABITUATION

We have recently isolated heritable variants in which the habituated state is not erased during shoot regeneration and is even transmitted to seed (Meins and Lutz, 1980; Meins, in press, b). Plants with different phenotypes were regenerated from leaf and cortex cells derived from what now appears to be a unique plant. Leaf tissues from leaf-derived plants exhibited the normal phenotype, i.e. the tissues were CDF requiring. This phenotype was also transmitted to all the progeny obtained by self-crossing leaf-derived plants. On the other hand, leaf tissues from cortex-derived plants exhibited the habituated phenotype and, in this regard, resembled the cortex cells from which they were derived. This habituated leaf (HL) phenotype segregated in progeny obtained from the self-cross. Three classes of progeny were obtained: 1) plants exhibiting the normal phenotype that breed true; 2) plants exhibiting the HL phenotype that breed true; and 3) plants exhibiting the HL phenotype that, when selfed, gave the same three classes as obtained in the original cross. Pure lines of HL and normal plants, when crossed, gave an F_1 generation in which all the plants scored were intermediate in degree of habituation between the parental types. The same result was obtained in the reciprocal cross indicating that HL is a semi-dominant trait inherited in a nuclear fashion. When the F_1 generation was back-crossed to the normal parent, about 50% of the progeny were normal and about 50% of the progeny exhibited the intermediate HL phenotype. About 25% of the plants in the F_2 generation were normal; the remainder exhibited phenotypes ranging from intermediate to fully HL. These results are consistent with the interpretation that HL is a single, semi-dominant, Mendelian trait and that the original cortex-derived plants were heterozygous for this trait. HL is not tissue specific in its action; pith tissue from homozygous HL plants also exhibit the habituated phenotype in culture. Thus, it appears that, with regard to CDF requirement and competence to habituate, HL converts pith and leaf cells into cortex cells.

CONCLUSIONS

The experimental evidence favors the hypothesis that genes transferred from the Ti plasmid to the plant cell alter the regulation of growth factor production and this, in turn, provides the transformed cell with the capacity for autonomous growth. The fact that teratoma cells can differentiate, participate in organogenesis, and remain in a non-dividing state even though they retain the T-DNA sequences shows that expression of the tumor state is ultimately regulated by the normal developmental mechanisms of the host plant.

CDF autotrophy, an important attribute of the transformed state, is not unique to crown gall. It arises in specific tissue types during development and can be induced in tissue culture. Different states of habituation result from epigenetic changes rather than from rare, random mutations of the classical type. Although the mechanisms for these changes affecting competence are not known, expression of the habituated phenotype appears to be maintained by a positive feedback loop involving CDF.

The HL "mutant" most closely mimics the behavior of the suppressed teratoma state. In both cases the autotrophic state is suppressed in organized structures of the plant, expressed in tissue culture, and lost in a stable fashion during meiosis. This suggests that a primary event in crown gall transformation is a change in the inducibility of the feedback loop stabilizing the CDF autotrophic state. This change affects auxin autotrophy as well. Whereas normal pith tissues of tobacco do not become auxin autotrophic after auxin treatment, pith tissues from suppressed teratoma shoots (Braun and Wood, 1976) or from tumor-prone hybrid Nicotiana plants (Cheng, 1972) are induced by this treatment.

ACKNOWLEDGEMENTS

Original work from the author's laboratory at the University of Illinois was supported by grant No. 78-10203 from the National Science Foundation and grant No. CA 20053 from the U.S. Public Health Service, National Cancer Institute.

REFERENCES

Binns AN (1979). "Habituation of Tobacco Pith Cells for Factors Promoting Cell Division", Ph.D. Dissertation. Princeton University, Princeton NJ, USA.
Binns AN, Meins F Jr (1979). Cold-sensitive expression of cytokinin habituation by tobacco pith cells in culture. Planta 145:365.
Binns AN, Wood HN, Braun AC (in press). Suppression of the tumorous state in crown gall teratomas of tobacco: A clonal analysis. Differentiation.
Braun AC (1958). A physiological basis for autonomous growth of the crown-gall tumor cell. Proc Natl Acad Sci USA 44:344.
Braun AC (1959). A demonstration of the recovery of the crown-gall tumor cell with the use of complex tumors of single-cell origin. Proc Natl Acad Sci USA 45:932.
Braun AC (1969). "The Cancer Problem: A Critical Analysis and Modern Synthesis", New York and London: Columbia University Press.

Braun AC, Wood HN (1976). Suppression of the neoplastic state with the acquisition of specialized functions in cells, tissues, and organs of crown gall teratomas of tobacco. Proc Natl Acad Sci USA 73:496.

Cheng T-Y (1972). Induction of indole-acetic acid synthetases in tobacco pith explants. Plant Physiol. 50:723.

Davey MR, Cocking EC, Freeman J, Pearce N, Tudor I (1980). Transformation of Petunia protoplasts by isolated Agrobacterium plasmids. Plant Sci Lett 18:307.

Meins F Jr (1974). Mechanisms underlying tumor transformation and tumor reversal in crown gall, a neoplastic disease of higher plants. In King TJ (ed): "Developmental Aspects of Carcinogenesis and Immunity", New York-San Francisco-London: Academic Press, p.23.

Meins F Jr (in press, a). Habituation of cultured plant cells. In Schell J, Kahl G (eds):"Molecular Biology of Plant Tumors", New York: Academic Press.

Meins F Jr (in press, b). The nature of the cellular, heritable change in cytokinin habituation. In Earle E (ed): "Proceedings of the NSF-CNRS Workshop on Plant Tissue Culture", New York: Praeger.

Meins F Jr, Binns AN (1978). Epigenetic clonal variation in the requirement of plant cells for cytokinins. In Subtelny S, Sussex IM (eds): "The Clonal Basis of Development", New York: Academic Press, p. 185.

Meins F Jr, Lutz J (1979). Tissue-specific variation in the cytokinin habituation of cultured tobacco cells. Differentiation 15:1.

Meins F Jr, Lutz J (1980a). The induction of cytokinin habituation in primary pith explants of tobacco. Planta 149:402.

Meins F Jr, Lutz J (1980b). Epigenetic changes in tobacco cell culture: Studies of cytokinin habituation. In Rubenstein I, Phillips RL, Green CE, Gengenbach B (eds): "Emergent Techniques for the Genetic Improvement of Crops", Minneapolis: University of Minnesota Press, p. 220.

Turgeon R, Wood HN, Braun AC (1976). Studies of the recovery of crown gall tumor cells. Proc Natl Acad Sci USA 73:3562.

Wood HN, Binns AN, Braun AC (1978). Differential expression of oncogenicity and nopaline synthesis in intact leaves derived from crown gall teratomas of tobacco. Differentiation 11:175.

Yang F, Montoya AL, Merlo DJ, Drummond MH, Chilton M-D, Nester EW, Gordon MP (1980). Foreign DNA sequences in crown gall teratomas and their fate during the loss of the tumorous traits. Molec Gen Genetics 177:707.

Yang F, Simpson RB (in press). Revertant seedlings from crown gall tumors retain a portion of the bacterial Ti plasmid DNA sequences. Proc Natl Acad Sci USA.

Zambryski P, Holsters M, Kruger K, Depicker A, Schell J, Van Montagu M, Goodman HM (1980). Tumor DNA structure in plant cells transformed by A. tumefaciens. Science 209:1385.

DNA INTEGRITY IN PLANT EMBRYOS AND THE IMPORTANCE OF DNA REPAIR

Daphne J Osborne

Developmental Botany, Weed Research Organization, Oxford OX5 1PF, U.K.

INTRODUCTION

There is an implicit assumption that physically or chemically induced lesions in the DNA molecules of prokaryotes and eukaryotes can be converted to heritable mutations. These may persist with different degrees of success through successive generations of the species. It is also understood that not every lesion is automatically converted to a mutation. Both prokaryotes and eukaryotes possess enzymic DNA repair systems which in a series of events first recognise a lesion, then excise and repair the damaged region and resynthesise a new patch that is joined back into the existing DNA molecule. This is "repair" in the strict Oxford Dictionary sense of the word. It is therefore a mechanism of mutation avoidance which when fully functional and error-free protects the integrity of the genome from genetic defects and mutagenic change. Clearly, however, either inability to recognise the presence of a lesion or loss of fidelity of the processes of repair can result in failure to re-establish the initial status of the genome. Subsequent replication and recombination events can then fix the initial modification and favour secondarily induced changes in DNA molecules so that the altered loci become potential sites for heritable mutagenesis.

Much of the information we have concerning pathways of DNA repair derives from work on prokaryotes; some but not all the mechanisms appear to operate similarly in eukaryotes and the steps involved in mammalian systems or in yeast are perhaps the best understood. In higher plants biochemical evidence for repair of DNA lesions has only relatively recently been established (see repair of single strand breaks induced in barley seeds by alkylating agents, Velemínský, Zadražil and Gichner, 1972; the unscheduled DNA synthesis in gamma-irradiated barley embryos, Yamaguchi, Tatara and Naito, 1975; and light and dark monomerization or excision of UV-induced pyrimidine dimers in DNA of wild carrot, Howland, 1975). In the few examples of plant cancers, (as distinct from the hormonally regulated differentiation of callus tissue) the relationship between foreign DNA, the genetic status of the cell and its tumour potential is currently being resolved (see the plasmid conveyed genetic information responsible for tumour induction in plants infected by Agrobacterium tumefaciens, Yadav et al, 1980).

Although mutation and cancer development in plants has received considerable attention over the years most questions have been orientated differently from those in bacteria and mammalian tissues and relatively few have been concerned with processes of DNA repair. Perhaps the earliest speculation that phenotypic expression was regulated by changes in heritable information was set out by De Vries in his 2 volume work "Die Mutationstheorie" (1901, 1903). He discusses how old seed which has been stored in the dry state for a number of years produces seedlings with a higher number of morphological variants than freshly harvested seed. In the 1930s there was much interest in these matters and many reports correlated the duration of seed storage time with the number of chromosomal aberrations to be found in the first round of mitosis in root tips from the old seed: the numbers of aberrations reflecting the frequency of the subsequent abnormal growth morphology in the seedlings. Navashin (1933) for example, found that 80% of plants produced from 6-7 year old Crepis seeds showed chromosomal aberrations in the root tips whilst only 0.1% of such mutations were seen in 1 year old seed. Stubbe (1935) found 14% gene mutations in 10 year old Antirrhinum seedlings and only 1% amongst seedlings from 1 year old seed. In interesting,and in the present context important, studies Cartledge and Blakeslee (1934, 1935) describe the higher numbers of pollen abortion mutations in seedlings from seed aged dry compared with those from seed aged imbibed in the soil: 7% in 10 year laboratory stored Datura seed compared with only 1.8% in 22 year old soil stored material. This and other early evidence indicated that chromosome disturbances tend to accumulate more readily in non-hydrated cells than in those cells that are hydrated and metabolically active. Of considerable interest therefore is Nichols' (1941) report that high levels of chromosomal aberrations were produced in root tips from seeds of Allium cepa stored dry, but no aberrations could be found in the root tips produced from regenerating bulbs. Nichols himself suggests that dehydration plays some part in inducing chromosomal damage and that the hydrated conditions in the bulb are less conducive to chromosome breakage. He also considered the likelihood of a differential survival of damaged cells during development, and pointed out how few cells with gross chromosomal disturbances are found in reproductive tissue. Although Nichols does not preclude the possibility that the broken ends of chromosomes must heal, he favoured the view that cells containing chromosomal abnormalities would tend to be eliminated in competition with cells containing the normal gene complement. Supporting evidence was based upon an analysis of chromosomal aberrations seen at mitosis in root tips of Allium cepa at successive stages of development (see his Table 5, presented here as Table 1).

Table 1

Chromosome aberrations in Allium cepa root tips at successive stages of root development. (After Nichols, 1941)

LENGTH OF ROOTS mm	TOTAL CELLS ANALYZED	PERCENTAGE ABERRATIONS
2 - 5	450	10.4
7 - 9	58	6.6
10 - 12	138	5.8
25 - 30	190	2.1
80 - 100	116	1.7

However, a root that is 2 - 5 mm long has already been hydrated for many hours before the first round of DNA replication and the first mitoses. We now know that most other biochemical events of early germination are initiated within minutes of imbibition (Sen, Payne and Osborne, 1975; Sen and Osborne, 1977). There is therefore a "window" of some hours in which changes can occur in the status of DNA between that present in the dry seed and that at first prophase.

This manuscript concerns certain lesions in DNA that we know to occur naturally in dry seeds. It describes some very early repair processes in the embryos of one species (rye, Secale cereale) and how these might facilitate rehabilitation of the damaged genome during the earliest stages of rehydration before the advent of the first mitotic prophase. It includes reflections on the role of these repair processes in mutagenesis.

Embryo material and biochemical methods are given in detail in Cheah and Osborne, 1978; Osborne, Sharon and Ben-Ishai, 1980).

THE INTEGRITY OF NUCLEAR DNA IN DRY SEEDS

In the root tips of rye embryos excised from dry seed, initiation of RNA and protein synthesis occurs within minutes of imbibition of water. The replication of DNA, however, is always a relatively late event in seed germination and is not detectable in embryos of freshly harvested rye seed until 3 - 4 hours after imbibition at the earliest. With increasing time of seed storage a number of degradative changes occur in the dry seeds such that although percentage germination may not be reduced, the rates of early protein and RNA synthesis become progressively slower and the lag to the first round of DNA replication is progressively delayed (Osborne, 1980). Loss of membrane integrity, impairment of enzyme function (including mitochondrial dehydrogenases, aminoacyl-tRNA ribosome binding transferases), loss of integrity of high molecular weight ribosomal RNA (26S and 18S) and fragmentation of nuclear DNA all appear to be contributory factors in the slow germination typical of aged seeds. Although the numbers of chromosomal aberrations apparent at first prophase, or metaphase reflect the numbers of morphogenic abnormalities in the seedlings, the fact that major chromosome damage takes place and accumulates in the dry seed has only relatively recently been established (Cheah and Osborne, 1978). Both natural ageing and gamma-irradiation lead to increased fragmentation of the DNA. The autoradiographic results of Table 2 show the higher number of single stranded breaks detectable in nuclear DNA in preparations of fixed and sectioned dry embryos isolated from long stored seed (0% germination), freshly harvested 95% germination stocks and 95% embryos after 500 kR gamma-irradiation in the dry state.

Table 2

Incorporation of ^{3}H-dCTP in the presence of calf thymus 3'OH terminal deoxynucleotidyl transferase (EC 2.7.7.31) into nuclei of 3 μm sections cut from dry embryos*

Embryo Samples (% germination stocks)	Grains per nucleus					
	0-20	20-40	40-60	60-80	80-100	>100
	Numbers of nuclei					
95%	3	39	39	8	1	2
0%	1	13	53	30	7	1
95% (irradiated 500 kR)	0	2	32	42	29	10

* Range of duplicates <10%, 200 nuclei scanned per embryo. For further details see Osborne, Sharon and Ben-Ishai (1980). Method of Modak and Bollum (1972).

Clear differences in integrity of DNA can also be demonstrated by electrophoretic fractionation on polyacrylamide denaturing gels of DNA isolated from nuclei of embryos of 95% and non-viable seed stocks. As seen in Fig 1a DNA of 95% embryos is predominantly high molecular weight and little low molecular weight material enters the gel. DNA from the non-viable embryos (Fig 1b) shows a large proportion of low molecular fragments visible as a heterodisperse splurge of DNA throughout the gel. Of considerable interest is the enhanced degradation of DNA when the non-viable embryos are imbibed in water for 18 h before isolation of the nuclei (Fig 1c). Microdensitometry of Feulgen stained

Fig 1. Fractionation of DNA isolated from nuclei of dry embryos of a) freshly harvested 95% germinating seed, b) stored, non-viable seed. c) Non-viable embryos imbibed in water for 18 h before isolation of nuclei. DNA (20 μg) fractionated under denaturing conditions in 4% polyacrylamide slab gels containing 7M urea. Gels run at 25 mA and 35 volts for 16 h at room temperature. Gels stained with ethidium bromide. For details of method see Cheah and Osborne, 1978.

sections of the embryo demonstrates no change in the total DNA content per nucleus during the 18 h imbibition period and the progressive fragmentation of the DNA with time is attributable to a continuous endodeoxyribonuclease activity associated with the chromatin (Cheah and Osborne, 1978).

EVIDENCE FOR DNA REPAIR AT GERMINATION

High levels of incorporation of ^{3}H-thymidine into non-replicating DNA have been observed in the embryos of seeds of a number of species following irradiation or treatment with chemical mutagens (Soyfer and Ciemenis, 1974; Yamaguchi, Tatara and Naito, 1975).

Such unscheduled DNA synthesis is deemed repair synthesis for several reasons. In barley, alkaline sucrose density gradient analysis of DNA from nuclei isolated from gamma-irradiated embryos shows a shift from a low overall molecular weight to a higher value following the period of non-replicative DNA synthesis, indicating that a repair of single stranded breaks in DNA occurs before the replicative step (Tano and Yamaguchi, 1977). Autoradiographic techniques used by Gudkov and Grodzinsky (1976) have shown increasing levels of unscheduled DNA synthesis unassociated with mitosis in nuclei of germinating pea seeds which reflect increasing doses of gamma-irradiation to dry seed. When seed that has been irradiated, or treated with chemical mutagens is exposed to conditions that permit repair, but not replication (eg storage imbibed at 30% moisture content), there is a reduction in the number of chromosomal aberrations subsequently observed in root tips at germination and fewer seedlings subsequently show phenotypic mutations (see review by Velemínský and Gichner, 1978). Similar recoveries have been obtained by Villiers and Edgcumbe (1975) for irradiated dormant lettuce seed subsequently held imbibed but not germinating.

Evidence from rye (Table 3) indicates that an early non-scheduled DNA synthesis is a normal occurrence even in freshly harvested 95% germination embryos and in embryos of aged seed (52% germination) this synthesis is initially significantly higher (but see also Fig 3). Fractionation of the extracted DNA on benzoylated-naphthoylated DEAE cellulose columns following imbibition in ^{3}H-thymidine for 30 minutes confirms incorporation into double stranded (repaired) DNA eluted at 1M NaCl, but little incorporation into the fractions containing single stranded fragments or replicative forks. Fig 2a.

Table 3

Incorporation of ^{3}H-methyl-thymidine into TCA-insoluble material during early imbibition by embryos of rye from 52% and 96% germination stocks.

Embryo samples (% germination)	Imbibition min	Total Uptake cpm	Incorporation (TCA insoluble material)
96%	20	3310	547
		2220	399
	30	-	1110
	40	5470	1039
	60	10020	1820
		5300	1436
52%	30	2790	6001
		2135	3311
		-	4420

Results of Dell'Aquila and Osborne.

That the embryos of seeds possess an active DNA repair system seems in little doubt, but when the DNA of dry seed is damaged by physical or chemical mutagen-inducing treatments, the repair systems operate only upon rehydration of the embryo (Veleminský and Gichner, 1978). In metabolically active hydrated tissues of higher plants efficient repair mechanisms appear to be continually in operation. For example, carrot protoplasts exposed to gamma-irradiation repair single-stranded breaks in DNA, as judged by alkaline sucrose density gradient analysis, such that 50% repair is achieved within 5 min and complete repair within 1 h (Howland,

Hart and Yette, 1975). Similar active repair processes probably account for the absence of chromosomal disturbances in root tips produced from onion bulbs (but not from dry seeds) as reported by Nichols (1941) and to the low level of aberrations in seeds stored imbibed in the soil (Cartledge and Blakeslee, 1934, 1935). Since no detectable synthetic events occur in dry seeds, any lesions or structural damage to DNA accumulate during the dry state and requires a reactivation of the enzymic repair processes at imbition to initiate repair. The effectiveness and fidelity of the repair system then depends upon the extent of degradation of the repair enzyme proteins during storage and the overall cellular integrity and organisation of the embryo tissue on rehydration.

Impaired protein synthesis and reduced rates of transcription of all classes of RNA are typical symptoms shown by embryos of aged or irradiated seeds during the early hours of imbibition (Sen and Osborne, 1977). Although long lived mRNA is present in seeds, its physiological importance remains speculative and the transcription of new mRNA molecules is essential to maintain germination and seedling growth. Unrepaired lesions in the DNA can preclude the synthesis of RNA coded at these loci. If activity of any repair enzymes is attenuated during storage in the dry state and damage to DNA arises at those cistrons that code for repair enzyme proteins (particularly if they are single copy genes) then an irreversible cellular senescence programme will have been initiated. The following preliminary studies suggest that a failure of DNA repair enzyme function can occur during seed storage and could be a major controlling factor in the loss of synthetic metabolic activity and slow germination on imbibition of low viability seeds.

FAILURE OF DNA REPAIR PROCESSES AT GERMINATION

Embryos isolated from freshly harvested (96% germination) and stored (52% germination) rye seeds have been used to test whether the capacity of the DNA repair system declines in the dry state. Some embryos were gamma-irradiated with 100 kR from a Gravatom Industries RX30/55M cobalt source and all were imbibed for different periods of time in ^{3}H-methyl-thymidine. The scans in Fig 2a and b show the elution profiles of radioactivity from the DNA isolated from irradiated and non-irradiated embryos, fractionated on

Fig 2. Elution profiles of DNA fractionated on benzoylated-naphthoylated DEAE cellulose columns to separate double stranded repair DNA (1M NaCl) from single strand containing forked, or replicating DNA. Embryos imbibed in ^{3}H-methyl-thymidine for the times shown, before extraction and fractionation of the DNA.
a) Embryos from 96% and 52% germination stocks.
b) Embryos from 96% and 52% stocks exposed to gamma-irradiated (100 kR) in the dry state.
Results of Dell'Aquila and Osborne.
Method of Scudiero, Henderson, Norin and Strauss, 1975.
See also Osborne, Sharon and Ben-Ishai, 1980.

2a

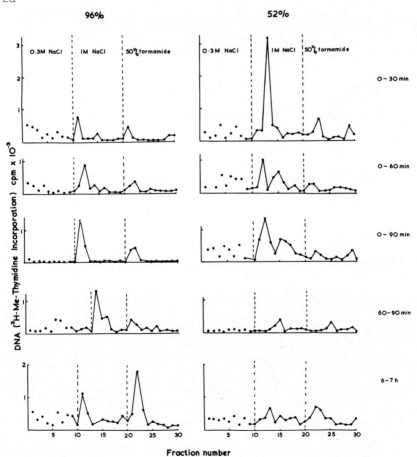

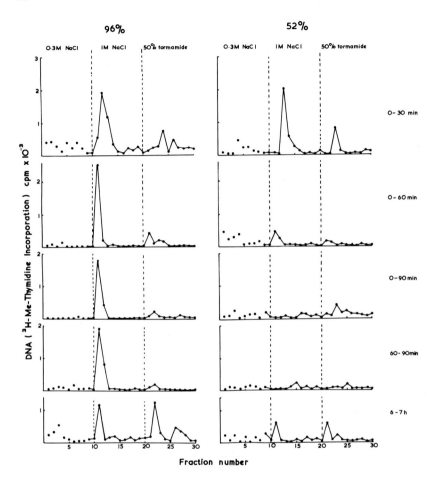

columns of benzoylated naphthoylated DEAE cellulose to separate double stranded (repaired) DNA from single stranded, forked, or replicative DNA. From Fig 2a it is evident that during the first 30 min of imbibition high incorporation occurs into the double stranded DNA fraction (eluted with

1M NaCl) in the 52% embryo samples, indicating an initial repair event considerably in excess of that in 96% embryos (see also Table 3). Incorporation into the repair fraction of 96% embryos increases with time, but repaired DNA of 52% embryos is unstable and radioactivity is subsequently lost from the double stranded DNA fraction during the next 60 minutes with little new incorporation in the period 60-90 minutes. With irradiated 96% embryo samples (Fig 2b), incorporation into the repair fraction is considerably increased above the non-irradiated samples (Fig 2a) and the ^{3}H-thymidine is stable and retained indicating a functional DNA repair system. In 52% embryos, however, irradiation reduces the overall incorporation of ^{3}H-thymidine. The subsequent loss of incorporated radioactivity is even greater than in non-irradiated samples, suggesting that the DNA repair mechanism is further impaired by the radiation treatment. This loss of radioactivity from ^{3}H-methyl-thymidine initially incorporated into the DNA of 52% embryos is evident also from the time course of in vivo incorporation into TCA-insoluble material (Fig 3). The results in Fig 2 are not therefore attributable to degradation during extraction and fractionation of the DNA.

An important feature of the rye results (Fig 2) is the low incorporation by 52% embryos between hours 6 - 7; a period when DNA replicative synthesis and mitotic activity are already fully initiated in the 96% material. Failure to reinstate the DNA integrity of the genome in 52% embryos offers a possible explanation towards understanding the prolonged lag before DNA replication that occurs in all old seed. Also, failure to effect full repair before replication is initiated may be a major cause of the high incidence of chromosomal aberrations (including sister chromatid exchanges) and phenotypic mutations in seedlings produced from old seed.(The experiments of Schvartzman and Gutierrez (1980) with Allium cepa root meristems have demonstrated more sister chromatid exchanges in 5-bromodeoxy-uridine substituted nuclei when the tissue is exposed to light irradiation at progressively closer times to DNA replication.

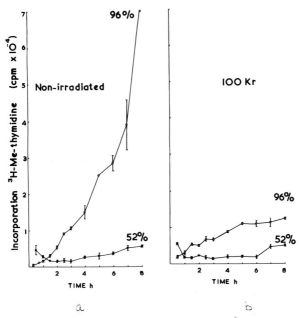

Fig 3. Time course of incorporation of ^{3}H-methyl-thymidine into TCA-insoluble material of embryos from 96% and 52% germination stocks.
a) Non-irradiated embryos.
b) Embryos exposed to gamma-irradiation (100 kR) in the dry state.
Results of Dell'Aquila and Osborne.
For details of method see Sen and Osborne, 1977.

RELATIONSHIP BETWEEN MUTAGENESIS AND DNA REPAIR AT GERMINATION

The reasons for failure of DNA repair in embryos of aged seed are at present speculative. Considering the nature of the DNA damage, an excision repair and ligation of modified bases of single strand breaks should be operative. DNA polymerase activity and ATP availability in 52% embryos certainly permits high levels of early thymidine incorporation (three times that of 96% embryos at 30 min, Fig 3). One explanation could be that the polynucleotide ligase of old seed becomes functionally impaired during storage so that the 3' hydroxy and 5' phosphodiester ends of adjacent deoxyribonucleotides on either side of a break are not properly joined. 3'-5'-exonuclease activity of the polymerase might result in digestion of the newly synthesised patch and explain the loss of radioactivity observed in incorporation experiments. Such systems are not unknown in prokaryotes. For example, in E. coli mutants with only 4% of normal ligase activity, Pauling and Hamm (1969) showed that single strand gaps remained open for longer than usual and led to abnormally high frequencies of post replication recombination.

With a fully operative repair process, single strand lesions in DNA appear to be non-damaging in the long term (as in 96% embryos) and germination is successfully achieved with mutations as rare events. Monitoring for lesions and proof-reading of the repair can be seen as a mechanism by which integrity of the genome is reestablished in G_1 before the first round of DNA replicative synthesis at S phase. That the mechanism is not fully effective in embryos from old seed is evident from the high level of chromosomal aberrations seen in root tips at germination and from the large numbers of developmental abnormalities in the seedlings produced. Mutations passing to the germ line may then be heritable and the aberrations reported in pollen of plants from old seed (Cartledge and Blakeslee, 1934) may be of this type.

If sufficient numbers of single strand breaks in DNA occur in close proximity in old seeds they may be converted to double strand breaks. Coupled with a failing ligase activity, the likelihood of such conversion is increased. Whereas single strand breaks can be repaired with speed and may be error-free, presenting little potential hazard to genetic fidelity, double strand breaks may be lethal and their repair highly prone to error.

The embryos of aged seed therefore present a large potential source of DNA change for genetic mutagenesis, but because they germinate slowly and are poor competitors with their more vigorous rivals, their chances of producing progeny in the wild are low. Although it is difficult to judge what importance the spontaneous mutations of old seeds may have had during evolution, we can recognise the unusual possibilites they offer for modern genetic manipulation.

ACKNOWLEDGEMENTS

I am indebted to Mrs Moya Jones and to Mrs Sheila Dunford for their expert assistance in the preparation of this manuscript.

REFERENCES

Cartledge JL, Blakeslee AF (1934). Mutation rate increased by aging seeds as shown by pollen abortion. Proc Natl Acad Sci USA 20: 103-110.
Cartledge JL, Blakeslee AF (1935). Mutation rate from old Datura seeds. Science 81: 492-493.
Cheah KSE, Osborne DJ (1978), DNA lesions occur with loss of viability in embryos of ageing rye seed. Nature (Lond) 272: 593-599.
De Vries H (1901,1903). "Die Mutationstheorie" Band 1 and 2. Leipzig: von Veit and Comp.
Gudkov IN, Grodzinsky DM (1976). Induction by gamma-radiation of DNA synthesis in radicle cells of germinating seeds of Pisum sativum L. Int J Radiat Biol 29: 455-462.
Howland G (1975). Dark repair of ultra-violet induced pyrimidine dimers in the DNA of wild carrot protoplasts. Nature (Lond) 254: 160-161.
Howland GP, Hart RW, Yette ML (1975). Repair of DNA strand breaks after gamma irradiation of protoplasts isolated from cultured wild carrot cells. Mutation Res 27: 81-87.
Modak SP, Bollum F (1972). Detection and measurement of single-strand breaks in nuclear DNA in fixed lens sections. Exp Cell Res 75: 307-313.
Navashin M (1933). Origin of spontaneous mutations. Nature (Lond) 131: 436.
Nichols C (1941). Spontaneous chromosome aberrations in Allium. Genetics 26: 89-100.
Osborne DJ (1980). Senescence in seeds. In Thimann KV (ed): "Senescence in plants," Boca Raton: CRC Press, pp 13-37.
Osborne DJ, Sharon, R, Ben-Ishai R (1980/81). Studies on DNA integrity and DNA repair in germinating embryos of rye (Secale cereale). Israel J Bot 29: 259-272.
Pauling C, Hamm L (1969). Properties of a temperature sensitive, radiation sensitive mutant of Escherichia coli. II DNA replication. Proc Natl Acad Sci USA 64: 1195-1202.

Schvartzman JB, Gutierrez C (1980). The relationship between the cell time available for repair and the effectiveness of a damaging treatment in provoking the formation of sister-chromatid exchanges. Mutation Res 72: 483-489.

Scudiero D, Henderson E, Norin A, Strauss B (1975). The measurement of chemically induced DNA repair synthesis in human cells by BND-cellulose chromatography. Mutation Res 29: 473-488.

Sen S, Osborne DJ (1977). Decline in ribonucleic acid and protein synthesis with loss of viability during the early hours of imbibition of rye (Secale cereale L.) embryos. Biochem J 166: 33-38.

Sen S, Payne PI, Osborne DJ (1975). Early ribonucleic acid synthesis during the germination of rye (Secale cereale) embryos and the relationship to early protein synthesis. Biochem J 148: 381-387.

Soyfer VN, Ciemenis KGK (1974). Dark repair in higher plants. Proc Acad Sci USSR 215: 1261-1264.

Stubbe H (1935). Samenalter und Genmutabilität bei Antirrhinum majus L. Biol Zbl 55: 209-215.

Tano S, Yamaguchi H (1977). Repair of radiation-induced single-strand breaks in DNA of barley embryo. Mutation Res 42: 71-78.

Veleminský J, Gichner T (1978). DNA repair in mutagen-injured higher plants. Mutation Res 55: 71-84.

Veleminský J, Zadražil S, Gichner T (1972). Repair of single strand breaks in DNA and recovery of induced mutagenic effects during the storage of ethyl methanesulphonate treated barley seeds. Mutation Res 14: 259-261.

Villiers TA, Edgcumbe DJ (1975). On the cause of seed deterioration in dry storage. Seed Sci and Technol 3: 761-774.

Yadav NS, Postle K, Saiki RK, Thomashow MF, Chilton M-D (1980). T-DNA of a crown gall teratoma is covalently joined to host plant DNA. Nature (Lond) 287: 458-461.

Yamaguchi H, Tatara A, Naito T (1975). Unscheduled DNA synthesis induced in barley seeds by gamma rays and 4-nitroquinoline 1-oxide. Japan J Genet 50: 307-318.

THE BLASTOCYST IN CONTROL OF EMBRYONAL CARCINOMA

G. Barry Pierce, M.D. and Robert S. Wells, M.D.

Department of Pathology
University of Colorado School of Medicine
4200 E. Ninth Avenue
Denver, Colorado 80262

Teratocarcinomas are highly malignant tumors composed of a chaotic mixture of tissues representing the three embryonic germ layers which are intermingled with embryonal carcinoma cells (Pierce et al. 1959). These embryonal carcinoma cells resemble inner cell mass cells of early mouse embryos, and when they proliferate and differentiate the resulting cells are organized into the primary germ layers. Further differentiation from these layers results in ectodermal, mesodermal and endodermal tissues chaotically arranged, and intermingled with the rapidly proliferating embryonal carcinoma cells. The important point is that the differentiated tissues derived from the embryonal carcinoma cells are benign (Pierce et al. 1960, 1978).

This raises the possibility that enhancement of this naturally occurring differentiation could serve as an alternative to cytotoxic therapy for teratocarcinoma. Cytotoxic therapy for human teratocarcinomas is now extremely effective and a significant number of people with visceral metastases are cured with it. We have examined surgically removed lung metastases which, after chemotherapy, were reported to be slowly growing. No cancer cells were present in these metastases. Obviously the cancer cells had either been killed by the treatment or they had been directed in their differentiation to slowly-growing, benign-appearing tissues. Evidence can be mobilized to support either possibility.

Two experimental approaches to the direction or enhancement of differentiation in teratocarcinomas have

been taken. One employs chemicals such as retinoic acid
(Strickland & Sawey 1980) or dimethylacetamide (Speers et
al. 1979) in tissue culture or in vivo to cause endodermal
and other differentiations from nullipotent embryonal
carcinomas. The other approach is embryological and is
based on the observation that single or small numbers of
embryonal carcinoma cells injected into a blastocyst of
the mouse can be incorporated into the egg. They are
regulated and take part in normal development resulting in
chimeric mice (Brinster 1974, Mintz & Illmensee 1975,
Papaioannou et al. 1975, 1979). Chimerism was first
recognized by variations in coat color and by appropriate
immune responses (Brinster 1974) and subsequently by
biochemical analyses of isoenzyme markers (Mintz & Illmensee
1975). It was also observed that the blastocyst could
control only a few embryonal carcinoma cells: if more
were injected, mice were born with tumors (Papaioannou
1975). In this respect, mice were occasionally born with
tumors when only a single cell had been incorporated into
the blastocyst.

The fate of the injected embryonal carcinoma cells is
not known. In the situation in which the embryonal carcinoma cell and its progeny are incorporated into the
developing embryo to form a chimeric mouse, it is presumed
that they undergo differentiation. The formation of
chimeric mice occurs in a minority of situations, however,
and the fate of the cancer cells, which do not take part
in embryonic development, is not known.

What is known from the accumulated data is that the
blastocyst can regulate some embryonal carcinoma cells
with the incorporation of their progeny into embryos that
develop into functional mice. Would it be possible to use
embryonal carcinoma cells to probe early development of
the mouse embryo, with the expectation that elucidation of
the mechanism might be useful in clinical treatment of
cancer?

An assay was developed (Pierce et al. 1979) based on
the observation that when embryonal carcinoma cells are
cloned they either produce teratocarcinomas or nothing
(Kleinsmith & Pierce 1964). This suggests that when
embryonal carcinoma cells begin to differentiate they lose
their ability to produce tumors. Accordingly, the assay
compared the incidence of tumors produced when embryonal

carcinoma cells were cloned either alone or after incorporation into blastocysts. Embryonal carcinomas of strain 129 origin were employed. Recipient blastocysts were of Swiss-Webster origin. When injected into the intraperitoneum of strain 129 mice, these blastocysts underwent homograft rejection after 7-10 days, but the cancer cells were isologous with the host and were not harmed by the immune response to the blastocyst (Pierce et al. 1979).

Three lines of embryonal carcinoma were tested. 402A$\bar{x}$ is a so-called nullipotent line capable of some endodermal differentiation but devoid of major histocompatibility antigens, F-9 is a nullipotent line with some endodermal differentiation and 247 is a multipotent but extremely undifferentiated line (Pierce et al. 1981b).

In the original series of experiments, 402A$\bar{x}$ was not tumorigenic if the cell was placed in the blastocoele cavity. If placed in the perivitelline space, it produced tumors in from 25-60% of instances. This proved to be an important positive control used throughout the experiments. With further refinement in technique, and elimination of error, 402A$\bar{x}$ was found to be controlled at the 50% level, F-9 was never controlled, but 247 was controlled in almost absolute fashion. In other words, if five 247 cells were injected into a blastocyst which was then injected into the testes tumors developed in 8% of cases. This is considered to be within the experimental error for the technique. When five cells were injected into the testis alone or were incorporated into the perivitelline space of blastocysts, and then injected into the testis, they gave rise to tumors in 37% of cases (Pierce et al. 1981b).

In order to determine whether the control of tumor formation by the blastocyst was specific for embryonal carcinoma cells, B16 melanoma cells (Pierce et al. 1979), sarcoma 180 cells and L1210 leukemia cells were injected into Swiss-Webster blastocysts and assayed in the appropriate hosts (Pierce et al. 1981b). Single cells injected into the blastocyst produced tumors in comparable incidence to that obtained when single cells were injected into the host in the absence of blastocysts.

It was concluded that the blastocyst was capable of regulating certain lines of embryonal carcinoma cells and there appeared to be a specificity to the reaction because they did not abrogate tumorigenicity of the other three cell types tested. It may be inferred from these observations that an embryonic field can regulate only its closely related kind of cancer. Is there an appropriate embryonic field for each type of carcinoma?

Some regulation of C1300 neuroblastoma cells was observed (Pierce et al. 1981b). The degree of regulation was minimal, but reproducible. Two possibilities were considered to explain the observation because neurulation follows blastulation by about three days, the neuroblastoma cells may have persisted and been controlled at the time of neurulation, or the inductive events leading to neurulation may have occurred in late blastulation and abrogated tumor formation by some of the neuroblastoma cells at that time.

It was decided to examine how the blastocyst controlled tumorigenicity of embryonal carcinoma cells. The effect could be mediated by inner cell mass cells, trophectoderm or by blastocoele fluid, which is secreted by trophectoderm. In addition, the mechanism may be the result of a combination of cell-cell contact plus the effect of blastocoele fluid. In the initial experiments it was decided to incubate fragments of trophectoderm or inner cell mass with single embryonal carcinoma cells and test the ability of these preparations to form tumors (Pierce et al. 1981c). Inner cell masses were isolated by immunosurgery as described by Solter and Knowles (1975). Pieces of trophectoderm free of inner cell mass cells were dissected from blastocysts. Single embryonal carcinoma cells were stuck to each of these preparations, were incubated for either 2 or 24 hours and transferred into animals. No evidence was found for control of tumorigenicity when embryonal carcinoma cells were cultured in apposition to inner cell mass cells. Embryonal carcinoma cells attached to trophectoderm produced fewer tumors than the controls in which the cancer cells were cloned in the intraperitoneum free of either inner cell mass or trophectoderm. The data were not definitive (Pierce et al. 1981c). Accordingly, it was decided to inject embryonal carcinoma cells into trophectodermal vesicles and determine if the vesicle was capable of controlling colony formation of the cancer cell (Pierce et al. 1981a). A colony-forming assay was developed by Wells (1981) as a more efficient assay to measure the effect of the blastocyst on embryonal carcinoma cells.

In the colony assay embryonal carcinoma cells are grown in tissue culture either alone or after incorporation into the blastocoele cavity or in the perivitelline space. Blastocysts hatch and form a small patch of polyploid trophoblast with a clump of inner cell mass cells. If the blastocyst controls an embryonal carcinoma cell injected into its cavity, no colony of tumor cells is formed. If the cell is injected into the perivitelline space or is cloned alone a large colony of tumor cells forms within 6 days. The results of the assay using 247 cells corresponded closely to those obtained with the tumor-forming assay.

The experiment to determine the effect of trophectoderm was done in the following manner: single embryonal carcinoma cells were injected into mouse blastocysts, and after removal of the injection pipette, the tip of a 26-gauge needle attached to a micromanipulator head was used to guillotine off the inner cell mass. The guillotine action pressed the cut edges of the trophectoderm together and stuck them to the scar made by the needle in the bottom of the plastic petri dish. The inner cell mass and polar trophectoderm were removed by suction. The trophoblastic vesicle collapsed and upon re-expanding was examined to see if it contained a cell. Initially, during the perfection of the technique, and periodically during the course of the experiment, embryonal carcinoma cells labeled with fluorescent beads were employed (Pierce et al. 1981a). Samples of trophectodermal vesicles thought to contain cancer cells were examined under ultraviolet light to confirm the presence of a labeled cell. (The ultraviolet light kills the cell and these were not used in the experiments.) Only vesicles with easily visible embryonal carcinoma cells were used in the experiment.

It turned out that colony formation of embryonal carcinoma cells incorporated into trophectodermal vesicles was controlled in almost absolute fashion in comparison to colony formation of single 247 cells in vitro. This control occurred only when the embryonal carcinoma cell was placed on the secretory surface (in vesicles) rather than on the base of the trophectoderm (perivitelline space). Furthermore, when inner cell masses were isolated by immunosurgical destruction of trophectoderm, and cultured with attached embryonal carcinoma cells, colony formation occurred as frequently as it did when single embryonal carcinoma cells were cultivated alone (Pierce et al. 1981a). Although this evidence is negative, it is compatible with the idea that the inner cell mass does not control colony formation of single embryonal carcinoma cells.

During the course of these experiments, an interesting observation was made concerning the interaction of embryonal carcinoma cells attached to trophoblastic outgrowths of Swiss-Webster blastocysts (Pierce et al. 1981c). In these experiments blastocysts were allowed to hatch, attach and grow out on glass cover slips that were 3mm in diameter. Embryonal carcinoma cells were placed on the outgrowth and these coverslips with attached cells were placed in the intraperitoneum of mice. Little or no difference was noticed in tumor formation between these preparations and the ones in which the embryonal carcinoma cells were placed in the intraperitoneum in the absence of a substrate. Of interest, however, was the observation that the ascites of tumor cells grown on substrates was associated with metastases whereas single cells placed in the intraperitoneum were not. In our experience with germinal tumors we have never before observed metastases. The $402A\bar{x}$ cells employed had marker chromosomes which were doubled in number in the metastases indicating that the cells that metastasized were polyploid. This genetic change was not permanent because when cells from metastases were reinoculated into the intraperitoneum they did not remetastasize and their chromosome number reverted to that characteristic for $402A\bar{x}$ cells (Pierce et al. 1981c). Yolk sac carcinoma cells were also tested for their ability to metastasize from substrates. Whereas single yolk sac carcinoma cells injected into the peritoneal cavity also gave rise to tumors that did not metastasize, those attached to substrates metastasized frequently. At this point it is not known if it is the substrate that is important or the feeder layer of cells (trophectoderm or mouse embryo fibroblast) which promote metastases. In the experiments of Boone (1975) 3T3 cells when injected into appropriate hosts did not develop tumors, but when they were attached to glass beads and then injected into hosts, they formed tumors. The important point to be made is that metastases may have an epigenetic component. This comes as no surprise to developmental biologists who know that an appropriate environment is required for a phenotypic expression to occur.

In summary, if one cellular environment can abrogate tumorigenicity or colony formation of malignant cells and another can enhance malignant expression as evidenced by induction of metastases, we must determine how the environment of cells regulates phenotypic expression. Heppner et al. (1978) has shown that mixtures of drug-sensitive and drug-resistant cells mutually interact and that the population becomes more drug-sensitive than expected. Similarly, Gray & Pierce showed that melanomas composed of mixtures of cells with a four-fold difference in growth rate did not segregate out into fast-growing portions of the tumor and slow-growing portions. Rather, it appeared as if the slow-growing tumor cells inhibited the fast-growing ones (Pierce et al. 1964).

Studies of microenvironments may well offer biological means of controlling the growth of tumor cells.

ACKNOWLEDGEMENT:

This work was supported in part by a gift from R. J. Reynolds Industries, Inc. and by Grants #CA15823 from the National Cancer Institute and #CD-81 from the American Cancer Society.

REFERENCES:

Boone CW (1975). Malignant hemangioendotheliomas produced by subcutaneous inoculation of Balb/3T3 cells attached to glass beads. Science 188:68.
Brinster RL (1974). The effect of cells transferred into the mouse blastocyst on subsequent development. J Exp Med 140:1049.
Gray JM, Pierce GB (1964). Relationship between growth rate and differentiation of melanoma in vivo. J Natl Cancer Inst 32:1201.
Heppner GH, Dexter DL, DeNucci T, Miller FR, Calabrisi P (1978). Heterogeneity in drug sensitivity among tumor cell subpopulations of a single mammary tumor. Cancer Res 38:3758.

Kleinsmith LJ, Pierce GB (1964). Multipotentiality of single embryonal carcinoma cells. Cancer Res 24:1544.

Mintz B, Illmensee K (1975). Normal genetically mosaic mice produced from malignant teratocarcinoma cells. Proc Natl Acad Sci USA 72:3585.

Papaioannou VE, Evans EP, Gardner RL, Graham CF (1979). Growth and differentiation of an embryonal carcinoma cell line (C145b). J Emb exp Morph 54:277.

Papaioannou VE, McBurney MW, Gardner RL, Evans RL (1975). Fate of teratocarcinoma cells injected into early mouse embryos. Nature 258:70.

Pierce GB, Dixon FJ (1959). Testicular teratomas. I. The demonstration of teratogenesis by metamorphosis of multipotential cells. Cancer 12:573.

Pierce GB, Dixon FJ, Verney EL (1960). Teratocarcinogenic and tissue forming potentials of the cell types comprising neoplastic embryoid bodies. Lab Invest 9:583.

Pierce GB, Shikes R, Fink LM (1978). "Cancer: A Problem of Developmental Biology." New Jersey: Prentice-Hall.

Pierce GB, Lewis SH, Miller GJ, Moritz E, Miller P (1979). Tumorigenicity of embryonal carcinoma as an assay to study control of malignancy by the murine blastocyst. Proc Natl Acad Sci USA 76:6649.

Pierce GB, Hood G, Wells RS (1981a). Trophectodermal control of embryonal carcinoma. In preparation.

Pierce GB, Pantazis C, Caldwell JE, Wells RS (1981b). Specificity of the control of tumor formation by the blastocyst. Submitted to Cancer Research.

Pierce GB, Pantazis CG, Caldwell JE, Wells RS (1981c). Embryologic control of malignancy. In Harris C, Ceruth P (eds): "Mechanisms of Chemical Carcinogenesis," in press.

Solter D, Knowles BB (1975). Immunosurgery of mouse blastocyst. Proc Nat Acad Sci USA 72:5099.

Speers WC, Birdwell CR, Dixon FJ (1979). Chemically induced bidirectional differentiation of murine teratocarcinoma cells in vitro. Am J Path 97:563.

Strickland S, Sawey MJ (1980). Studies on the effect of retinoids on the differentiation of teratocarcinoma stem cells in vitro and in vivo. Dev Biol 78(1):76.

Wells RS (1981). An in vitro assay to probe blastocyst control of embryonal carcinoma cells. Submitted to Science.

RECENT OBSERVATIONS ON THE PATHOGENESIS OF CANCER METASTASIS

Isaiah J. Fidler and Ian R. Hart

Cancer Metastasis and Treatment Laboratory
NCI Frederick Cancer Research Center
Frederick, Maryland 21701

The most devastating aspect of cancer is the propensity for malignant neoplasms to spread from their primary site of growth to distant organs where secondary tumors, i.e., metastases, can develop. Despite major advances in surgical techniques and in patient care as well as the development of aggressive adjuvant therapies, most deaths of cancer patients are caused by metastases. Obviously then, the major obstacle to the treatment of cancer is not the excision of the primary tumor mass but the elimination of metastases.

New approaches to the treatment of metastasis must be based on a better understanding of the basic biology of the process of metastasis (Fidler et al., 1978). Research done in many laboratories in the past few years is providing us with new insights into the mechanics of the metastatic process and the characteristics of metastatic tumor cells.

The process of metastasis can be subdivided, albeit somewhat arbitrarily, into a series of linked, sequential steps. Failure to complete any one of these steps leads to the elimination of the disseminating tumor cell (Fig. 1). Thus, malignant cells that eventually give rise to secondary tumor deposits have survived a series of potentially lethal interactions, the final outcome of which is dependent on both host responses and intrinsic tumor cell properties (Fidler et al., 1978; Hart and Fidler, 1980; Poste and Fidler, 1980). Because of the complexity of the pathogenesis of cancer metastasis, it is not surprising that the process is an inefficient one (Weiss, 1980). Tumor cells present in the circulation do not necessarily constitute a metastasis (Salsbury, 1975), for most cells released into the bloodstream are eliminated

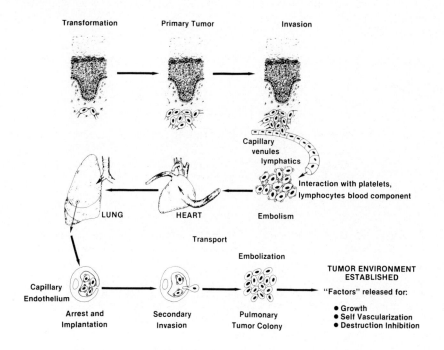

Figure 1. The pathogenesis of cancer metastasis.

rapidly (Butler and Gullino, 1975). Indeed, using radiolabeled murine B16 melanoma cells injected directly into the venous circulation, we have shown that less than 0.1% of the original inoculum survives to form secondary growths (Fidler, 1970).

Consideration of this finding has led us to question whether metastasis is a random or a selective process (Fidler and Kripke, 1977). That is, given the quantity of neoplastic cells that can be shed into the blood, do the eventual progenitors of distant metastatic foci represent the fortuitous survival of a few cells or the selection of a preexistent metastatic subpopulation from the parental tumor? It is unwise to generalize about any aspects of the metastatic process, and random or selective facets of the phenomenon are not mutually exclusive. However, with this caveat in mind, we believe that metastasis generally represents a highly selective, nonrandom event. At every step of the process, the rules of "survival of the fittest" apply with regard to the interplay of the tumor cells with their host. This belief is contrary to the widely accepted view that neoplastic dissemination is the ultimate expression of cellular anarchy. Rather, it indicates that we do not yet fully understand the regulatory mechanisms that control tumor spread. It is the purpose of this review to present a view of metastasis within this framework of cellular diversity and tumor progression.

At the time of diagnosis, many human and animal neoplasms are not composed of biologically uniform cells. Several alternative, but not necessarily mutually exclusive, mechanisms have been proposed to explain this finding. The heterogeneity in tumors could be attributed to their multicellular origin (Reddy and Fialkow, 1979) or to a tumor mass consisting of multiple primary tumors. Most studies, however, conclude that cancer probably results from the proliferation of a single transformed cell. Tumor evolution and progression, therefore, are probably the major factors responsible for the development of biological diversity. There are many examples showing that neoplasms are heterogeneous with regard to such phenotypes as karyotype, drug and hormone sensitivity, antigenicity, pigment production, growth rate, metabolic characteristics, enzyme markers and radiosensitivity (review Fidler, 1978; Poste and Fidler, 1980; Hart and Fidler, 1981). Experimental proof that metastatic subpopulations preexist within a primary tumor was obtained from studies of the B16 melanoma (Fidler and Kripke, 1977). Based upon the modified fluctuation assay of Luria and Delbrück (1943), it was shown that clones, each derived from individual cells, isolated from the parent tumor varied dramatically in their ability to form pulmonary nodules following i.v. inoculation into syngeneic recipient mice. Control subcloning procedures demonstrated that the observed diversity was not a consequence of the cloning procedure. Since this first demonstration, several investigators, using different tumor lines, have found a similar degree of metastatic heterogeneity in neoplasms that were either venerable or of more recent origin (Nicolson, 1978; Kripke et al., 1978; Dexter et al., 1978; Suzuki et. al., 1978; Talmadge et al., 1979; Reading et al., 1980). Except for the UV-2237 fibrosarcoma (Kripke et al., 1978), extensive analysis of tumor heterogeneity for metastasis has been done with serially propagated tumors such as the B16 melanoma. Because the B16 melanoma arose spontaneously in a

C57BL/6 mouse in 1954, repeated passages could have provided sufficient opportunity for "progression" to have occurred. Whether the findings we obtained with the B16 tumor were relevant to other melanomas of recent origin (primary melanomas) remained unclear. Recently, Kripke (1979) has described the induction and isolation of a new K-1735 melanoma syngeneic to the C3H/HeN mouse. Using this autochthonous neoplasm, we were able to determine whether a melanoma of recent origin would also exhibit preexistent heterogeneity for metastasis (Fidler and Kripke, 1980; Fidler et al., 1981). The K-1735 melanoma arose in a C3H mouse subjected to ten 1-hr exposures of UV radiation followed by the application of 2.5% croton oil in acetone to the skin of the scapular region for 2 years (Kripke, 1979). The primary tumor was removed and fragments were transplanted into immunodeficient recipient animals to obviate the possibility of reduction of heterogeneity by

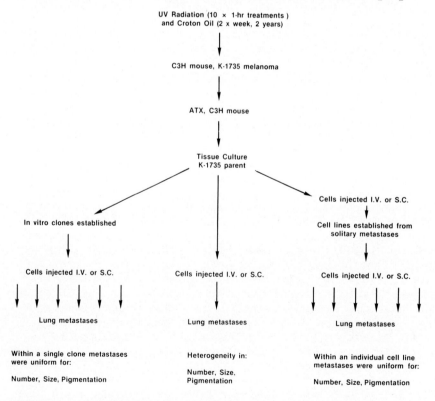

Figure 2. Experimental procedures used to investigate the metastatic heterogeneity of the K-1735 melanoma.

immunoselection (Fidler, 1978). Several weeks after transplantation, a tissue culture line was established and cells from the fifth passage were used to perform the procedures outlined in Fig. 2. Clones and tumor lines were stored frozen until expanded for in vivo work.

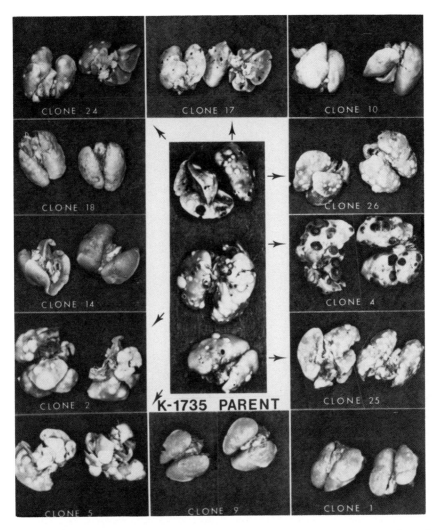

Figure 3. Gross appearance of pulmonary metastases produced by cells of the K-1735 melanoma (2X) and its cloned subpopulations 35 days after the i.v. injection of 100,000 cells (15X) (Fidler et al., 1981).

Intravenous injection of cells from the parental tumor resulted in the establishment of pulmonary nodules exhibiting marked phenotypic variability (Fig. 3). Lung tumor colonies varied in size, number and pigmentation; however, this striking pleomorphism was, to a large extent, eliminated when clones were injected i.v. Counts of lung tumors obtained under the dissecting microscope showed that clones differed significantly among themselves and from the parent tumor with regard to the median number of lung nodules formed. Furthermore, in contrast to the lung metastases produced by cells from the parent K-1735 melanoma, metastases established from the cloned lines tended to be much more uniform with regard to size and pigmentation (Table 1). Statistical analysis of the results indicated that only 2 of 22 K-1735 clones were indistinguishable from the parent tumor (Fidler et al., 1981). In the B16 melanoma system, 2 of 17 clones were indistinguishable from the parent tumor (Fidler and Kripke, 1977). With these figures as an indication of the degree of metastatic heterogeneity of the tumors, it is apparent that the K-1735 melanoma, which is of recent origin, is no less heterogeneous than the B16 melanoma, a tumor that arose in 1954. Again, control subcloning experiments carried out with the two tumor systems ruled out that the high degree of heterogeneity was caused by the process of in vitro cloning.

Although the K-1735 melanoma exhibits considerable cellular diversity for the metastatic phenotype, the source of this heterogeneity is unclear. It is possible that a physically induced tumor, such as the K-1735, has a multicellular origin (Reddy and Fialkow, 1979) and thus the different phenotypes represent differences between the progeny of several transformed cells. However, it is thought that most human tumors are unicellular in origin (Fialkow, 1976; Iannaccone, et al., 1978) and most human tumors are heterogeneous for a variety of characteristics by the time of clinical presentation (Baylin et al., 1978; Trope et al., 1975). In order to explain this multiformity, Nowell (1976) has suggested that acquired genetic variability within developing clones of tumor cells, coupled with the selection pressure exerted by host responses, leads to the emergence of new sublines with enhanced survival ability that is manifested by increased malignancy. Although we will discuss data supporting this hypothesis later in this review, it should be borne in mind that the continuous evolution of tumors does not always lead to the emergence of more malignant cells.

Table 1. Experimental metastasis in syngeneic C3H/HeN mice produced by the i.v. injection of cells from K-1735 melanoma parent tumor and cloned lines.

Source of cells	Pulmonary metastasis[a]			Extrapulmonary metastases
	Median	Range	p[b]	
Clone 16	0	(0-2)	<0.0001	None
Clone 13	0	(0-2)	<0.0001	None
Clone 11	0	(0-3)	<0.0001	None
Clone 9	0	(0-5)	<0.0001	None
Clone 10	0.5	(0-4)	<0.0001	1/15 heart
Clone 6	1	(0-1)	<0.0001	None
Clone 23	1	(0-6)	<0.0001	None
Clone 3	1	(0-6)	<0.0001	None
Clone 19	1	(0-7)	<0.0001	None
Clone 14	2	(0-10)	<0.0001	None
Clone 8	4	(0-8)	<0.0001	None
Clone 15	7	(0-17)	<0.0022	None
Clone 5	23.5	(4-73)	<0.98	None
Clone 18	55	(11-140)	<0.05	2/15 heart
Clone 24	56.5	(23-160)	<0.0195	3/15 brain
Clone 17	90	(70-156)	<0.0002	None
Clone 1	111.5	(44-187)	<0.0005	2/15 brain
Clone 2	123	(41-168)	<0.0009	2/15 brain
Clone 4	156	(101-196)	<0.0001	None
Clone 26	209	(48-383)	<0.0001	2/20 heart
Clone 25	257	(15-521)	<0.0015	3/20 heart, 3/20 skin
Clone 7	320	(82-375)	<0.0001	4/15 heart, 4/20 skin 3/15 kidney, 4/15 lymph node
Parent K-1735	33	(0-152)		6/35 brain, 3/15 brain 4/35 heart, 3/35 lymph node

[a]Five weeks after i.v. injection of 1×10^5 tumor cells (n = >15 mice/group).

[b]Probability of no difference from parent (Mann-Whitney U-test).

In order to determine whether tumors of unicellular origin can exhibit metastatic heterogeneity, we examined the in vivo behavior of murine embryo fibroblasts transformed by an oncogenic virus (Fidler and Hart, 1981). Six colonies of BALB/c embryo fibroblasts, each derived from a single cell, were infected in vitro with mouse sarcoma virus and then propagated as individual cell lines. Injection of 2×10^5 viable cells from each clone into the tail vein of recipient BALB/c mice resulted in marked differences among the clones with regard to the resultant number of lung nodules. Because each tumor line was derived from the progeny of a single, transformed cell, the data indicate that multicellular origin can engender metastatic heterogeneity. Equally, when the clones from two colonies (one of high and one of low experimental metastatic capacity) were further subcloned and evaluated in the same manner, both clones exhibited a pattern of metastatic heterogeneity. Interestingly, the clone with higher metastatic capacity exhibited a greater degree of variability than the clone with lower metastatic capacity. Thus, regardless of the unicellular origin of the tumor, by the time of assay, which was six weeks after transformation, clones contained subpopulations of cells with different metastatic properties. These data also demonstrate that the generation of metastatic heterogeneity in neoplasms does not require a prolonged latent period of months or even weeks (Fig. 4).

If tumors are heterogeneous with regard to metastasis such that the formation of secondary tumor deposits represents the enrichment or isolation of metastatic variants, then continued application of the correct selection pressure should lead to the emergence of a more metastatic population. Such is the case. The repeated cycling of B16 melanoma cells through i.v. injection, followed by harvesting of lung nodules, placing them in culture and then repeating the process led, after ten passages, to a cell line, designated B16-F10, which was more highly metastatic in the experimental metastasis assay than the initial tumor line which was designated B16-F1 (Fidler, 1973). Again, the K-1735 melanoma, though physically induced and of more recent origin than the B16 melanoma, exhibited a comparable pattern of behavior. Cells from the parental K-1735 line were injected i.m. into the footpads of adult C3H mice and at predetermined times after the amputation of the primary tumor, animals were killed and individual metastases were harvested from the lungs. These metastases, which varied both in size and pigmentation, were established in tissue

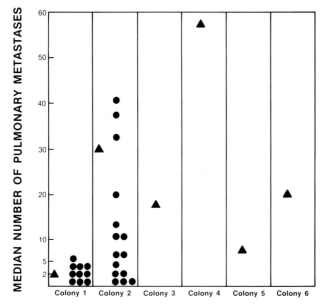

Figure 4. Experimental metastasis produced by cells of individual BALB/c 3T3 fibroblasts transformed in vitro with mouse sarcoma virus (▲). Subcloning experiments were performed within 2-3 weeks after the in vitro transformation (●). The median number of pulmonary metastases per mouse was derived from counts taken from 10 mice/group. Mice were injected i.v. with 2×10^5 viable cells.

culture and subsequently injected i.v. into groups of age- and sex-matched syngeneic mice. As illustrated in Table 2, all the cell lines derived from lung nodules gave a significantly greater median number of lung nodules ($p \leq 0.001$) than equal numbers of cells from the parental line. In the next set of experiments, we examined the production of spontaneous metastases (following i.m. implantation) by cells isolated from the parent K-1735 melanoma and several spontaneous metastases. The data, also shown in Table 2, clearly indicate that cells isolated from metastases produce a higher incidence of spontaneous metastasis than the parental tumor.

Table 2. Metastatic potential of cells from parent K-1735 melanoma and from lines established from spontaneous metastases.

Source of cells	Experimental metastasis[a]			Spontaneous metastasis[b]			
	Median number of pulmonary metastases (range)			Amputated[c]	Nonamputated	Total	
K-1735 parent	44 (4-265)			0/7	2/10	2/17	
K-1735-M1	252 (55->500)		p<0.001	1/7	6/10	7/17	p<0.01
K-1735-M2	376 (1->500)		p<0.001	0/8	9/9	9/17	p<0.01
K-1735-M3	143 (11-315)		p<0.001	3/7	6/8	9/15	p<0.01
K-1735-M4	91 (6->500)		p<0.01	5/10	7/9	12/19	p<0.01

[a]Median number of metastases produced 35 days after the i.v. injection of 10^5 viable cells.

[b]Metastases produced in lung and/or lymph nodes of mice implanted with viable tumor cells in the footpad.

[c]Legs were amputated when tumors were 12-15 mm in diameter.

If, as seems probable, metastases are generally more metastatic than unselected heterogeneous parental lines, why did it require ten passages to produce the B16-F10 line (Fidler, 1973) rather than a single selection cycle? In part, this may be attributable to the mixing of different lung nodules, each of which was derived from a cell of different metastatic potential. Alternatively, it may have been attributable to the progression and development of heterogeneity within a single clone since, as already pointed out, this phenomenon does not necessarily lead to the development of more metastatic populations. This latter possibility could be tested by comparing the metastatic variability of clones derived from small, newer metastases with the behavior of clones obtained from larger, older metastases. The expected result would be that clones derived from the smaller, newer metastases would show a more homogeneous spectrum of metastatic variability than those from larger and older tumors. Partial support for this hypothesis has come from some studies recently completed in our laboratory and in collaboration with Dr. G. Poste (Smith Kline and French Laboratories, Philadelphia, PA).

In vitro passage of individual clones isolated from heterogeneous parental populations rapidly resulted in the emergence of variant subclones with different metastatic properties (Poste, Doll and Fidler, 1981). Two clones exhibiting low metastatic capacity and two clones exhibiting higher metastatic capacity were isolated from the B16-F10 line. These clones were cultured in vitro for 10, 20 or 40 subcultivations (5, 10 or 20 weeks) and then subclones derived at each of these intervals were reassessed for their metastatic behavior. After only 10 subcultivations in vitro, many subclones were derived that differed significantly from their original parent clone. Continued cultivation introduced further variability such that by 20 and 40 subcultivations, the majority of the clones tested were significantly different from the parent clone. In contrast to other tumor lines, which will be discussed later, greater phenotypic shifts were detected in subclones from the low metastatic clones as compared to the high metastatic clones. Exactly comparable findings were obtained when the cloned lines were passaged in vivo by s.c. transplantation. In marked contrast to these results, the metastatic phenotype of the uncloned B16 melanoma lines, B16-F1 and B16-F10, remained remarkably stable over 60 in vitro and 30 in vivo passages. Maintenance of metastatic uniformity could be imposed upon different clones by mixing them together and cocultivating them as polyclonal populations (Poste et al., 1981). Identification of the appropriate clones was achieved by mixing together wild-type clones of known metastatic potential (low, medium or high) with clones of comparable metastatic capacity bearing stable biochemical markers. Different clones, after 10 or 20 in vitro subcultivations, were isolated according to their sensitivity to various drugs, and these subclones were subsequently tested for their metastatic capacity. The metastatic properties of subclones isolated from serially passaged, mixed clone populations resemble the metastatic characteristics of the original parent clones.

The obvious inference to be drawn from these studies is that different cloned subpopulations of tumor cells act to "stabilize" their relative proportions. Removal of the stabilizing effect by isolating clones or by the application of a strong selection pressure leads to the rapid generation of diversity in the resurgent population. Once diversity occurs it would result in the reimposition of the previously noted stabilizing effect.

As we have emphasized elsewhere in this chapter, the metastatic process is highly selective. It has been suggested that metastases could be clonal in their origin (Suzuki and Withers, 1979). Nowell's hypothesis for tumor progression (Nowell, 1976) seems to predict that increasing malignancy, i.e., increasing metastatic capacity, would be accompanied by increasing genetic instability. To test for this in our laboratory, we have examined the phenotypic stability of metastatic and nonmetastatic tumor lines for experimental metastasis. Concomitantly, the rates of mutation to ouabain and/or 6-thioguanine resistance of four metastatic tumor lines have been compared with their respective tumor lines of low metastatic capacity, isolated from the same neoplasms (Cifone and Fidler, 1981).

Clones with low metastatic potential (LMC) and clones with high metastatic potential (HMC) were isolated from the UV-2237 fibrosarcoma (Kripke, 1979). Classification of the metastatic capacity of these clones was based upon the ability to produce both spontaneous and experimental metastases (Kripke et al., 1978). The LMC and HMC were carried in vitro for 72 or 60 days, respectively. During the same time period the LMC and HMC were carried in vivo, in the subcutis of syngeneic mice, for 72 and 60 days. At the end of these time periods, cell cultures were established from these solid tumors, and one week later a series of subclones were isolated. The ability of these subclones to form experimental lung metastases was compared to that of subclones derived from the in vitro passaged clones and to that of subclones, obtained at the initial establishment of the LMC and HMC, which had been recovered from the freezer. The pattern of behavior of the subclones of the UV-2237 LMC was remarkably similar to that of the parent clone, regardless of whether they were derived at the time of isolation or after 72 days of continuous cultivation in vitro or in vivo. The behavior of the subclones derived from the HMC contrasted considerably with that of the parent clone. Considerable and significant diversity had been generated by 60 days after both in vitro and in vivo generation, suggesting that the metastatic phenotype of the HMC is unstable (Figs. 5,6). Although these findings seem to support Nowell's hypothesis, it should be noted that not all tumors follow this pattern. Studies with the B16 melanoma clones have shown that the clones with less metastatic capacity generated the greatest phenotypic diversity (Poste et al., 1981).

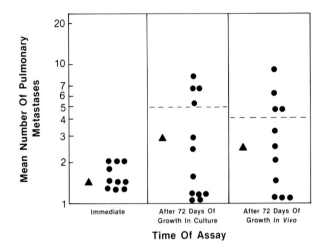

Figure 5. Experimental metastasis produced by cells of individual subclones (●) isolated from parent LMC of UV-2237 fibrosarcoma (▲). Subcloning experiments were performed at 72 days after the parent LMC was grown in culture or injected s.c. into syngeneic C3H mice. The median number of pulmonary metastases/mouse was derived from counts taken from at least 10 mice/group. Mice were injected i.v. with 10^5 cells, were killed 3 weeks later, and lung tumor colonies were counted under a dissecting microscope.

The rates of spontaneous mutation of cells from the two UV-2237 fibrosarcoma clones to resistance to 4 mM ouabain and 6-thioguanine (4 µg/ml) were determined using the fluctuation assay of Luria and Delbrück (1943). The mutation rates for ouabain and 6-thioguanine were 3- and 4.6-fold higher, respectively in the HMC than in the LMC. Similar differences in the rates of spontaneous mutation between high- and low-metastatic cells were found when these studies were extended to 3 other tumor systems. In these studies with the UV-2237 fibrosarcoma, the K-1735 melanoma

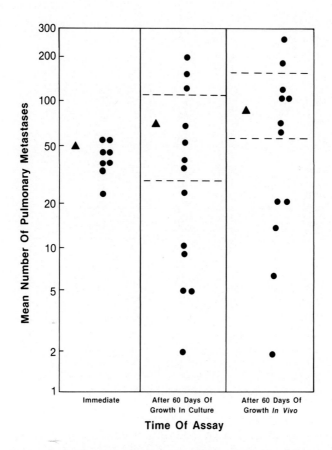

Figure 6. Experimental metastasis produced by cells of individual subclones (●) isolated from parent HMC of UV-2237 fibrosarcoma (▲). Subcloning experiments were performed 60 days after the parent HMC was grown in culture or injected s.c. into syngeneic C3H mice. For experimental details see footnote to Fig. 5.

and the SF-19 fibrosarcoma (spontaneous in origin), metastatic cells showed an increase in the rate of spontaneous mutation to ouabain resistance (4.6-, 6.5-, 7.0- and 5.8-fold, respectively) which correlated with the degree of metastatic capacity (Table 3).

Table 3. Rate of spontaneous mutation to ouabain resistance of tumor cells with low or high metastatic potential isolated from three different murine neoplasms.

Cell line	Metastatic potential[a]	Rate of mutation ($\times 10^6$ per cell generation)[b]	Fold increase HMC/LMC
UV-2237 fibrosarcoma LMC-1	low	0.158	
UV-2237 fibrosarcoma HMC-1	high	0.728	x4.61
UV-2237 fibrosarcoma LMC-2	low	0.0764	
UV-2237 fibrosarcoma HMC-2	high	0.502	x6.52
K-1735 melanoma LMC	low	0.0873	
K-1735 melanoma HMC	high	0.610	x6.98
SF-19	low	0.0178	
SF-19-UV-9	high	0.103	x5.78

[a]Metastatic potential was determined by the ability of tumor cells to produce spontaneous metastases from a s.c. tumor or lung metastases after i.v. injection.

[b]The rate of mutation was calculated by the equation originally described by Luria and Delbruck (1943) and Cifone and Fidler (1981).

These results seem to support the contention that tumor progression occurs as a result of acquired genetic alterations. The evolution of metastatic heterogeneity in the HMC did not progress in a unidirectional manner (some subclones were less metastatic than the parental clone), but this finding is not incompatible with Nowell's hypothesis (Nowell, 1976). It is possible that in the absence of selection pressure, such as under in vitro conditions or in the relatively benign milieu of the subcutis, increased rates of mutation lead to the emergence of less, as well as more, malignant variants. Recent mutagenesis experiments on other murine neoplasms support this contention. Mutagenesis of malignant cell lines has resulted in the appearance of clones incapable of progressive growth in normal hosts (Boon and Van Pel, 1978; Van Pel et al., 1979) or variants with malignant capacities less than or equal to the parental tumor (Kerbel, 1979).

SUMMARY AND CONCLUSIONS

The complexity of the metastatic process makes it difficult to provide generalized explanations. Results and hypotheses based upon a single tumor system (or a restricted range of tumors) are likely to be revised as more studies are carried out. Furthermore, results obtained with our tumor systems have not always been comparable to those obtained with other tumor models. However, bearing these limitations in mind and ignoring our own cautionary warnings, we have concluded the following:

1. The metastatic process selects from a heterogeneous starting population, the preexistent metastatic subpopulations.

2. Diversity for the metastatic phenotype may be a consequence of the multicellular origin of a neoplasm or it may be the result of continuous evolution and progression in tumors of unicellular origin. Either of these modes of diversification may be operative in the primary or the metastatic tumor.

3. Tumor cell subpopulations of clonal (unicellular) origin are phenotypically less stable than heterogeneous cell populations. Achievement of a certain degree of diversity imposes an equilibrium, with regard to the metastatic phenotype, which is presumably mediated by some sort of interaction among the different subpopulations.

4. Metastatic clones appear, in general, to be less stable than benign or nonmetastatic clones in that they generate a greater diversity and exhibit higher rates of spontaneous mutation. It is possible that in the presence of strong selection pressures, such as would occur during metastasis, these clones are more likely to survive and emerge as the progenitors of distant tumor foci.

The generation of this diversity and the sequelae of the selective nature of the metastatic process have profound implications both for studies on the biology of tumor spread and the design of therapeutic regimens to counter the intransigent aspects of neoplasia (Poste and Fidler, 1980; Hart and Fidler, 1981).

Research sponsored by NCI, DHHS, under Contract N01-CO-75380 with Litton Bionetics, Inc.

Baylin SD, Weisburger WR, Eggleston JC, Mendelsohn G, Beaven MA, Abeloff MD, Ettinger DS (1978). Variable content of histaminase, L-dopa decarboxylase and calcitonin in small-cell carcinoma of the lung. Biologic and clinical implications. N Engl J Med 299:105.

Boon T, Van Pel A (1978). Teratocarcinoma cell variants rejected by syngeneic mice: Protection of mice immunized with these variants against other variants and against the original malignant cell line. Proc Natl Acad Sci USA 75:1519.

Butler T, Gullino P (1975). Quantitation of cell shedding into efferent blood of mammary adenocarcinoma. Cancer Res 35:512.

Cifone MA, Fidler IJ (1981). Increasing metastatic potential is associated with increasing genetic instability of clones isolated from murine neoplasms. Proc Natl Acad Sci USA, in press.

Dexter DL, Kowalski HM, Blazar BA, Fligiel Z, Fogel R, Heppner GH (1978). Heterogeneity of tumor cells from a single mouse mammary tumor. Cancer Res 38:3174.

Fialkow PJ (1976). Clonal origin of human tumors. Biochim Biophys Acta 458:283.

Fidler IJ (1970). Metastasis. Quantitative analysis of distribution and fate of tumor emboli labeled with ^{125}I-5-iodo-2'-deoxyuridine. J Natl Cancer Inst 45:773.

Fidler IJ (1973). Selection of successive tumor lines for metastasis. Nature (New Biol) 242:148.

Fidler IJ (1978). Tumor heterogeneity and the biology of cancer invasion and metastasis. Cancer Res 38:2651.

Fidler IJ, Hart IR (1981). The origin of metastatic heterogeneity in tumors. Eur J Cancer 17:487.

Fidler IJ, Kripke ML (1977). Metastasis results from preexisting variant cells within a malignant tumor. Science 197:893.

Fidler IJ, Kripke ML (1980). Metastatic heterogeneity of cells from the K-1735 melanoma. In Grundmann E (ed): "Cancer Campaign Vol. 3: Metastatic Tumor Growth," Stuttgart: Gustav Fischer Verlag, p 71.

Fidler IJ, Gersten DM, Hart IR (1978). The biology of cancer invasion and metastasis. Adv Cancer Res 28:149.

Fidler IJ, Gruys E, Cifone MA (1981). Demonstration of multiple phenotypic diversity in a murine melanoma of recent origin. JNCI, in press.

Hart IR, Fidler IJ (1980). Cancer invasion and metastasis. Q Rev Biol 55:121.

Hart IR, Fidler IJ (1981). The implications of tumor heterogeneity for studies on the biology and therapy of cancer metastasis. Biochim Biophys Acta, in press.

Iannaccone PM, Gardner RL, Hans H (1978). The cellular origin of chemically-induced tumors. J Cell Sci 29:249.

Kerbel RS (1979). Immunologic studies of membrane mutants of a highly metastatic murine tumor. Am J Pathol 97:609.

Kripke ML (1979). Speculations on the role of ultraviolet radiation in the development of malignant melanoma. JNCI 63:541.

Kripke ML, Gruys E, Fidler IJ (1978). Metastatic heterogeneity of cells from an ultraviolet light-induced murine fibrosarcoma of recent origin. Cancer Res 38:1962.

Luria SE, Delbruck M (1943). Mutations of bacteria from virus sensitivity to virus resistance. Genetics 28:491.

Nicolson GL (1978). Experimental tumor metastasis: Characteristics and organ specificity. Bioscience 28:441.

Nowell PC (1976). The clonal evolution of tumor cell subpopulations. Science 194:23.

Poste G, Fidler IJ (1980). The pathogenesis of cancer metastasis. Nature 283:139.

Poste G, Doll J, Fidler IJ (1981). Interactions between clonal subpopulations affect the stability of the metastatic phenotype in polyclonal populations of B16 melanoma cells. Proc Natl Acad Sci USA, in press.

Reading CL, Belloni PN, Nicolson GL (1980). Selection and in vivo properties of lectin-attachment variants of malignant murine lymphosarcoma cell lines. JNCI 64:1241.

Reddy AL, Fialkow PF (1979). Multicellular origin of fibrosarcomas in mice induced by the chemical carcinogen 3-methylcholanthrene. J Exp Med 150:878.

Salsbury AJ (1975). The significance of the circulating cancer cell. Cancer Treatment Rev 2:55.

Suzuki N, Withers HR (1979). Lung colony formation: A selective cloning process for lung-colony-forming ability. Br J Cancer 39:196.

Suzuki N, Withers R, Koehler MW (1978). Heterogeneity and variability of artificial lung colony forming ability among clones from mouse fibrosarcoma. Cancer Res 38:3349.

Talmadge JE, Starkey Jr, Davis WC (1979). Characteristics of induced metastatic variants from a cloned metastatically homogeneous cell line. J Supramol Struct 9 (Suppl. 3):183.

Trope C, Hakansson L, Dencker U (1975). Heterogeneity of human adenocarcinomas of the colon and the stomach as regards sensitivity to cytostatic drugs. Neoplasma 22:423.

Van Pel A, Georlette M, Boon T (1979). Tumor cell variants obtained by mutagenesis of a Lewis lung carcinoma cell line: Immune rejection by syngeneic mice. Proc Natl Acad Sci USA 76:5282.

Weiss L (1980). Cancer cell traffic from the lungs to the liver: An example of metastatic inefficiency. Int J Cancer 25:385.

GONIAL CELL NEOPLASM OF GENETIC ORIGIN AFFECTING BOTH
SEXES OF DROSOPHILA MELANOGASTER

Elisabeth Gateff

Biologisches Institut I, Albert-Ludwigs-Universität

78 Freiburg i. Br., Schänzlestr. 1, F.R.G.

INTRODUCTION

In Drosophila melanogaster mutated developmental genes induce benign, lethal-benign and malignant neoplasms in various embryonic, larval and adult tissues (Gateff, 1978a,b,c). In search for mutants, causing tumors in adult tissues, a new autosomal, recessive mutant was found which develops benign gonadal tumors in both sexes. The mutant is designated as benign gonial cell neoplasm (bgcn; Gateff, 1981).

Numerous mutants have been described which affect either the male or the female germ line (King and Mohler, 1975; Lindsley and Tokuyasu, 1980). bgcn, in contrast, is active in the germ line of both sexes. It represents the first case of a mutant in which the germ cell precursors in the male as well as in the female fail to differentiate and grow in a tumorous fashion. Moreover, the bgcn allele seems to act at an earlier developmental stage than any of the known ovarian tumor mutants.

This paper describes the mutant phenotype, the histology and the developmental capacities of the neoplastic female and male gonial cells.

MATERIALS AND METHODS

The autosomal, recessive mutation benign gonial cell neoplasm (bgcn) is located between dumpy and black on chromosome 2 (Gateff, 1981; for the characterization of the genetic markers see Lindsley and Grell, 1968). It was induced in wild-type males with ethyl methan sulfonate (EMS; Lewis and Bacher, 1968), and was isolated by a standart procedure for

for the detection of mutations on the second chromosome. In order to prevent crossing over, the bgcn chromosome was complexed with the balancer chromosome SM5 (Linsley and Grell, 1968). bgcn/SM5 animals are fertile and thus, serve to continue the strain. The bgcn/bgcn individuals develop the gonadal tumors and are sterile. SM5/SM5 animals die at the end of embryonic life. Thus, in the larval and adult population the relation between bgcn/bgcn and bgcn/SM5 animals is 1:2.

A wild-type laboratory strain, designated as Oregon-R was used for anatomical and histological comparisons with the mutant and in transplantation experiments (see below). Larvae of a yellow (y) strain (see Lindsley and Grell, 1968) were also used as hosts in the transplantation experiments. Standart culture methods were applied. Animal age in hours after egg laying (h AEL).

To study the developmental capacities of the tumorous gonial cells in the wild-type environment, the in vivo transplantation method of Hadorn (1963) was applied. Ovaries of freshly eclosed, mutant flies were dissected in a drop of Ringer's solution and implanted into wild-type female flies. The implants were recovered from the hosts after 14 days, and their morphology and cellular content were examined in phase contrast.

The mode of development of mutant ovaries from 3rd instar larvae in a wild-type environment was also tested. Ovaries from 100 h old larvae of the mutant strain were implanted into female larvae of comparable age which were homozygous for the cuticle color gene yellow (y). The host larvae were allowed to metamorphose and complete adult development (for further details see Results).

For histology, the mutant and wild-type animals were fixed in Carnoy's solution, embedded in paraffine, sectioned at 5 mμ and stained with hematoxylin and eosin.

RESULTS

Comparative anatomy and histology of mutant and wild-type female and male gonads

bgcn female and male flies are sterile but perfectly viable. Except for their gonads they show no other deviations from the wild-type. We will compare first the anatomy and histology of the bgcn and wild-type female gonads.

Each of the two wild-type ovaries consist of approximately 15 ovarioles. The distal most region of each ovariole is occupied by a germarium, while the proximal portion represents the vitellarium (Fig. 1).

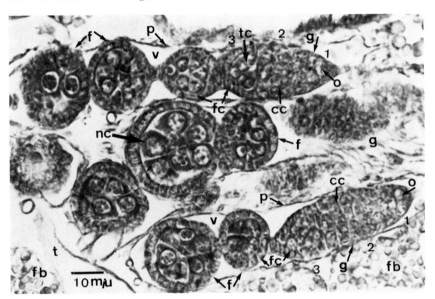

Figure 1. Longitudinal section through three ovarioles from a freshly eclosed wild-type fly. Each ovariole exhibits a germarium (g) and a vitellarium (v) in which two to three follicles (f) can be seen. Centrally located in the follicles are the nurse cells (nc) which are enveloped by a monolayer of follicle cells (fc). The ovarioles are surrounded by the peritoneal sheath (p), fat body (fb) and a trachea (t). Note the three regions in the germaria (1,2,3). Region 1 contains the mitotically active oogonia (o) and cystoblasts. Region 2 is inhabited by cystocytes (cc) in various stages of cyst formation. Region 3 contains terminal cysts (tc) surrounded by follicle cells (fc). Hematoxylin and eosin.

In the apical region of the germarium, oogonia divide giving rise to cystoblasts (Fig. 1). They, in turn, replicate and become cystocytes, which produce in four consecutive mitosis cysts of 16 interconnected cystocytes (Fig. 1). The cysts receive an envelop of follicle cells at the posterior end of the germarium and enter the vitellarium were oogenesis takes place. One of the 16 cystocytes becomes the oocyte while the remaining 15 differentiate into nurse cells (King, 1970; Mahowald and Kambysellis. (1980).

In newly emerged mutant flies the germaria are filled with numerous single, round cells while the vitellaria are empty (Fig. 2a). Neither cyst-nor follicle formation can be observed. Upon rupturing the peritoneal sheath surrounding the germarium, the released cells form a single cell suspension (Fig. 2b). In phase contrast as well as in histological preparations the cells appear cystoblast-like. They can clearly be distinguished from the smaller, cuboidal follicle cells, present at the posterior end of the germarium (Fig. 2a).

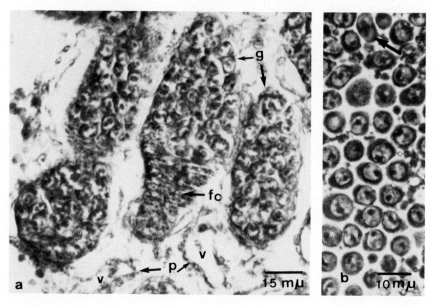

Figure 2a. Longitudinal section through three germaria from a freshly eclosed bgcn fly. Note the empty vitellaria (v) deliniated by the peritoneal sheath (p), and the germaria (g) filled with oogonia (o) and cystoblasts. In the posterior region of one of the three germaria cuboidal follicle cells (fc) can be seen. b. Phase contrast micrograph of a group of cells derived from a bgcn germarium. Note the large nuclei and the mitosis (arrow). Hematoxylin and eosin.

As the flies age, the cystoblasts from the germaria enter the vitellaria where cell division continues. The walls of the vitellaria become lined with follicle cells, while the inner space contains millions of neoplastic cystocytes (Fig. 3a,b).

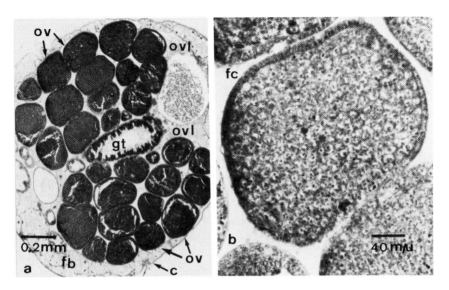

Figure 3a. Cross section through the abdomen of a seven day old mutant fly showing the two ovaries (ov), each with 15 ovarioles (ovl) surrounding the gut (gt) and embedded in fat body (fb). b. Cross section through an ovariole (enlargement from a). A single layer of cuboidal follicle cells (fc) lines the ovariole. The entire lumen is filled with cystoblasts. Hematoxylin and eosin.

Thus, the above histology shows that the bgcn allele inhibits the differentiation of the female germ cells at the cystoblast stage. As a result, cystoblasts engage in a neoplastic mode of growth and accumulate by the millions in the ovarioles.

In contrast to all known ovarian tumor mutants, which affect exclusively the female germ line (King, 1970), the bgcn causes neoplastic growth also of the male germ line. In the following paragraphs we will compare the histology of the bgcn and wild-type testies.

Wild-type spermatogenesis procedes as follows: primary spermatogonia, located at the distal end of the testis, divide and give rise in four consecutive mitosis to cysts of 16 interconnected primary spermatocytes. After a period of growth they undergo meiosis and form cysts of 64 syncitial spermatids (Fig. 4; Lindsley and Tokuyasu, 1980).

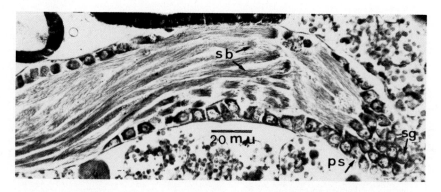

Figure 4. Longitudinal section through the distal end of a wild-type testis. Various stages of spermatogenesis, such as, spermatogonia (sg), primary spermatocytes (ps) and spermatid bundles (sb) can be observed. Hematoxylin and eosin.

In contrast to the wild-type, bgcn testes exhibit exclusively primary spermatocytes in their lumen (Fig. 5). The primary spermatocytes do not enter the growth phase, but continue to divide forming clusters of many hundreds of cells.

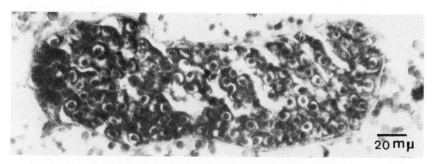

Figure 5. Oblique section through a bgcn testis showing only primary spermatocytes. Hematoxylin and eosin.

In conclusion, the histological studies demonstrate that in the bgcn mutant oogenesis as well as spermatogenesis are arrested at the cystoblast respectively primary spermatocyte stage, while the basic endowment of the cells to divide remains. This activity is persued by the neoplastic cells until the ovarioles and the testes become completely filled (Fig. 6). The neoplastic growth is non-invasive and not lethal to the animals.

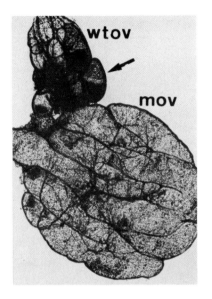

Figure 6. Wholemount of wild-type (<u>wtov</u>) and mutant ovaries (<u>mov</u>). Note the vitellogenic follicles in the wild-type ovary (arrow) and the lack of such structures in the mutant ovary, the ovarioles of which are filled with cystoblasts. X 80.

Developmental capacities of the bgcn ovary

The response of the <u>bgcn</u> cystoblasts from young adults to the wild-type environment was studies by implantation of ovaries from freshly eclosed flies into wild-type female hosts (see Materials and Methods). After 14 days of growth in the hosts 15 implants were recovered. Each implanted ovary had grown into a number of spheres which contained millions of cells resembling the cystoblasts found in situ in the <u>bgcn</u> ovarioles (Fig. 7).

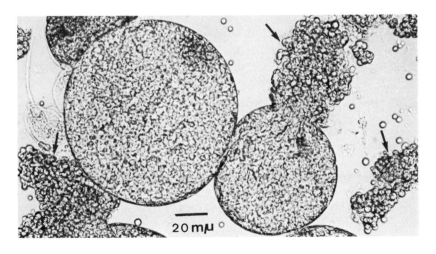

Figure 7. Spheres, containing cystoblasts derived from a young mutant ovary after in vivo culture for 14 days in a wild-type host. Some of the spheres have brocken and release their cystoblasts (arrows).

To test for the developmental capacity of the gonial cells in the bgcn larval ovary, we implanted 100 h old larval ovaries into female larvae of comparable age, homozygous for the X-linked body color allele y (see Materials and Methods). The later larvae show normal fertility. Since the larval population in the bgcn strain consists of 1/3 homozygous bgcn/bgcn and 2/3 heterozygous bgcn/SM5 individuals (see Materials and Methods), and since at this age no distinction can be made between the two genotypes, we collected female larvae at random and injected their ovaries into the above host larvae. The implanted ovaries were permitted to go through metamorphosis and adult development with their hosts, where the implants usually connect to the genital tracts of the host. After the completion of adult development, the host flies were crossed in pairs to y males.

If the implanted ovary was of the bgcn/bgcn genotype (see above), two modes of development are possible, depending whether the mutant trait is autonomous or non-autonomous. In the first case a gonial cell tumor will develop in the implant, just like in situ. Thus, in the progeny of such a host only y/y individuals will be found. If however, the second possibility would apply, in the implanted ovary oogenesis will procede normally, and among the progeny of the host there will be, in addition to y/y, also phenotypically wild-type bgcn/+ individuals.

As mentioned above, 2/3 rd of the implanted ovaries are expected to be genotypically bgcn/SM5. In these cases oogenesis should take place normally. The progeny of such pairs will show three phenotypes, (i) y/y, (ii) bgcn/+ (wild-type) and (iii) SM5/+ marked by Curly wings (for Curly see Lindsley and Grell, 1968). Table 1 shows the results of this experiment. The five hosts yielding exclusively y/y progeny showed after dissection a third ovary, which had connected to the host's oviducts. Similarely to the mutant ovary in situ, their ovarioles were filled with large numbers of cystocytes. The above results demonstrate that the $bgcn^+$ allele acts in the gonial cells at some stage between pole cell formation in the embryo and the middle of the third larval instar.

Table 1. Implantation of 100 h old larval ovaries from the bgcn strain into y/y female larvae of equal age

No. of implanted hosts	No. of pairs with y/y progeny	No. of pairs with y/y and bgcn/+ progeny	No. of pairs with y/y, bgcn/+ and SM5/+ progeny
16	5	-	11

DISCUSSION

The above results demonstrate that the cystoblasts in the female, respectively the primary spermatocytes in the male fail to differentiate and engage in an autonomous, non-invasive and non-lethal growth typical of benign tumors. Including bgcn, six ovarian tumor mutants are presently known (Table 2).

Table 2. Ovarian tumor mutants of Drosophila

Designation of mutant	Mutation affects	References
female sterile (2) of Bridges	early cystocyte division	Koch and King, 1964; King, 1970
female sterile (1) 231 G and 231 M (alleles)	early cystocyte division	Gans et al., 1975; King et al., 1978; King, 1979; King and Buckles, 1978; Mohler, 1977
narrow (5 alleles)	early cysrocyte division	King, 1970; King and Bodenstein, 1965
fused (7 alleles)	late cystocyte division	King et al., 1957; Smith and King, 1966; King, 1970;
female sterile (1) 1621	late cystocyte division	Gollin and King, 1981
benign gonial cell neoplasm	differentiation of cystoblast into cystocyte, resp., primary into secondary spermatocyte	Gateff, 1981; present paper

With the exception of the bgcn mutant, all other ovarian tumor mutants listed in table 2 affect only the female germ line. A thorough literature search for germ line mutants affecting both sexes, revealed no mutant with the bgcn characteristics (King and Mohler, 1975; Romrell, 1975; Mahowald and Kambysellis, 1980; Lindsley and Tokuyasu, 1980).

The female and male germ lines seem to diverge very early during development. Thus, genes affecting the fertility of either of the two sexes are comparatively rare. And indeed Watanabe and Lee (1977), found 80 male and 60 female sterility loci on the second chromosome, but none affecting both sexes. This rare class of genes must act either during pole cell formation, or before the entry of the pole cells into the presumptive gonads. The bgcn gene is apparently a representative of this class of germ line genes.

After the entry of the primary gonial celss, the pole cells, into the presumptive gonads female respectively male specific gene sets become activated. The five ovarian tumor mutants, shown in table 2 and studied by King and his coworkers, are exemples of this second class of sex-specific germ line mutants. In these mutants, cyst formation is interrupted either at the early or late cystocyte divisions (Table 2). Thus, in the tumorous egg chambers, in addition to numerous cystocytes, cysts composed of variable numbers of cells are also present.

Compared to the above ovarian tumor mutants, the bgcn mutant shows neither cyst- nore follicle formation. The ovaries contain millions of cystoblasts, which represent a differentiation step prior to that of the cystocytes.

Similarly, in the bgcn male germ cell differentiation is also arrested at an early, premeiotic stage. The cell type which is developmentally arrested is the primary spermatocyte. The only mutant, which approximates the bgcn male phenotype is the X-linked, male sterile mutant $l(1)55$ (Lifschytz, 1978). However, the female seems to be fertile.

In summary: the autosomal mutation benign gonial cell neoplasm (bgcn) of Drosophila melanogaster arrests germ cell differentiation at the cystoblast- respectively the primary spermatocyte stage. The above cells continue to divide and give rise to autonomous, benign neoplasms in the male and female gonads. bgcn is the first case of a mutant causing gonadal tumors in both sexes. Furthermore, the bgcn mutation arrests germ cell differentiation at an earlier stage than any previously known mutant.

Acknowledgments. This work was supported by the Deutsche Forschungsgemeinschaft (SFB 46). Gratitude to D. Willer and I. Brillowski for able technical assistance.

REFERENCES

Gans M, Audit C, Masson M (1975). Isolation and characterisation of sex-linked female-sterile mutants of Drosophila melanogaster. Genetics 81:683.

Gateff E (1978a). Malignant neoplasms of genetic origin in Drosophila melanogaster. Science 200:1448.

Gateff E (1978b). Malignant and benign neoplasms of Drosophila melanogaster.In Ashburner M, Wright TRF (eds): "The Genetics and Biology of Drosophila," Longon New York: Academic Press, vol 2b, p 181.

Gateff E (1978c). The genetics and epigenetics of neoplasms in Drosophila. Biol Rev 53:123.

Gateff E (1981). bgcn: benign gonial cell neoplasm. Drosophila Inform Serv 56:191.

Gollin SM, King RC (1981). Studies of fs(1621, a mutation producing ovarian tumors in Drosophila melanogaster. Develop Gen 2:203.

King RC (1970). "Ovarian Development in Drosophila melanogaster." New York: Academic Press, p.33.

King RC (1979). Aberrant fusomes in the ovarian cystocytes of the fs(1)231 mutant of Drosophila melanogaster Meigen (Diptera:Drosophilidae), Int J Insect Morphol Embryol 8:297.

King RC, Bodenstein D (1965). The transplantation of ovaries between genetically sterile and wild-type Drosophila melanogaster. Z Naturforsch 20b:293.

King RC, Mohler JD (1975). The genetic analysis of oogenesis in Drosophila melanogaster. In King RC (ed): "Handbook of Genetics vol. 3," New York: Plenum Press, p 757.

King RC, Buckles BD (1980). Three mutations blocking early steps in Drosophila oogenesis: fs(4)34, fs(2)A16 and fs(1)231M. Drosophila Inform Serv 55:74.

King RC, Burnett RG, Staley NA (1957). Oogenesis in adult Drosophila melanogaster. IV. Hereditary ovarian tumors. Growth 21:230.

King RC, Bahns M, Horowitz T, Larramendi P (1978). A mutant that affects female and male germ cells differentially in Drosphila melanogaster Meigen (Diptera:Drosophilidae).

Int J Insect Morphol Embryol 7:359.
Koch EA, King RC (1964). Studies on the fes mutant of Drosophila melanogaster. Growth 28:325.
Lewis E, Bacher C (1968). Method of feeding ethyl methan sulfonate (EMS) to Drosophila males. Drosophila Inform Serv 43:193.
Lifschytz E (1978). Uncoupling of gonial and spermatocyte stages by maens of conditional lethal mutants of Drosophila melanogaster. Develop Biol 66:571.
Lindsley DL, Grell EH (1969). The mutants of Drosophila melanogaster. Washington: Carnegie Institution of Washington Publ 627.
Lindsley DL, Tokuyasu KT (1980). Spermatogenesis. In Ashburner M, Wright TRF (eds): "The Genetics and Biology of Drosphila," London:Academic Press, Vol 2d, p 225.
Mahowald AP, Kambysellis MP (1980). Oogenesis. In Ashburner M, Wright TRF (eds): "The Genetics and Biology of Drosophila Vol 2d," London: Academic Press, p 141.
Mohler JD (1977). Developmental Genetics of the Drosophila egg. I. Identification of 59 sex-linked cistrons with maternal effects on embryonic develoment. Genetics 85:259.
Romrell LJ (1975). Mutations influencing male fertility in Drosphila melanogaster. In King RC (ed): "Handbook of Genetics Vol 3," New York: Plenum Press, p 735.
Smith PA, King RC (1966). Studies on fused, a mutant gene producing ovarian tumors in Drosophila melanogaster. J Natl Cancer Inst. 36:455.
Watanabe TK, Lee WH (1977). Sterile mutations in Drosphila melanogaster. Genetic Res Camb 30:107.

PROLIFERATING RAT HEPATOCYTES IN CULTURE CONTAIN mRNA CODING
FOR ALPHA-FETO-PROTEIN AND ALBUMIN.

Hans H. Arnold, H.J. Ruthe and Dieter Paul

Department of Toxicology, University of Hamburg
Medical School.
Grindelallee 117, 2000 Hamburg 13, FRG.

Alpha fetoprotein (AFP) (m.w. 72,000 daltons) is the major serum component of the developing fetus (Ruoslahti and Seppälä 1979). At early developmental stages AFP is synthesized by the yolk sac, while at later stages it is made primarily by the liver. After birth serum AFP levels decrease and become barely detectable in non-pregnant adults (Ruoslahti and Settälä 1979). While serum AFP levels decline around birth, albumin (m.w. 68,000 daltons) levels, which are very low in the fetus become increasingly prominent until albumin is the major constituent of the adult serum (Peters 1975. Ruoslahti and Settälä 1979). It is believed that AFP and albumin serve similar physiological functions including the maintenance of the osmotic pressure in body fluids and the efficient binding of a variety of low molecular weight metabolites (Peters 1977) such as estradiol, which binds with high affinity to murine AFP (Aussel and Massayaeff 1977).

In the adult animal AFP synthesis resumes in the liver in response to a variety of insults including partial hepatectomy, intoxication by compounds such as CCl_4, carcinogens (Sell et al. 1974)and infection with hepatitis virus (Bassendine et al. 1980). It is also produced in some hepatomas (Becker et al. 1973. Sell and Wepsic 1975) and in teratocarcinomas (Adamson 1977). Under those circumstances where AFP synthesis has been reactivated in adult liver, albumin levels, although somewhat lower than in normal liver, remain relatively high (Sala -Trepat et al.1979a). Therefore, synthesis of albumin and AFP in the adult animal are not necessarily inversely regulated as seems to be the case during fetal development. The sequential expression of AFP and albumin genes in the developing animal at a defined stage after fertilization is part of a developmental program and is presumably similar to the switch from fetal hemoglobin (Hgb F) to adult hemoglobin (Hgb A) in fetal erythropoiesis (Wood and Weatherall 1973).

An intimate evolutionary relationship between AFP and albumin has been demonstrated (Gorin et al.1981) based on the aminoacid sequences of the two proteins and on identical organization of the genes coding for these proteins (Gorin et al.1981). Both genes in the mouse consist of 15 coding segments interrupted by 14 intervening sequences (Kioussis et al.1981). The gene coding for albumin is located 13.5 kb upstream of the AFP gene in the same orientation (Ingram et al. 1981). It has been shown that 1-2 albumin genes and approximately 2-3 AFP genes are present per haploid genome in the rat (Sala-Trepat et al. 1979b).

The rates of synthesis of AFP and albumin in the mouse and the rat seem to be correlated with cellular levels of the corresponding mRNA (Sala-Trepat et al.1979a).Therefore their rate of synthesis is most likely controlled by the rate of transcription of the two genes, whereas translational control or gene amplification (Sala-Trepat et al.1979) do not appear to play major roles for the expression of either gene. The switching of AFP gene transcription to transcription of the albumin gene around birth can be assumed to be regulated, at least in part, by humoral or hormonal signals.

AFP gene transcription in the adult liver increases after partial hepatectomy (Sell et al.1979). It is not known, however, whether all hepatocytes which start dividing during the regeneration process are engaged in AFP production. Although it has been suggested by immunocytochemical methods that AFP production (but not that of albumin) occurs predominantly at late G_1 and S in cultured fetal rat hepatocytes (Sell at al.1975) it remains unclear whether AFP synthesis occurs in all cells that have been recruited to enter the cell cycle in response to growth stimulation and whether AFP and albumin expression are indeed cell cycle dependent. Furthermore, it is not clear whether synthesis of AFP or albumin is affected by agents such as CCl_4 or carcinogens in resting cells or only in cells that have entered the cell cycle. Increased AFP production in response to drugs or as the result of the commitment of cells to divide might be mediated by different mechanisms (Sell and Becker 1978) as suggested by the observation that treatment of adult rats with CCl_4 leads to a ten-fold higher AFP production than partial (68%) hepatectomy even though the rate of DNA synthesis in the liver is similar under both conditions (Watanabe et al. 1977).

To investigate cellular aspects of AFP and albumin gene expression in detail we have begun studies using well defined cell culture systems of fetal or newborn rat hepatocytes which we have previously shown to multiply in selective arginine-free medium (Leffert and Paul 1972. Paul and Walter 1975) and of adult rat hepatocytes which proliferate in culture under defined conditions (unpublished results). Here we report some initial observations showing that AFP gene expression in cultured fetal rat hepatocytes is dependent on the growth state of the cells whereas that of albumin does not appear to be linked to cell multiplication.

MATERIAL AND METHODS

Cells. Fetal rat hepatocytes were cultured essentially as described (Leffert and Paul 1972). Briefly, livers of late term (day 21) Wistar rat fetuses were treated with collagenase (Sigma) (0.5 mg/ml). The resulting cell suspension was cultured in arginine free medium supplement with 10% dialyzed fetal calf serum (Gibco). Routinely, 10^6 cells were plated per 10 cm dish (Falcon). Cells were harvested in 0.8 mM EDTA solution buffered with Tris/saline (pH 7.4). Autoradiography of cultures treated with ^{3}H-thymidine cultures (5 µCi/dish) was performed using Kodak AR-10 stripping film in formalin-fixed cultures. Morris 7777 hepatoma was kindly provided by Dr. Sala -Trepat (Gif sur Yvette, France) and was transplated subcutanously on Buffalo rats. Tumor tissue was dispersed with collagenase (1 mg/ml) and cultured as single cell suspension in medium containing 10% calf serum. Mycoplasma free cloned cell lines were stored in liquid nitrogen. For the studies described here MH 7777 clone 2 was used.

Flow microfluorometry (FMF). Cell cycle analysis of cells was conducted as described (Crissman and Tobey 1974).

Probes for AFP and albumin. A recombinant plasmid containing 950 bp of rat AFP and cDNA in pBR322 (Innis et al. 1979) was a gift from Dr. D.L. Miller (Nutley, N.J.). A recombinant plasmid containing 800 bp of rat albumin cDNA in pBR322 was kindly provided by Dr. A.Alonso (Heidelberg) (Capetanaki et al. 1981). Antisera against rat AFP or against rat albumin were kindly provided by Drs. Kuhlmann (Heidelberg) and Heinrich (Freiburg), respectively.

Extraction of RNA from cells and tissues. RNA was prepared using the guanidinium thiocyanate method (Chirgwin et al. 1979) modified as follows: cells 1-1.5 x 10^7) or tissue (0.5-2 g) were homogenized in 6 ml of freshly prepared guanidinum thiocyanate (4 M) and the resulting homogenate centrifuged (11,000 rpm, 10 min) in a Sorvall SS34 rotor at 20°C. Ethanol (0.75 vol) was added to the supernatant and the mixture kept at -20°C over night. Precipitated nucleic acids were pelleted and dissolved in 1 ml guanidinium-HCl (7 M). A typical experiment using 10^7 cells yielded 200 - 450 µg total cellular RNA.

Cell free translation in rabbit reticulocyte lysates and immunoprecipitation of AFP and albumin synthesized in vitro. RNA samples were translated in vitro in a reticulocyte cell free system as described (Pelham and Jackson 1976). For immunoprecipitation aliquots were preincubated with 1 µl of non-immune rabbit serum and unlabeled AFP for 20 min. at 0°C. Staph A protein (25°C was diluted 1:10 with adult rat liver extract 25,000 x g supernatant and mixed with the sample and incubated for 10 min. The complex was pelleted and the supernatant mixed with anti-AFP rabbit serum which had been pre-incubated with adult rat liver extract. After incubation at 0°C (1 hr) Staph A protein (50 µl) was added and incubation continued for 10 min. The immunoprecipitate formed was centrifuged, the pellet washed 3 times with buffer (100 µl) containing 20 mM Tris (pH 7.6), 0,1 M NaCl, 1 mM EDTA, 2,5 M K Cl and 0.5% NP40. The final pellet was extracted with 60 µl of protein sample buffer and aliquots run on SDS-PAGE.

Analysis of cell-free translation products by SDS-PAGE and Fluorography. 35-S-methionine labeled translation products were analyzed on 10% SDS-polyacrylamide slab gels (Laemmli 1970), and treated for fluorography as described (Bonner and Laskey 1974).

Electrophoresis of RNA on Agarose Gels under denaturing conditions. Electrophoresis of RNA in 1.5% agarose gels (12 x 10 cm) containing 10 mM methylmercury hydroxide was as described (Bayley and Davidson 1976) (16 h at 20 mA). Gels were stained with ethidium bromide after reduction of methylmercury hydroxide with ß-mercaptoethanol and transferred to ABM paper using sodium acetate buffer (pH 4,0). Following the transfer, covalently bound RNA was hybridized to nick-translated plasmid DNA probes at 42°C (16 hr) in buffer containing 50% formamide, 0.75 M NaCl, 75 mM sodium citrate, 0.02% polyvinylpyrrolidone, 0.02% bovine serum albumin, 0.02% ficoll, 100 mM sodium phosphate (pH 6.5), 500 µg/ml denatured calf thymus DNA and $1-3 \times 10^6$ cpm of ^{32}P-plasmid DNA.

Immunoprecipitation of ^{35}S-methionine labeled AFP and albumin synthesized in cells. Fetal rat liver cultures were incubated in methionine-free arginine deficient medium in the presence of 10% dialyzed fetal calf serum and of ^{35}S-methionine (NEN) (50 µCi/ml) for 6 hrs. Cells were harvested after rinsing, and immediately frozen in liquid nitrogen. Immunoprecipitation was conducted as described above.

RESULTS AND DISCUSSION

When fetal rat hepatocytes are cultured in selective arginine-free medium the cells multiply (Leffert and Paul 1972. Paul and Walter 1975) provided serum is present (Fig.1). When cells have reached the final cell density DNA synthesis is shut off (Fig.1) and cells arrest in G_1 as shown by flow microfluorometry (Fig.2). Low density cultures arrested by serum deprivation can be reinduced to enter the cell cycle by the addition of fresh serum (Fig.1).

It has been shown previously that rat hepatocytes in primary cultures produce AFP and albumin (Leffert and Paul 1972). Data summarized in Fig. 3 show the presence of immuno-precipitable AFP in extracts of ^{35}S-methionine labeled cells obtained at various times after plating. Some additional immunoprecipitable material found in cell extracts (Fig.3, lanes E-H) presumably represents breakdown products of AFP still recognizable by the antibody. The relative amount of intracellular AFP seems to decrease with increasing cell density, in accord with previously published observations (Sell et al.1975). In similar experiments we have determined the presence of albumin within the cells (Fig.4).Taken together, the results indicate that primary cultures of FRL cells do express the liver specific markers albumin and AFP thus confirming our previous observation (Leffert and Paul 1972) and that of others (Sell et al. 1975).

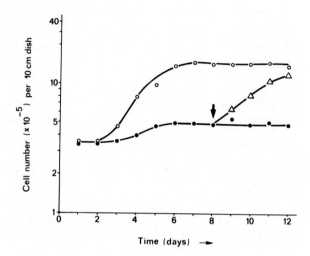

Fig. 1 Multiplication of fetal rat hepatocytes in primary Cells derived from term fetuses (10^6/10 cm dish) were cultured as described in Material and Methods. Cell number was determined daily using a Coulter Counter: o---o: 10% serum. ●---●: 2% serum. △---△ : 2% serum and addition of fresh serum (10%) at day 8 (arrow). The fraction of labeled nuclei after 24 hrs of incubation with (^{3}H) thymidine (autoradiography) for day 3, 5, 7 and 10 cultures was 36%, 31%, 3% and 2% respectively.

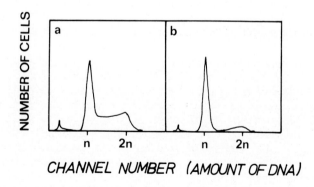

Fig. 2 Flow microfluorometic analysis showing the distribution of cellular DNA on populations of fetal rat hepatocytes at day 5 (panel A) and day 8 (panel 3) after plating the cells in 10% serum.

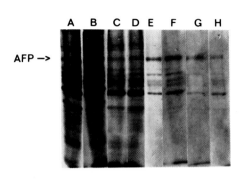

Fig. 3 ^{35}S-methionine labeled proteins and immunoprecipitated AFP from fetal rat liver cells in culture. Cultures were incubated with ^{35}S-methionine (see Materials and Methods). After extraction of total cell proteins aliquots were immunoprecipitated with anti AFP serum and analyzed on 10% SDS-PAGE and subsequently subjected to autoradiography. Migration of authentic AFP is indicated by arrow. Lanes A-D: Total protein extracts of cells labeled at days 4, 6, 8 and 10 after plating. Lanes E-H: Specific AFP immunoprecipitate of protein extracts of cells at 4, 6, 8 and 10 days after plating.

In an attempt to gain insight into the regulation of expression of AFP in cultured rat hepatocytes, we have isolated total poly(A) containing RNA and used it for two sets of experiments. First, to determine the amount of specific AFP sequences, RNA was subjected to agarose gel electrophoresis under strongly denaturing conditions and blotted onto ABM paper according to Alwine et al. (1979). Nick translated ^{32}P-pA5 plasmid DNA (AFP) was used as a specific hybridization probe. Secondly, to determine the amount of biologically active mRNA we have translated total cellular RNA in a cell free system and AFP synthesized in vitro was subsequently immunoprecipitated and analyzed on SDS-PAGE. As shown in figure 5

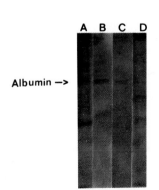

Fig. 4 ^{35}S-methionine labeled proteins and immunoprecipitated albumin from fetal rat liver cells in culture. Experiments were conducted as described in legend to Fig. 3 except that proteins were immunoprecipitated with anti-albumin antiserum. Migration of authentic albumin is indicated by arrow. Lane A: Total protein extract. Lane B-D: Specific albumin immunprecipitation of protein extracts of albumin 2, 5 and 10 days after plating.

the antiserum used specifically precipitates AFP from translation products of mRNA isolated from fetal rat livers (Lanes D and E) whereas no AFP is brought down from translation products directed by mRNA of adult rat livers (Lane A and B).

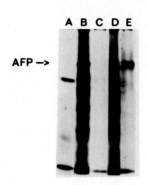

Fig. 5
Immunoprecipitation of cell free translation products of RNA isolated from adult and fetal rat livers with specific anti-AFP serum.
RNA was isolated and translated in rabbit reticulocyte lysate as described in Materials and Methods. Aliquots were subjected to immunoprecipitation and analyzed on 10% SDS-PAGE. Lane A: No mRNA added. Lane B: 10 μg of RNA from adult rat livers. Lance C: Immunoprecipitation of lane B. Lane D: 10 μg of RNA from fetal rat liver. Lane E: Immunoprecipitate from lane D.

Levels of translatable AFP mRNA in cells extracted at different times after plating increase markedly during logarithmic growth (day 5) and drop to lower levels from day 6 on when the growth curve (Fig. 1) is attaining saturation. At day 8, when most cells (>94%) are arrested in G_1 AFP mRNA levels are reduced 7 fold compared to the highest levels at day 5 (Fig. 6 and 7).

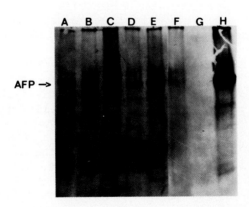

Fig. 6 Immunoprecipitation of AFP from <u>in vitro</u> translation products followed by SDS-PAGE and autoradiography.

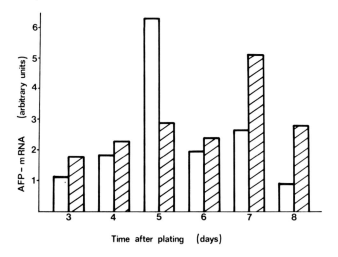

Fig. 7 Estimation of AFP mRNA in cultured hepatocytes. Optical density scans of autoradiograms of immunoprecipitated AFP (Fig. 6) are shown in arbitary units (ordinate) (OPEN BARS). Autoradiograms of Northern blots of total cytoplasmic RNA electrophoresed on 1.5% agarose hybridized to ^{32}P-PA5' plasmid (AFP) are shown in arbitrary units (ordinate) (HATCHED BARS).

The results obtained in the Northern blot experiments indicate also that increasing levels of specific AFP mRNA are present in the cultured cells until day 7 and then decline sharply at day 8. Although the time course of the appearance of AFP mRNA in this experiment does not exactly parallel the levels of translatable mRNA, it nevertheless shows the same pattern of decreasing concentrations of AFP mRNA when cells reach confluency (Fig. 7).

To compare the relative amounts of AFP mRNA in cultured rat hepatocytes with those in fetal rat liver tissue and Morris 7777 hepatomas in vivo and in culture AFP mRNA levels were determined in these tissues by in vitro translation and Northern blotting as described above. The levels in fetal rat liver cells in culture appear to be 2 fold or 4 fold lower

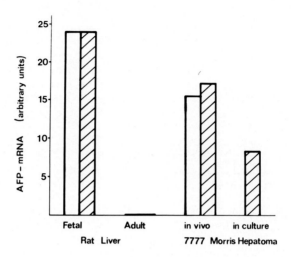

Fig. 8 Presence of translatable AFP mRNA in fetal and adult rat liver tissues. Total cytoplasmic RNA (10 µg) from fetal and adult rat liver, from Morris 7777 hepatoma tissue and from tumor cell cultures was translated and analyzed as described in legend to Fig. 6. In vitro translation products were immunoprecipitated using anti-AFP serum, subjected to SDS-PAGE and subsequently to autoradiography (OPEN BARS). Northern blotting of RNA after electrophoresis as described in legend to Fig. 7 (HATCHED BARS). Optical density scans of autoradiograms are shown in arbitary units (ordinate).

than in Morris 7777 hepatoma cells in culture or of tumor tissue, respectively. Fetal rat liver tissue contains about 6 fold higher levels of AFP mRNA than cultured hepatocytes (Fig. 8).

It appears that AFP mRNA levels are modulated during the growth process of cells in culture. Previous observations have suggested that AFP levels present in the culture medium of fetal rat liver cells (as determined by RIA) (Sell et al. 1975) might be cell cycle dependent. Our observations suggest the possibility that AFP gene expression might be controlled by hormones/growth factors which are involved in stimulating the multiplication of fetal rat liver cells in culture. Experiments to investigate the cell cycle dependent expression of AFP are currently under investigation.
Numerous laboratories have dealt with problems regarding the

stability of differentiated functions in normal liver cells in culture. The results described here lend support to the idea that least some liver specific functions (AFP, albumin) do not deteriorate in fetal rat liver cells during a minimum period of one week under the described culture conditions. It is possible that the observed decreasing cellular levels of AFP mRNA with time (Fig. 7) might reflect the differentiation process observed <u>in vivo</u>, which is characterized by decreasing hepatic AFP gene expression after birth. Therefore, the system described here appears to be useful to investigate the regulation of AFP gene expression during the cell cycle which is known to be regulated by multi-endocrine controls. Also, the system might be of interest to study the molecular basis of the switch from AFP to albumin gene expression in pure hepatocyte populations in culture.

Acknowledgments: We thank Ms.A.Piasecki for outstanding technical assistance. We are grateful to Drs. D.L.Miller and A.Alonso for providing plasmids containing rat AFP and albumin sequences, Drs. Heinrich and Kuhlmann for antibodies. We thank Ms. B.Strunck for expert assistance in the preparation of the manuscript and Dr. H.Marquardt for generous support.

REFERENCES

Adamson ED,Evans MJ,Magrane GG(1977).Biochemical Markers of the Progress of Differentiation in Cloned Teratocarcinoma Cell Lines. Eur J Biochem 79:607

Alwine JC,Kemp DJ,Parker BA,Reiser J,Renart J,Stark GR,Wahl GM (1979). Meth Enzymol 68:220

Aussel L,Massayaeff R (1977).Binding of Estrogens to Molecular Variants of Rat α-Fetoprotein.FEBS Letters 81:363

Bassendine MF,Arborgh BAM,Shipton U,Sherlock S(1980).Hepatitis B Survase Antigen and AFP Secreting Human Primary Liver Cell Cancer in Athymic Mice. Gastroenterol 79:528

Bayley IM,Davidson N (1976).Methylmercury as a Reversible Denaturing Agent for Agarose Gel Electrophoresis.Anal Biochem 70:75

Becker FF,Klein FM,Wolman SR,Asofsky R,Sell S(1973).Characterization of Primary Hepatocellular Carcinomas and Initial Transplant Generations. Cancer Res 33:3330

Bonner WM,Laskey RA(1974).A Film Detection Method for Tritium-Labelled Proteins and Nucleic Acids in Polyacrylamide Gels. Eur J Biochem 46:83

Capetany Y,Flytzanis CN,Alonso A(1981).Repression of the Albumin Gene in Novikoff Hepatoma Cells.Submitted for publication.

Chirgwin JM,Przybyla A,MacDonald RJ,Rutter WJ(1979).Isolation of Biologically Active RNA from Sources Enriched in Ribonuclease. Biochemistry 18:5294

Crissman HA,Tobey RA(1974).Cell Cycle Analysis in 20 minutes. Science 184:1297

Gorin MB,Cooper DL,Eiferman F,v d Rijn P,Tilghman SM(1981). The Evolution of α-Fetoprotein and Albumin. J Biol Chem 256:1954

Ingram RS,Scott RW,Tilghman SM(1981).The α-Fetoprotein and Albumin Genes are in Tandem in the Mouse Genome.Proc Nat Acad Sci USA 78:4694

Innis MA,Harpold MM,Miller DL(1979).Amplification of α-Fetoprotein Complementary DNA by Insertion into a Bacterial Plasmid. Arch Biochem Biophys 195:128

Kioussis D,Eiferman F,v d Rijn P,Gorim MB,Ingram RS,Tilghman SM (1981).The Evolution of α-Fetoprotein and Albumin.J Biol Chem 256:1960

Laemmli VK(1970).Clearage of Structural Proteins during Assembly of the Head of Bacteriophage T_4. Nature 227:680

Leffert HL,Paul D(1972).Studies on Primary Cultures of Differentiated Fetal Liver Cells. J Cell Biol 52:559

Pelham HRB,Jackson RS(1976).An Efficient mRNA-Dependent Translation System from Reticulocyte Lysates.Eur J Biochem 67:247

Paul D,Walter S(1975).Growth Control in Primary Fetal Rat Liver Cells in Culture.J Cell Physiol 85:113

Peters T(1975).Serum Albumin.In Putnam FW(ed)Vol I:"The Plasma Proteins",New York:Academic Press,p 131

Peters T(1977).Scrum Albumin.Recent Progress in the Understanding of its Structure and Biosynthesis.Clin Chem 23:5

Ruoslahti E,Seppälä M(1979).α-Fetoprotein in Cancer and Fetal Development.Adv Cancer Res 29:275

Sala-Trepat JM,Devor J,Sargent TD,Thomas K,Sell S(1979a).Changes in Expression of Albumin and α-Fetoprotein Genes During Rat Liver Development and Neoplasia.Biochem 18:2167

Sala-Trepat JM,Sargent TD,Sell S,Bonner J(1976). α-Fetoprotein and Albumin Genes of Rats.Proc Nat Acad Sci USA 76:695

Sell S,Wepsic HT (1975a).α-Fetoprotein.In Becker FF(ed)"The Liver:Normal and Abnormal Functions",New York:Dekker Inc.p.773

Sell S,Leffert HL,Skelly H,Müller-Eberhard V,Kida S(1975b). Relationship of the Biosynthesis of AFP,Albumin,Hemopexin and Haptoglobin to the Growth State of Fetal Rat Hepatocyte Cultures.Ann NY Acad Sci 259:45

Sell S,Becker FF(1978).α-Fetoprotein.J Nat Cancer Inst 60:19

Sell S,Thomas K,Michaelsen M,Sala-Trepat JM,Bonner J(1979). Control of Albumin and AFP Expression in Rat Liver and in some Transplantable Hepatocellular Carcinomas.Biochem Biophys Acta 564:173

Watanabe A, Miyazaki M, Taketa T(1976). Differential Mechanisms of Increased AFP Production in Rats Following CCl_4 Injury and Partial Hepatectomy. Cancer Res 36:2171

Wood WG, Weatherall DJ(1973). Haemoglobin Synthesis during Human Fetal Development. Nature 244:162

DIFFERENCES IN THE ADHESIVE PROPERTIES OF OSTEOSARCOMA-
DERIVED CLONAL VARIANTS; ENHANCED BY PROTEASE TREATMENT

D. DE MARTELAERE, J.J. CASSIMAN, F. VAN LEUVEN
and H. VAN DEN BERGHE
Division of Human Genetics, University of Leuven
Minderbroedersstraat 12
B-3000   Leuven, Belgium.

ABSTRACT

The sensitivity of the adhesive properties of human osteosarcoma derived clonal cell lines to various proteases was
examined. A protocol to dissociate cell layers with minimal
concentrations of proteases in EDTA, without affecting the
ability of the cells to aggregate rapidly in suspension was
developed. Using this protocol, 0.5 μM dispase in 0.02%
EDTA, cell suspensions of clone I_6, II_{11} and III_2 were
obtained. These cell suspensions have presumably lost most
of the extracellular material which maintains the cells in a
spread form on the substratum. Treatment of these cell
suspensions with nanomolar concentrations of chymotrypsin
(and also papain) resulted in the loss of the aggregative
ability of clone I_6 only. The other clones were resistant
to these proteases even at micromolar concentrations. Moreover, the sensitivity of clone I_6 to the proteases was
modulated by the presence of Ca^{++}. An attempt to identify
the cell surface components affected by protease treatment
of metabolically labeled cells, was unsuccessful.
Nevertheless, the results indicate that the aggregation of
the three clonal cell lines is the result of differences in
their cell surface properties.

INTRODUCTION

The adhesive properties of malignant cells are thought to
play an important role in tumor metastasis (Nicolson, 1978).
To measure adhesive properties, however, the nature of the
assay and the dissociation procedure must be taken into

account. Indeed, transformed cell lines suspended with EDTA have a higher rate of initial aggregation compared to normal cell lines (Cassiman & Bernfield, 1975; Wright et al. 1977). Also, using the single cell-cell layer binding assay (Walther et al. 1973), normal cells are distinguishable from transformed cells depending on the dissociation procedure : the differences in the rate of attachment of trypsinized normal and transformed single cells to homotypic preformed cell layers disappears completely when EDTA-dissociated cells are used (Brugmans et al. 1978). Moreover, the adhesivity of a particular cell line, normal as well as transformed, can be modified using different concentrations of proteases in the presence or absence of calcium (Takeichi, 1977; Urushihara et al. 1977; Grunwald et al. 1980; Magnani et al. 1981; Brackenbury et al. 1981).

The differences between normal and transformed cell lines depend also on the heterogeneity of the tumor cell population (Hart & Fidler, 1981). Therefore, a human osteosarcoma derived clonal system was established and clones were characterized for their adhesive properties in three different in vitro adhesion assays (De Martelaere et al. 1981). Clonal variation in the expression of some cell surface markers was demonstrated and their variation was indicative for an altered cell surface organisation.

In this report, the sensitivity of the adhesive properties of representative clones to protease treatments was examined. Different proteases were screened for their effect on the initial aggregation and for their ability to disperse the cell layer into single cells. Based on these results, a procedure of sequential protease treatments was developed in an attempt to digest cell surface molecules responsible for the adhesive behavior. This approach was extended to other representative clones taking advantage of the variation within the tumor-derived clonal system. It will be shown that the clones not only expressed variation in their adhesive and cell surface properties as described (De Martelaere et al. 1981), but also differed in their sensitivity to protease dissociation and in their susceptibility to alter their adhesivity by a double protease treatment.

RESULTS AND DISCUSSION

1. Effect of proteases on the dissociation of cell layers and on the subsequent aggregation of clone I_6 cells

Proteases with different substrate specificity were compared for their capacity to disperse cell layers into single cells, without affecting the ability of the cells to aggregate rapidly in suspension. Clone I_6 was chosen for this screening, since the extent of its initial aggregation (65% single cells after 30min) would allow us to detect stimulatory as well as inhibitory effects of the enzyme treatment. The cell layers were rinsed three times with 0.02% EDTA followed by incubation at 37°C with the appropriate enzyme at 0.05, 0.5 and 5 μM in 0.02% EDTA until complete dissociation of the cell layers; the maximum incubation time was 20min. The conditions in which the cell layers were incubated were standardized for comparison. Since EDTA without proteases was used as the control treatment, all enzyme solutions contained the same amount of EDTA (0.02%). As a consequence, the effect of certain proteases, such as dispase II, thermolysin and collagenase, whose activities require the presence of divalent cations, might have been underestimated.

At the lowest concentration tested (0.05 μM) only thrombin treatment yielded well-dispersed cell suspensions without requiring the application of extensive mechanical shear or careful filtration of the dispersed cell layers to remove cell clumps. At 0.5 μM thrombin still yielded the best single cell suspension within the shortest incubation time. The other proteases however, dissociated the cell layers readily within 20 min, requiring much less mechanical shear than the EDTA treatment. At 5 μM the dissociation of the cell layers was rapid and complete for all the proteases. The effect of the protease digestion on the extent of the initial aggregation of clone I_6 was concentration dependent for most proteases (Fig. 1); stimulation was never observed. To avoid interference with the aggregation of proteases remaining associated with the cell suspension, even after extensive washing (Brugmans et al. 1979; Cassiman et al. 1981b), α_2Macroglobulin (α_2M) was included at 10^{-3} μM in all suspensions before aggregation. α_2M has been shown to inhibit all the peptidases used in this study (Starkey & Barrett, 1977) but by itself had no effect on the extent of the aggregation (68% single cells). Bovine serum albumin, added at concentrations similar to the proteases, affected aggregation only marginally (73% single cells).

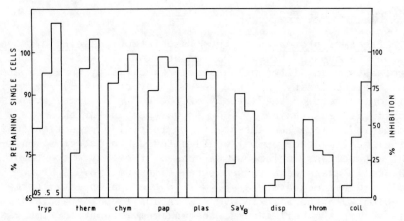

Fig. 1 : *Extent of initial aggregation of clone I_6 following dissociation with proteases in EDTA.*
Cell layers were rinsed three times with EDTA, incubated during 10 to 20min at 37°C with different proteases at 0.05 µm (A), 0.5 µM (B) and 5 µM (C) in 0.02% EDTA. The cells were washed three times and aggregated in full medium (DMEM). Aggregation was carried out as described (De Martelaere et al., 1981). The extent of the initial aggregation is expressed as the percent single cells after 30min of aggregation compared to time 0. Cell viability was measured before each assay. Each value represents the mean of at least one duplicate experiment. (a) trypsin; (b) thermolysin; (c) α-chymotrypsin; (d) papain; (e) plasmin; (f) S. aureus V_8; (g) dispase II; (h) thrombin; (i) collagenase.

There was no correlation between the capacity of the proteases to disperse cell layers and their ability to inhibit aggregation. Indeed, whereas only poor dissociation was obtained with chymotrypsin, papain and plasmin at 0.05 µM, these proteases abolished the subsequent aggregation of suspended cells of clone I_6 almost completely (more than 90% remaining single cells). Thrombin, the most effective protease for the dissociation of cell layers affected aggregation slightly. Its effect was variable but independent of the concentration. The most striking result was obtained with dispase II. At 5 µM about 30% inhibition was observed, while at 0.5 and 0.05 µM its effect on aggregation was negligible. Nevertheless, an adequate dissociation of the cell layer was obtained at 0.5 µM.
We have previously demonstrated that 100 µM dispase II-EDTA

dissociated human fibroblast cell layers rapidly and that the cell suspensions expressed surface properties different from those of trypsin-EDTA or EDTA treated cells (Cassiman et al. 1981a,b). In the present study, we show that the human osteosarcoma derived clone I_6 requires much less enzyme for dissociation than the human fibroblast cell layer. At these micromolar concentrations, dispase II in EDTA maintains nevertheless proteolytic activity towards ^{3}H-acetylated hemoglobin (results not shown). Again it is clear that this enzyme treatment yields cell suspensions, which in contrast to cells treated with other proteases, have maintained important surface properties including those responsible for rapid aggregation. Apparently, the surface components involved in the initial aggregation are extremely sensitive to most proteases while being resistant to dispase II and thrombin under the conditions described.
The differential effect of thrombin on cell dissociation and on aggregation is interesting. It would indicate that the surface components involved in initial aggregation are not thrombin sensitive, in contrast to components responsible for cell-substratum interactions. Whether, the interaction of thrombin with cell surface nexin (Baker et al. 1980), might have contributed to the release of the cells from the substratum remains to be determined.
The screening of proteases has allowed us to elaborate a procedure -0.5 µM dispase II in 0.02% EDTA- which yields cell suspensions that have presumably lost a large amount of their extracellular material (Glimelius et al. 1978) and possibly some of their cell surface components. The surface molecules required for their initial reaggregation are apparently unaffected by this treatment.

2. Protease digestion of dispase suspended cells
In the previous section we have demonstrated that clone I_6 cells can be dissociated from the cell layer with proteases without loosing their ability to rapidly form stable intercellular bonds. In an attempt to study the nature of the molecules involved in intercellular adhesion, the sensitivity of the aggregative behavior for protease treatment in suspension was examined. Cell layers were dissociated with dispase-EDTA and the cell suspensions were treated either with EDTA or with chymotrypsin-EDTA at different concentrations (Fig. 2). The extent of aggregation of these cells was measured in serum-free medium after 30 min on the gyrotory shaker.
Aggregation was nearly completely inhibited by treating the

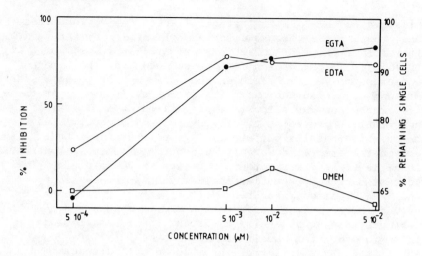

Fig. 2 : *Aggregation of dispase/EDTA dissocated cells of clone I_6 retreated with chymotrypsin.*
Cell layers were dissociated with 0.5 μM dispase II in 0.02% EDTA, washed, treated with chymotrypsin for 5min at the indicated concentrations in 0.02% EDTA, in 0.02% EGTA and in DMEM, washed and aggregated in full serum-free medium for 30min. Each value is the mean of one duplicate experiment.

cells with 5 nM or higher concentrations of chymotrypsin in EDTA. This is in contrast with the effect of chymotrypsin on cell layers where a 10 fold higher concentration (5.10^{-2} μM) is required before any inhibitory effect of the protease digestion can be measured (results not shown). Presumably, the removal of large amounts of extracellular material, the release of the cells from close contacts with the substratum and the more homogeneous distribution of the cells in solution, provide more easy acces of the enzyme to its substrate at the cell surface. Similar results were obtained using papain instead of chymotrypsin (results not shown).

In order to better understand the mechanism of action of chymotrypsin, similar incubations were performed in EGTA, in serum-free medium (Fig. 2) and in Mg^{++}-free phosphate buffered saline. The results indicate that only in EDTA or EGTA the adhesive properties of dispase suspended cells were lost during chymotrypsin treatment. In serum-free medium or Mg^{++}-free PBS no effect of chymotrypsin was observed even at 5.10^{-2} μM. Higher concentrations of the enzyme resulted

in extensive lysis of the cells.
The inhibitory effect of chymotrypsin-EDTA treatment was not due to EDTA since no inhibition was observed when dispase-dissociated cells were incubated in EDTA (68% single cells). Moreover, the proteolytic activity of chymotrypsin was comparable in EDTA, in DMEM and in MF.PBS as measured towards 3H-acetylated hemoglobin according to Van Leuven et al. (1981) (results not shown). Our results thus indicate that the presence of Ca^{++} during the protease treatment, protected the adhesive properties of dispase suspended cells against proteolysis by chymotrypsin. The removal of cell surface Ca^{++} by EDTA might render cell surface components, responsible for their adhesive properties, susceptible to proteolytic attack as suggested earlier by Takeichi (1977).

3. Protease sensitivity of the aggregation of other MG-63 clones

To determine whether the susceptibility of aggregation of clone I_6 to the different proteases was an exclusive property of this clone or was shared with other cell lines, two other clones of the MG-63 cell line were examined. The two clones

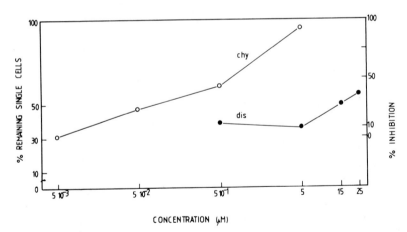

Fig. 3 : Extent of initial aggregation of dispase/EDTA and chymotrypsin/EDTA dissociated cells of clone III_2.
Cell layers were rinsed three times with 0.02% EDTA, dispersed by chymotrypsin or dispase II at the indicated concentrations in 0.02% EDTA, washed and aggregated in full medium for 30min. Each point is the mean of one duplicate. Similar results were obtained for clone II_{11}.

II_{11} and III_2 were shown to aggregate extensively within 30 min (35 and 30% remaining single cells, respectively). Cell layers of these clones were dissociated with dispase or chymotrypsin in EDTA at the concentrations used to disperse clone I_6 cells. The aggregation of the cells in serum-free medium was inhibited in a concentration dependent fashion. However, compared to clone I_6 much higher concentrations of the enzymes were required to inhibit their initial aggregation completely (5 µM chymotrypsin i.o. 0.05 µM) (Fig. 3). After dispase dissociation, treatment with chymotrypsin in EDTA was ineffective in inhibiting the aggregation of the cells (Table 1), even at increased concentrations of either dispase or chymotrypsin. At the highest concentration of chymotrypsin (5×10^{-2} µM) significant cell lysis was observed.

4. Identification of the adhesive molecules sensitive to proteases

The results presented thusfar indicated that chymotrypsin treatment of dispase-dissociated clone I_6 cells, abolished their ability to aggregate, whereas other clones were resistant to this treatment. An attempt was therefore made to

Table 1 : Effect of chymotrypsin on the aggregation of dispase-dissociated clone II_{11} and III_2 cells.

Clone	Dispase II (µM)	Chymotrypsin (µM)	% Remaining single cells
III_2	0.5	-	41
		5×10^{-4}	37
		5×10^{-3}	28
		5×10^{-2}	51
	5	-	28
		5×10^{-3}	26
		5×10^{-2}	24
II_{11}	0.5	-	53
		5×10^{-2}	40
	5	-	53
		5×10^{-2}	51

Cell layers were dissociated with dispase II in 0.02% EDTA, washed in full medium, redigested with chymotrypsin for 5min in 0.02% EDTA and aggregated in full medium (DMEM) for 30min.

determine whether cell surface molecules removed by chymotrypsin could be identified. Clone I_6 cells were metabolically labeled during 18hrs with ^{35}S-methionine. The cell layers were dissociated with dispase-EDTA and the cell suspensions were treated with EDTA with or without 5 and 0.5nM chymotrypsin for 5min; as shown in section 2 the highest concentration abolished the adhesivity while the lowest concentration was without effect. After adding 10 μg α_2M, to arrest further proteolysis, the suspensions were filtered (0.45 μm Millipore Filter) to remove the cells, and the clear supernatant was treated with 10% TCA for 30min on ice. The results indicate (Table 2) that a substantial

Table 2 : Quantitative analysis of the cellular components released by chymotrypsin.

Treatment	Total cpm[1] x 10^{-3}	Supernatant[2] cpm x10^{-3}	TCA precipitable cpm[3] x10^{-3}
EDTA exp 1	5,400	383 (7.1)	339 (83)
exp 2	6,052	268 (4.4)	245 (77)
Chymotrypsin			
5 nM exp 1	5,671	344 (6.1)	349 (82)
exp 2	5,931	405 (6.8)	402 (82)
0.5 nM exp 1	5,772	417 (7.2)	383 (82)
exp 2	5,342	383 (7.2)	356 (80)

Cell layers were labeled for 18hrs with 2.5 μCi/ml ^{35}S-methionine (1000 Ci/ mmol; Radiochemical Centre Amersham) in methionine-free DMEM supplemented with 10% Newborn Calf Serum, washed twice, dissociated with 0.5 μM dispase in EDTA, washed twice and treated with EDTA or chymotrypsin in EDTA at the indicated concentration for 5min at 37°C.
(1) total cpm of the dispase-dissocated cells
(2) cpm recovered in the clear supernatant after centrifugation of the treated cells and filtration through a 0.45 μm diameter filter. The percent of the total counts is given in parenthesis.
(3) the clear supernatant was precipitated with 10% TCA for 30 min at 0°C and the precipitate was washed extensively on Millipore filter (HA, pore size 0.45 μm).
Both the TCA precipitable and soluble counts were examined. The percent precipitable counts of the total (precipitable and soluble) is given in parenthesis.

fraction (+80%) of the label is precipitated from the supernatant but that no significant quantitative differences between EDTA and the two chymotrypsin treatments were found. Attempts to detect qualitative differences on 6-20% gradient SDS polyacrylamide gels, by coomassie staining or after autoradiography, equally failed.

One of the major problems with these experiments is that by metabolic labeling, the contribution of cell lysis to the label in suspension is substantial and might mask a specific enrichment after chymotrypsin treatment. Similar experiments using cell surface specific markers might overcome this problem.

CONCLUSION

The results of these investigations demonstrate that clonal variants, which showed quantitative differences in their extent of aggregation (De Martelaere et al. 1981) can be further distinguished on the basis of the susceptibility of their aggregative behavior to protease treatment. Cell layers of the three clones examined could be dispersed with low concentrations of the same protease (dispase II). However, only the aggregation of clone I_6 was susceptible to further protease treatment. Moreover, this susceptibility was modulated by the presence of Ca^{++} in the dissociation medium. These results then suggest that the surface properties of clone I_6, responsible for its initial aggregation differ from those of the two other clones.

The nature of these differences remains to be determined. Chymotrypsin and papain hydrolyse many peptide bonds. Therefore the three clones should have shown a similar susceptibility to these proteases if the same molecules on their surface had been involved in their aggregation. The fact that this is not the case suggests that qualitative differences, rather than quantitative ones are responsible for the differences in their aggregative behavior.

REFERENCES

Baker JB, Low DA, Simmer RL, Cunningham DD (1980) Protease-nexin : a cellular component that links thrombin and plasminogen activator and mediates their binding to cells. Cell 21:37.

Brackenbury R, Rutishauser U, Edelman GM (1981) Distinct calcium-independent and calcium-dependent adhesion systems of chicken embryo cells. Proc Natl Acad Sci USA 78:387.

Brugmans M, Cassiman JJ, Van den Berghe H (1978) Selective adhesion and impaired adhesive properties of transformed cells. J Cell Sci 33:121.

Brugmans M, Cassiman JJ, Van Leuven F, Van den Berghe H (1979) Quantitative assessment of the amount and the activity of trypsin associated with trypsinized cells. Cell Biol Intern Reports 3:257.

Cassiman JJ, Bernfield MR (1975) Transformation-induced alterations in fibroblast adhesion : masking by trypsin treatment. Exp Cell Res 91:31.

Cassiman JJ, Brugmans M, Van den Berghe H (1981a) Growth and surface properties of dispase dissociated human fibroblasts. Cell Biol Intern Reports 5:125

Cassiman JJ, Marynen P, Brugmans M, Van Leuven F, Van den Berghe H (1981b) Role of the a_2M-receptor in attachment and spreading of human fibroblasts. Cell Biol Intern Reports 5:901.

De Martelaere D, Brugmans M, Cassiman JJ, Van Leuven F, Verlinden J, Van den Berghe H (1981) Variation in the adhesive properties and cell surface molecules of clones derived from a human osteosarcoma. Intern J Cancer Submitted.

Glimelius B, Norling B, Westermark B, Wasteson Å (1978) Turnover of cell surface associated glycosaminoglycans in cultures of human normal and malignant glial cells. Exp Cell Res 117:179.

Grunwald GB, Geller RL, Lilien J (1980) Enzymatic dissection of embryonic cell adhesive mechanisms. J Cell Biol 85:766.

Hart IR, Fidler IJ (1981) The implications of tumor heterogeneity for studies on the biology and therapy of cancer metastasis. Biochem Biophys Acta 651:37.

Magnani JL, Thomas WA, Steinberg MS (1981) Two distinct adhesion mechanisms in embryonic neural retina cells. I. A kinetic analysis. Devel Biol 81:96.

Nicolson GL (1978) Cell and tissue interactions leading to malignant tumor spread (metastasis). Amer Zool 18:71.

Starkey PM, Barrett AJ (1977) in 'Proteinases in mammalian cells and tissues' (Barrett AJ, ed). North Holland, Amsterdam, pp. 663.

Takeichi M (1977) Functional correlation between cell adhesive properties and some cell surface proteins. J Cell Biol 75:464.

Urushihara H, Ueda MJ, Okada TS, Takeichi M (1977) Calcium-dependent and -independent adhesion of normal and transformed BHK cells. Cell Struct Funct 2:289.

Van Leuven F, Cassiman JJ, Van den Berghe H (1981) Functional modifications of α_2-Macroglobulin by primary amines. II. Inhibition of covalent binding of trypsin to α_2M by methylamine and other primary amines. J Biol Chem 257: 9023.

Walther BT, Öhman R, Roseman S (1973) A quantitative assay for intercellular adhesion. Proc Natl Acad Sci USA 70: 1569.

Wright TC, Ukena TE, Campbell R, Karnovsky MJ (1977) Rates of aggregation, loss of anchorage dependence, and tumorigenicity of cultured cells. Proc Natl Acad Sci USA 74:258.

Index

Acetylcholine, 404, 473–481
Acetylcoline transferase, 491
N-Acetylglucosamine, 185–191
N-Acetylglucosaminidase, 173
Acid phosphatase, active and inactive, 271–281
Acrosin and fertilization in human, 61
Acrosome, sperm, 59–60, 62
α-Actin, 393–399, 406, 412
    anti-actin antibodies, 347–348
    in egg cortex, 45, 50
    locomotion, 341–354
    and mouse myoblast differentiation, 391–399
    and muscle assembly, 318–322, 331–333
α-Actinin, 45–52, 319–328, 348–351
Actinomycin D, 21–22, 37
Adenosine 5′-monophosphate (AMP), cyclic (cAMP) 167–181, 288, 330–336
Adhesion
    in *Dictyostelium discoideum*, 183–191
    fibronectin, in developing mouse cerebellum, 509–519
    and osteosarcoma-derived clonal variants, 647–656
Adrenergic differentiation, 491–492
Agglutinin, wheat germ, 535–543
Aggregation factor
    in marine sponges, 193–208
    in sea urchin embryonic cells, 211–220
Agrobacterium tumefaciens, 567–574, 578
Alanine transaminase assay, 185–186
Albumin, 288
    maize, 546–549
    rat, mRNA coding for, 633–643
Alkaline phosphatase, 263–267
Allantois, 262–268
Allium cepa, 579, 588
AMP. *See* Adenosine 5′-monophosphate (AMP)
β-Andrenergic antagonists, 330

Androgenesis, in mice, 88–99
Animal dimple, 45–52
Anthocyanin biosynthesis, 555–565
Anthomedusae, 295
Antibodies
    anti-actin, 45, 347–348
    ferritin-coupled, for laminin, 155–164
    monoclonal, 326–327, 368, 475
    polyclonal, 326–327
    polyclonal, rabbit, against chicken skeletal muscle, 385
Antigens, histocompatibility, 145–152
Antirrhinum, 578
Antiserum
    anti-β_2-microgbulin, 145–146
    anti-fibronectin, 512–513
Arbacia lixula, 213–219
Ascites, tumor-cell, 598
Aspergillus, 169
Assembly, bird muscle, 317–336
Astroglia cells, 509–519

Bacteria, crown gall, 567–574
Bacterial plasmid vecton (pBR322), 98, 104–109, 112–119, *see also* Plasmid recombinants
Barley, 578
Basal lamina, mesenchyme differentiation, 249–258
Basement membrane glycoproteins, 155–164
Biogenesis, membrane, 125–132
Birds
    integument specialization, 309–315
    Z-disc assembly, 317–336
Blastocyst
    control in embryonal carcinoma, 593–599
    human, differentiation of, 69–84
    mouse, 87–99
Blastoderm, preorganogenetic, 223–232
Blastomere
    ambhibian embryonic, 235–244

Rana pipiens, 238
Bombina orientalis, 239
Brush-border enzymes, 261-268
Brush border, intestinal, 348
Butanol, 212-220

C perfringens, 480
Caenorhabditis, 521
Calcium, 404, 474
    and fertilization in sea urchin egg, 11-18
    in cell aggregation, 647-656
    in egg fertilization, 59-60
Cancer. *See* Metastasis; Tumor; Tumorigenesis; specific neoplasias
Capacitation of sperm, 58-64
Carbonic anhydrase-c
    cellular localization, 461-469
    in chick retina development, 459-469
    in lens development, 463
Carcinogenesis. *See* Neoplastic transformation; Tumor; Tumorigenesis; specific neoplasias
Carcinogen stimulation of AFP synthesis, 633
Carcinoma, embryonal, 593-599
Carrots, 522, 577-578, 594
Catecholamine, 365-366, 473-481
Catecholamine-producing cells, 488-491
Cell activation, receptors in, 145-152
Cell adhesion
    amphibian, 235-244
    Cephalopod eggs, 223-232, 223-232
    in *Dictyostelium discoideum*, 183-191
    marine sponges, 193-208
    in metastasis, 647-656
    sea urchin embryonic cells, 215
Cell aggregation
    in marine sponges, 193-208
    in sea urchin embryonic cells, 211-220
    species specificity, 193-208, 211-220
Cell communication in Cephalopod eggs, 223-232
Cell contact, 167-181, 445-455
Cell culture, 383-386
    chick embryo muscle, 403-412
    parsley, 556-565
    rat hepatocytes, 633-643
    sympathetic neurons, 473-481
Cell differentiation, 136-142
Cell differentiation in slime mold, 283-291
Cell division factor, 568, 571-574
Cell interaction, 223-232, 280-281
    in Cephalopod eggs, 223-232
    at cleavage in sea urchin embryos, 21-28
Cell junctions, 76-79, 174
Cell layer, Purkinje, 514
Cell lines
    mouse teratocarcinoma, differentiated, 155-164
    plant, 521-532
    teratocarcinoma and retinoic acid, 315
Cell mass, inner, 69-84
Cell migration, neural crest, in quail, 359-373
Cell motility
    chick embryo, 433-443
    muscle, bird, and Z-disc assembly, 317-336
Cell recognition, 183-191
    amphibian blastomeres, 235-244
    marine sponges, 193-208
Cell signaling, 178
Cell-surface components and protease treatment, 647-656
Cell-surface proteins, 145-152
    p30, 146-149, 152
    p45, 146-159
Cellobiose, 185-191
Cells
    astroglia, in developing mouse cerebellum, 509-519
    Bergmann glial, 518
    catecholamine-producing, 488-491
    choanocytes, 206
    clone, neuronal, 486-495
    contraction state, 433-443
    discarded embryonic, 71-84
    embryonic sea urchin, aggregation factors, 211-220
    eukaryotic, 103, 125-132
    eukaryotic and heat-shock proteins, 36
    glial, 318, 364-371
    gliocyle-derived, 450, 452
    gonial tumor, 621-631
    granule, migration, 509-519
    heart, 486, 491
    hepatocyte, chicken, 187
    hepatocyte, rat, 187

hepatocyte, rat, cultured, 633–643
leukemic, mouse, 136
marrow, mouse 136–137
mechanical behavior, 433–443
mesenchymal, 249–258, 261–268
Mueller glial, 445–455, 459–469
neural crest, migration, 497–507
neuronal clone, 486–495
peripheral cephalic crest, 500–507
photoreceptor, 449, 462
pigment, 489–491
pinacocytes, 205
pluripotent neural crest, 486–495
pole, 630
precursor granule, 512
prokaryotic and cAMP, 180–181
Purkinje, 514
Schwann sheath, 485
stem, mouse pluripotent, 135–142
tobacco teratoma, 568
trunk crest, 500–505
yeast, 168
Cephalopod eggs, 223–232
Cerebellum development, mouse, 509–519
Chalcone synthase gene, 555–565
Chaotrope KI, 319–322
Chemotherapy, 593
Chicken
 chorioallantoic membrane, 310
 embryo, mechanical behavior of cells, 433–443
 fibroblast locomotion, 342, 351–354
 gizzard, 320, 322, 326, 332, 347
 gizzard endoderm in embryo, 261–268
 integument specialization, 309–315
 mouse-chick recombinants, 261–268
 myogenesis, 403–412
 myomesin, reactivity against rabbit polyclonal antibodies, 385
 neural retina development in chick, 459–469
Cholinergic
 differentiation, 488–491
 factor, pH stability, 473–481
Chondroitin sulfate, 361
Chorioallantoic membrane, chick, 310
Chorioallantois, 427, 429
Chromosome aberrations in plants, 577–590
Chromosomes

SM5 in *Drosophila*, 621
T6 in mice, 94–96
X- and gene expression, 88
Citrus, 523
Cliona, 197, 201
Clomiphene citrate, 56, 70
Clonal variants of osteosarcoma, 647–656
Cloning
 cDNA, 392–399
 embryonic carcinoma cells, 594–599
 neuronal cells, 486–495
Cnidaria, 305
Colcemid
 sensitivity, 325–327, 336
 treatment, 405
Collagen, 361–373
 in neural crest cell migration pathways, 498
 type IV, 254–255
Colony-formation assay, 136–142
Compartmentalization, glutamine synthetase, 460, 468
Concanavalin A, 217, 238–244
Conditioned medium
 from nonneuronal cells, 473–481
 heart cell, 486, 491
Corn. *See* Maize
Corpus luteum
 human, 58
 monkey, 58
Cortical reaction and calcium in sea urchin egg, 15, 17
Corticosteroids, 460–469; *see also* specific steroids
Corticosterone, 455
Cortisol, 460–467
Cortisol receptors, cytoplasmic, 445–455
Creatine kinase, 404
 transformation from BB to MM, 381–386, 396
Creatine phosphokinase, 395
Crepis seeds, 578
Crown gall bacterium, 567–574
Culture, cell. *See* Cell culture
Culture, embryo, 69–84
Currents, electrical, transcellular, 235, 242–244
Cyclic AMP, 167–181, 288, 330–336
Cytochalasin B, 89, 242, 403–404
Cytokinesis, 90–91

Cytokinins, 568–574
Cytoplasm in sea urchin embryos, 23–28
Cytoskeleton, chick, 341–354

Datura, 578
Dentine extracellular matrix, 249–258; *see also* Extracellular matrix
Deoxyribonucleic acid (DNA)
  at germination, 590
  cDNA cloning, 392–399
  cDNA pLF56, 563–565
  cDNA probes, 171, 175
  expression in differentiation, 169
  integrity in plants, 577–590
  interferon cDNA, 112
  microinjection, 103–109, 111–123
  recombinant, 103–109, 111–123,
  repair in plants, 577
  synthesis, 211, 219
  synthesis blockers, 206
  see also Recombinants
Dermal-epidermal recombination, 310, 313–314
Desmin, 317–336
Development
  neural crest in birds, 473–481
  regulatory processes, 381–386
Dictyostelium discoideum, 439, 521
  adhesion, 183–191
  gene expression in differentiation, 168–181
  pattern formation, 283–292
Dictyostelium mucoroides, 283
  see also Slime mold
Differentiation,
  adrenergic in quail, 485–495
  amoebic, 167
  behavioral, 271–281
  cell, 136–142
  cell-line, 155–164
  cell, in slime mold, 283–292
  cholinergic in quail, 488–491
  environmental cues, 361–373
  epidermal, 309–315
  from embryonal carcinoma cells, 593–599
  gene expression during, in Dictyostelium discoideum, 167–181
  melanogenic in quail, 486, 492–494
  mesenchymal, 249–258

  mesenchymal cell, 261–268
  morphogenetic, 271–281
  mouse myoblast, 391–399
  neural crest in quail, 359–373
  procambium, 528
Discoidin I and II, 190–191
Drosophila, 521
  and heat-shock proteins, 36, 39
  melanogaster, 621–631
Dynein, 50

Eggs
  amphibian and α-actinin, 45–52
  animal dimple, 45–52
  Cephalopod, 223–232
  invertebrate, 223–232
  mouse, and transplanted human insulin gene, 103–109
  mouse, enucleation in, 87–99
  mouse, transplantation of human insulin gene into, 111–123
  yolk syncytium, 255–272
Elastase, 126–131
Electrical currents, transcellular, 235, 242–244
Embryo culture, 69–84
Embryogenesis, mutant plant, 521–532
Embryonal carcinoma, 593–599
Embryonic cells, aggregation factors, sea urchin, 211–220
Embryos
  chick, gizzard endoderm, 261–268
  chick, mechanical behavior of cells, 433–443
  mouse, 87–99, 103–109
  neural crest in quail, 485–495
  rye, 580–590
Endoderm
  gizzard, in chick embryo, 261–268
  ultrastructure, 69–84
Endonuclease Pru II, 105–108
Endoplasmic reticulum, rough, 74–82, 125–132
Endosperm, maize, 546–552
Endothelium, glomerular, 425–429
Enzyme localization, retinal, 459–469
Enzymes, intestinal brush-border, 261–268; *see also* specific enzymes
Epidermal-mesenchymal interaction, 249–258

Index / 663

Epigenetic changes, 571-574
Erythropoiesis, mouse, 135-142
Escherichia coli, 561, 590
Estradiol, 633
Estradiol and human corpus luteum function, 58
Estrogen, 57-58
Ethyl methan sulfate, 621; *see also* Tumor promoters; Tumorigenesis
Eukaryotes, higher 111
Evolutionary relation of AFP and albumin genes, 634
Extracellular matrix, 497-507
    dentine, 249-258
    glycosaminoglycans in, 498
Eye, embryonic chick, 445-455, 459-469

Fab fragment, 214
Fallopian tube, human, 60-62
Feather formation, retinoic acid in, 309-315
Ferritin, 155-164
Fertilization
    and calcium in sea urchin egg, 11-18
    effects of temperature on, 12-15
    envelope, 15
    human, in vitro, 69-84
    mammalian, in vitro, 55-64
    potassium in, 70
α-Fetoprotein, rat mRNA coding for, 633-643
Fetuin, 237
Fibroblasts
    chicken, locomotion, 342, 351-354
    cultured chick, 403-405
    quail, 364
Fibronectin, 166-164, 254, 361-367, 498-507
    adhesion in developing mouse cerebellum, 509-519
    and malignancy, 504
    human plasma, 510-519
    quail, 486, 488-491
    role in neural crest migration, 510-519
Fibrosarcoma
    SF-19, 614-615
    UV-2237, 603, 612-615
Filaments
    intermediate, in bird muscle assembly, 317-336
    microfilaments, 341-354
Filamin, 321-322, 333, 350-351
Fimbrin, 348
Flavonoid biosynthesis in parsley, 555-565
Focal contacts, 341, 351-354

Galactose, 186-188, 190-191, 272
Gamete collection, 56
Gels, polyacrylamide derivatized, 183-191
Gene activity in myogenesis, 382-386
Gene expression
    X-linked, 88
    actin, in mouse myoblast differentiation, 391-399
    coordinated and chalcone synthase gene, 555-565
    in differentiation in *Dictyostelium discoideum,* 167-181
    myosin, in mouse myoblast differentiation, 391-399
Gene transcription regulation, 122, 167-181
Gene transfer
    DNA-mediated, 103-109
    human insulin into fertilized mouse eggs, 111-123
    in mouse, 87-99
Genes
    albumin and AFP, 634
    chalcone synthase, 555-565
    β-globin human, 98, 104, 123, 111-123
    IgG, heavy-chain, 145-146
    insulin, human, 87-99, 103-109, 111-123
    insulin, mouse, 103-109
    preproinsulin, 104
    thymidine kinase, 111-123
Genetic instability, 606-616
Gentiobiose, 188
Germ layer, 224-228
Germ line, 109
    sequence transmission, 111-123
Germarium, 621
Germination, rye seed, 580-590
Gizzard, chicken, 261-268, 320, 322, 326, 332, 347
Globulins, maize, 545-552
Glomerular endothelium, 425-429
Glucagon, 473

Glucose, 185-191, 288
Glucosephosphate isomerase, 96
Glucuronic acid, 204
Glutamine synthetase, 459-469
　cellular localization, 488-450
　hormonal induction, 445-455
　induction, 460
Glycocalyx, amphibian embryonic, 240
Glycoconjugates, cell-surface, 236-244
Glycogen phosphorylase, 179
Glycoproteins, 361
　basement-membrane, 155-164
　cell-surface, 271-281
　see also specific glycoproteins
Glycosaminoglycans, 498
Glycosidases, wheat germ, 536, 540-543
Glycosides, 187, 189
Golgi apparatus, 73-75, 125
Gonadotropin
　human, 56-58
　human chorionic, 70
Gonial tumor cells, 621-631
Grafts
　chick-quail, 427, 429
　interspecies, 427, 429
　mouse-quail, 427, 429
　see also Recombinants
Granule-cell migration, 509-519
Granule cells, precursor, 512
Granulopoicsis, mouse, 135-142
Growth factor requirements, 571-574
Gynogenesis, in mice, 88-99

Habituation to cell-division factor, 571-574
Halichondria, 198; see also Sponges marine
Heart contractions, in chick embryo 433-443
Hemagglutination assays, 538-542
Hemagglutinin, 236-237
Hematopoesis, mouse, 135-142
Hematopoietic islands, 426-430
Hensen node 438-439, 441-443
Heparin sulfate, 361
Hepatocytes, rat, cultured, 633-643
Hepatomas
　expression of AFP, 633
　Morris 7777, 641-642
Heterochromatin, 73

Hexose, 200-201
Hind III site, 563-565
Histone messenger, 40-43
Homing of neural crest cells, 498-507
Hormonal induction of glutamine synthetase, 445-455
Hormones
　adrenal corticosteroid, 446-455, 460
　luteinizing, 57
　see also specific hormones
Hyaluronic acid, 361-362, 366, 498, 506
Hybridization, Southern blot 108, 112-115
Hybridomas, 368
Hydra, 305-306
11-β-Hydroxycorticosteroids, 455
Hydrozoa, 295
Hypothesis, signal, 125-132

Immobilization
　ligand, 190-191
　sugar, 190-191
Immunoglobulin
　anti-spore, 283-289
　IgG 385-386
　　goat anti-rabbit, 449
　　heavy-chain gene, 145-146
　　light-chain, 127
　　rabbit, 130
Immunolocalization of α-actinin in amphibian eggs, 45-52
Immunological techniques, 211-220
Induction, 223-232
　fetal liver, 427
　fetal spinal cord, 427
　hormonal, of glutamine synthetase, 445-455
　in kidney development, 425-430
Inner cell mass (ICM), 69-84, 206, 593, 596-597
Insulin
　gene, human, 87-99
　human, 123
Integument specialization, bird, 309-315
Interferon
　cDNA, 112
　expression, human, 123
　mRNA activation and, 176
Intestinal brush border, 348
Invertase, 237
Isoprotein switching, 381-386

Isoproterenol, 330

Junctions, intercellular, 21–28

Keratin, 155–164, 318
Kidney development, 425–430
Kinases 330–334; *see also* specific kinases

Lactose, 188
Lamella, leading, 341, 352–354
Laminin, 155–164, 254, 361
Laparoscopy, 56, 58
Lectin, 475
    endogenous cell-surface, 236–244
    function, endogenous, 535–543
    localization in wheat germ, 536–543
Letruce seed, 583
Leucine, 31
Leukemia
    cells, in mouse, 136–137
    LIZIO, 595
Leukoagglutinin, 145, 151–152
Ligands, immobilized, 190–191
Localization. *See* Enzyme localization; Gene localization; Lectin
Locomotion. *See* Actin; Fibroblasts; Motility
Loligo pealei, 224, 229
Loligo vulgaris, 223–232
Lymphocyte, T-, membrane, 145–152
Lysine, maize, 545

M-line, 383–386
Macroglobins, 61
Maize globulins in development, 545–552
Malignancy and fibronectin, 504; *see also* Tumor
Maltase, 263–267
Maltose, 185–189
Mannan, 237
Mannose, 185–199, 272
Manubrium, *Podocoryne*, 297–303
Marine sponges. *See* Sponges, marine
Matthiola incana (Brassicaceae), 555–565
Medusae, *Podocoryne*, 296
Melanin, 364, 367, 491
Melanogenesis, quail, 364, 367
Melanogenic differentiation, 486, 492–494
Melanoma
    B16, 595, 599

    B16, mouse, 602–612
    K-1735, 604–615
Membrane
    basement, glycoproteins, 155–164
    biogenesis, 125–132
    epithelial (polarity), 235–244
    plasma (structure), 235–244
    polarity, transcellular, 235, 242–244
    proteins, 125–132, 211–220
Mendelian inheritance ratios, 111–123
Merocyanine, 540, 135–142
Mesenchyme
    differentiation, 249–258
    metanephric, 425–430
Metamorphosis, organogenesis during, in marine sponge, 193, 205–208
Metanephros, 425–430
Metastasis
    and cell adhesion, 647–656
    organ-specific, 193
    pathogenesis of, 601–616
    tumor-cell ascites and, 598
    see also Tumor; Tumorigenesis; specific neoplasias
Methionine, maize, 545
Mice, transgenic, 112–123
Microciona prolifera
    aggregation, 193–208
    organogenesis, 193–208
    see also Sponges, marine
Microfilaments, 341–354
β_2-Microglobulin, 145–146
Micromeres, as pacemakers for mitotic activity, 21–28
Migration of granule cells, 509–519
Mitochondria in human blastocysts, 73–80
Mitotic pattern in sea urchin embryos, 21–28
Monkey, rhesus, 58, 61
Monoclonal antibodies. *See* Antibodies, monoclonal
Morphogenesis of feather, and retinoic acid, 309–315
Mosaicism in mouse, 119–122
Motility
    actinin, 341–354
    cell, 342–354, 433–443
Mouse
    astroglia cells in developing cerebellum, 509–519

blastocyst, 87-99
eggs, 103-109
eggs, transplantation of human insulin gene into, 103-109, 111-123
embryo, 103-109
fetus, 87-99
mosaicism in, 119-122
mouse-chick recombinants, 261-268
mouse-quail recombinants, 249-258
myoblast differentiation, 391-399
neural retina, 468
nuclear transplantation in, 87-99
placenta, 99, 105-109
transcriptional regulation in, 122
tumorigenesis, 594-599
Mucin, 237
Mueller glial cells, 459-469
Muscle, cardiac, 317-336
Muscle-cell assembly, 317-336
    skeletal, 317-336
    smooth, 317-336
Mutagens. *See* Tumor promotors; Tumorigenesis
Mutants
    autosomal recessive, 621-631
    plant, and temperature, 521-532
    temperature-sensitive, 521-532
Mutation-avoidance mechanisms, 577
Mutations induced by mutated developmental genes, 621-631
Mycale fusca, 201
Myofibrillogenesis, 381-386; see also Myogenesis
Myogenesis, 317-336
    chicken, 381-386, 403-412
    mouse, 391-399
    see also Muscle assembly
Myomesin, 381-386
Myosin, 318-322, 384, 412
    heavy- and light-chain, 391-399
    in egg cortex, 45, 50
    in nonmuscle cell, 341, 350

Neoplastic transformation in plants, 567-575
Neural crest
    cell migration and homing, 497-507
    development in birds, 473-481
    development in quail, 359-373
    embryonic quail, 485-495
    migration and fibronectin, 510-519
Neural retina, chick embryo, 445-455, 459-469
Neuraminidase, 147, 480
Neuroblastoma, Cl300, 596
Neurogenesis, quail, 485-495
Neuronal differentiation, 359-373
Neurons
    amacrine, 459-469
    cultured sympathetic, in neurotransmitter choice, 473-481
Neurotransmitter choice, 473-481
Nicotiana, 574
Nuclear envelopes in sea urchin embryos, 23
Nuclear transplantation, mouse, 87-99

Odontoblast, 249-258, 485
Onion, 585
Oocytes
    Podocoryne 299-302
    human, and in vitro fertilization, 55-64
    mouse, fertilized, 112
Oogenesis in *Drosophila*, 622-629
Organogenesis
    marine sponge, 193, 205-208
    tooth, 249-258
Orthophosphate, 271, 273, 280
Osteosarcoma clonal variants, 647-656
Ouabain, 612-613
Ovalbumin, 237
Ovariole, 622-627
Ovomucoid, 237
Ovulation, induced, 57-58, 60-62

Paracentrotus lividus, 21-28, 31-43, 212-218; see also Sea urchin
Paranemin, 331-336
Parsley, 555-565
Parthenogenesis, 63
Pattern formation in *Dictyostelium discoideum*, 283-292
Perikarya, 463, 465
Petroselinum hortense (Apiaceae) 555-565
Phenylalanine ammonia-lyase (PAL), 555-565
Phorbol esters as tumor promotors, 403-412, 486, 492-494
Phosphatase, acid, 271-281
Phosphorylation, 325, 327, 330, 334

Photosensitization, merocyanine 540-mediated, 135–142
Physcomitrella, 521
Phytohemagglutinin, 145, 151–152
Pigment cells, 489–491
  cortical melanin, 236
  quail, 364–367
Placenta, mouse, 99, 105–109
Plasmid recombinant
  pIf, 114–121
  pST6, 112–121
  pST9, 115, 121
  see also Bacterial plasmid vector
Plasmids
  muscle actin-coding, 393–394
  tumor-inducing, 568–574
Plasminogen activator, 404
Podocoryne carnea, 29–306
Polarity, membrane, and transcellular electrical currents, 235, 242–244
Polyacrylamide gels, derivatized, 183–191
Polyclonal antibodies. *See* Antibodies, polyclonal
Polyps, *Podocoryne*, 295–306
Polysomes, 37–43
Polyspermy, block to, 16–18
Positional information, 283–292
Potassium, 70, 474
Primitive streak, 438–439, 441–442
Procambium differentiation, 528
Progesterone, and human corpus luteum function, 58
Proline, maize, 547
Pronase, 480
Protease treatment and osteosarcoma adhesion, 647–656
Protein regulation in myogenesis, 317–336
Protein synthesis
  chick embryo muscle cell cultures, 403–412
  myogenesis, 381–386
  transcriptional control of, 39–43
  translational control of, 39–43, 88
Proteins
  cell-surface, 145–152
  cell-surface glycoproteins, 271–281
  contractile, 45–52
  α-fetoprotein in rat, 633–643
  glycoproteins, 361
  heat-shock, 31–43, 319

M-line, 383–386
  membrane, 125–132
  membrane, sea urchin embryonic cells, 211–220
  regulation in myogenesis, 317–336
  transport of, 125–132
  triton-insoluble, 145–152
  see also specific proteins
Proteoglycans, 254
  fragments, 194
  glycosaminoglycans (GAGs), 361–363

Quail
  mouse–quail recombinants, 249–258
  neural crest development, 359–373
  neural quest in embryos, 485–495

Rabbit polyclonal antibodies against chicken myomesin, 385
Rana pipiens blastomere, 238
Rat
  albumin and mRNA, 633–643
  α-fetoprotein and mRNA, 633–643
Reassembly, 295–306
Receptors
  carbohydrate-binding, 183–191
  cell-activation, 145–152
  cell-membrane, in *Dictyostelium discoideum*, 183–191
  cortisol, cytoplasmic, 445–455
  plasma membrane, 411
Recombinant DNA. *See* Deoxyribonucleic acid (DNA), recombinant
Recombinants
  dermal–epidermal, 310, 313–314
  heterologous
    mouse–chick, 261–268
    mouse–quail, 249–258
  see also Grafts; Deoxyribonucleic acid (DNA), recombinant
Restriction enzyme analysis, 112–119, 555–565
Reticulum, endoplasmic, rough, 74–82
Retina
  embryonal chick, 445–455
  mouse neural, 468
Retinoic acid in feather morphogenesis, 309–315
Ribonucleic acid (RNA)
  aggregation-specific, 168–181

constitutive, 168–181
mRNA, 40–43, 87–99, 127
mRNA precursor, 37–43
muscle-specific, 391–393
myoblast differentiation, 395–396
particles, 31–43
polyadenylated, 169–179, 391, 395
stability, 168–181
synthesis, 22
synthesis blockers, 206
Ribosomes in human blastocysts, 73, 79–83
Rye, 580–590

180 Sarcoma, 595
Scale formation, 309–315
Scatchard analysis, 202–203
Schizosaccharomyces
 cerevisiae, 272–273, 279, 281
 pombe, 217–281
 see also Yeast
Schwann sheath cells, 485
Sea urchin
 aggregation factors in embryonic cells, 211–220
 calcium in egg fertilization, 11–18
 embryo
  cells, aggregation factors, 211–220
  cytoplasm, 23–28
  and heat shock, 31–43
 mitotic pattern in embryos, 21–28
Secale cereale, 580–590
Secretion, protein, in eukaryotic cells, 125–132
Seed development in maize, 545–552
Seeds
 Crepis, 578
 lettuce, 583
 storage and morphological variance, 578–590
Self-nonself recognition in *Podocoryne* stolons, *296–306*
Signal hypothesis, 125–132
Signaling, cell, 178
Slime mold, cellular, 283–292
 see also specific organisms
Southern blot analysis, 99
Sperm
 acrosome, 59–60, 62
 acrosome-reacted, 11–18

human, and in vitro fertilization, 58–64
 motility, 61
Spermatocytes, *Podocoryne*, 299–302
Spermatogenesis in *Drosophila*, 626
Sponges, marine, cell–cell recognition in, 193–208
Squid, 223–232
Staphylococcus aureus, 326
Stem cells, mouse pluripotent, 135–142
Steroids. See Corticosteroids
Stress fibers, 341–354
Substratum and differentiation, 359–373
Sucrase, 263–267
Sucrose, 188
Sugars, immobilized, 190–191

T-lymphocyte. See Lymphocyte
Temperature-sensitive mutants, 521–532
Temporal emzyme changes, 459–469
Teratocarcinoma
 and AFP, 633
 mouse, PYS-2, differentiated cell line, 155–164
 therapy for, 593–599
 See also Tumors
Teratoma, tobacco, 568
Testis
 Drosophila, 626
 tumors of, in mouse, 595
Thioquanine, 612–613
Thrombin, 649–651
Thryoglobulin, 237
Tissue
 culture of parsley cells, 556–565
 elasticity, 433–443
 incompatibility, 295–306
 interaction in Cephalopod eggs, 223–232
 viscosity micromeasurement, 433–443
Tobacco teratoma cells, 568
Trans-β-retinoic acid. See Retinoic acid
Transcriptional regulation in mouse, 122
Transferrin, 237
Transgenic mice, 112–123
Transplantation
 heterotopic, 366
 gene in mouse, 87–99
 gene, human insulin, into fertilized mouse eggs, 111–123
 nuclear, in mouse, 87–99

Trehalose, 188
Triticum aestivum, 536
Trophectoderm, 95-96
  mouse, 596
  ultrastructure, 69-84
Tropomin, 319
Tropomyosin, 50, 319, 350-351
Tryptophan, maize, 545
Tubulin, 50, 331
Tumor
  heterogeneity, 603-616
  testicular, mouse, 595
  -inducing plasmid, 568-574
  metastasis, pathogenesis of, 606-616
  progression, 603-616
  promotion in chick embryo muscle cells, 403-412
  reversal implants, 567-574
  see also Metastasis; specific neoplasias
Tumor promotors
  ethyl methan sulfate, 621
  phorbol esters, 403-412, 486, 492-494
Tumorigenesis in mouse, 594-599

Uranic acid, 200-201

Valency, 202-204
Vascularization in kidney development, 425-430

Vimentin, 163-164, 317-336
Viruses
  hepatitis and AFP, 633
  herpes simplex, 98, 104, 112
  mouse sarcoma, 609-616
  oncogenic, 568, 608-609
  Sendai, 94
  simian 40 (SV40), 98, 104, 112-119
Vitellarium, *Drosophila*, 622-629
Vitellus, 63
Volvox, 521

Wheat germ agglutinin, 535-543

Yeast, 577
  and heat-shock proteins, 36

Z-disc assembly in bird muscle, 317-336
Zein, maize, 545, 549
Zona pellucida
  chicken, 436, 439-440
  human, 71, 76-84
  penetration, 61, 63
Zonula
  adherens, 46
  occludens, 242

PROGRESS IN CLINICAL AND BIOLOGICAL RESEARCH

Series Editors
Nathan Back
George J. Brewer

Vincent P. Eijsvoogel
Robert Grover
Kurt Hirschhorn

Seymour S. Kety
Sidney Udenfriend
Jonathan W. Uhr

Vol 1: **Erythrocyte Structure and Function**, George J. Brewer, *Editor*

Vol 2: **Preventability of Perinatal Injury**, Karlis Adamsons and Howard A. Fox, *Editors*

Vol 3: **Infections of the Fetus and the Newborn Infant**, Saul Krugman and Anne A. Gershon, *Editors*

Vol 4: **Conflicts in Childhood Cancer: An Evaluation of Current Management**, Lucius F. Sinks and John O. Godden, *Editors*

Vol 5: **Trace Components of Plasma: Isolation and Clinical Significance**, G.A. Jamieson and T.J. Greenwalt, *Editors*

Vol 6: **Prostatic Disease**, H. Marberger, H. Haschek, H.K.A. Schirmer, J.A.C. Colston, and E. Witkin, *Editors*

Vol 7: **Blood Pressure, Edema and Proteinuria in Pregnancy**, Emanuel A. Friedman, *Editor*

Vol 8: **Cell Surface Receptors**, Garth L. Nicolson, Michael A. Raftery, Martin Rodbell, and C. Fred Fox, *Editors*

Vol 9: **Membranes and Neoplasia: New Approaches and Strategies**, Vincent T. Marchesi, *Editor*

Vol 10: **Diabetes and Other Endocrine Disorders During Pregnancy and in the Newborn**, Maria I. New and Robert H. Fiser, *Editors*

Vol 11: **Clinical Uses of Frozen-Thawed Red Blood Cells**, John A. Griep, *Editor*

Vol 12: **Breast Cancer**, Albert C.W. Montague, Geary L. Stonesifer, Jr., and Edward F. Lewison, *Editors*

Vol 13: **The Granulocyte: Function and Clinical Utilization**, Tibor J. Greenwalt and G.A. Jamieson, *Editors*

Vol 14: **Zinc Metabolism: Current Aspects in Health and Disease**, George J. Brewer and Ananda S. Prasad, *Editors*

Vol 15: **Cellular Neurobiology**, Zach Hall, Regis Kelly, and C. Fred Fox, *Editors*

Vol 16: **HLA and Malignancy**, Gerald P. Murphy, *Editor*

Vol 17: **Cell Shape and Surface Architecture**, Jean Paul Revel, Ulf Henning, and C. Fred Fox, *Editors*

Vol 18: **Tay-Sachs Disease: Screening and Prevention**, Michael M. Kaback, *Editor*

Vol 19: **Blood Substitutes and Plasma Expanders**, G.A. Jamieson and T.J. Greenwalt, *Editors*

Vol 20: **Erythrocyte Membranes: Recent Clinical and Experimental Advances**, Walter C. Kruckeberg, John W. Eaton, and George J. Brewer, *Editors*

Vol 21: **The Red Cell**, George J. Brewer, *Editor*

Vol 22: **Molecular Aspects of Membrane Transport**, Dale Oxender and C. Fred Fox, *Editors*

Vol 23: **Cell Surface Carbohydrates and Biological Recognition**, Vincent T. Marchesi, Victor Ginsburg, Phillips W. Robbins, and C. Fred Fox, *Editors*

Vol 24: **Twin Research, Proceedings of the Second International Congress on Twin Studies,** Walter E. Nance, *Editor*
Published in 3 Volumes:
  Part A: **Psychology and Methodology**
  Part B: **Biology and Epidemiology**
  Part C: **Clinical Studies**

Vol 25: **Recent Advances in Clinical Oncology,** Tapan A. Hazra and Michael C. Beachley, *Editors*

Vol 26: **Origin and Natural History of Cell Lines,** Claudio Barigozzi, *Editor*

Vol 27: **Membrane Mechanisms of Drugs of Abuse,** Charles W. Sharp and Leo G. Abood, *Editors*

Vol 28: **The Blood Platelet in Transfusion Therapy,** G.A. Jamieson and Tibor J. Greenwalt, *Editors*

Vol 29: **Biomedical Applications of the Horseshoe Crab (Limulidae),** Elias Cohen, *Editor-in-Chief*

Vol 30: **Normal and Abnormal Red Cell Membranes,** Samuel E. Lux, Vincent T. Marchesi, and C. Fred Fox, *Editors*

Vol 31: **Transmembrane Signaling,** Mark Bitensky, R. John Collier, Donald F. Steiner, and C. Fred Fox, *Editors*

Vol 32: **Genetic Analysis of Common Diseases: Applications to Predictive Factors in Coronary Disease,** Charles F. Sing and Mark Skolnick, *Editors*

Vol 33: **Prostate Cancer and Hormone Receptors,** Gerald P. Murphy and Avery A. Sandberg, *Editors*

Vol 34: **The Management of Genetic Disorders,** Constantine J. Papadatos and Christos S. Bartsocas, *Editors*

Vol 35: **Antibiotics and Hospitals,** Carlo Grassi and Giuseppe Ostino, *Editors*

Vol 36: **Drug and Chemical Risks to the Fetus and Newborn,** Richard H. Schwarz and Sumner J. Yaffe, *Editors*

Vol 37: **Models for Prostate Cancer,** Gerald P. Murphy, *Editor*

Vol 38: **Ethics, Humanism, and Medicine,** Marc D. Basson, *Editor*

Vol 39: **Neurochemistry and Clinical Neurology,** Leontino Battistin, George Hashim, and Abel Lajtha, *Editors*

Vol 40: **Biological Recognition and Assembly,** David S. Eisenberg, James A. Lake, and C. Fred Fox, *Editors*

Vol 41: **Tumor Cell Surfaces and Malignancy,** Richard O. Hynes and C. Fred Fox, *Editors*

Vol 42: **Membranes, Receptors, and the Immune Response: 80 Years After Ehrlich's Side Chain Theory,** Edward P. Cohen and Heinz Köhler, *Editors*

Vol 43: **Immunobiology of the Erythrocyte,** S. Gerald Sandler, Jacob Nusbacher, and Moses S. Schanfield, *Editors*

Vol 44: **Perinatal Medicine Today,** Bruce K. Young, *Editor*

Vol 45: **Mammalian Genetics and Cancer: The Jackson Laboratory Fiftieth Anniversary Symposium,** Elizabeth S. Russell, *Editor*

Vol 46: **Etiology of Cleft Lip and Cleft Palate,** Michael Melnick, David Bixler, and Edward D. Shields, *Editors*

Vol 47: **New Developments With Human and Veterinary Vaccines,** A. Mizrahi, I. Hertman, M.A. Klingberg, and A. Kohn, *Editors*

Vol 48: **Cloning of Human Tumor Stem Cells,** Sydney E. Salmon, *Editor*

Vol 49: **Myelin: Chemistry and Biology,** George A. Hashim, *Editor*